HORIZONTAL WELL TECHNOLOGY

HORIZONTAL WELL TECHNOLOGY

S. D. JOSHI, Ph.D.
Joshi Technologies International, Inc.
Tulsa, OK, U.S.A.

PENNWELL PUBLISHING COMPANY
TULSA, OKLAHOMA

PennWell Corporation
1421 South Sheridan Road
Tulsa, Oklahoma 74112-6600 USA

800.752.9764
+1.918.831.9421
sales@pennwell.com
www.pennwellbooks.com
www.pennwell.com

Marketing Manager: Julie Simmons
National Account Executive: Barbara McGee
Director: Mary McGee
Production Manager: Sheila Brock

Library of Congress Cataloging-in-Publication Data
Joshi, Sada D.
Horizontal well technology / Sada D. Joshi
p.cm.
Includes bibliographical references and index
ISBN 0-87814-350-5
ISBN13 978-0-87814-350-4
1. Horizontal oil well drilling. I. Title.
TN871.25.J67 1991
622'.3382—dc20 90–27822
CIP

Printed in the United States of America

2 3 4 5 6 11 10 09 08 07

This book is humbly dedicated to my mother, Sumati D. Joshi, my father, Dattatray M. Joshi and my wife, Claudette J. Joshi.

I am grateful to my parents for providing me with education, inspiration, and confidence. I am also indebted to my wife who provided the encouragement, fortitude, and extraordinary understanding which enabled me to steal many hours from our family while writing this book.

CONTENTS

CHAPTER 7
Pseudo-Steady State Flow

CHAPTER 8
Water and Gas Coning in Vertical and Horizontal Wells

CHAPTER 9
Horizontal Wells in Gas Reservoirs

PREFACE

The major purpose of writing this book is to summarize the state-of-the-art of horizontal well technology. Recent advances in drilling and completion have resulted in a rapid increase in the number of horizontal wells drilled each year around the world. A horizontal well, to some extent, is different from a vertical well because it requires an interdisciplinary interaction between various professionals, such as geologists, reservoir engineers, drilling engineers, production engineers, and completion engineers. Because of the large volume of literature that is available in different disciplines, I have decided to divide this book into two parts. The first part (Volume 1), is presented here. This first volume mainly deals with reservoir and production engineering.

In this book, I have included published literature available as of June 1990. Additionally, I have included example problems to illustrate the use of various theoretical solutions. Wherever possible, I have not only discussed practical difficulties that one may encounter while using theoretical solutions, but I have also listed some of the methods that one can use to obtain the desired information. I have included descriptions on field histories wherever they were available. The available field histories that I have chosen not only represent successes of horizontal well technology but also include some economic failures.

To some extent, writing this book was difficult because of the interdisciplinary nature of horizontal well technology. The book is mainly directed to the practicing professionals who make engineering calculations and decisions on horizontal well applications. This book can also be used as a graduate level textbook. For managers, the book helps to review the present state of the art. I have also outlined some of the gaps in technology that exist today. These gaps in technology will be useful for research engineers and research professionals to determine the areas of future research.

Many solutions which are presented are based upon my personal experiences dealing with various vertical well and horizontal well field projects around the world. I am thankful to the many companies with whom I had the opportunity to work on the field projects. I am also thankful to all the people who have suffered through my teaching of horizontal well classes.

Our class discussions and their suggestions were very valuable in making this book useful to a practicing engineer.

Chapter 1 of this book is an overview of horizontal well technology and is a general introduction to the technology from a reservoir, drilling, and completion standpoint.

Chapter 2 mainly looks at the reservoir engineering concepts and their application for horizontal wells. The chapter also includes a discussion on well spacing of horizontal wells.

Chapter 3 includes steady state solutions and their applications. It also includes discussions on formation damage problems in horizontal wells. In addition to horizontal wells, it also contains a discussion of slant wells. There are cases where slant wells may be more beneficial than horizontal wells.

Chapter 4 deals with the influence of well eccentricity on productivity of a horizontal well. Well eccentricity represents a vertical distance between the horizontal well location and the center of the pay zone. Though influence of the well eccentricity on productivity of a well is minimal, it will have a strong influence on the ultimate reserves for a horizontal well drilled in reservoirs with top gas or bottom water.

Chapter 5 compares horizontal and fractured vertical wells. This chapter discusses practical aspects of hydraulic fracturing of a vertical well, its advantages, and the limitations. The chapter also includes reasons for stimulating horizontal wells and calculation of productivities for fractured horizontal wells.

Chapter 6 focuses on transient well testing. In general, transient well testing is a highly mathematical subject. At the same time, it is one of the most important and useful subjects to understand the well behavior in a given reservoir. To make the chapter complete, I have included all the necessary mathematics and many concepts which are essential to interpret the behavior of a horizontal well.

Chapter 7 deals with pseudo-steady state solutions. In this chapter, I have listed various solutions for vertical wells, fractured vertical wells, and horizontal wells. I have also included available solutions for the partially perforated or partially open horizontal wells. The chapter also describes the performance of horizontal wells completed in solution gas drive reservoirs.

Chapter 8 examines water and gas coning in vertical and horizontal wells. It outlines many of the available solutions for calculating water and gas coning behavior in horizontal and vertical wells. It also contains discussion of available field histories. The histories not only show successes but also the failure of horizontal wells in minimizing water and gas coning. The chapter also outlines benefits and risks associated with production testing of vertical wells to estimate the potential of horizontal wells.

Chapter 9 looks at the application of horizontal wells in gas reservoirs. In my opinion, horizontal wells are highly suitable for low permeability as well as high permeability gas reservoirs.

Chapter 10 deals with the pressure drop through a horizontal well and how important it is in the estimation of horizontal well performance.

To make the book complete, I have included *Appendix A* which refers to fluid properties. *Appendix B* includes data on gas compressibility. *Appendix C* contains various conversion factors. (I have included Appendix C because the book is written in U.S. field units, and Appendix C will be helpful to convert the examples to different field units.) *Appendix D* includes a discussion about various pseudo-skin factors and their definitions. *Appendix E* consists of tables of recovery factors that one can expect from various types of reservoirs and under different types of drive mechanisms. *Appendix F* is a glossary of the terms that are used in this book. I believe this glossary will be useful for people who are not familiar with reservoir and production engineering terminology.

To the readers, I would very much be interested in any comments, suggestions, or questions you may have about the contents of the book. Please feel free to contact me directly:

Sadanand D. Joshi, Ph.D.
Joshi Technologies International, Inc.
5801 E. 41st St., Suite 603
Tulsa, OK 74135
(918) 665-6419

I consider myself to be a student of this technology. After writing this book, reading many published papers and working on various field projects, I realized more than ever that there are many more things which I need to learn before I will ever know all the answers.

Tulsa, Oklahoma
Sept. 6, 1990

ACKNOWLEDGEMENTS

Several people were instrumental in helping me to complete this book. First of all, I would like to acknowledge my family which has been very encouraging and understanding while I spent time away from them to write this book. I am also thankful to three engineers at our company, JTI; George Saville, Pralhad N. Mutalik, and Dr. Wenzhong Ding, who also spent many hours proofreading and solving some of the examples which have been included in the book. Without their help, it would not have been possible to complete this book. I am especially thankful to Dr. Wenzhong Ding, whose attention to details helped to eliminate many typographical errors in the book.

I have also been privileged to work with many professionals in the oil industry who have taught me many things and helped me grow and develop as an engineer. I have been fortunate to be associated with Dr. R. Raghavan who taught me how to think rationally. I am also thankful to: W. B. (Ben) Lumpkin, who taught me many skills in practicing reservoir engineering; Thomas B. Reed of the Department of Energy, who persuaded me in 1980 to start studying horizontal wells; M. J. Fetkovich and R. B. Needham of Phillips Petroleum Company and P. H. Doe of Shell Research Company, who taught me how to use reservoir engineering.

I am indebted to Frank J. Schuh, an outstanding drilling engineer, who introduced me to the drilling part of horizontal well technology. By teaching classes with Frank and working with him on field projects, I have learned about horizontal drilling operations and design. I will be eternally indebted to him for teaching me the practical part of the business. I would also like to thank R. V. Westermark of Phillips Petroleum Drilling Department, for working with me and educating me in this technology from day one. I would like to thank Dr. W. M. Maurer of Maurer Engineering, for all the encouragement and assistance that he provided. I am also thankful to many companies who were generous in providing the field histories and data which were used in the book. In particular, I would like to thank T. O. Stagg of British Petroleum, Alaska, and Barry Anderson of Western Mining Ltd., Australia, for being generous in providing all the necessary information to make this book complete. I am also thankful to Dr. D. K. Babu of Mobile Research who was kind enough to read Chapter 7 and provide many valuable suggestions.

There are numerous other individuals who have been very helpful in teaching me the various aspects of horizontal well technology. If I omitted any names, it is simply an oversight on my part.

CHAPTER 1

Overview of Horizontal Well Technology

INTRODUCTION

In the last few years, many horizontal wells have been drilled around the world.[1–27] The major purpose of a horizontal well is to enhance reservoir contact and thereby enhance well productivity. As an injection well, a long horizontal well provides a large contact area, and therefore enhances well injectivity, which is highly desirable for enhanced oil recovery (EOR) applications.

In general, a horizontal well is drilled parallel to the reservoir bedding plane. Strictly speaking, a vertical well is a well which intersects the reservoir bedding plane at 90°. In other words, a vertical well is drilled perpendicular to the bedding plane (see Fig. 1–1). If the reservoir bedding plane is vertical, then a conventional vertical well will be drilled parallel to the bedding plane and in the theoretical sense it would be a *horizontal well*. As shown in Figure 1–2, even in the reservoirs with vertical bedding plane, it is still possible to

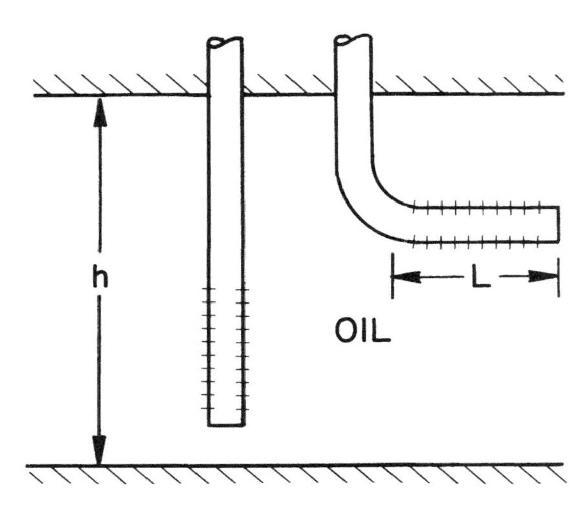

Figure 1–1 A Schematic of a Vertical Well Drilled Perpendicularly to the Bedding Plane, and a Horizontal Well Drilled Parallel to the Bedding Plane.

drill down vertically and then drill sideways. The objective here is to intersect multiple pay zones. (In some instances, from the drilling standpoint, it may be easier to stay in one zone to have effective control on well trajectory.) In the mid-continental region and in the Gulf Coast region of the United States, some reservoir bedding planes are almost vertical. Similarly, in California some reservoirs are steeply deepening. Thus, while analyzing hori-

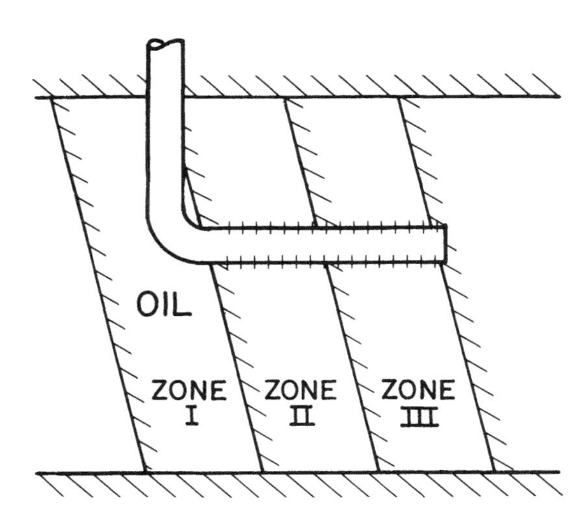

Figure 1–2 A Horizontal Well in a Reservoir With Vertical Bedding Planes.

zontal well performance, geometric configuration of the reservoir bedding planes should be considered.

A typical horizontal well project is different from a vertical well project because productivity of a well depends upon the well length. Moreover, the well length depends upon the drilling technique that is used to drill the well (See Table 1–1).[6] Therefore, it is essential that reservoir and drilling engineers work together to choose the appropriate drilling technique which will give the desired horizontal well length.

The other important consideration is well completion scheme: one can either have an open hole, insert a slotted liner, insert a liner with external casing packers, or case the hole and perforate the casing, depending upon local completion needs and experience. The type of completion affects horizontal well performance, and certain types of completions are possible only with certain types of drilling techniques. Thus, well length, the well's physical location in the reservoir, the tolerance in drilling location, and the type of completion that can be achieved strongly depend upon the drilling method. Therefore, it is very important for reservoir engineers to understand different drilling techniques, their advantages and disadvantages. Similarly, drilling engineers, completion engineers, production engineers, and geologists

TABLE 1–1 HORIZONTAL WELL LENGTHS[6]

TYPE	HOLE DIAMETER (in.)	RADIUS (ft)	RECORDED† (ft)	EXPECTED (ft)
Ultrashort*		1–2		100–200
Short**	$4\frac{3}{4}$	30	425	250–350
(Rotory)	6	35	889	350–450
Short**	$4\frac{3}{4}$	40	—	—
(Mud motors)	$3\frac{3}{4}$	40	—	—
Medium	$4\frac{1}{2}$	300	1300	500–1000
	6	300	2200	1000–2000
	$8\frac{1}{2}$	400–800	3350	1000–3000
	$9\frac{7}{8}$	300	—	—
Long	$8\frac{1}{2}$	1000	4000	1000–3000
	$12\frac{1}{4}$	1000–2500	1000	—

* several radials can be drilled from a single vertical well

** several drainholes at different elevations can be drilled from a single vertical well

† In early 1990, over 4500 ft long, medium radius horizontal wells were drilled.

should also understand and appreciate different factors that influence a horizontal well's performance. Hence, cooperation and teamwork of different professionals is essential to ensure the successful horizontal well project. A horizontal well project requires a multidisciplinary approach for an economic success.

LIMITATIONS OF HORIZONTAL WELLS

As noted earlier, the major advantage of a horizontal well is a large reservoir contact area. Currently, one can drill as long as 3000- to 4000-ft-long wells, providing significantly larger contact area than a vertical well. The major disadvantage is that only one pay zone can be drained per horizontal well. Recently, however, horizontal wells have been used to drain multiple layers. This can be accomplished by two methods: 1) one can drill a "staircase" type well where long horizontal portions are drilled in more than one layer, and 2) one can cement the well and stimulate it by using propped fractures. The vertical fractures perpendicular to the wells could intersect more than one pay zone and thereby drain multiple zones. It is important to note that in some cases due to strength of each pay zone and intermediate barriers, it may not be possible to interconnect the zones at different elevations by fracturing horizontal wells.

The other disadvantage of horizontal wells is their cost. Typically, it costs about 1.4 to 3 times more than a vertical well, depending upon drilling method and the completion technique employed. The incremental cost of drilling horizontal wells over vertical wells has reduced significantly over the last 10 years. Some of the early horizontal well projects listed in Tables 1–2 and 1–3 show that in the late seventies and the early eighties, horizontal well costs were six to eight times more than vertical well costs. By the mid-eighties and late eighties, typical drilling costs were two to three times more than vertical well costs.

An additional factor in cost determination is drilling experience in the given area. Typically, a first horizontal well costs much more than the second well. As more and more wells are drilled in the given area, an incremental drilling cost over a vertical well is reduced. Thus, there is a learning curve.

Field experience and published results of horizontal well costs in Cold Lake, Canada;[7] Prudhoe Bay, Alaska;[8] offshore Indonesia;[9] offshore The Netherlands;[10] Austin chalk formation in the United States;[11] and Bakken formation in North Dakota, U.S.A,[12] show a significant reduction in drilling costs over time and with experience. In these projects, a typical first horizontal well cost was two to four times more than a vertical well, but after drilling a few wells, a typical horizontal well cost is only about 1.4 times the vertical well cost. In some cases, with extensive drilling experiences, the horizontal well costs are reported to be almost the same or even lower than vertical well costs.[7] This tells us that for an economic success, the preferred option is to undertake a multiwell rather than a single well horizontal drilling program.

Costs for horizontal drilling and completions for 16 wells in Prudhoe Bay,[8] Alaska, are shown in Figures 1–3a and 1–3b. The figures show that drilling costs have decreased initially over time and have remained constant over the last two years. However, completion costs have remained constant over a four-year period. As shown in Figures 1–4a and 1–4b, similar cost

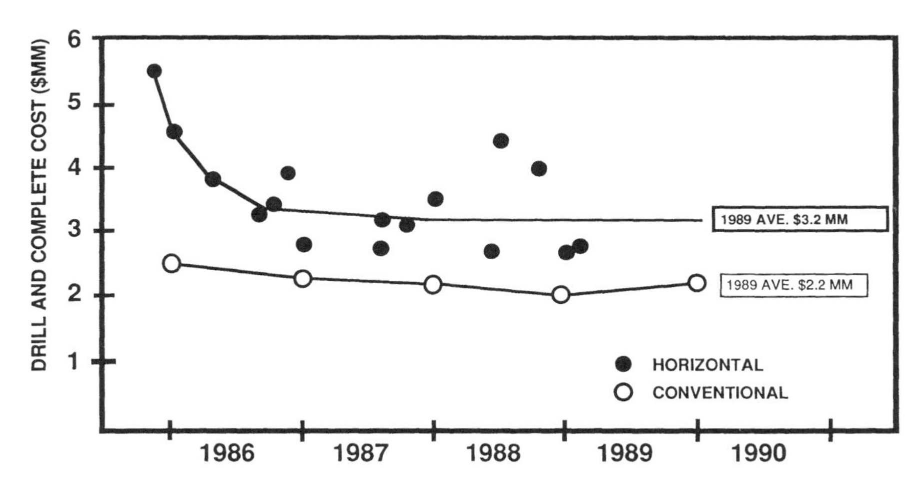

Figure 1–3a A Comparison of Horizontal and Vertical Well Costs for Prudhoe Bay, Alaska (CIM/SPE 90–124, Broman et al.).

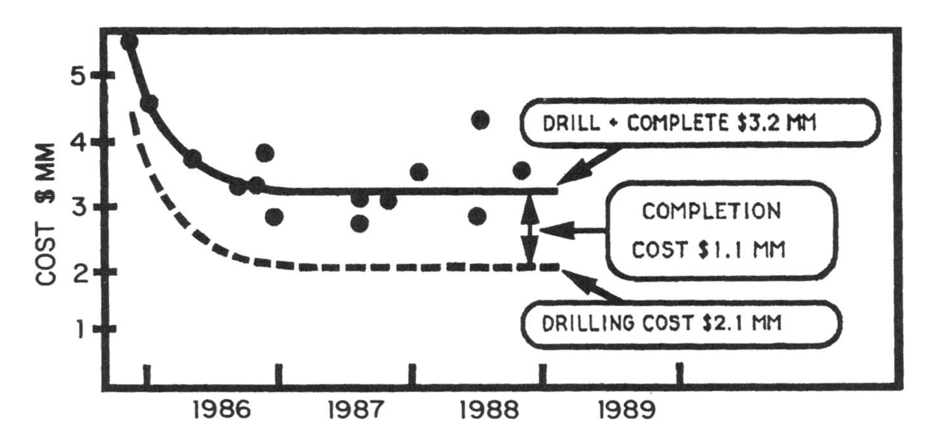

Figure 1–3b A Comparison of Drilling and Completion Costs of Horizontal Wells in Prudhoe Bay, Alaska (CIM/SPE 90-124, Broman et al.).

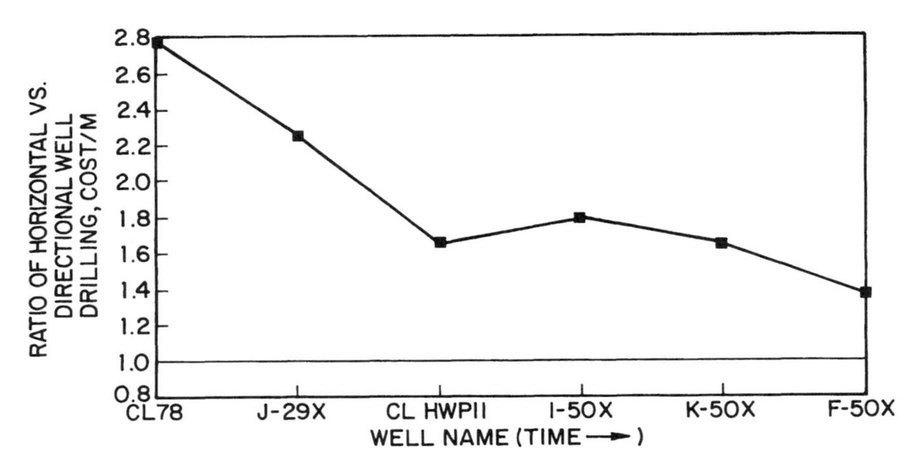

Figure 1–4a A Comparison of Horizontal and Slant Well Costs, Norman Wells, Canada.[7]

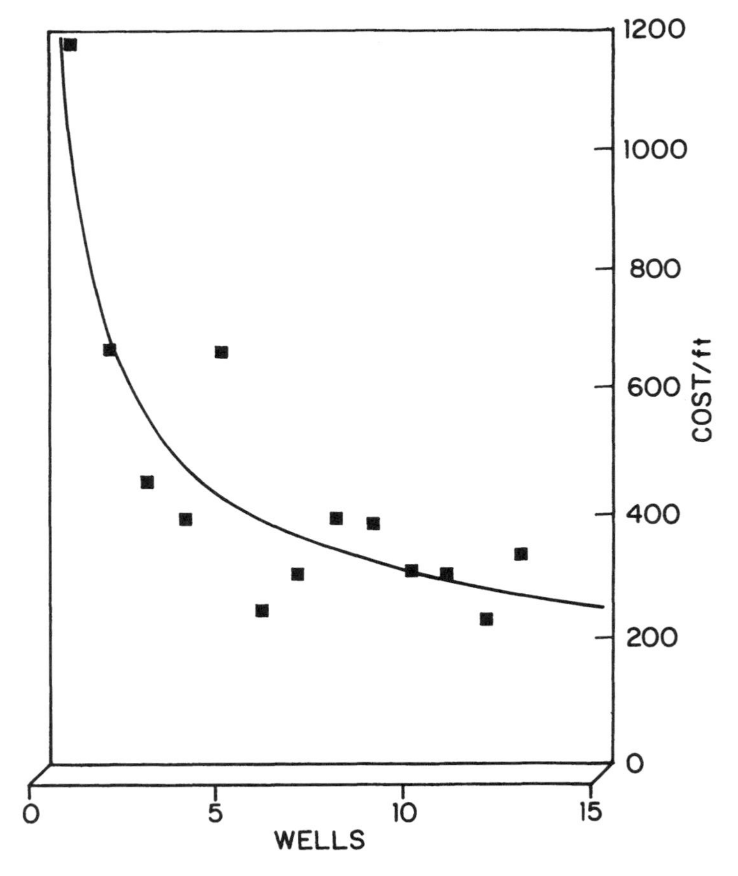

Figure 1–4b Horizontal Well Drilling Costs (Horizontal Section) in the Austin Chalk Formation, Texas, (*Petroleum Engineer,* pp. 22, April 1990).

trends are also seen in Canada and in Austin chalk wells in Texas, respectively.[7,8a]

As noted above, horizontal well costs are 1.4 to 3 times more than a vertical well. Hence, for an economic success, producible reserves from a horizontal well not only have to be proportionately larger, but they should also be produced in a shorter time span than a vertical well. If one assumes that for horizontal and vertical wells, the percentage recovery of original oil in place is the same (i.e., barrels recovered per acre-ft are the same), then to achieve larger producible reserves, horizontal wells will have to be drilled with a larger well spacing than vertical wells.

HORIZONTAL WELL APPLICATIONS

Horizontal wells have been used effectively in the following applications (a list of some of the early projects is given in Tables 1–2 and 1–3):[2,3]

1. In naturally fractured reservoirs, horizontal wells have been used to intersect fractures and drain them and the reservoir effectively (examples: Bakken formation,[12,13] North Dakota, U.S.A.; Austin chalk formation,[14,15] Texas, U.S.A.; and Devonian Shale,[16,17] West Virginia, U.S.A.).
2. In reservoirs with water and gas coning problems, horizontal wells have been used to minimize coning problems and enhance oil production (example: Rospo Mare field, offshore Italy;[18,19] Helder Field, offshore The Netherlands;[20,21] Bima Field, Indonesia;[9,22] Prudhoe Bay, Alaska, U.S.A.;[23] and Empire Abo Unit, New Mexico, U.S.A.[24,25,26]).
3. In gas production, horizontal wells can be used in low-permeability as well as in high-permeability reservoirs. In low-permeability reservoirs, horizontal wells can improve drainage area per well and reduce the number of wells that are required to drain the reservoir. In high-permeability reservoirs, where near-wellbore gas velocities are high in vertical wells, horizontal wells can be used to reduce near-wellbore velocities. Thus, horizonal wells can be used to reduce near-wellbore turbulence and improve well deliverability in high-permeability reservoirs. A recent application of a horizontal well in the Zuidwal gas field in The Netherlands confirms the effectiveness of horizontal wells in reducing near-wellbore turbulence.[10] A detailed discussion on this is included in Chapter 9.
4. In EOR applications, especially in thermal EOR, horizontal wells have been used (see a brief project list in Table 1–3). A long horizontal well provides a large reservoir contact area and therefore enhances injectivity of an injection well. This is especially beneficial in EOR applications where injectivity is a problem. Horizontal wells have also been used as producers.

TABLE 1–2 HORIZONTAL WELLS DRILLED

YEAR	COMPANY	FIELD	NO. OF WELLS
1937	—	Yarega, USSR	many
1939–41	Leo Ranney, et al.	McConnesville, Ohio	6
1942		Franklin Henry Field, Venago County, Pennsylvania	4 4
1942(?)	—	Midway Sunset, San Joaquin Valley, California	2
1946		Round Mountain Field, Kern County, California	9
1946	New Tech Oil, Malta, Ohio	—	117
1952		San Joaquin Valley, California (Midway Sunset)	1
1952	Venezuelan Oil Concessions, Ltd.	La Pas Field, Western Venezuela	?
1952	Long Beach Oil Development Co.	Los Angeles Basin Area (Wilmington Field)	8
1957		USSR	1
1967		China	1
1968		Marcovo, East Siberia, USSR	1
1978	Esso, Canada	Cold Lake, Alberta	1(?)
1979	Conoco	Tisdale, Wyoming	6
1979	Texaco	Fort McMurry, Alberta	3

TABLE 1–2 (CONTINUED)

DEPTH (ft)	RESERVOIR CONTACT LENGTH (ft)	(HORIZONTAL/ VERTICAL) COST	COMMENTS
	1000 max	—	Mine-assisted steam, production at about 1000 BOPD
	1000	—	Drilled through a tunnel
388	1000	Very Expensive	Drilled through a tunnel
388	600	—	
1100	70	—	Drilled through existing vertical well using downhole motor and flexible drill pipe
1650	56	—	Drilled through existing vertical well using downhole motor and flexible drill pipe
—	200	—	Drilled through a tunnel
1200	50	—	Six drainholes from a well
3700	50	—	Six drainholes from a well
10,000	50–10(?)	—	Drilled to reduce gas coning
3500 to 4800	50	—	In each well 6 to 8 drainholes were drilled. In some wells several drainholes were drilled at depth to reach many zones
1000	300	—	Production well
3600	1600	—	Produced 5 to 10 times more fluid than a vertical well, but collapsed in seven days
7200	1800	—	—
1558	1000	8–12	Drilled from the surface into unconsolidated sand, steam stimulation
—	1700 max	—	Drilled from tunnel
415	1000	5–6	Steam stimulation. Presently shut down

TABLE 1–2 HORIZONTAL WELLS DRILLED

YEAR	COMPANY	FIELD	NO. OF WELLS
1979	Esso, Canada	Normal wells under Mckenzie River,	2
		Alberta, Canada	1
1980–81	Elf-Aquitaine	Lacq Field, Southwest France	1
	Elf-Aquitaine	Lacq Field, Southwest France	1
1981–83	Elf-Aquitaine	Rospo Mare, Offshore Italy	1
	Elf-Aquitaine	Casterla Lou, South France	1
1980–84	ARCO	Empire Abo Unit, New Mexico	2
1981–84	ARCO	Empire Abo Unit, New Mexico	8
1984–84	Preussag	Lehrte Field, W. Germany	1
1985	Esso, Canada	Cold Lake, Alberta, Canada	1
1985	Petrobras	Fazenda Belam Field	1
1985	Sohio	McMullen Co., Texas	1
1985	Sohio	Glassock Co., Texas	1
1985–87	Sohio	Prudhoe Bay, Alaska	4
1985–86	Trendwell Oil	Niagaran reef trend reservoir, Muskegan County, Michigan	1
1985–86	Texas Eastern Skyline	Grassy Trail, Utah	2

TABLE 1–2 (CONTINUED)

DEPTH (ft)	RESERVOIR CONTACT LENGTH (ft)	HORIZONTAL/ VERTICAL COST	COMMENTS
1603	1860	—	—
—	4013	—	—
2195	330	4.3	Drilled from the surface
4100	1214	3.5	Drilled from the surface
4500	1988	2.1	Drilled into fractured limestone. Produced 15 times better than a vertical well. Reduced water coning
9500	1300	2.1	Not only increased production, but reduced water coning in a limestone formation
6200	300–400	2	Drilled new wells from the surface
6200	300–400	$250,000/ well	Drilled through existing vertical well
6300	1200	—	Drilled in a salt dome
1558	3330	—	Cyclic steam stimulation
1000	554	—	Produces five times better than the surrounding vertical wells
10,300	1908	—	Long horizontal well
—	295	—	Short drainhole
8989	1400	2–1.4	Three-fold increase in oil production rates over conventional well
3550	263	—	Short-radius drainhole, cost $350,000. Initial flow 629 BOPD
3900	250	—	Short-radius drainholes, up to three drainholes from a single vertical well

TABLE 1–2 HORIZONTAL WELLS DRILLED

YEAR	COMPANY	FIELD	NO. OF WELLS
1985–86	Liapco	Java Sea, Rama 1–7	1
1985–86	ARCO	Austin chalk, Rockwell County, Texas	2
1985–86	ARCO	Spraberry trend, Texas	1
1985–86	ARCO Indonesia	Bima field	9
1985–86	DOE/BDM	Wayne County, West Virginia	1
1985–86	USSR	Salym field, West Siberia	1

A proper orientation of horizontal wells, especially in naturally fractured reservoirs, could enhance sweep efficiency in EOR applications. Recently, horizontal wells have been used in waterflood, in polymer flood applications to improve sweep efficiency,[27] and also in miscible flood.[27a]

Other applications of horizontal wells are mainly related to overcoming the drilling and drilling related cost problems. In offshore wells, in remote locations, and in environmentally sensitive areas, where the project cost can only be reduced by minimizing the number of wells that are required to drain the given reservoir volume, horizontal wells are highly desirable. In these cases, horizontal wells provide unique advantages. For example, in offshore wells, platform costs are proportional to the number of slots, i.e., the number of wells that can be drilled from a platform. Long horizontal wells can be used not only to reduce the number of wells that are required to drain the given reservoir volume, but they can also increase reservoir volume that can be drained from a single platform, and reduce offshore project costs significantly. Similarly, in environmentally sensitive areas and reservoirs under cities, horizontal wells can be employed to drain a large reservoir volume with a minimum surface disturbance.

DRILLING TECHNIQUES

Before discussing various drilling techniques, it is important to define two terms: horizontal well and drainhole.

TABLE 1–2 (CONTINUED)

DEPTH (ft)	RESERVOIR CONTACT LENGTH (ft)	HORIZONTAL/ VERTICAL COST	COMMENTS
2650	1180	—	Long-radius horizontal well
1380	1344	—	Medium-radius wells, experimental drilling study in a non-productive zone
7871	1000	—	Medium-radius well, 891 ft exposed to the productive formation
2650	1200–2100	—	Medium-radius wells to reduce coning problems
6016	2200	Very Expensive	Gas well productive interval is about 800 ft
9000	1214	—	Found no commercial oil or gas

1. *Horizontal well:* A horizonal well is a new well drilled from the surface. The length usually varies from 1000 to 4500 ft.
2. *Drainhole:* Drainholes, which are also called laterals, are normally drilled from an existing well. The length usually varies from 100 to 700 ft.

The drilling techniques to drill horizontal wells and drainholes are classified into four categories, depending upon their turning radius. Turning radius is the radius that is required to turn from the vertical to the horizontal direction. The four drilling categories are:

1. *Ultrashort:* turning radius is 1 to 2 ft.; build angle is 45° to 60°/ft.
   In this technique, 100- to 200-ft-long drainholes are drilled using water jets.[28,29] As shown in Figure 1–5, the drainholes are drilled through a 7- to 10-foot-long, under-reamed zone, which is approximately two feet in diameter. The drainhole tubing diameter varies from 1¼ to 2½ inches, depending upon the drilling system used. After drilling, the tubing is perforated or gravel-packed. Then the tubing is severed and the next drainhole is drilled at the same elevation. It is possible to drill several drainholes, like bicycle spokes, at a given elevation.
2. *Short:* Turning radius is 20 to 40 ft.; build angle is 2° to 5°/ft.
   In this technique, drainholes are drilled either through a cased or through an uncased vertical well. In cased holes, a window, about 20 ft. long, is

TABLE 1–3 HORIZONTAL WELL, STEAM FIELD PROJECTS

COMPANY AND YEAR	LOCATION	REFERENCE	NUMBER	LENGTH FT
Yarega	USSR	[62]	Many	300 max
Signal Oil Co. (1965–66)	Sunny Side, Carbon County Utah	[63, 64]	3	370 ft, but 50 ft open to pay zone
Petro-Canada & other companies MAISP-I (1978)	Fort McMurry, Alberta, Canada	[65]	1 2	1083 394
Hopco (1982–83)	Kern County, California	[66, 68]	4	430 700
Esso–I (1978) Esso–II (1985)	Cold Lake, Alberta, Canada	[45, 46]	1 1	840 3330
Texaco (1979)	Fort McMurray, Alberta, Canada	[42, 43, 69]	3	1000
Texaco (1986)	California	—	1	—
Petrobras (1985)	Fazenda Belem Field, Brazil	[70]	1	544
UTF–AOSTRA (1985–87)	Athabasca Tar Sands, Alberta, Canada	[71]	3	230
Sceptre Resources & Murphy Oil (1987–88)	Saskatchewan, Canada	[72]	1	1640

milled to kick-off laterally. The earlier drilling system versions used surface rotary drilling to drill holes. In addition to surface rotation, flexible drill collar joints (see Fig. 1–6) are used to facilitate drilling. Normally, an angle build-up assembly is used to drill off the whipstock into the formation up to about 85°; a second stabilized assembly drills the rest of the hole. As noted in Table 1–1, it is possible to drill either 4¾- or 6-in. holes with expected length of about 250 to 450 ft. A maximum drilled length using

TABLE 1–3 (CONTINUED)

STEAM PROCESS	CUM. OIL PRODUCTION (bbls)	SOR	STATUS AND COMMENTS
—	—	—	Active/Mine Assist.
cyclic and drive	556	23	Inactive, 9–10° API oil, drilled from an outcrop side
cyclic and drive	5350	16	Inactive, ~10° API oil, drilled from an outcrop. Tests of horizontal wells, drilled from a mine
cyclic and drive	46,984	17	Inactive, 13° API oil, drilled from a shaft
cyclic steam & steam-assisted gravity drainage	—	—	Active, ~10° API oil Active, ~10° API oil
cyclic and drive	—	—	Inactive, ~10° API oil, drilled from the surface
cyclic	—	—	Active, 13° API oil
(?)	75 BOPD	—	14° API oil
—	—	—	Mine/shaft project, 10° API oil
steam-assisted gravity drainage	—	—	Vertical injectors and a horizontal producer, 13–14° API oil

this technique is 889 ft. One of the limitations of this system has been limited directional control. Recently developed systems which used downhole mud motors provide good directional control. This new system uses specially designed short mud motors. An angle-build motor is used to drill a 40-ft radius curve section of the wellbore and an angle-hold motor is used to drill the horizontal well section. With this new system, it is possible to drill a $4\frac{3}{4}$-in. diameter horizontal wellbore through a vertical well with a minimum $6\frac{1}{8}$-in. diameter. A $3\frac{3}{4}$-in. horizontal sec-

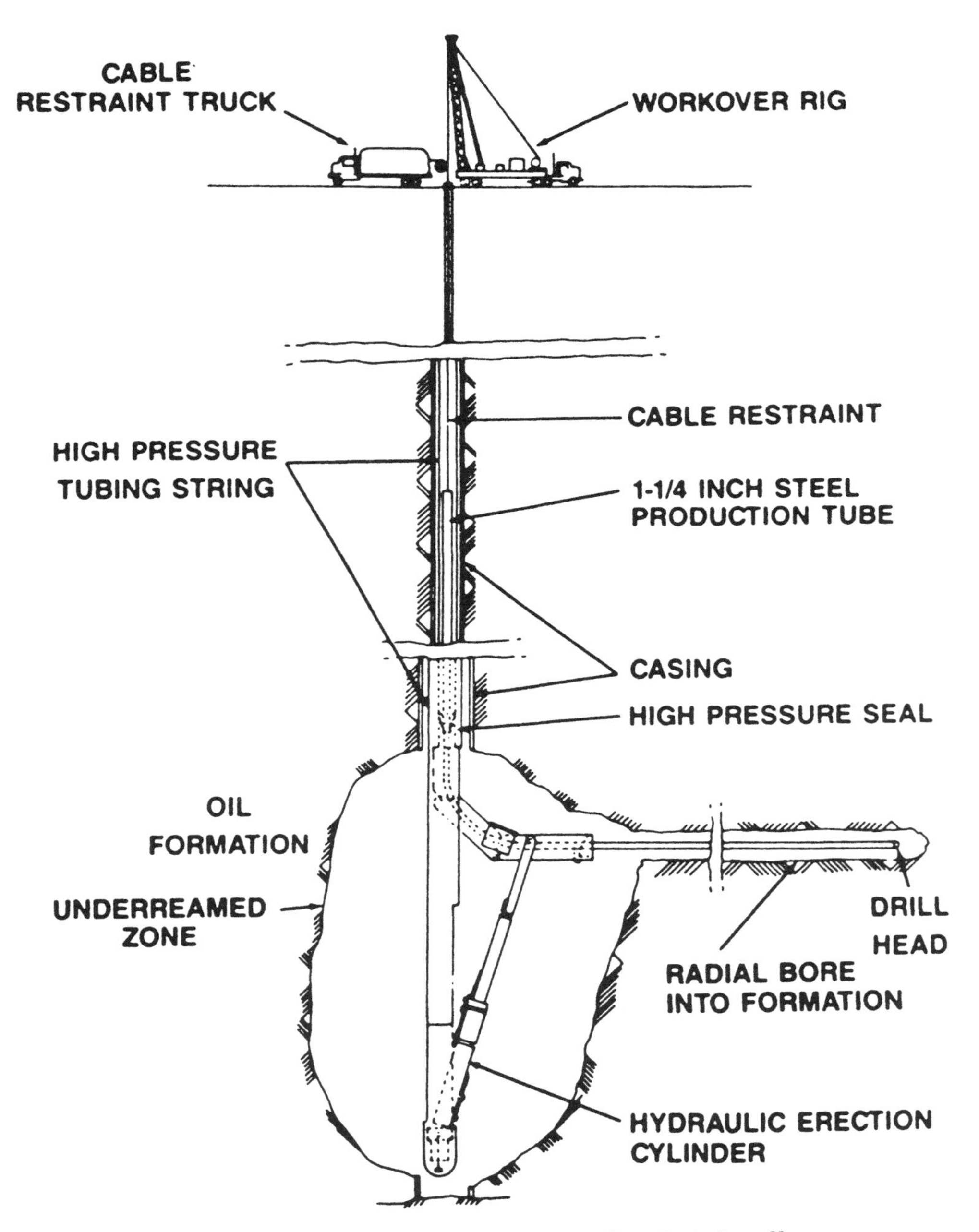

Figure 1–5 A Schematic of Water Jet Drilling Technique.[29]

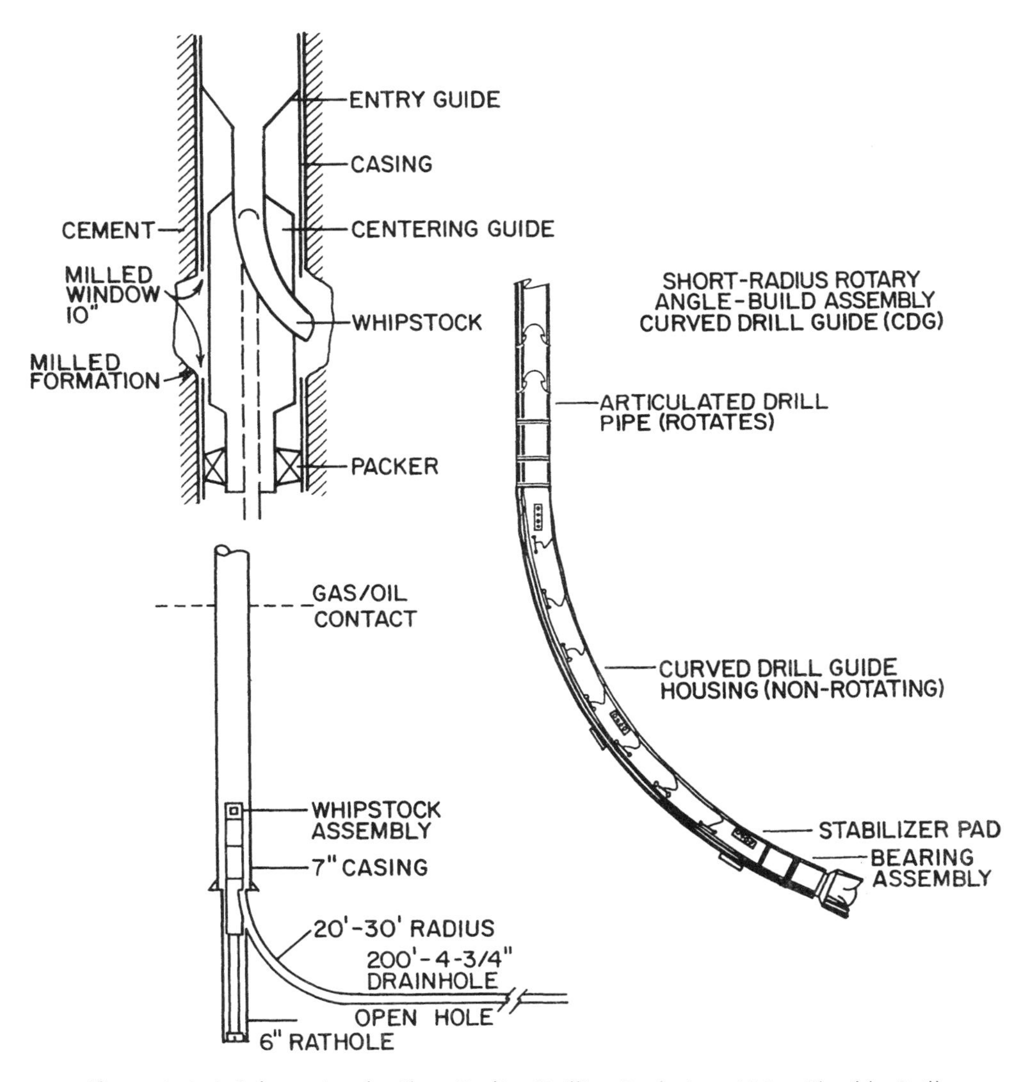

Figure 1–6 A Schematic of a Short-Radius Drilling Technique Using Flexible Drill Collar Joints (Eastman Christensen).

tion can be drilled through a vertical well with a minimum diameter of $4\frac{7}{8}$ in. As described later, these wells can be completed as an open hole or with slotted liners. Additionally, using this new technique, it is possible to drill about 1000-ft-long wells.

3. *Medium:* Turning radius is 300 to 800 ft; build angle is 6° to 20°/100 ft. This is becoming a predominant method to drill horizontal wells. Because

of the generous turning radius, it is possible to use most of the conventional oil field tools in the hole.

Specially designed downhole mud motors are used to drill horizontal wells. An angle-build motor is used to build angle and an angle-hold motor is used to drill the horizontal well section. It is possible to drill very long, 2000- to 4000-ft-long wells. Additionally, it is possible to complete them as open hole, with slotted liners, with liners and external casing packers, or it is possible to cement and perforate these wells. These types of wells have been cored, and it is possible to fracture stimulate these wells effectively.

4. *Long:* Turning radius is 1000 to 3000 ft; build angle is 2° to 6°/100 ft. This technique uses a combination of rotary drilling and downhole mud motors to drill these wells. Similar to conventional directional drilling, bent subs are used to kick-off and build angle. The horizontal position is drilled using downhole mud motors. Very long wells can be drilled using this technique. It is possible to core these wells. In addition, as noted in Table 1–4, several completion options are also available with these holes.

The above four methods are also summarized in Figure 1–7. Table 1–1 lists expected lengths and recorded lengths to date using different drilling techniques.[6] The details of various drilling methods can be found in References 28 to 47.

HORIZONTAL WELL LENGTHS BASED UPON DRILLING TECHNIQUES AND DRAINAGE AREA LIMITATIONS

Figure 1–7 and Table 1–1 show that we can drill a horizontal well with a turning radius as short as 1 to 2 ft and as long as 1000 to 3000 ft.

TABLE 1–4[1]

METHOD	COMPLETION	LOGGING
Ultrashort Radius	Perforated tubing or gravel pack	No
Short Radius	Open hole or slotted liner	No
Medium Radius	Open hole, slotted liner or cemented and perforated liner	Yes
Long Radius	Slotted liner or selective completion using cementing perforation	Yes

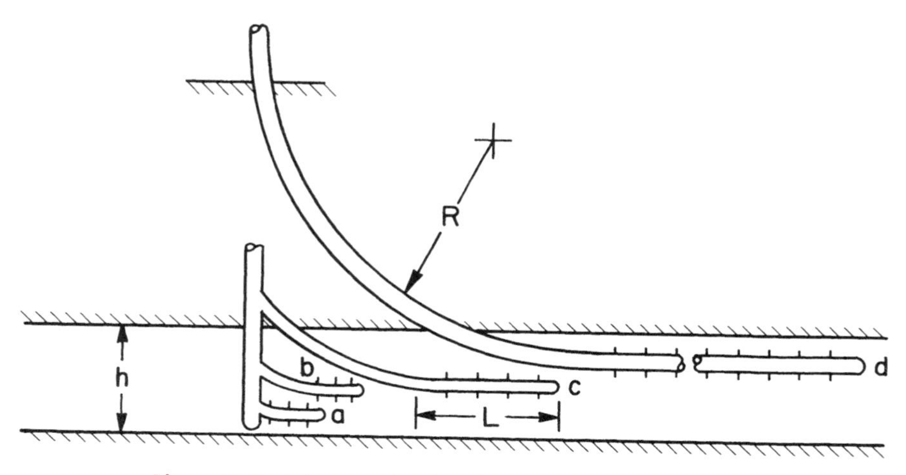

Figure 1–7 A Schematic of Different Drilling Techniques
a) Ultra short radius, $R = 1\text{–}2$ ft, $L = 100\text{–}200$ ft
b) Short radius, $R = 20\text{–}40$ ft, $L = 100\text{–}800$ ft
c) Medium radius, $R = 300\text{–}800$ ft, $L = 1000\text{–}4000$ ft
d) Long radius, $R \geqslant 1000$ ft, $L = 1000\text{–}4000$ ft

From the reservoir standpoint, turning radius is very important because certain acreages and lease sizes restrict the use of certain techniques. For example, in a 40-acre lease, it is difficult to drill a 2000-ft turning radius well.

EXAMPLE 1–1

In a 40-acre lease size, what is the maximum length of a well located centrally in the area that can be drilled using different drilling techniques?

Solution

Let us assume that the 40-acre lease is a square.

$$\text{Then lease area} = 40 \text{ acre} \times 43{,}560 \text{ ft}^2/\text{acre} = 1{,}742{,}400 \text{ ft}^2$$

and each side of a square is 1320 ft. Under most state laws in the U.S.A., one cannot cross the lease boundary. Moreover, in some states, one may have to stop the well at a certain distance from the lease boundary (typically 150 ft), specified by the state regulatory body. As shown in Figure 1–8, the maximum well length that we can drill along either the x or the y axis is

$$1320 - (150 \times 2) = 1020 \text{ ft}$$

The diagonal length of the square is 1867 ft. Hence the maximum possible well length along the diagonal is 1442 ft.

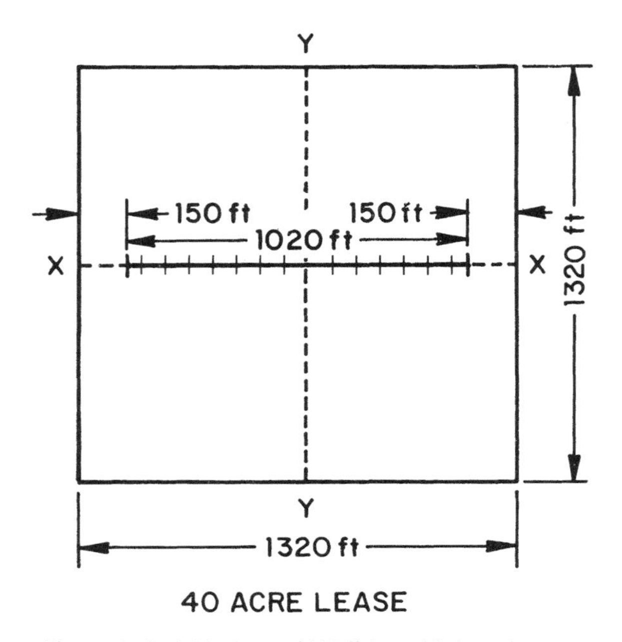

Figure 1–8 A Horizontal Well in a 40-Acre Lease.

1. *Ultrashort radius,* turning radius = 1 to 2 ft.
   - Maximum well length is maximum possible length of the drilling technique: 200 ft.
2. *Short radius,* turning radius = 20 to 40 ft.
   - As shown in Table 1–1, the generally expected length from the drilling technique is 450 ft. (Note that longer lengths are possible.)
3. *Medium radius,* turning radius = 300 ft.
   - If we locate a rig on one lease boundary, then we can drill a maximum well length of 720 ft for a centrally located well in the *x-y* drainage plane. The tips of this well would be 300 ft away from the lease boundaries which are perpendicular to the well.
4. *Long radius,* turning radius = 1000 ft.
   - It may be difficult to drill a centrally located well along the *x-y* drainage plane. However, we can drill a 665-ft-long well along the diagonal drainage plane. One well tip will be 1000 ft away from the ends of the diagonal, while the other well tip will be 212 ft away from the other end of the diagonal. Thus, this well will not be equally spaced from all the boundaries.

EXAMPLE 1–2

What are the maximum lengths of a horizontal well that can be drilled using different drilling techniques in 60-acre and 80-acre leases?

Solution

The maximum length of a horizontal well that can be drilled in a given acre spacing is determined by the x and y dimensions of the drainage area and the length of the diagonal. We assume a square drainage area in the calculations. The possible horizontal well lengths for different drilling techniques are shown in Figures 1–9 and 1–10 and are summarized below:

Drainage Area =		60 acre	80 acre
Maximum x or y dimension		1617 ft	1867 ft
Maximum length of diagonal		2286 ft	2640 ft
Drilling Method	**Turning radius, ft**	**Maximum well length, ft**	**Maximum well length, ft**
Ultrashort Radius	1 to 2	200	200
Short Radius	20 to 40	450	450
Medium Radius	300 to 800	1017*	1267*
Long Radius	1000 to 2500	1074**	1428**

* Centrally located well along the x-x or y-y axis of the drainage plane

** Well along the diagonal of the drainage plane but not centrally located in the drainage plane

COMPLETION TECHNIQUES

As noted earlier, it is possible to complete horizontal wells as open hole, with slotted liners, liners with external casing packers (ECPS), and cemented and perforated liners. The choice of completion method can have a significant influence on well performance. The various completion options and their advantages and disadvantages are summarized below.[48–61] Additionally, various issues that need to be considered before choosing completions are also briefly discussed. More details about the completion are included in Volume II.

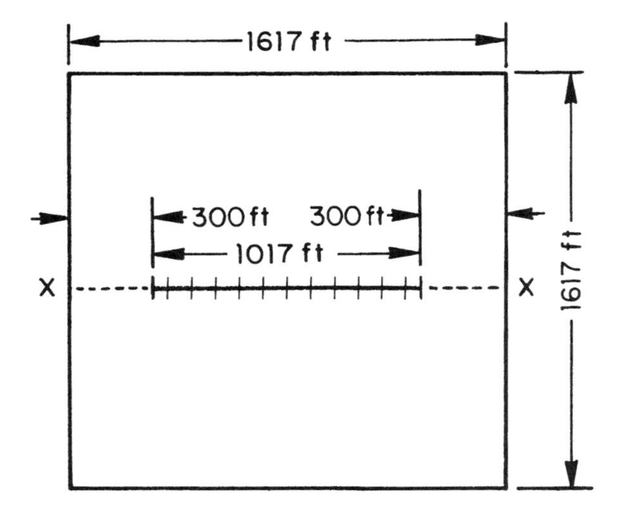

Figure 1–9 A Medium-Radius Horizontal Well in a 60-Acre Lease.

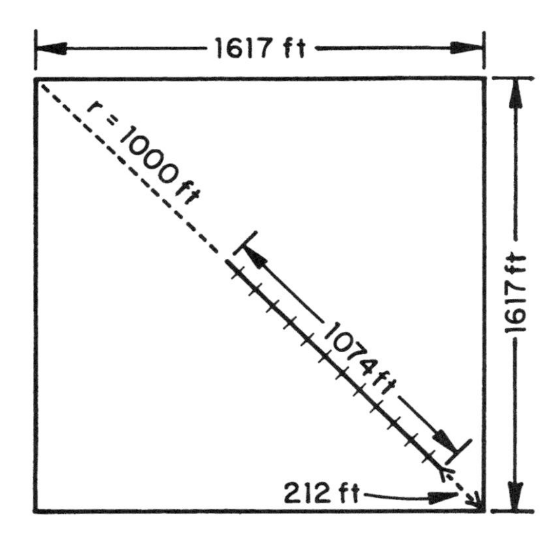

Figure 1–10 A Long-Radius Horizontal Well in a 60-Acre Lease.

COMPLETION OPTIONS

Figure 1–11 shows a schematic diagram of various completion options for horizontal wells. These completion aspects are described below:

1. *Open Hole:* Open-hole completion is inexpensive but is limited to competent rock formations. Additionally, it is difficult to stimulate open-hole

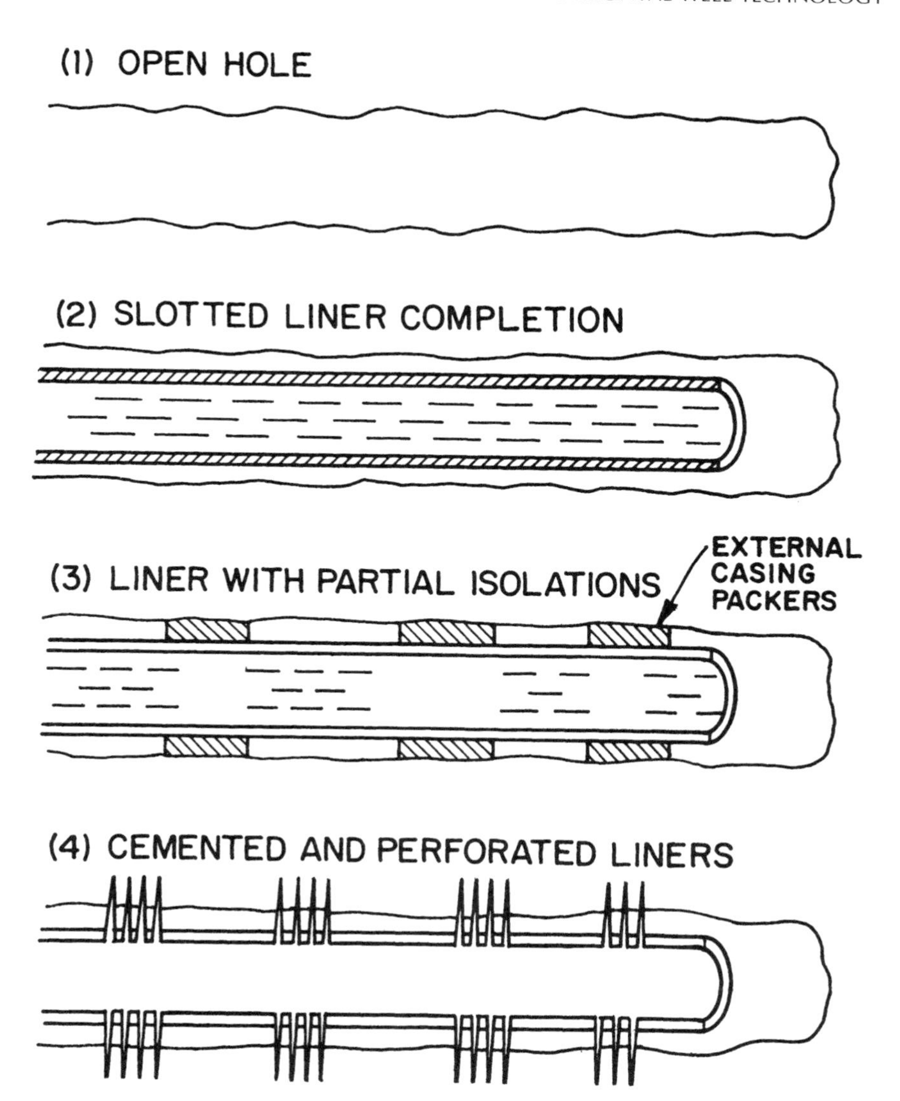

Figure 1–11 A Schematic of Various Completion Techniques for Horizontal Wells.

wells and to control either injection or production along the well length. A few early horizontal wells have been completed open hole but the present trend is away from using open hole completions, except in formations such as Austin Chalk.

2. *Slotted Liner Completion:* The main purpose of inserting a slotted liner in a horizontal well is to guard against hole collapse. Additionally, a liner provides a convenient path to insert various tools such as coiled tubing in a horizontal well. Three types of liner have been used:
   a. perforated liners, where holes are drilled in the liner;
   b. slotted liners, where slots of various width and depth are milled along the liner length; and
   c. prepacked liners.

Slotted liners provide limited sand control by selecting hole sizes and slot width sizes. However, these liners are susceptible to plugging. In unconsolidated formations, wire wrapped slotted liners have been used effectively to control sand production. Recent literature indicates the successful use of gravel packing for effective sand control in horizontal wells.

The main disadvantage of a slotted liner is that effective well stimulation can be difficult, due to the open annular space between the liner and the well. Similarly, selective production and injection is difficult.

3. *Liner with Partial Isolations:* Recently, external casing packers (ECPS) have been installed outside the slotted liner to divide a long horizontal wellbore into several small sections (Fig. 1–11). This method provides limited zone isolation, which can be used for stimulation or production control along the well length.

A recent Department of Energy (DOE) well in the Devonian shale formation in West Virginia used a liner with port collars.[14,15] Additionally, external casing packers were also used to divide a long horizontal well into several sections. The well was stimulated successfully using foam fracturing in several of the selected zones.

Normally, horizontal wells are not horizontal; rather, they have many bends and curves. In a hole with several bends, occasionally it may be difficult to insert a liner with several external casing packers.

4. *Cemented and Perforated Liners:* It is possible to cement and perforate medium and long radius wells. As noted earlier, at the present time, it is not economically possible to cement short radius wells. Cement used in horizontal well completion should have significantly less free water content than that used for vertical well cementing. This is because in a horizontal well, due to gravity, free water segregates near the top portion of the well and heavier cement settles at the bottom. This results in a poor cement job. To avoid this it is important to conduct a free water test for cement at least at 45°, in addition to or instead of, the conventional API free water test, which is conducted in the vertical position.

COMPLETION CONSIDERATIONS

Several items need to be considered before selecting an appropriate completion scheme. A brief discussion of each item is given below:

1. *Rock and Formation Type:* If an open-hole completion is considered, then it is important to ensure that the rock is competent and the drilled hole will be stable. Several early horizontal wells drilled in competent limestone formations have been completed as open holes. Present field experience reported in the literature indicates that when the horizontal wells are drilled along the direction of least horizontal stress, they exhibit excellent hole stability.
2. *Drilling Method:* As noted before, with short radius only an open-hole or a slotted-liner completion is possible. With a medium- and long-radius well, one can complete them either as open hole, open hole with slotted liner, or with cement and perforate them.
3. *Drilling Fluid/Mud Cleanup:* Formation damage during horizontal drilling is a serious problem in many wells, especially for wells drilled in low permeability reservoirs. Horizontal drilling takes significantly longer time than drilling a vertical well, and the producing formation is exposed to drilling fluid for a longer time period than in a vertical well. Thus, the possibility of mud invasion and related formation damage in a horizontal well is higher than that in a vertical well. Therefore, a method must be devised for well cleanup. Although not impossible, it is difficult to clean up a horizontal well completed as an open hole or with slotted liners. If the well has a large turning radius then swab tools can reach at least up to the end of the curve. For sharp turning radius wells, swab tools cannot reach beyond the vertical well portion.

To minimize damage while drilling horizontally, one can drill under balance. It is also possible to use a special mud, for example, a polymer mud with either minimal or no solids. However, these types of muds may have problems with shale caving and sloughing. Moreover, some mud systems may have a limited capacity to carry solid cuttings. This may result in cutting accumulation in the horizontal well portion.

Another alternative for dealing with formation damage is to cement and perforate the horizontal well, the way we do for vertical wells. Perforations may extend past the drilling damage. Then, one can either break down or do a limited fracture job on a horizontal well to regain the lost productivity due to drilling and cementing. The objective of stimulating here is to achieve well productivity at least the same as that for an undamaged open-hole horizontal well.

It is important to note that many horizontal wells, especially in offshore Europe and Asia, have been successfully completed using slotted liners. In many of these wells, flow rates exceed a few hundred to thousands of barrels per day. At a high flow rate the well has a better chance of self cleanup than at low flow rates.

4. *Stimulation Requirements:* A cemented horizontal well is preferred if the well is to be fractured. The well can be isolated in several zones along its length by using bridge plugs and each zone can be fractured independently. Recently, several wells have been completed by inducing multiple fractures along the well length. From a mechanical standpoint, it is preferable to fracture various zones along the well length in stages. It is prudent to use reservoir engineering criteria to design the number of fractures required along the well length to maximize recovery and minimize fracturing cost.

Recently, several papers have been published which discuss the orientation of induced fractures with respect to drilled direction of a horizontal well.[58-60] In principal, for primary production one would like to obtain fractures perpendicular to the drilled wellbore direction and to enhance drainage. However, such a configuration is not desirable for secondary recovery applications. In a developed pattern, induced fractures perpendicular to the well direction would cause short-circuiting of the injection fluid, resulting in an early breakthrough and poor sweep efficiency. Thus, if secondary recovery is contemplated, inducing a long fracture perpendicular to a horizontal well's direction should be avoided. In fact, for secondary recovery, an induced fracture parallel to the horizontal well's direction may be desirable.

It is difficult to fracture wells completed as an open hole or with a slotted liner. This is because of large leak-off occurring along the well length. Similarly, uniform acidization along the length of a well completed as open hole or with slotted liner is also difficult. The difficulty in uniform acidization along the well length can be reduced by using coiled tubing. To ensure uniform acid distribution along the well length, the coiled tubing may have to be moved up and down the hole while spraying the acid. Another alternative is to use a staged acidization treatment where each stage is separated by a chemical diverter. (The chemical diverting agent breaks down over time.)

5. *Production Mechanism Requirements:* In some wells, especially those drilled in a fractured reservoir with a bottom water drive, water may break through in a certain portion of the long horizontal well. Similarly, in an enhanced recovery application, the injected fluid, such as water, may

show a premature breakthrough along a small portion of a long producing horizontal well. In such cases, one may have to plug off a certain portion of a long well.

The effective way to plug off well length is to isolate the zone where undesirable fluids are entering the well and squeeze that zone off using cement. A completion plan should include design considerations for such contingencies.

In reservoirs with gas caps, it is important to obtain effective well isolation from the gas cap. One can either use packers or cemented liners to isolate production tubing from the gas cap. Literature indicates that some horizontal wells were not able to meet their expectations due to premature gas breakthrough in the well portion located in the proximity of the gas cap.

Rarely are horizontal wells truly horizontal; rather, they wander up and down in the vertical plane. In low rate wells, well shapes can have significant impact on well productivity, especially when multiphase flow is involved. For example, water may accumulate in a low portion of the wellbore, and it may be difficult to displace it. Similarly, there is a possibility of gas lock occurring near hook-shaped well portions. In such situations, gas anchors can be used to mitigate the problem. However, the best way to handle this completion problem is to design a well path slightly up-dipping or down-dipping, depending upon the reservoir mechanism. This will facilitate fluid segregation along the well length and reduce problems due to gas blocking in oil wells and liquid loading in gas and condensate wells.

6. *Workover Requirements:* Before selecting a completion option, workover requirements must be considered, but they are also difficult to anticipate. For example, consider completing a medium radius horizontal well in a competent but fractured limestone reservoir with a bottom water drive. One can anticipate a possibility of water breakthrough along a small portion of the horizontal well sometime during the well life. The following three completion scenarios are possible:
   a. One can insert a slotted liner and pull it out later when water breaks through or water cut gets high. After pulling the liner out, one can insert a casing and cement it. This will stop water production. However, how risky is it to pull a slotted liner out of a horizontal hole?
   b. One can cement the well and perforate it. Once the water breaks through, production logging can be used to locate the high water production zone. Later, one can squeeze the zone off using cement.
   c. One can complete the well as an open hole and wait until water breakthrough occurs to design a course of action.

   Each of these options has costs and risks associated with it. The completion choice should be based on local operating experience and the operator's willingness to assume a degree of risk.

Currently, in the ultrashort-radius technique, tubing is severed once the hole is drilled. Therefore, it is not possible to reenter the horizontal section of the wellbore. In a short-radius well it is possible to reenter by using coiled tubing. With coiled tubing, it is probably safer to reenter a hole completed with a slotted liner than to reenter an open hole. In medium- and long-radius wells, reentry is not very difficult. In these wells either coiled tubing or drill pipe conveyed tools can be used.

7. *Abandonment Requirements:* At the present time no special regulations are in effect for abandoning a horizontal well. However, an operator should anticipate these needs and design well completion so that the well can be abandoned safely.

COMPLETION SUMMARY

Proper well completion is essential to ensure a successful horizontal well project. Based on the completion needs, one can choose an appropriate drilling technique. For example, if the well is to be cemented, ultrashort and short-radius drilling techniques cannot be used. In contrast, medium- and long-radius wells can be cemented and perforated. Table 1–4 includes a summary of possible completion options and logging options for difficult drilling methods.

SUMMARY

Chapter 1 gives a brief review of horizontal well technology which includes a list of early horizontal well projects for primary recovery and enhanced oil recovery (EOR) applications, a discussion of horizontal well applications, and shows a substantial benefit in naturally fractured formations such as Austin Chalk, Texas, and Bakken formation in North Dakota. Similarly, the benefit of horizontal wells is also seen in gas and water coning applications, such as Rospo Mare field, offshore Italy; Prudhoe Bay, Alaska; offshore The Netherlands; offshore Indonesia; and in Empire Abo Unit, New Mexico.

The applications of horizontal wells are based on reservoir needs and drilling needs. Many times in offshore fields, arctic fields, or in environmentally sensitive areas, horizontal wells can improve the project economics by reducing the number of wells that are required to drain the reservoir. A reduction in the number of wells results in a substantial cost savings.

This chapter also includes a brief discussion about various drilling methods and the expected horizontal well lengths that can be drilled using these methods, and provides a description of various completion methods, listing several reservoir, drilling and production aspects that must be considered before deciding on the completion scheme.

REFERENCES

1. Joshi, S. D.: "A Review of Horizontal Well and Drainhole Technology," paper SPE 16868, presented at the 1987 Annual Technical Conference, Dallas, Texas. A revised version was presented in the 1988 SPE Rocky Mountain Regional Meeting, Casper, Wyoming, May 1988.
2. Joshi, S. D.: "A Review of Thermal Oil Recovery Using Horizontal Wells," *In Situ,* vol. 11, no. 1 of 3, pp. 211–259, 1987.
3. Bosio, J. C., Fincher, R. W., Giannesini, J. F., and Hatten, J. L.: "Horizontal Drilling—A New Production Method," presented at the 12th World Petroleum Congress, Houston, Texas, April 1987.
4. Moritis, G.: "Worldwide Horizontal Drilling Surges," *Oil & Gas Journal,* pp. 53–63, Feb. 27, 1989.
5. Moritis, G.: "Horizontal Drilling Scores More Success," *Oil & Gas Journal,* pp. 53–64, Feb. 26, 1990.
6. Schuh, Frank J.: "Horizontal Drilling: Where We've Been and Where We're Going," *Drilling,* pp. 10–14, May/June 1989.
7. Gust, D.: "Horizontal Drilling Evolving from Art to Science," *Oil & Gas Journal,* pp. 43–52, July 24, 1989.
8. Personal Communication with Ted Stagg, B. P. Petroleum, Prudhoe Bay, Alaska, Feb. 6, 1990. See also paper by Broman, W. H., Stagg, T. O. and Rosenzweig, J. J.: "Horizontal Well Performance Evaluation at Prudhoe Bay," CIM/SPE paper No. 90–124, 1990.

8a. Moore, Steven D. "Oryx Develops Horizontal Play," *Petroleum Engineer International,* pp. 16–22, April 1990.

9. Barrett, S. L., Lyon, R. G.: "The Navigation Drilling System Proves Effective in Drilling Horizontal Wells in the Java Sea," paper IADC/SPE 17238, presented at 1988 IADC/SPE Drilling Conference, Dallas, Texas, Feb. 28–March 2, 1988.
10. Celier, G. C. M. R., Jouault, P., de Montigny, O. A. M. C.: "Zuidwal: A Gas Field Development With Horizontal Wells," paper SPE 19826, presented at the 64th Annual Technical Conference and Exhibition of the Society of Petroleum Engineers, San Antonio, Texas, Oct. 8–11, 1989.
11. "Horizontal Drilling," *Oil & Gas Journal,* p. 105, July 31, 1989.
12. "Horizontal Drilling Grows in Williston," *Oil & Gas Journal,* pp. 22–25, Nov. 6, 1989.
13. "Horizontal Drilling," Equity Research, *First Boston.*
14. Sheikholeslami, B. A., Schlottman, B. W., Siedel, F. A., Button, D. M.: "Drilling and Production Aspects of Horizontal Wells in the Austin Chalk," paper SPE 19825, presented at the 64th Annual Technical Conference, San Antonio, Texas, Oct. 8–11, 1989.
15. Stang, C. W.: "Alternative Electronic Logging Technique Locates Fractures in Austin Chalk Horizontal Well," *Oil & Gas Journal,* pp. 42–45, Nov. 1989.
16. Yost, II, A. B., Overbey, Jr., W. K., Wilkins, D. A., Locke, C. D.: "Hydraulic Fracturing of a Horizontal Well in a Naturally Fractured Reservoir: Case Study for Multiple Fracture Design," paper SPE 17759, presented at the SPE Gas Technology Symposium, Dallas, Texas, June 13–15, 1988.
17. Yost, II, A. B.: "Horizontal Gas Well Promises More Devonian Production," *The American Oil & Gas Reporter,* pp. 29–32, July 1988.
18. Reiss, L. H.: "Production From Horizontal Wells After 5 Years," *Journal of Petroleum Technology,* pp. 1411–1416, November 1987.
19. Reiss, L. H.: "Producing the Rospo Mare Oil Field By Horizontal Wells," pre-

sented at Seminar on Recovery from Thin Oil Zones, Stavanger, Norway, April 21–22, 1988.
20. Murphy, P. J.: "Performance of Horizontal Wells in the Helder Field," paper SPE 18340, presented at the SPE European Petroleum Conference, London, United Kingdom, Oct. 16–19, 1988.
21. Stewart, C. D., Williamson, D. R.: "Horizontal Drilling Aspects of the Helder Field Redevelopment," paper SPE 17886, presented at the 20th Annual OTC, Houston, Texas, May 2–5, 1988.
22. Cooper, R. E., Troncoso, J. C.: "An Overview of Horizontal Well Completion Technology," paper SPE 17582, presented at the SPE International Meeting on Petroleum Engineering, Tianjin, China, Nov. 1–4, 1988.
23. Sherrard, Dave W., Brice, Bradley W., MacDonald, David G.: "Application of Horizontal Wells at Prudhoe Bay," *Journal of Petroleum Technology*, pp. 1417–1425, November 1987.
24. Dech, J. A., Wolfson, L.: "Advances in Horizontal Drilling," 7th Bi-Annual Petroleum Congress of Turkey, Ankara, Turkey, April 1987.
25. Stramp, R. L.: "The Use of Horizontal Drainholes in Empire Abo Unit," paper SPE 9221, presented at the 1980 Annual Meeting, Dallas, Texas, Sept. 21–24.
26. Detmering, T. J.: "Update on Drainhole Drilling—Empire Abo," Proceeding of 31st Annual Southwestern Petroleum Short Course, Lubbock, Texas, pp. 25–41, April 1984.
27. Bruckert, Louis: "Horizontal Well Improves Oil Recovery From Polymer Flood," Technology, *Oil & Gas Journal*, pp. 35–39, Dec. 18, 1989.
27a. Adamache I., et al.: "Horizontal Well Application in a Vertical Miscible Flood," CIM paper No. CIM/SPE 90-125, presented at the joint CIM-SPE meeting in Calgary, Canada, June 10–13, 1990.
28. Dickinson, W., and Dickinson, R. W., Petrolphysics Ltd.: "Horizontal Radial Drilling System," SPE 13949, presented at the SPE 1985 California Regional Meeting, Bakersfield, California, March 27–29, 1985.
29. Dickinson, W., Petrolphysics Ltd., Anderson, R. R., Bechtel National Inc., and Dickinson, R. W., Petrolphysics Ltd.: "A Second-Generation Horizontal Drilling System," IADC/SPE 14804, presented at the 1986 IADC/SPE Drilling Conference, Dallas, Texas, Feb. 10–12, 1986.
30. Mall, T., Fincher, R.: "Michigan Operator Salvages Well Using Lateral Drilling," *Oil & Gas Journal*, pp. 33–38, June 1986.
31. Fincher, R. W.: "Short-Radius Lateral Drilling: A Completion Alternative," *Petroleum Engineering International*, pp. 29–35, February 1987.
32. Parson, R. S. and Fincher, R. W.: "Short-Radius Lateral Drilling: A Completion Alternative," paper SPE 15943, presented at the 1986 Eastern Regional Meeting, Columbus, Ohio, Nov. 12–14.
33. "Short-Radius Motor System: Advanced Technology for High Performance Drilling," brochure by Eastman Christensen Company, Houston, Texas, May 1989.
34. Karlsson, H., Cobbley, R., and Jaques, G. E., Eastman Christensen: "New Developments in Short-, Medium-, and Long-Radius Lateral Drilling," SPE/IADC 18706, presented at the 1989 SPE/IADC Conf., New Orleans, Louisiana, Feb. 28–March 3, 1989.
35. Kerr, D., and Lesley, K., Smith Intl.: "Mechanical Aspects of Medium Radius Well Design," SPE 17618, presented at the SPE International Meeting on Petroleum Engineering, Tianjin, China, Nov. 1–4, 1988.
36. "Horizontal Re-Completion Techniques Refined," *DRILLING CONTRACTOR*, pp. 28–29, Dec. 1988–Jan. 1989.

37. Greener, J. M., Sauvageau, K. A., Pasternack, I., Amoco Production Co.: "Horizontal Drilling of Eolian Nugget Sandstone in the Overthrust Belt," SPE 17527, presented at the SPE Rocky Mountain Meeting, Casper, Wyoming, May 11–13, 1988.
38. Edlund, P. A.: "Application of Recently Developed Medium-Curvature Horizontal Drilling Technology in the Spraberry Trend Area," paper SPE 16170, 1987 SPE/IADC Drilling Conference, New Orleans, Louisiana, March 15–18.
39. Briggs, G. M.: "How to Design a Medium-Radius Horizontal Well," *Petroleum Engineer International,* pp. 26–37, September 1989.
40. Prevedel, B.: "New Techniques in Horizontal and Drainhole Drilling Optimization: Lehrte 41 Lateral Drilling Project," paper SPE 15694, presented at the 1987 SPE Middle East Oil Show, Manama, Bahrain, March 7–10, 1987.
41. Jourdan, A.: "Drilling of Horizontal Wells," *Petrole et Techniques,* No 294, pp. 37–39, December 1982.
42. Pugh, G. E.: "Drilling of Three Horizontal Hole Pattern, Fort McMurray, Alberta," paper CIM 82–33–68, presented at the 1982 Annual Meeting of the Petroleum Society of CIM, Calgary, Alberta, June 6–9, 1982.
43. Loxam, D. C.: "Texaco Canada Completes Unique Horizontal Drilling Program," *Petroleum Engineering International,* pp. 40–52, September 1982.
44. Baldwin, D. D., Royal, R. W., and Gill, H. S.: "Drilling High Angle Directional Wells," Proceedings of the Eleventh World Petroleum Congress, vol. 3, pp. 15–24, London, 1983.
45. Bezaire, G. E. and Markiw, I. A.: "Esso Resources Horizontal Hole Project at Cold Lake," paper CIM 79–30–10, presented at the 1979 Annual Meeting of the Petroleum Society of CIM, Banff, Canada, May 8–11, 1979.
46. MacDonald, R. R.: "Drilling the Cold Lake Horizontal Well Pilot No. 2," paper SPE 14428, presented at the 1985 Annual Meeting, Las Vegas, Nevada, Sept. 22–25, 1985.
47. Markle, R. D.: "Drilling Engineering Considerations in Designing a Shallow Horizontal Well at Norman Wells, N.W.T., Canada," paper SPE/IADC 16148, presented at the 1987 SPE/IADC Drilling Conference, New Orleans, Louisiana, March 15–18, 1987.
48. Zaleski Jr., T. E., Spatz, Edward: "Horizontal Completions Challenge for Industry," *Oil & Gas Journal,* May 2, 1988.
49. Austin, Carl, Zimmerman, Chris, Sullaway, Bob, and Sabins, Fred, Halliburton Services: "Fundamentals of Horizontal Well Completions," *Drilling,* May/June 1988.
50. Spreux, A., Lessi, J.: "Most Problems in Horizontal Completions are Resolved," *Oil & Gas Journal,* June 13, 1988.
51. Joshi, Dr. S. D.: "Proper Completion Critical For Horizontal Wells," *The American Oil & Gas Reporter,* December 1989.
52. Dickinson, W., Anderson, R. R., and Dickinson, W., Petrolphysics Ltd., and Dykstra, H.: "Gravel Packing of Horizontal Wells," SPE paper 16931, presented at the 62nd Annual Technical Conference and Exhibition of the Society of Petroleum Engineers, Dallas, Texas, Sept. 27–30., 1987.
53. Elson, T. D., Darlington, R. H., Mantooth, M. A.: "High-Angle Gravel Pack Completion Studies," paper SPE 11012, presented at 57th Annual Fall Technical Conference and Exhibition of the Society of Petroleum Engineers of AIME, New Orleans, Louisiana, Sept. 26–29, 1982.
54. Austin, C. E. and Rose, R. E.: "Simultaneous Multiple Entry Hydraulic Fracture Treatments of Horizontally Drilled Wells," paper SPE 18263, presented at the

63rd Annual Technical Conference and Exhibition of the Society of Petroleum Engineers, Houston, Texas, Oct. 2–5, 1988.
55. Giannesini, J. F.: "Production Technology for Horizontal Wells Takes New Direction," *World Oil,* May 1989.
56. Weirich, J. B., Zaleski Jr., T. E., Mulchay, P. M., Baker Sand Control: "Perforating the Horizontal Well: Designs and Techniques Prove Successful," SPE 16929, presented at the 62nd Annual Technical Conference and Exhibition of the Society of Petroleum Engineers, Dallas, Texas, Sept. 27–30, 1987.
57. Zimmerman, J. C., Winslow, D. W., Hinkie, R. L., and Lockman, R. R.: "Selection of Tools for Stimulation in Horizontal Cased Hole," SPE 18995, presented at the SPE Joint Rocky Mountain Regional/Low Permeability Reservoirs Symposium and Exhibition, Denver, Colorado, March 6–8, 1989.
58. El Rabaa, W.: "Experimental Study of Hydraulic Fracture Geometry Initiated from Horizontal Wells," paper SPE 19720, presented at the 64th Annual Technical Conference and Exhibition of the Society of Petroleum Engineers, San Antonio, Texas, Oct. 8–11, 1989.
59. Dowell Schlumberger brochure, "Horizontal Wells."
60. Economides, M. J.: "Formation Conditions Dictate Stimulation," *The American Oil & Gas Reporter,* December 1988.
61. Personal communication with C. White and A. Mullins of Baker Oil Tools, Houston, Texas, August, 1990.
62. Butler, R. M.: "The Potential of Horizontal Wells—Some Introductory Comments," Proceedings of AOSTRA's 5th Annual Advances in Petroleum Recovery and Upgrading Technical Conference, Calgary, Canada, June 14–15, 1984.
63. Marchant, L. C.: "U.S. Tar Sand Oil Recovery Projects—1984." Proceedings of Expl. for Heavy Crude Oil and Bitumen—AAPG/UNITAR Joint Conference, Santa Maria, California, Vol. II, October 29–November 2, 1984.
64. Combs, J. E.: "Signal Oil and Gas Company Summary Test Data for Horizontal Well Tests in Sunnyside Area," Carbon County, Utah, 1967.
65. Towson, D. E.: "The Mine Assisted In-Situ Project," Proceedings of Expl. or Heavy Crude Oil and Bitumen—AAPG/UNITAR Joint Conference, Santa Maria, California, Vol. II, October 29–November 2, 1984.
66. "Kern River 'Hotplate' Project Launched," *Oil & Gas Journal,* August 23, 1982.
67. Rintoul, B.: "Hot Plate Cooking in California," *Drilling Contractor,* pp. 46–54, October 1982.
68. Dietrich, J. K.: "The Kern River Horizontal—Well Steam Pilot," SPE Paper No. 16346, presented at the 1987 California Regional Meeting, Ventura, California, April 8–10, 1987.
69. "Texaco Mothballs Oil Sands Pilot," *Oil Week,* p. 7, March 17, 1986.
70. "Petrobras Steams Fields in Potiguar Basin," *EOR Week,* Washington, DC, February 5, 1985.
71. Best, D. A., Cordell, G. M., and Haston, J. A.: "Underground Test Facility: Shaft and Tunnel Laboratory for Horizontal Well Technology," SPE Paper No. 1433, presented at the 60th Annual Technical Conference, Las Vegas, Nevada, September 22–25, 1985.
72. "Canadian Combine Steam Drive, Horizontal Well," *Oil & Gas Journal,* p. 22, June 1, 1987.

CHAPTER 2

Reservoir Engineering Concepts

INTRODUCTION

Figure 2–1 shows a drainage area for a vertical well and a horizontal well. A vertical well drains a cylindrical volume, whereas a horizontal well drains an ellipsoid, a three-dimensional ellipse. In general, we expect a horizontal well to drain a larger reservoir volume than a vertical well.

Figure 2–2 shows a fractured vertical well. The well is drilled in the reservoir of height h. The well is fractured and the fracture is fully penetrating, i.e., it covers the entire reservoir height. The fracture half-length is equal to x_f. Moreover, we assume that the fracture has an infinite conductivity, which means that pressure drop within the fracture is negligible. In other words, pressure in the vertical wellbore and at every point within the fracture is the same. This represents an ideal or desired fracture for a vertical well. If the height of this fracture is reduced, one would obtain a horizontal well (see Fig. 2–3). A horizontal well represents a limiting case of an infinite-

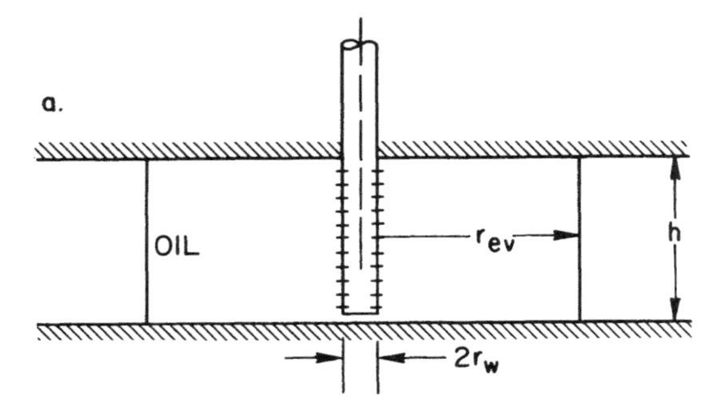

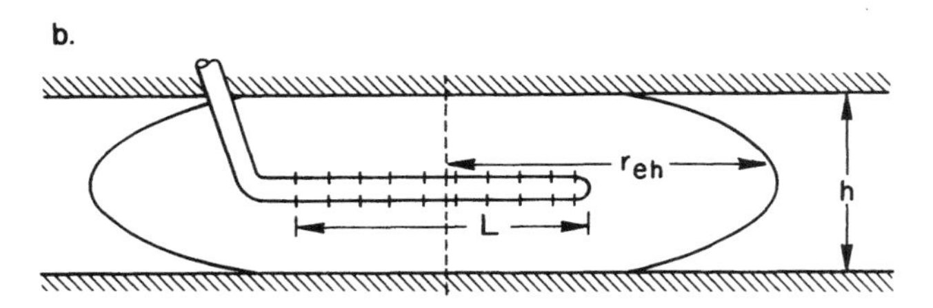

Figure 2–1 A Schematic of Horizontal and Vertical Well Drainage Areas.

conductivity fracture where the fracture height is equal to the wellbore diameter. This also tells us that the hole diameter of a horizontal well would have an influence on its performance. For example, instead of drilling a $4\frac{1}{2}''$ diameter hole, a $9\frac{7}{8}''$ hole is drilled; then the hole size has more than doubled. This increases the wellbore area open to the flow.

As noted in Chapter 1, a drainhole is normally drilled from a vertical well (see Fig. 2–4). It is also called a reentry. The turning radius here is about 20 to 40 ft. The well length varies from 100 to several hundred feet. The

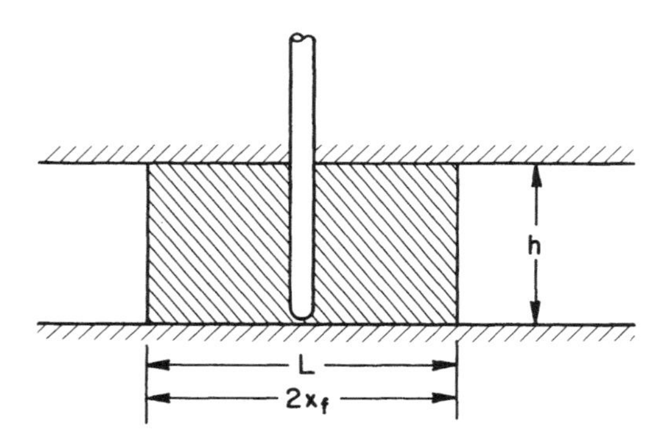

Figure 2–2 A Schematic of a Fractured Vertical Well.

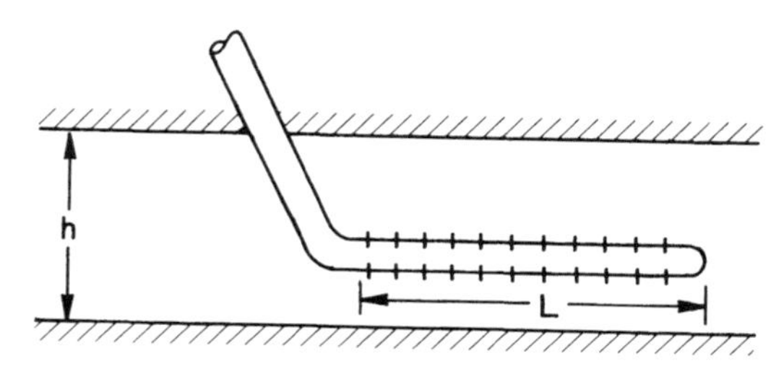

Figure 2–3 A Schematic of a Horizontal Well.

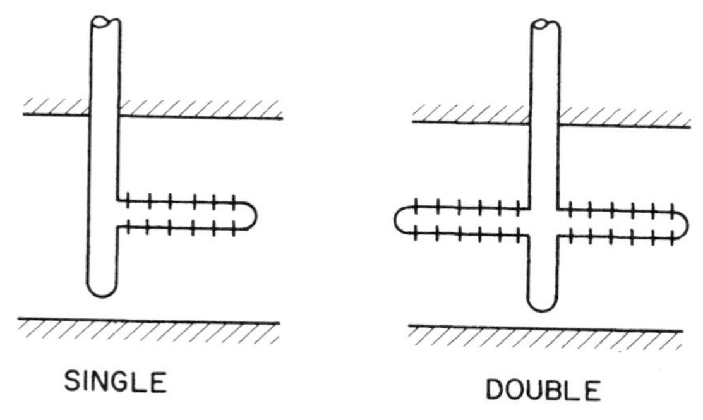

Figure 2–4 A Schematic of Single and Double Drainholes.

drainhole also represents a limiting case of an infinite-conductivity fracture. Thus horizontal wells, as well as drainholes, represent a limiting case of a fully penetrating infinite-conductivity vertical fracture.

The above discussion is for a single horizontal well or drainhole. However, using some drilling techniques, it is possible to drill several drainholes through one vertical well. In some cases one can drill several drainholes at a fixed elevation, i.e., multiple drainholes at a given elevation (see Fig. 2–5). Some drilling techniques facilitate drilling drainholes at different el-

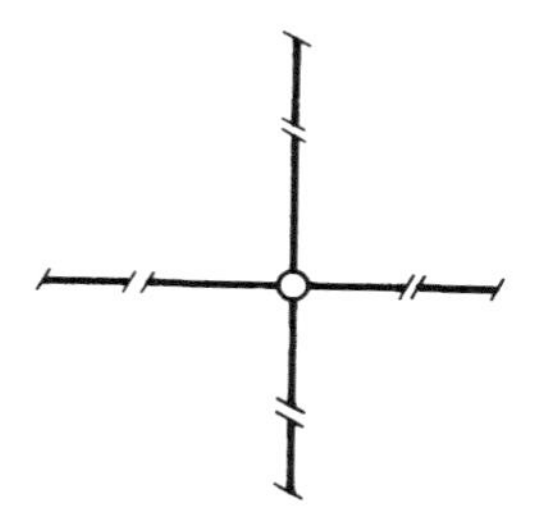

Figure 2–5 A Schematic of Multiple Drainholes at a Single Elevation.

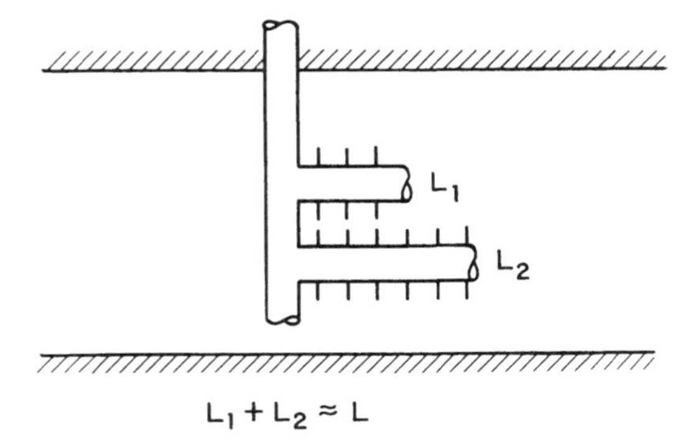

Figure 2–6 A Schematic of Multiple Drainholes at Different Elevations.

evations through a single vertical well (see Fig. 2–6). Each of these multiple drainholes individually represents a limiting case of a fully penetrating fracture. Evaluation of the effect of multiple drainholes on productivity is discussed later.

Before discussing horizontal well productivity, basic reservoir engineering concepts are reviewed in the following sections. Their applications to horizontal wells are also discussed.

SKIN FACTOR

Van Everdingen and Hurst introduced the idea of a skin factor to the petroleum industry.[1,2,3] They noticed that for a given flow rate, the measured bottom-hole flowing pressure was less than that calculated theoretically. This indicated to them that there was an additional pressure drop over that calculated theoretically. Moreover, this pressure response was found to be independent of time. They attributed this pressure drop to a small zone of changed or reduced permeability around the wellbore. While drilling a well, it may be necessary to have a positive pressure differential acting from the wellbore into the formation to prevent the flow of reservoir fluids into the wellbore. This results in a limited flow of mud in the formation, and particles suspended in drilling mud may plug some pore spaces around the wellbore. This results in a zone of reduced permeability around the wellbore.

Van Everdingen and Hurst called this "invaded" zone, or damaged zone, a skin zone, and the associated pressure drop as a skin factor effect. It is important to note that Van Everdingen and Hurst discussed only *damaged* wells, which means the pressure drop is higher than the theoretical pressure drop. The skin pressure drop is given as:

$$s = \frac{kh(\Delta p)_{skin}}{141.2\, q\mu_o B_o} \qquad (2\text{–}1)$$

In general, the skin factor in wells can vary from +1 to +10, and even higher values are possible. (Note that skin factor s is dimensionless.) This thin skin concept works very well in damaged wells, but this concept has some mathematical and physical difficulties when the skin factor is negative. (In Eq. 2–1, a negative skin factor represents flow from the wellbore to the formation.) To overcome this problem of thin skin, Hawkins[4] introduced a concept of thick skin. He showed that the skin factor for a skin zone of radius r_s with permeability k_s in a formation with permeability, k, and wellbore radius, r_w, can be shown to be (see Fig. 2–7):

$$s = [(k/k_s) - 1] \ln(r_s/r_w) \tag{2–2}$$

Similar to Van Everdingen and Hurst, Hawkins also assumed a steady state flow through the skin region. (Strictly speaking, the skin concept is only valid for steady and pseudo-steady states.) Because of Hawkins' modification, a trend was set to represent fractured or stimulated vertical wells with a negative skin factor. Under such an assumption, the producing pressure in the wellbore is higher than the expected pressure; i.e., the well is producing a desired flow rate at a significantly higher bottom hole flowing pressure than the theoretically calculated value (see Fig. 2–8). Stimulated vertical wells could have negative skin factors as high as −6. It will be shown later that, theoretically, horizontal wells can also be represented as a vertical well with a large negative skin factor, i.e., theoretically horizontal wells can be represented as highly stimulated vertical wells.

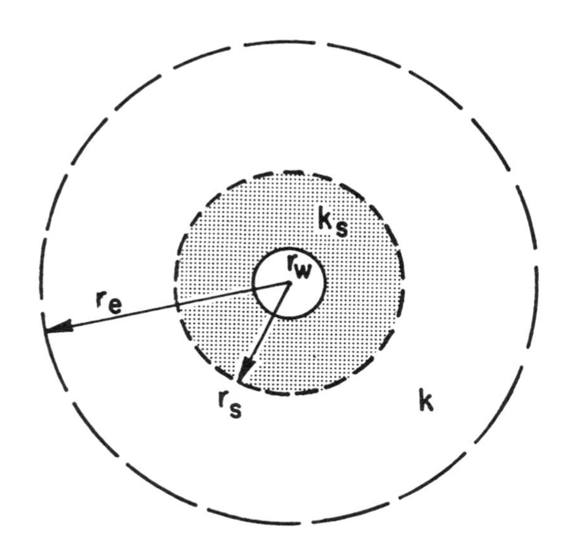

Figure 2–7 A Schematic of a Well with a Damaged Zone (Skin Damage).

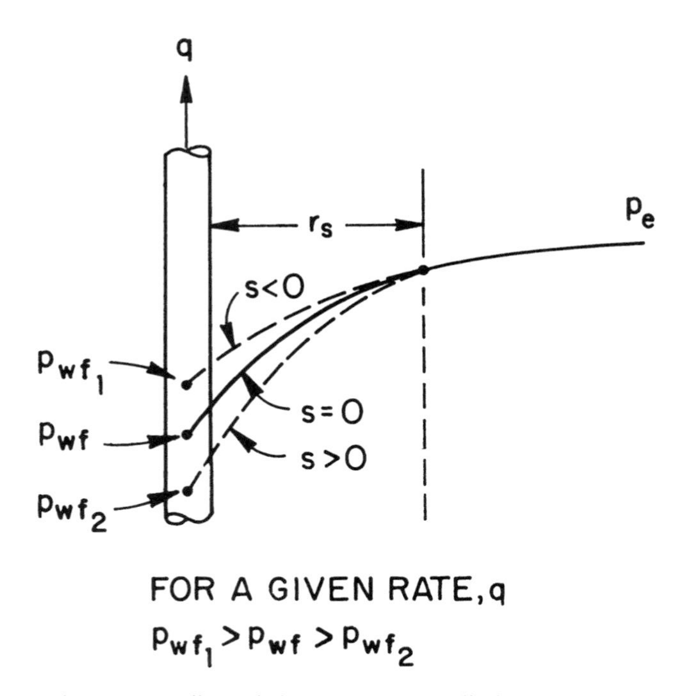

Figure 2–8 Effect of Skin Factor on Well Flow Pressure.

SKIN DAMAGE IN LOW- AND HIGH-PERMEABILITY FORMATIONS

As noted earlier, drilling fluid invasion causes a zone of reduced permeability around the wellbore. In general, a high-permeability reservoir exhibits a higher invasion zone thickness than a low-permeability formation. However, the percentage loss in permeability in a high-permeability zone is smaller than that in a low-permeability zone. In some cases, depending upon drilling mud used, one may see a limited mud invasion zone even in high permeability formations. High-permeability formations probably have larger pore throat sizes than the low-permeability formations. These large pore throats are probably not completely blocked by solids in drilling fluids and by mud invasion, resulting in a small percentage loss of the initial formation permeability. In contrast, in low-permeability formations, where pore throat sizes are small, solids in drilling fluids and mud invasion probably block these pore throats, resulting in a large reduction in permeability around the wellbore. This reduced permeability around the wellbore minimizes fluid loss and therefore minimizes radius of the mud-invaded zone.

A careful observation of Equation 2–2 tells us that the invasion zone thickness is less important than the permeability change in the invaded zone.

This is because invasion zone radius r_s is in logarithmic terms in Equation 2–2, therefore, exercising less influence on the skin factor than the invaded zone permeability, k_s. As noted above, in general high-permeability formations exhibit a high mud invasion zone thickness, but they also exhibit a small loss in initial reservoir permeability in the invasion zone, giving a small value of positive skin factor s. In contrast, in a low-permeability formation, the mud invasion zone thickness is small, but a loss in the reservoir permeability in the invaded zone is high, resulting in large near wellbore damage, or a large positive skin factor after drilling. Thus, damage in high-permeability zones is relatively smaller than that in low-permeability zones.

Normally, skin factors are estimated using drill stem testing (DST) or pressure build-up testing. Knowing the skin factors, pressure drops across the damaged zones can be estimated. Reformulating Equation 2–1 to calculate vertical well pressure drops in the skin region,

$$(\Delta p)_{skin} = s(141.2\ \mu_o B_o/k)(q/h) \qquad (2\text{–}3a)$$

Thus, $(\Delta p)_{skin}$ depends upon q/h, i.e., rate of fluid entry per unit length of the wellbore. Hence, for horizontal wells we can approximate skin pressure drop as

$$(\Delta p)_{skin} = s(141.2\ \mu_o B_o/k)(q/L) \qquad (2\text{–}3b)$$

Comparison of Equations 2–3a and 2–3b demonstrates that for the same positive skin factor s, the pressure loss in the skin region in a horizontal well is smaller than that in a vertical well (see Figs. 2–9a and 2–9b). This is because the rate of fluid entry into wellbore per unit length of a horizontal well is much smaller than that for a vertical well. This is shown in Example 2–1. It is important to note that in the horizontal well calculations in Equation 2–3b, k represents effective reservoir permeability, ($k = \sqrt{k_v k_h}$, where k_v is vertical permeability and K_h is horizontal permeability).

EXAMPLE 2–1

Calculate the pressure drops in the skin zones, i.e., damaged zones, in vertical and 1000-ft-long horizontal wells. The well tests show skin factor +1 for vertical as well as horizontal wells. The following reservoir properties are given:

$s = +1$	$k_v = k_h = 10$ md
$h = 25$ ft	$q_v = 1000$ BOPD
$q_h = 2500$ BOPD	$\mu_o = 0.8$ cp
$B_o = 1.06$ RB/STB	

Note q_h is horizontal well rate and z_v is vertical well rate.

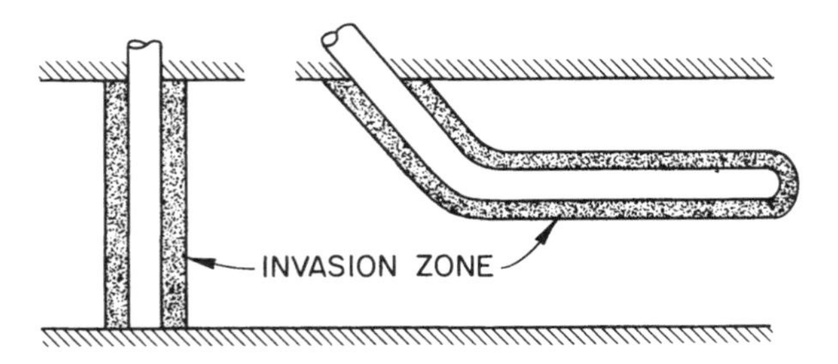

Figure 2–9a Ideal Mud Damage Zones for Horizontal and Vertical Wells.

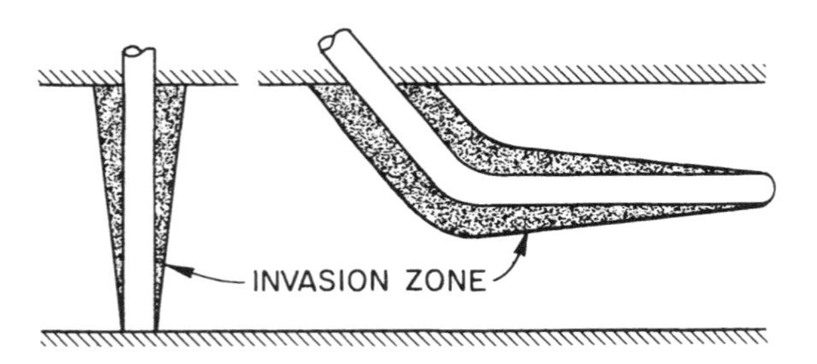

Figure 2–9b Conical Shaped Damaged Zones.

Solution

From Equation 2–1

$$s = \frac{kh(\Delta p)_{skin}}{141.2\ q\mu_o B_o}$$

or

$$(\Delta p)_{\text{skin,vertical}} = \frac{s(141.2\ q\mu_o B_o)}{kh}$$

$$(\Delta p)_{\text{skin,vertical}} = \frac{1 \times 141.2 \times 1000 \times 0.8 \times 1.06}{10 \times 25}$$

$$= 479 \text{ psi}$$

To calculate the $(\Delta p)_{skin}$ for a horizontal well, substituting the horizontal well rate and horizontal well length L for the reservoir height h gives:

$$(\Delta p)_{\text{skin,horiz}} = \frac{1 \times 141.2 \times 2500 \times 0.8 \times 1.06}{10 \times 1000}$$

$$= 29.9 \text{ psi} = 30 \text{ psi}$$

SKIN DAMAGE FOR HORIZONTAL WELLS

Example 2–1 clearly indicates that for a given positive skin factor, pressure drop in the skin region for a horizontal well is considerably smaller than that for a vertical well. (Note that Example 2–1 assumes a uniform fluid entry rate along the well length.) This shows that for a given skin damage the stimulation treatment to remove near-wellbore damage would have less effect on the productivity of a horizontal well than on the productivity of a vertical well. Therefore, before deciding to stimulate a horizontal well, it is important to estimate pressure loss in the skin zone and compare it with the overall pressure drop from the reservoir to the wellbore pressure. This comparison can be used to determine a need for horizontal well stimulation.

The minimum influence of near-wellbore damage on horizontal well productivity in a high-permeability reservoir also explains the reasons for many successful horizontal well field projects in high-permeability reservoirs. In these reservoirs, horizontal wells were drilled and put on production, in some cases even without acidizing. These wells had minimum damage which had no significant influence on horizontal well productivity. In contrast, in low-permeability reservoirs, the influence of damage on horizontal wells can be severe. Therefore, some horizontal well field projects have been uneconomical due to severely damaged horizontal wells. This issue is examined further in the following paragraphs.

In many reservoirs, especially in low-permeability reservoirs, after drilling vertical wells, the vertical wells are cemented and perforated. Prior to production, these wells are stimulated using propped or unpropped fractures. Without fracturing these wells do not produce at economic rates. In these types of reservoirs, vertical well drilling probably causes severe damage, but it is overcome by fracture stimulation. If a horizontal well is drilled in such a reservoir, the damage due to horizontal wells will be larger than that in the vertical well. This is because horizontal drilling takes a longer time than vertical drilling, resulting in a conical-shape damage zone (Figure 2–9b). This damage zone can significantly reduce productivity of a horizontal well. As shown in Figure 2–10a, skin damage could vary along the well length. Based on the experience with vertical well skin damage, and vertical well drilling time, one can attempt to estimate damage along the length of a horizontal well. Availability of core data for damage or drill stem test data on vertical wells will be useful in constructing such plots, as shown in Figure 2–10a. These plots can be used to estimate expected average damage from a horizontal well. It is important to note that in a horizontal well test, the calculated damage value would represent an average value for the entire well. Based on the expected damage value, a proper stimulation and near-wellbore formation cleanup procedure needs to be critically reviewed. As noted in Chapter 1, a well completed as an open hole or with a slotted liner may be difficult to clean and special cleanup procedures may have to be

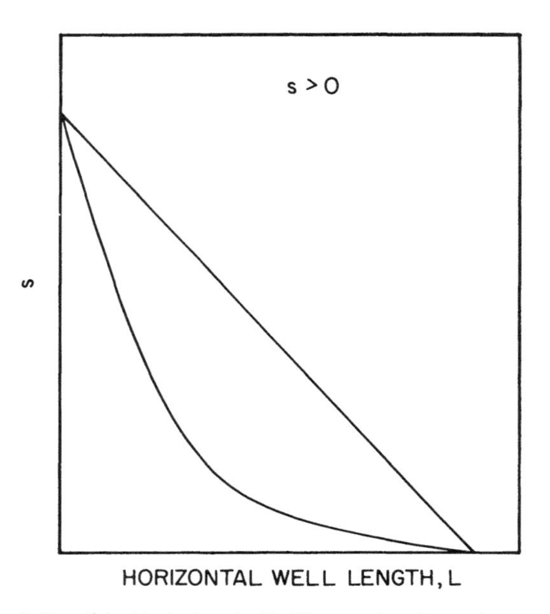

Figure 2–10a A Possible Variation in Drilling Related Mechanical Skin Factor along the Well Length.

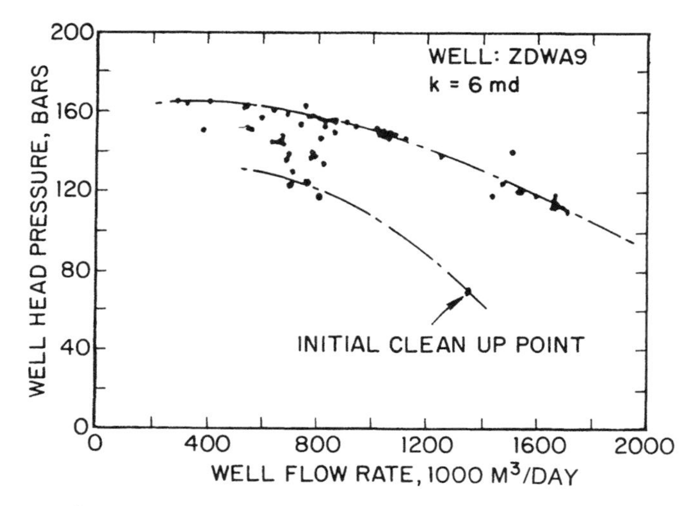

Figure 2–10b Mud Clean-Up Effect in a Horizontal Gas Well, Zuidwal Field, Netherlands.[5]

devised. Swabbing the well is one alternative, but it can be time-consuming and may be inefficient to clean up long horizontal wells. Moreover, depending upon well turning radius, a swab tool may reach only up to the top of the curve section of the wellbore, and it is difficult to reach with a swab tool in the horizontal well portion. Another option, where severe damage is expected, is to consider cementing and perforating horizontal wells. Small stimulation treatments in the perforated zones can be designed to overcome near-wellbore damage.

SUMMARY OF SKIN FACTOR DISCUSSION

In general, drilling-related damage in a high-permeability reservoir is smaller than that in a low-permeability reservoir. For the similar skin damage value, the influence of damage on horizontal well productivity is not as detrimental as in a vertical well. Thus, horizontal wells can sustain more damage than vertical wells without a significant loss of well productivity. However, due to additional drilling time incurred in horizontal wells, horizontal wells may show much more near-wellbore damage than a vertical well and a proper procedure must be adopted to either minimize this severe damage or to clean up this damage. Figure 2–10b shows a clean-up history of a horizontal gas well, completed with a slotted liner, in Zuidwal Field, Netherlands.[5] After initial flow, as time progressed due to clean-up effects, the well exhibited higher productivity. Similar behavior is observed in many wells, especially those drilled in high permeability reservoirs.

Recently, mathematical equations have been proposed to estimate the influence of mud damage on productivity of a horizontal well. These equations are included in Chapter 3.

EFFECTIVE WELLBORE RADIUS, r'_w

The effective wellbore radius concept is used to represent the well which is producing at a rate different than that expected from the calculations based upon a drilled wellbore radius. Effective wellbore radius is the theoretical well radius required to match the observed production rate. Thus, stimulated wells will have an effective wellbore radius greater than the drilled wellbore radius, and damaged wells will have an effective wellbore radius smaller than the drilled wellbore radius. A steady-state equation with effective wellbore radius can be written as

$$q = 0.007078\, kh\, \Delta p / [\mu_o B_o \ln(r_e / r'_w)] \tag{2–4}$$

where

q = oil rate, STB/day
k = permeability, md

h = reservoir thickness, ft
μ_o = viscosity, cp
B_o = formation volume factor, RB/STB
r_e = drainage radius, ft
r'_w = effective wellbore radius, ft
Δp = pressure drop from the drainage radius to the wellbore, psi

EXAMPLE 2–2

We have the following reservoir parameters:

$k = 10$ md $\quad h = 40$ ft
$p_e = 2{,}000$ psi $\quad p_{wf} = 500$ psi
$\mu = 0.5$ cp $\quad B_o = 1.25$ RB/STB
$r_e = 1{,}000$ ft

where:

p_e = reservoir pressure at drainage radius, r_e
and p_{wf} = bottom hole well flowing pressure.

Calculate steady-state flow rates for $r'_w = 0.35$, 0.1, and 20 ft. The drilled wellbore diameter is 0.35 ft.

Solution

The steady-state flow equation is

$$q = \frac{0.007078\, kh\, \Delta p}{\mu_o B_o \ln(r_e/r'_w)} \tag{2–4}$$

In the above equation, substituting reservoir parameters gives:

$$q = \frac{0.007078 \times 10 \times 40 \times (2000 - 500)}{0.5 \times 1.25 \ln(1000/r'_w)}$$

$$q = 6794.9/\ln(1000/r'_w)$$

If $r'_w = 0.35$ ft, $q = 854$ BOPD
If $r'_w = 0.1$ ft, $q = 738$ BOPD
If $r'_w = 20$ ft, $q = 1737$ BOPD

Thus, a change in flow rate can be represented as a change in an apparent wellbore radius. Hence, the effective wellbore radius is an apparent wellbore radius that represents the flow rate.

Mathematically, we can show that

$$r'_w = r_w \exp(-s) \tag{2–5}$$

substituting Equation 2–5 into Equation 2–4

$$q = \frac{0.007078\, kh\, \Delta p/(\mu_o B_o)}{\ln(r_e/r_w) + s} \tag{2–6}$$

EXAMPLE 2–3

What are the skin factors for $r'_w = 0.35$, 20, and 0.1 in Example 2–2 if the drilled wellbore radius is 0.35 ft?

As noted earlier, a positive skin factor indicates a damaged wellbore, and a negative skin factor indicates a stimulated wellbore. The drilled radius is $r_w = 0.35$ ft. Substituting new values in Equation 2–5 yields

$$r'_w = 0.35 \exp(-s) \tag{2–7}$$

$$-s = \ln(r'_w/0.35) \tag{2–8}$$

Thus, for

$$r'_w = 0.35 \text{ ft}, \; s = 0$$
$$r'_w = 0.1 \text{ ft}, \; s = 1.25$$
$$r'_w = 20 \text{ ft}, \; s = -4.05$$

EXAMPLE 2–4

The following parameters apply to an oil well having an estimated drainage area of 80 acres.

$k = 1$ md, $h = 60$ ft, $r_w = 0.3$ ft.
$p_e = 2500$ psi, $p_{wf} = 900$ psi, $\mu_o = 0.6$ cp and
$B_o = 1.25$ RB/STB

Calculate (1) the steady-state flow rates and (2) the skin factors, for effective wellbore radii of 0.3, 0.05, and 30 ft.

Solution

Assuming a circular drainage area A, drainage radius r_e is calculated as

$$\text{Drainage radius } r_e = \sqrt{80 \times 43{,}560/3.142} = 1053 \text{ ft}$$

Steady-state flow rate can be calculated for $r'_w = 0.3$ ft using Equation 2–4 as

$$q = \frac{0.007078kh\,(p_e - p_{wf})}{\mu_o B_o \ln(r_e/r'_w)}$$

$$q = \frac{0.007078 \times 1 \times 60 \times (2500 - 900)}{0.6 \times 1.25 \ln(1053/r'_w)}$$

$$q = \frac{906.24}{\ln(1053/r'_w)} = \frac{906.24}{\ln(1053/0.3)}$$

$$q = 111 \text{ BOPD}$$

Skin factor $s = -\ln(r'_w/r_w)$

$$s = -\ln(0.3/0.3)$$
$$s = 0$$

The calculated steady-state flow rates and skin factors for different values of r'_w are summarized in the following table.

Effective Wellbore Radius, ft. (*m*)	**Steady-State Flow Rate, BOPD (m^3/D)**	**Skin Factor s**
0.30 (.09)	111 (17.63)	0
0.05 (.015)	91 (14.46)	+1.79
30.00 (9.14)	255 (40.51)	−4.6

PRODUCTIVITY INDEX, *J*

The Well Productivity Index, J, is defined as $q/\Delta p$, where q is the flow rate and Δp is pressure drop. The units of J are bbl/day/psi or m^3/day/kPa. Initially, for simplicity, we will assume that we have constant pressure at the reservoir boundary and at the well center. This will give us the steady-state productivity index. It is important to note that this assumes a constant pressure being maintained at the drainage radius and also at the wellbore. In actual reservoir operations, for primary recovery, this pressure drop would change over time as more and more fluid is withdrawn from the reservoir; therefore, this constant pressure drop Δp will reduce over time. By definition, productivity index for steady state is calculated by reformulating Equation 2–4 as

$$J = q/\Delta p = \frac{0.007078\, kh/(\mu_o B_o)}{\ln(r_e/r'_w)} \tag{2–9}$$

The productivity index can be used to compare productivities of two different wells in the same reservoir as

$$J_1/J_2 = [\ln(r_e/r'_w)]_2/[\ln(r_e/r'_w)]_1 \qquad (2\text{–}10)$$

where subscripts 1 and 2 refer to the first and second well. One can also rewrite Equation 2–10 in terms of skin factors as

$$J_1/J_2 = [\ln(r_e/r_w) + s]_2/[\ln(r_e/r_w) + s]_1 \qquad (2\text{–}11)$$

Thus, concepts of skin factor, effective wellbore radius, and productivity index are used to represent well productivity. As shown above, it is fairly straightforward to convert skin factors into either effective wellbore radii or productivity indices.

EXAMPLE 2–5

Calculate the ratio of productivities of two wells (1 and 2) in the same reservoir drilled at two different well spacings

Well 1: Area = 40 acres, r'_w = 20 ft.
Well 1: Area = 80 acres, r'_w = 4 ft.

Solution

The ratio of productivities of two different wells in the same reservoir is

$$J_1/J_2 = [\ln(r_e/r'_w)]_2/[\ln(r_e/r'_w)]_1 \qquad (2\text{–}10)$$

For well 1, drainage radius, $r_{e1} = \sqrt{40 \times 43{,}560/\pi} = 745$ ft.

For well 2, drainage radius, $r_{e2} = \sqrt{80 \times 43{,}560/\pi} = 1053$ ft. Substituting drainage and effective wellbore radii in Equation 2–10

$$J_1/J_2 = \ln(1053/4)/\ln(745/20)$$
$$= 1.54$$

FLOW REGIMES

So far all discussion has been restricted to the steady-state flow, which assumes constant pressure at the reservoir boundary and at the wellbore. As noted earlier, in a reservoir there is an initial pressure drop between the wellbore and the boundary. This pressure drop decreases over time and similarly the production rate from a well decreases over time. This can be explained by considering various flow regimes.

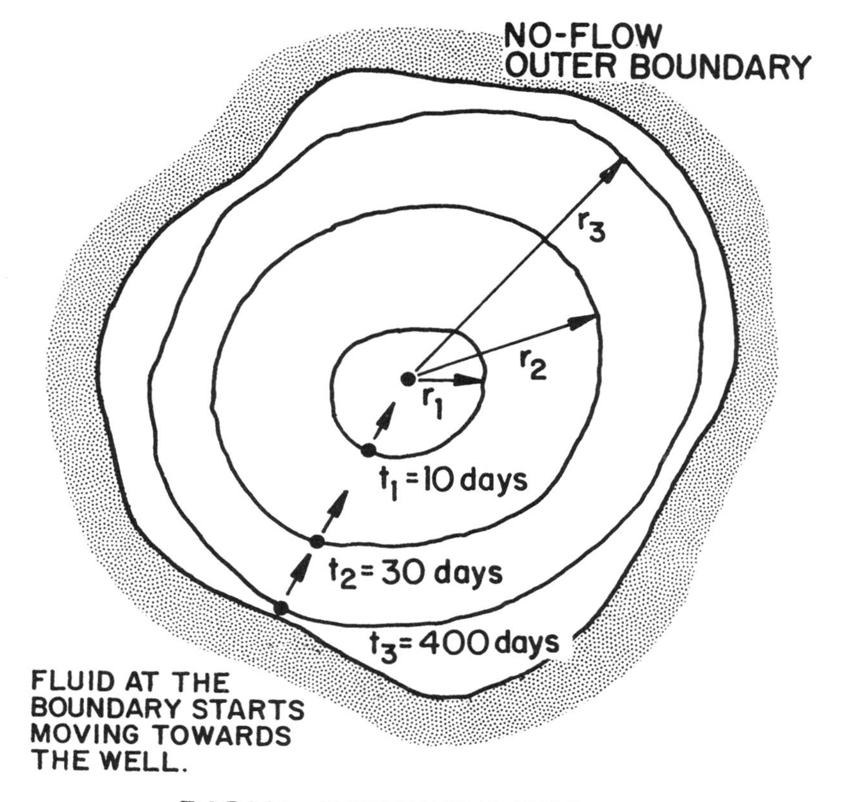

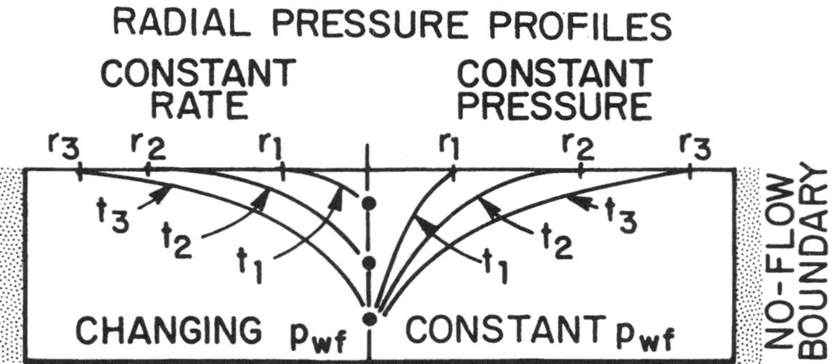

Figure 2–11 Pressure Profile as a Function of Time.[6]

CONCEPT OF STEADY STATE AND PSEUDO-STEADY STATE

During the operational life, the reservoir passes through various stages. These stages are classified into two categories: (1) infinite acting or transient state, and (2) pseudo-steady state or steady state, depending upon the boundary condition. To describe what happens in reservoir life consider the example below:

> Drop a stone in a calm pond. As soon as the stone is dropped in the pond, one can see circular waves going outward (see Fig. 2–11).

A similar phenomenon occurs when a well is put on production in a given drainage area. The pressure disturbance caused by the well transmits outward radially; fluid away from the wellbore experiences the pressure gradient, and it starts moving toward the wellbore. The rate of pressure propagation outward from a wellbore depends upon reservoir permeability. The higher the permeability, the faster is the rate of pressure propagation. As time progresses, this circular disturbance travels outward and eventually the disturbance reaches the "drainage boundary." At this point, fluids located at the drainage boundary start moving toward the wellbore. The time period that is required for pressure disturbance to reach the boundary of the circular area is called *infinite acting time*.[6,7] Once the disturbance has reached the boundary of the circular area, then as time progresses, and as more and more fluids are withdrawn from the reservoir, the average reservoir pressure starts decreasing over time. This occurs in a reservoir with a *no-flow boundary* condition, i.e., in a reservoir where there is no flow across the well's drainage boundary. This is called a *pseudo-steady state;* this is also called a *depletion state* or *semisteady state*. If the boundary condition is such that we have a fixed pressure at the boundary, then the well would achieve a *steady state*.

TRANSITION FLOW

The two flow periods, infinite acting and pseudo-steady state or steady state describe flow periods for circular and square drainage areas sufficiently. However, in the case of rectangular drainage areas, as soon as the disturbance is created, the well will see the closest boundary first (see Fig. 2–12). Then it will see the second boundary, and then it will start the depletion state. The time required for the well to reach the first boundary is called the end of an infinite acting period. Up to this point, the well has seen no boundary. A time period where the well has seen one boundary, but not the other, is called the *transition region* or transition time. The last stage, again, when both boundaries are seen, is called the pseudo-steady state or steady state, depending on the boundary condition. These concepts are very important in actual practice. The time periods required to reach the pseudo-steady state for different vertical well locations in the drainage plane and for various configurations of the drainage areas are listed in Table 2–1.[7]

FLOW EQUATIONS FOR DIFFERENT FLOW REGIMES (OR FOR STEADY AND PSEUDO-STEADY STATE)

Flow equations for steady state and pseudo-steady state for wells located centrally in the areal drainage plane are summarized in Table 2–2. It is

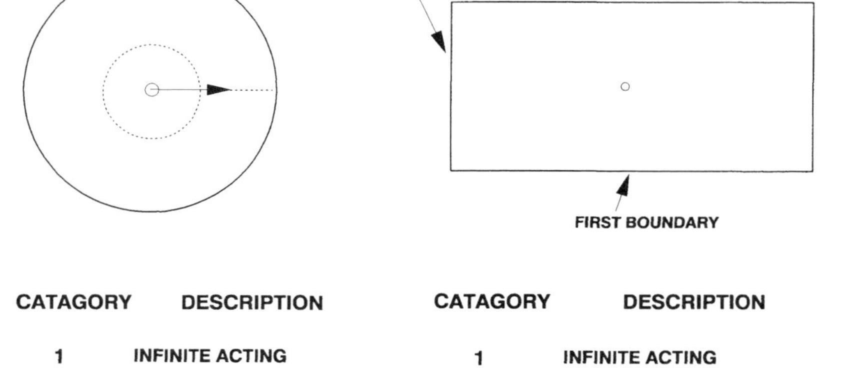

CATAGORY	DESCRIPTION
1	INFINITE ACTING
2	PSEUDO-STEADY OR STEADY STATE

CATAGORY	DESCRIPTION
1	INFINITE ACTING
2	TRANSITION (SEE FIRST BOUNDARY BUT NOT SECOND)
3	PSEUDO-STEADY OR STEADY STATE

Figure 2–12 Flow Regimes Encountered for Circular and Rectangular Drainage Areas.

important to note that the equations are based either on drainage boundary pressure p_e or average reservoir pressure $\overline{p}$. In practice, reservoir pressure is estimated by using either a DST test, a pressure bomb test, or a buildup test. These methods estimate average reservoir pressure $\overline{p}$.

TIME TO REACH PSEUDO-STEADY STATE

Dimensionless time t_D, which is used to define various flow regimes, is given as

$$t_D = \frac{0.000264\ kt}{\phi\ \mu\ c_{ti}\ r_w^2} \tag{2–12}$$

and area-based dimensionless time is defined as

$$t_{DA} = t_D(r_w^2/A) \tag{2–13}$$

Thus,

$$t_{DA} = \frac{0.000264\ kt}{\phi\ \mu\ c_{ti}\ A} \tag{2–14}$$

TABLE 2–1 SHAPE FACTORS AND TIME TO REACH STEADY OR PSEUDO-STEADY STATE FOR DIFFERENT VERTICAL WELL CONFIGURATIONS[7]

IN BOUNDED RESERVOIRS	C_A	$\ln C_A$	$1/2 \ln\left(\frac{2.2458}{C_A}\right)$	EXACT FOR $t_{DA}>$	LESS THAN 1% ERROR FOR $t_{DA}>$	USE INFINITE SYSTEM SOLUTION WITH LESS THAN 1% ERROR FOR $t_{DA}<$
	31.62	3.4538	−1.3224	0.1	0.06	0.10
	31.6	3.4532	−1.3220	0.1	0.06	0.10
	27.6	3.3178	−1.2544	0.2	0.07	0.09
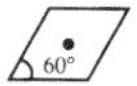 60°	27.1	3.2995	−1.2452	0.2	0.07	0.09
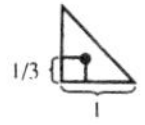 1/3; 1	21.9	3.0865	−1.1387	0.4	0.12	0.08
 4	0.098	−2.3227	+1.5659	0.9	0.60	0.015
	30.8828	3.4302	−1.3106	0.1	0.05	0.09
	12.9851	2.5638	−0.8774	0.7	0.25	0.03
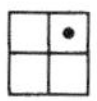	4.5132	1.5070	−0.3490	0.6	0.30	0.025
	3.3351	1.2045	−0.1977	0.7	0.25	0.01
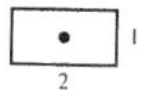 1; 2	21.8369	3.0836	−1.1373	0.3	0.15	0.025

TABLE 2–1 (Continued)

	C_A	$\ln C_A$	$1/2 \ln \left(\frac{2.2458}{C_A} \right)$	EXACT FOR $t_{DA} >$	LESS THAN 1% ERROR FOR $t_{DA} >$	USE INFINITE SYSTEM SOLUTION WITH LESS THAN 1% ERROR FOR $t_{DA} <$
	10.8374	2.3830	−0.7870	0.4	0.15	0.025
	4.5141	1.5072	−0.3491	1.5	0.50	0.06
	2.0769	0.7309	+0.0391	1.7	0.50	0.02
	3.1573	1.1497	−0.1703	0.4	0.15	0.005
	0.5813	−0.5425	+0.6758	2.0	0.60	0.02
	0.1109	−2.1991	+1.5041	3.0	0.60	0.005
	5.3790	1.6825	−0.4367	0.8	0.30	0.01
	2.6896	0.9894	−0.0902	0.8	0.30	0.01
	0.2318	−1.4619	+1.1355	4.0	2.00	0.03
	0.1155	−2.1585	+1.4838	4.0	2.00	0.01
	2.3606	−0.8589	−0.0249	1.0	0.40	0.025

TABLE 2–2 SUMMARY OF FLOW EQUATIONS

	STEADY STATE	SEMI-STEADY STATE*
General relationship between p and r	$p - p_{wf} = \frac{q\mu}{2\pi kh} \ln \frac{r}{r_w}$	$p - p_{wf} = \frac{q\mu}{2\pi kh} \left(\ln \frac{r}{r_w} - \frac{r^2}{2r_e^2} \right)$
Inflow equations expressed in terms of $p = p_e$ at $r = r_e$	$p_e - p_{wf} = \frac{q\mu}{2\pi kh} \ln \frac{r_e}{r_w}$	$p_e - p_{wf} = \frac{q\mu}{2\pi kh} \left(\ln \frac{r_e}{r_w} - \frac{1}{2} \right)$
Inflow equations expressed in terms of the average pressure	$\bar{p} - p_{wf} = \frac{q\mu}{2\pi kh} \left(\ln \frac{r_e}{r_w} - \frac{1}{2} \right)$	$\bar{p} - p_{wf} = \frac{q\mu}{2\pi kh} \left(\ln \frac{r_e}{r_w} - \frac{3}{4} \right)$

* Wells located centrally in drainage plane

To express in field units (stb/d, psi, mD, ft) the term $q\mu/(2\pi kh)$ should be replaced by $141.2 q\mu B_o/(kh)$, in each of the above equations.

As an alternative, the skin factor can be accounted for in the inflow equation by changing the wellbore radius. For example, including the skin factor

$$\bar{p} - p_{wf} = \frac{q\mu}{2\pi kh} \left[\ln \left(\frac{r_e}{r'_w} \right) - \frac{3}{4} \right]$$

in which

$$r'_w = r_w e^{-s}$$

a) damaged well, $s > 0$
b) stimulated well, $s < 0$

where

k = permeability, md
t = time, hours
ϕ = porosity in fraction, dimensionless
μ = viscosity, cp
c_{ti} = initial total compressibility, psi^{-1}
A = area, ft^2
r_w = wellbore radius, ft.

As shown in Table 2–1, for a vertical well located at the center of a drainage circle or a square, the time to reach pseudo-steady state is $t_{DA} = 0.1$. Substituting this in Equation 2–14,

$$t_{DA} = 0.1 = \frac{0.000264\, kt}{\phi\, \mu\, c_{ti} A} \tag{2–15}$$

$$t_{pss} = \frac{379\, \phi\, \mu\, c_{ti} A}{k} \tag{2–16}$$

t_{pss} = time to reach pseudo-steady state in hours

$$t_{pdss} = \frac{15.79\, \phi\, \mu\, c_{ti} A}{k} \tag{2–17}$$

t_{pdss} is the time to reach pseudo-steady state in days.

Generally, oil wells are developed on 40-acre spacing and gas wells are developed on 160-acre spacing. Hence

$$40 \text{ acres} = 40 \times 43{,}560 \text{ ft}^2/\text{acre} = 1.7424 \times 10^6 \text{ ft}^2 \tag{2–18}$$
$$160 \text{ acres} = 160 \times 43{,}560 \text{ ft}^2/\text{acre} = 6.9696 \times 10^6 \text{ ft}^2 \tag{2–19}$$

Substituting these areas into Equation 2–17 gives

for a 40-acre well

$$t_{pdss} = \frac{27.512 \times 10^6\, \phi\, \mu\, c_{ti}}{k} \tag{2–20}$$

for a 160-acre well

$$t_{pdss} = \frac{110.05 \times 10^6\, \phi\, \mu\, c_{ti}}{k} \tag{2–20}$$

Equations 2–15 and 2–16 show that transient time depends on the basic reservoir properties, such as permeability, porosity, and compressibility. Time to reach pseudo-steady state does not depend on well stimulation. In the case of oil wells, time to reach pseudo-steady state normally is on the order of a few days to months. In contrast, for gas wells in low-permeability reservoirs, time to reach pseudo-steady state could be very long, in some cases as long as a few years.

EXAMPLE 2–6

For an oil well drilled at 40-acre spacing, calculate time to reach pseudo-steady state.

Given

$\phi = 10\%$, $c_{ti} = 0.00005$, psi^{-1}
$A = 40$ acres, $k = 35$ md
$\mu = 4.2$ cp (shallow well, dead oil)

Solution

Using Equation 2–16, time to reach pseudo-steady state is calculated as

$$t_{pss} = \frac{379.0 \times 0.1 \times 4.2 \times 0.00005 \times A}{35}$$
$$= 0.0002274\,A$$
$$= 0.0002274\,(40 \times 43{,}560)$$
$$= 396 \text{ hours}$$
$$t_{pss} = 16.5 \text{ days}$$

EXAMPLE 2–7

Calculate the time required to reach pseudo-steady state for a gas well drilled at either 20- or 160-acre spacing in a reservoir with an initial pressure of 1450 psi. The following reservoir properties are given.

$\phi = 7\%$, $\mu = 0.015$ cp
$p_i = 1450$ psi, $A = 20$ and 160 acres
$k = 0.03$ md, $c_{ti} = 0.000690$ psi^{-1}

Solution

$$t_{pss} = \frac{379\,\phi\mu\,c_{ti}\,A}{k} \qquad (2\text{–}1b)$$

$$= \frac{379 \times 0.07 \times 0.015 \times 0.000690 \times A}{0.03}$$

$$= 0.00915A$$

For 20 acres

$$t_{pss} = 7974 \text{ hours}$$
$$= 332 \text{ days}$$
$$= 0.91 \text{ years}$$

For 160 acres

$$t_{pss} = 63{,}772 \text{ hours}$$
$$= 2657 \text{ days}$$
$$= 7.3 \text{ years of infinite-acting period}$$

As noted in Example 2–7, for gas wells (or oil wells) drilled in low permeability reservoirs, especially gas wells in reservoirs with permeability less than 0.1 md, it can take years for the transient state to end. In such tight reservoirs, it is very difficult to drain the reservoir economically. In those cases, we need a method to accelerate reservoir drainage. Infill drilling and horizontal drilling provide alternatives to drain the reservoir effectively.

Results in Table 2–1 tell us that the dimensionless time to reach pseudo-steady state is $t_{DA} = 0.1$, as long as the well is centrally located in a drainage plane, i.e., when the well is at the center of a circle or a square ($x_e/y_e = 1$). When the drainage area becomes rectangular, the time to reach pseudo-steady state increases. For example, when one side of a drainage rectangle is five times larger than the other side ($x_e/y_e = 5$) the dimensionless time to reach pseudo-steady state is $t_{DA} = 1.0$, i.e., 10 times longer than a vertical well located centrally in the drainage plane. Thus, vertical wells are unable to drain effectively rectangular drainage areas in uniform-permeability reservoirs. As shown in Figure 2–13, a long horizontal well can drain rectangular areas much more rapidly than a vertical well. As shown in Example 2–6, a 40-acre spacing vertical well reaches pseudo-steady state in 16 days. By the same principle, as shown in Figure 2–13, a 2000-ft-long well would reach pseudo-steady state in a 101-acre area in 16 days. Table 2–1 tells us that to reach pseudo-steady state using a vertical well draining a rectangle with $x_e/y_e = 2.5$ (shown in Figure 2–13) would take at least three times longer than it takes for a 2000-ft-long horizontal well. Thus, horizontal wells can be utilized to drain a large reservoir volume in a small time frame. This becomes very important in low permeability reservoirs where close vertical well spacing is required to drain the reservoir effectively. Therefore, in a low permeability reservoir, horizontal wells can be used to enhance drainage volume per well in a given time period.

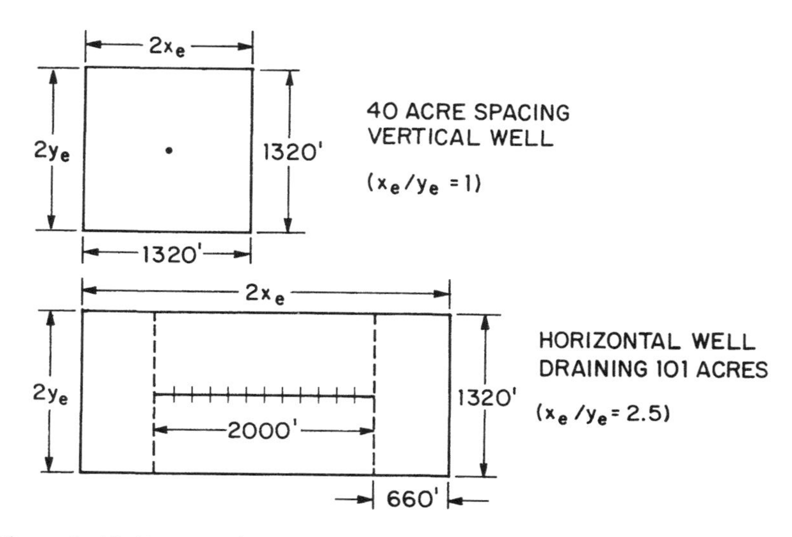

Figure 2–13 Horizontal and Vertical Well Drainage Areas for a Given Time.

INFLUENCE OF AREAL ANISOTROPY

The discussion so far was restricted to reservoirs with homogeneous areal permeability, namely $k_x = k_y$ (see Fig. 2–14a). In naturally fractured reservoirs, the permeability along the fracture trend is larger than in a direction perpendicular to fractures. In such cases, a vertical well would drain more length along the fracture trend. The derivation shown below can be used to estimate each side of a drainage area in an areally anisotropic reservoir. Assuming a single phase, steady-state (time independent) flow through porous formation, one can write the following equation.

$$\frac{\partial}{\partial x}\left(k_x \frac{\partial p}{\partial x}\right) + \frac{\partial}{\partial y}\left(k_y \frac{\partial p}{\partial y}\right) = 0 \qquad (2\text{–}21)$$

Assuming constant values of k_x and k_y in x and y directions, respectively, Equation 2–21 is rewritten as

$$k_x \frac{\partial^2 p}{\partial x^2} + k_y \frac{\partial^2 p}{\partial y^2} = 0 \qquad (2\text{–}22)$$

multiplying and dividing throughout by $\sqrt{k_x k_y}$, Equation 2–22 becomes

$$\sqrt{k_x k_y}\left[\sqrt{\frac{k_x}{k_y}}\frac{\partial^2 p}{\partial x^2}+\sqrt{\frac{k_y}{k_x}}\frac{\partial^2 p}{\partial y^2}\right]=0 \qquad (2\text{–}23)$$

This can be transformed into

$$\sqrt{k_x k_y}\left[\frac{\partial^2 p}{\partial x^2}+\frac{\partial^2 p}{\partial y'^2}\right]=0 \qquad (2\text{–}24)$$

where

$$y' = y\sqrt{k_x/k_y} \qquad (2\text{–}25)$$

and

$$y = y'\sqrt{k_y/k_x} \qquad (2\text{–}26)$$

Thus, an areally anisotropic reservoir would be the equivalent of a reservoir with an effective horizontal permeability of $\sqrt{k_x k_y}$ and the length along the high-permeability side is $\sqrt{k_y/k_x}$ times the length along a low-permeability side. Thus, if permeability along the fracture trend is 16 times

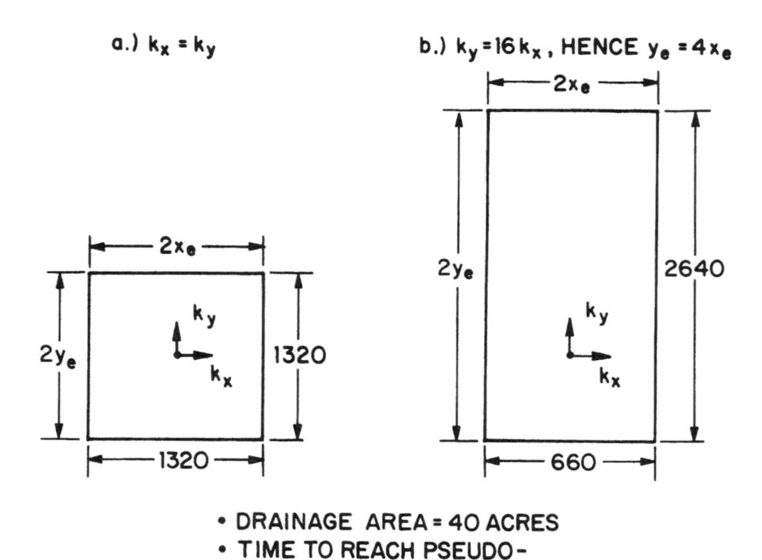

Figure 2–14 Drainage Areas of a Vertical Well in Isotropic and Anisotropic Reservoirs.

larger than perpendicular to it, then drainage length along the fracture is four times larger than the length perpendicular to the fracture (see Figure 2–14b). In such areally anisotropy reservoirs, it is difficult to drain larger reservoir length in the low-permeability direction using vertical wells. A horizontal well drilled along the low-permeability direction has the potential to drain a significantly larger area than a vertical well, resulting in a larger reserve for horizontal wells than vertical wells. Thus, horizontal wells are highly beneficial in areally anisotropic reservoirs. It is obvious that in naturally fractured formations, horizontal wells drilled in a direction perpendicular to the natural fractures would be highly beneficial. The success of horizontal wells in naturally fractured reservoirs, such as Austin Chalk formation in Texas, U.S.A., and Bakken Shale formation in North Dakota, U.S.A., illustrates the advantage of horizontal drilling in areally anisotropic formations.

So far the discussion has been mainly centered on vertical wells. For fractured vertical wells, limited results are available to calculate the time to reach the pseudo-steady state in square drainage boundaries (see Table 2–3).[7,8] Khan has obtained results for fractured vertical wells in rectangular areas.[8] Recently, similar results were also obtained for horizontal wells.[10–12] Tables 2–4 and 2–5 list the time to reach the pseudo-steady state for fractured vertical wells and horizontal wells in rectangular drainage areas, respectively.[10]

It is important to note that there is some discrepancy in calculation of time to start pseudo-steady state. This is because of the way the time to start pseudo-steady is calculated, as explained below. For a single-phase flow in a homogeneous reservoir, the relationship between dimensionless pressure

TABLE 2–3 SHAPE FACTORS AND TIME TO REACH PSEUDO-STEADY STATE FOR FRACTURED VERTICAL WELLS IN A SQUARE DRAINAGE AREA[7]

x_f/x_e*	C_A	$\ln C_A$	$1/2 \ln\left(\frac{2.2458}{C_A}\right)$	EXACT FOR $t_{DA}>$	LESS THAN 1% ERROR FOR $t_{DA}>$
0.1	2.6541	0.9761	−0.0835	0.175	0.08
0.2	2.0348	0.7104	+0.0493	0.175	0.09
0.3	1.9986	0.6924	+0.0583	0.175	0.09
0.5	1.6620	0.5080	+0.1505	0.175	0.09
0.7	1.3127	0.2721	+0.2685	0.175	0.09
1.0	0.7887	−0.2374	+0.5232	0.175	0.09

* x_f represents half length (one wing) of a fracture and x_e represents half length of a side of a square drainage area.

TABLE 2–4 TIME FOR START OF PSEUDO-STEADY STATE FOR FULLY PENETRATING INFINITE-CONDUCTIVITY FRACTURES WITH VARIOUS FRACTURE PENETRATION RATIOS (x_f/x_e) AND DIFFERENT (x_e/y_e) RATIOS[10]

	x_e/y_e					
x_f/x_e	1	2	3	5	10	20
0.01	0.2	0.3	0.45	0.7	2.0	3.0
0.05	0.2	0.3	0.45	0.7	2.0	3.0
0.10	0.2	0.3	0.45	0.7	2.0	3.0
0.20	0.2	0.3	0.45	0.7	2.0	3.0
0.40	0.2	0.3	0.45	0.7	2.0	3.0
0.50	0.2	0.3	0.45	0.7	2.0	3.0
0.70	0.2	0.3	0.45	0.7	2.0	3.0

TABLE 2–5 TIME FOR START OF PSEUDO-STEADY STATE FOR HORIZONTAL WELLS WITH VARIOUS WELL PENETRATION RATIOS ($L/(2x_e)$) AND DIFFERENT (x_e/y_e) RATIOS[10]

	x_e/y_e		
$L/(2x_e)$	1	2	5
0.2	0.15	0.2	0.6
0.4	0.15	0.2	0.6
0.6	0.15	0.2	0.6
0.8	0.15	0.2	0.6
1.0	0.10	0.15	0.3

and dimensionless time for a well producing at a constant rate in a bounded reservoir (i.e., reservoir with a fixed drainage area) is given, as

$$p_D = A' + 2\pi t_{DA} \tag{2–27}$$

where A' is a constant. Taking the derivative of Equation 2–27 gives

$$m = dp_D/dt_{DA} = 2\pi \tag{2–28}$$

Thus, in a single-phase flow calculation, pseudo-steady begins when slope m becomes 2π. Some engineers assume that when m reaches within 10% of 2π value, pseudo-steady state begins, while others use 5% criteria and a few others use 1% criterion. Depending upon the criterion used, one can estimate different values for beginning time of pseudo-steady state. The criterion used can give significantly different values of beginning time of pseudo-steady state as noted in References 13 and 14. At the present time, there is no consensus about the criterion, but most engineers accept $t_{DA} = 0.1$ as a dimensionless time to start a pseudo-steady state for a vertical well located centrally in either a circular or square drainage area. Reference 7 does not include information about criterion used to calculate $t_{DA} = 0.1$, probably because these results were obtained using a numerical simulator. The results of Reference 10 for calculation of pseudo-steady state for horizonal wells are probably conservative because they used a slope requirement of 5% within the value of 2π.

The above discussion indicates that before using any dimensionless time to reach pseudo-steady state, it is important to critically review the criterion that has been used. This is especially important in determining well spacing in leases where the lease concession lasts only for a few years, say less than 10 years. In these reservoirs, time for beginning of pseudo-steady state becomes important to drain a reservoir effectively in a limited time period.

HORIZONTAL WELL DRAINAGE AREA

Due to longer well length, in a given time period under similar operating conditions, a horizontal well would drain a larger reservoir area than a vertical well. If a vertical well drains a certain reservoir volume (or area) in a given time, then this information can be used to calculate a horizontal well drainage area. A horizontal well can be looked upon as a number of vertical wells drilled next to each other and completed in a limited payzone thickness. Then, as shown in Figures 2–13 and 2–15, each end of a horizontal well would drain either a square or a circular area, with a rectangular drainage area at the center. This concept implicitly assumes that the reservoir thickness is considerably smaller than the sides of the drainage area. It is possible to calculate the drainage area of a horizontal well by assuming an elliptical drainage area in the horizontal plane, with each end of a well as a foci of drainage ellipse. In the following examples, the methods to estimate drainage areas of horizontal wells are outlined. In general, different methods give fairly similar results. As a rule of thumb, a 1000-ft-long horizontal well can drain twice the area of a vertical well while a 2000-ft-long well can drain three times the area of a vertical well in a given time. Thus, it is important to use larger well spacing for horizontal well development than that used for vertical well development. The following examples show drainage area

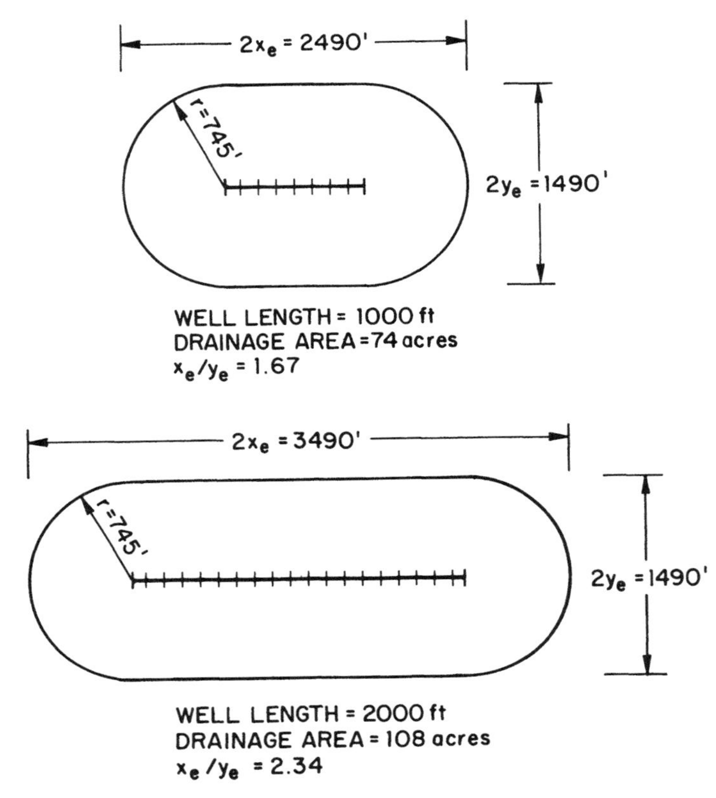

Figure 2–15 Drainage Areas of 1000 and 2000-ft-long Horizontal Wells.

calculations for areally isotropic and anisotropic reservoirs. In a fractured reservoir, where permeability in one direction is higher than the other, then the well would accordingly drain a larger length in a high-permeability direction by a factor of $\sqrt{k_y/k_x}$ where k_y represents higher permeability in the horizontal plane, and k_x represents lower permeability in the horizontal plane (see Figure 2–16).

EXAMPLE 2–8

A 400-acre lease is to be developed using 10 vertical wells. An engineer suggested drilling either 1000- or 2000-ft-long horizontal wells. Calculate the possible number of horizontal wells that will drain the lease effectively. Assume that a single vertical well effectively drains 40 acres.

Solution

A 40-acre vertical well would drain a circle of radius 745 ft. If r_{ev} is a drainage radius of a vertical well, then

$$\text{Area of a circle} = \pi r_{ev}^2 = 40 \text{ acres} \times 43{,}560 \text{ sq ft/acre}$$

$$r_{ev} = 745 \text{ ft}$$

Two methods can be employed to calculate horizontal well drainage area on the basis of a 40-acre drainage area of a vertical well.

Method I

As shown in Figure 2–15, a 1000-ft-long well would drain 74 acres. The drainage area is presented as two half circles at each end and a rectangle in the center. Similarly, as shown in Figure 2–15, a 2000-ft-long well would drain 108 acres.

Method II

If we assume that the horizontal well drainage area is an ellipse in a horizontal plane, then

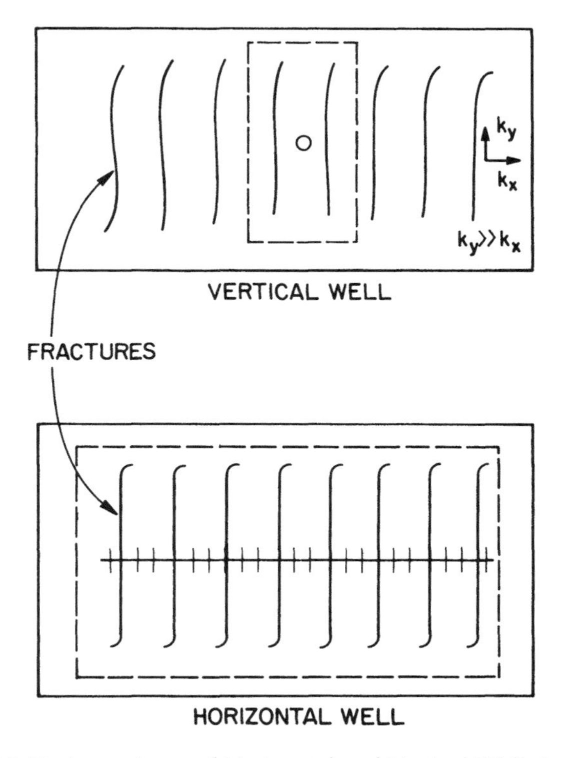

Figure 2–16 Drainage Areas of Horizontal and Vertical Wells in a Fractured Reservoir.

for a 1000-ft-long well,

$$a = \text{half major axis of an ellipse} = (L/2) + 745$$
$$= (1000/2) + 745 = 1245 \text{ ft}$$
$$b = \text{half minor axis of an ellipse} = 745 \text{ ft}$$
$$\text{Drainage area} = \pi ab/43{,}560$$
$$= \pi\,(1245 \times 745)/43{,}560 = 67 \text{ acres}$$

for a 2000-ft-long well,

$$a = \text{half major axis of an ellipse} = (L/2) + 745$$
$$= (2000/2) + 745 = 1{,}745 \text{ ft}$$
$$b = \text{half minor axis of an ellipse} = 745 \text{ ft}$$
$$\text{Drainage area} = \pi ab/43{,}560$$
$$= \pi\,(1745 \times 745)/43{,}560$$
$$= 94 \text{ acres}$$

As we see, two methods give different answers for drainage area. If we take average areas using two methods, a 1000-ft well would drain 71 acres and a 2000-ft well would drain 101 acres. Thus, a 400-acre field can be drained by 10 vertical wells; 6 1000-ft-long wells; or 4 2000-ft-long wells. Thus, horizontal wells seem very appropriate for offshore and hostile environment applications where a substantial upfront savings can be obtained by drilling long horizontal wells. This is because a large area can be drained by using a reduced number of wells. This reduces the number of slots that are required on offshore platforms, and thereby significantly reduces the cost of these platforms.

EXAMPLE 2–9

A 600-acre lease is to be developed with 10 vertical wells. Another alternative is to drill 500-ft-, 1000-ft-, or 2000-ft-long horizontal wells. Estimate the possible number of horizontal wells that will drain the leases effectively.

Solution

A 60-acre vertical well would drain a circle of radius r_{ev} of 912 ft.

$$\text{Area of a circle} = \pi r_{ev}^2 = 60 \text{ acres} \times 43{,}560 \text{ sq ft/acre}$$
$$r_{ev} = 912 \text{ ft}$$

Method I

We calculate the area drained by a horizontal well by two methods. In the first method, the drainage area is represented as two half circles of radius r_{ev} at each end and a rectangle, of dimensions $L \times 2r_{ev}$, in the center.

	Horizontal Well Length		
	500 ft	1000 ft	2000 ft
Area of 2 half circles (acres)	30 + 30 = 60	30 + 30 = 60	30 + 30 = 60
Dimensions of central rectangle (acres)	20.9	41.9	83.7
Total drainage area (acres)	80.9	101.9	143.7

Method II

This method assumes that a horizontal well drains an ellipse with minor axis a and major axis b.

The drainage area of an ellipse in acres is

$$A = \pi ab/43{,}560$$

	Horizonal Well Length		
	500 ft	1000 ft	2000 ft
a = half major axis, ft	250 + 912 = 1162	500 + 912 = 1412	1000 + 912 = 1912
b = half minor axis, ft	912	912	912
Drainage area, acres	76.4	92.9	125.8

Now, averaging drainage areas of the two methods we have

	Horizontal Well Length		
	500 ft	1000 ft	2000 ft
Average drainage area (acres) Methods I + II	79	98	135
Number of wells for 600-acre field	8	6	4 or 5

Thus, a 600-acre field can be effectively drained either by 10 vertical wells, eight 500-ft-long horizontal wells, six 1000-ft-long wells, or five 2000-ft-long wells.

EXAMPLE 2–10

A 360-acre lease shown in Figure 2–17 is to be developed using nine vertical wells. How many 1000-ft-long horizontal wells could drain this reservoir effectively? How many 2000-ft-long horizontal wells could drain this effectively? Suggest a development pattern.

Solution

As shown in Example 2–8, if a vertical well drains 40 acres effectively, 1000-ft and 2000-ft-long horizontal wells would drain 80 and 120 acres, respectively.

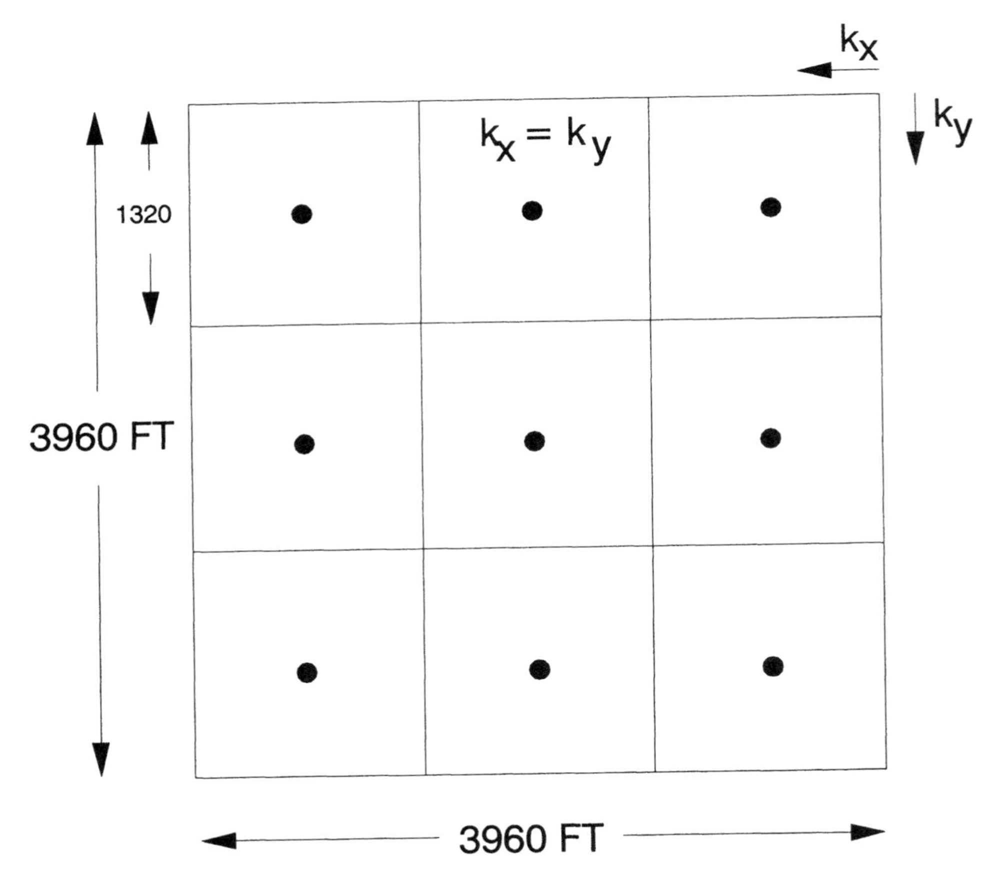

Figure 2–17 Nine Vertical Wells Draining a 360-Acre Lease.

With 1000-ft-long wells, the 360-acre lease could be developed using either (1) four horizontal wells and one vertical well or (2) three horizontal wells and three vertical wells. The possible configurations are shown in Figures 2–18 and 2–19.

As quoted in Example 2–8, a 2000-ft-long horizontal well could drain 120 acres. As shown in Figure 2–20, a 360-acre lease can be developed using three 2000-ft-long horizontal wells.

EXAMPLE 2–11

A well, Harris-1, drains approximately 40 acres in a 35-ft-thick naturally fractured reservoir. Pressure tests conducted between Harris-1 and the well to the east between Harris-1 and the well to the north indicate permeability difference along the two directions. The permeability along the east-west,

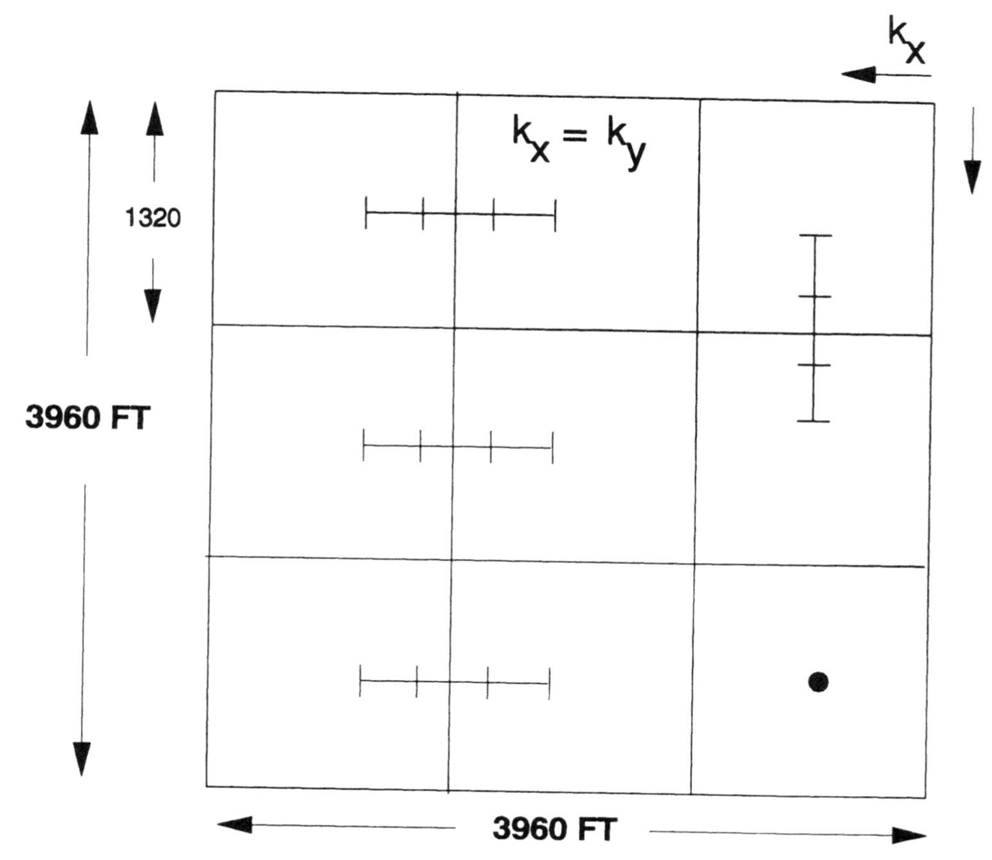

Figure 2–18 Four Horizontal and One Vertical Well Draining a 360-Acre Lease.

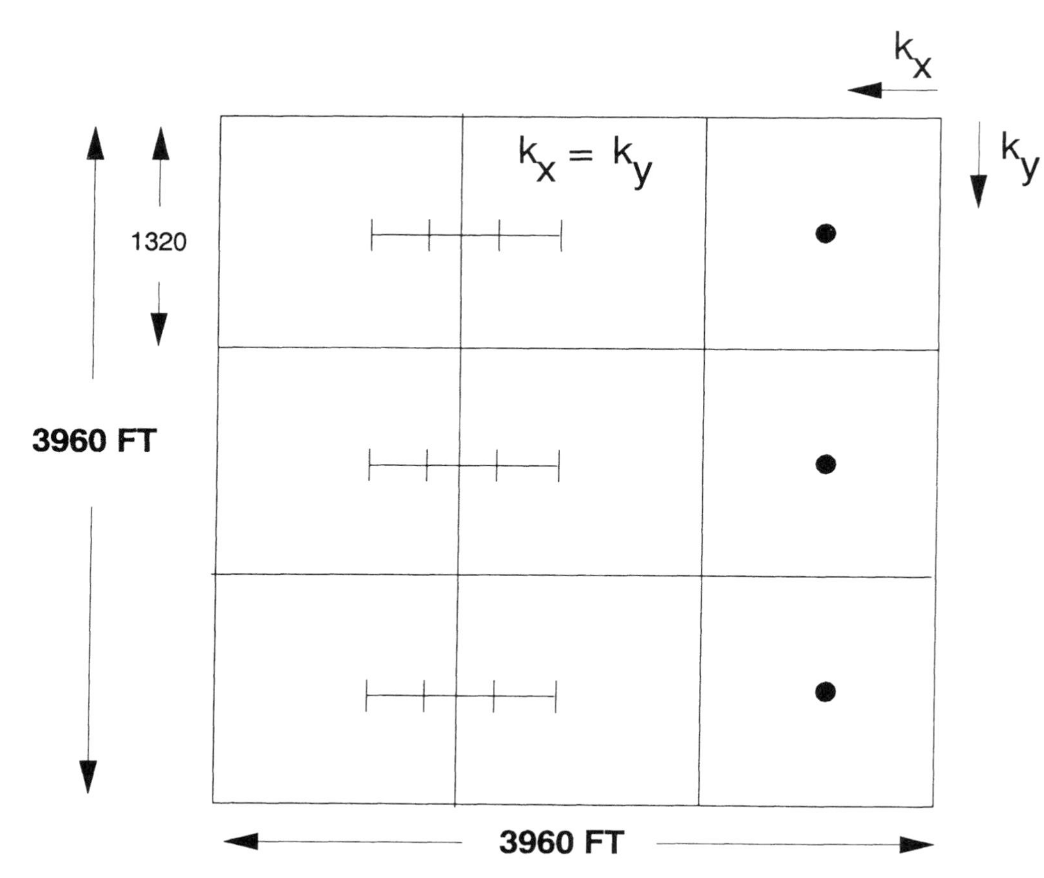

Figure 2–19 Three Horizontal and Three Vertical Wells Draining a 360-Acre Lease.

k_x, is 0.5 md, while the permeability along the north-south direction, k_y, is 4.5 md. An engineer proposed to drill a 2000-ft-long horizonal well along the east-west direction. Estimate the drainage area and dimensions of each drainage area side.

Solution

Let us assume that the vertical well, Harris-1, drains a rectangle area due to anisotropy. If the reservoir has uniform permeability, then the well would drain a 40-acre square with each side being

$$2x_e = 2y_e = \sqrt{40 \times 43{,}560} = 1320 \text{ ft}$$

The reservoir has non-uniform permeability in the areal plane with

$$k_x = 0.5 \text{ md} \qquad \text{and} \qquad k_y = 4.5 \text{ md}$$

hence, $k_y/k_x = 4.5/0.5 = 9$
and $\sqrt{k_y/k_x} = 3$

If the drainage rectangle has sides $2x_e$ and $2y_e$, and if we assume that Harris-1 drains only 40 acres,

$$(2x_e) \times (2y_e) = 40 \times 43{,}560$$

additionally, due to anisotropy,

$$2y_e/2x_e = 3$$

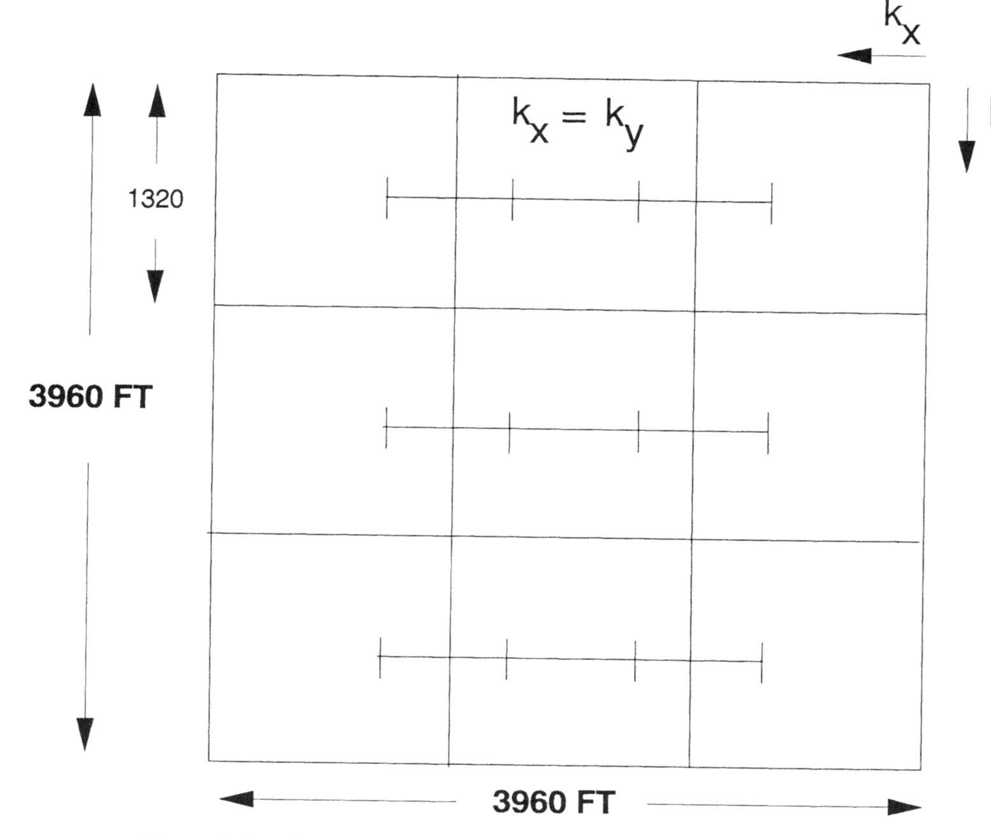

Figure 2–20 Three 2000-ft Horizontal Wells Draining a 360-Acre Lease.

Solving the above two equations simultaneously,

$$2x_e = 762 \text{ ft and } 2y_e = 2286 \text{ ft}$$

Thus, for a vertical well, the drainage length along the east-west direction, which is a low permeability direction, is 762 ft. Vertical well drainage along the high permeability north-south direction, $2y_e$, is three times longer than the east-west direction, $2x_e$.

Assuming the well tips of a horizontal well drain half of a vertical well, for a 2000-ft-long horizontal well drilled along the east-west direction, the drainage length along this direction is $2x_e = 2000 + 762 = 2762$ ft. Similarly, drainage length along the north-south direction will be the same as that for a vertical well which is $2y_e = 2286$ ft. Therefore, well spacing should be at least 2286 ft. along the north-south direction. Along the east-west direction, the horizontal well tips should be spaced at least 762 ft apart.

Thus, well spacing requirements for vertical, as well as horizontal wells, are different in isotropic and anisotropic reservoirs.

REFERENCES

1. van Everdingen, A. F., and Hurst, W.: "The Application of the Laplace Transformation to Flow Problems in Reservoirs," Trans., AIME, vol. 186, pp. 305–324, 1949.
2. van Everdingen, A. F.: "The Skin Effect and Its Influence on the Productive Capacity of a Well," *Trans.*, AIME, vol. 198, pp. 171, 1953.
3. Hurst, William: "Establishment of the Skin Effect and Its Impediment to Fluid Flow into a Well Bore," *The Petroleum Engineer*, October 1953.
4. Hawkins, M. F.: "A Note on the Skin Effect," *Trans.*, AIME, vol. 207, pp. 356–357, 1956.
5. Celier, G. C. M. R., Joualt, P. and de Montigny, O. A. M. C.: "Zuidwal: A Gas Field Development with Horizonal Wells," paper SPE 19826, presented at the 64th Annual Technical Conference and Exhibition of the Society of Petroleum Engineers, San Antonio, Texas, October 8–11, 1989.
6. Golan, M. and Whitson, C. H., *Well Performance*, International Human Resources Development Corporation, Boston, 1986.
7. Earlougher, R. C., Jr., "Advances in Well Test Analysis," Monograph Vol. 5 of the Henry L. Doherty Series in Society of Petroleum Engineers of AIME, 1977.
8. Gringarten, A. C., Ramey, H. J., Jr., and Raghavan, R.: "Unsteady-State Pressure Distribution Created by a Well with a Single Infinite-Conductivity Vertical Fracture," *Society of Petroleum Engineers Journal*, pp. 347–360, August 1974.
9. Khan, A. "Pressure Behavior of a Vertically Fractured Well Located at the Center of a Rectangular Drainage Region," M.S. Thesis, The University of Tulsa, 1978.
10. Mutalik, P. N., Godbole, S. P., and Joshi, S. D.: "Effect of Drainage Area Shapes on the Productivity of Horizontal Wells," paper SPE 18301, presented at the 63rd Annual Technical Conference, Houston, Texas, October 2–5, 1988.
11. Daviau, F., Mouronval, G, Bourdarot, G., and Curutchet, P.: "Pressure Analysis for Horizontal Wells," *SPE Formation Evaluation*, pp. 716–724, December 1988.

12. Goode, P. A. and Thambyanyagam, R. K. M., "Pressure Drawdown and Buildup Analysis of Horizontal Wells in Anisotropic Media," *SPE Formation Evaluation,* pp. 683–697, December 1987.
13. Onur, M. and Reynolds, A. C., "A New Approach for Constructing Derivative Type Curves for Well Test Analysis," *SPE Formation Evaluation,* pp. 197–206, March 1988.
14. Vongvuthipornchai, S., and Raghavan, R., "A Note on the Duration of the Transitional Period of Responses Influenced by Wellbore Storage and Skin," *SPE Formation Evaluation,* pp. 207–214, March 1988.
15. Dake, L. P.: *Fundamentals of Reservoir Engineering,* Elsevier Scientific Publishing Co., New York, 1978.

CHAPTER 3

Steady-State Solutions

INTRODUCTION

The steady-state analytical solutions are the simplest form of horizontal well solutions. These equations assume steady state, i.e., pressure at any point in the reservoir does not change with time.

In practice, very few reservoirs operate under steady-state conditions. In fact, most reservoirs exhibit change in reservoir pressure over time. In spite of this, steady-state solutions are widely used because (1) they are easy to derive analytically; (2) it is fairly easy to convert steady-state results to either transient and pseudo-steady state results by using concepts of expanding drainage boundary over time and effective wellbore radius and shape factors, respectively; and (3) steady-state mathematical results can be verified experimentally by constructing physical models in a laboratory. This is explained below.

From the standpoint of physics, Fourier's law of heat conduction, Ohm's law for flow of electricity, and Darcy's law for flow through porous media are similar.

$$\text{Fourier's law:} \qquad q = -kA\frac{\Delta T}{\Delta x} \tag{3–1}$$

where

q = heat transfer rate, BTU/hr
k = thermal conductivity, BTU/(hr-ft-°F)
A = cross-sectional area, ft^2
ΔT = temperature difference, °F
Δx = distance, ft

$$\text{Ohm's law:} \quad I = \frac{V}{R} \tag{3–2}$$

where

I = current, amperes
V = voltage, volts
R = resistance, ohms

$$\text{Darcy's law:} \quad q = \frac{-kA}{\mu}\frac{\Delta p}{\Delta x} \tag{3–3}$$

where

q = flow rate, cm^3/sec
k = permeability, darcy
A = cross-sectional area of flow, cm^2
μ = viscosity, cp
Δp = pressure drop, atmospheres
Δx = distance, cm

A comparison of Ohm's law and Darcy's law yields

$$I = q, \qquad V = -\Delta p, \qquad \text{and} \qquad R = (\mu\Delta x)/(kA). \tag{3–4}$$

Since the early days of petroleum engineering, many steady-state equations have been verified by using electrical models.[1,2,3] For example, well productivity based on perforation density was estimated using electrical analog experiments. Thus, steady-state analytical expressions offer a distinct advantage that these mathematical expressions can be checked out by laboratory experiments. As noted in Chapter 2, it is straightforward to convert steady-state expressions into pseudo-steady state expressions (see Table 2–2).

STEADY-STATE PRODUCTIVITY OF HORIZONTAL WELLS

Several solutions are available in the literature to predict the steady-state flow rate in a horizontal well. Borisov,[4] Merkulov,[5] Giger,[6] Giger et al.,[7] Renard and Dupuy,[8] and Joshi[9,10] have reported similar solutions. These solutions in generalized units are given below.

Borisov[4]

$$q_h = \frac{2\pi k_h h \Delta p/(\mu_o B_o)}{\ln[(4r_{eh}/L)] + (h/L)\ln[h/(2\pi r_w)]} \tag{3–5}$$

Giger[6]

$$q_h = \frac{2\pi k_h L \Delta p/(\mu_o B_o)}{(L/h)\ln\left(\dfrac{1+\sqrt{1-[L/(2r_{eh})]^2}}{L/(2r_{eh})}\right) + \ln[h/(2\pi r_w)]} \tag{3–6}$$

Giger, Reiss & Jourdan[7]

$$J_h/J_v = \frac{\ln(r_{ev}/r_w)}{\ln\left[\dfrac{1+\sqrt{1-[L/(2r_{eh})]^2}}{L/(2r_{eh})}\right] + (h/L)\ln[h/(2\pi r_w)]} \tag{3–7}$$

Renard and Dupuy[8]

$$q_h = \frac{2\pi\, k_h\, h\Delta p}{\mu_o B_o}\left[\frac{1}{\cosh^{-1}(X) + (h/L)\ln[h/(2\pi r_w)]}\right] \tag{3–8}$$

$X = 2a/L$ for ellipsoidal drainage area (3–9)

a = half the major axis of drainage ellipse (see Eq. 3–11)

Joshi[9,10]

$$q_h = \frac{2\pi k_h h \Delta p/(\mu_o B_o)}{\ln\left[\dfrac{a + \sqrt{a^2 - (L/2)^2}}{L/2}\right] + (h/L)\ln[h/(2r_w)]} \qquad (3\text{–}10)$$

$$a = (L/2)\left[0.5 + \sqrt{0.25 + (2r_{eh}/L)^4}\right]^{0.5} \qquad (3\text{–}11)$$

In Equations 3–5 through 3–10, L represents horizontal well length, h represents reservoir height, r_w represents wellbore radius, and r_{ev} and r_{eh} represent drainage radius of vertical and horizontal wells, respectively. Additionally, μ_o is oil viscosity, B_o is oil formation volume factor, Δp is pressure drop from the drainage boundary to the wellbore, and q_h is flow rate of a horizontal well. The productivity index J_h can be obtained by dividing q_h by Δp.

Note that all the above solutions are for isotropic reservoirs ($k_h = k_v$). *These equations can be modified to the practical field units by replacing 2π in the numerator by 0.007078. For example, Equation 3–10 can be rewritten in U.S. oil field units as*

$$q_h = \frac{0.007078\, k_h h \Delta p/(\mu_o B_o)}{\ln\left[\dfrac{a + \sqrt{a^2 - (L/2)^2}}{L/2}\right] + (h/L)\ln[h/(2r_w)]} \qquad (3\text{–}12)$$

In Equation 3–12, q_h is oil flow rate in STB/day, k_h is horizontal permeability in md, h is reservoir thickness in ft, Δp is pressure drop from the drainage radius to the wellbore in psi, μ_o is oil viscosity in cp, B_o is formation volume factor in RB/STB, L is horizontal well length in ft, and r_w is wellbore radius in ft. (Note that gas equations are included in Chapter 9.)

Relationships between $L/(2r_{eh})$, $L/(2a)$ and a/r_{eh} are given in Table 3–1. The detailed derivation of Equation 3–10 is given in Reference 9. The derivation of Equation 3–8 is given in Reference 8.

As shown in Figure 3–1, the three-dimensional horizontal well problem is divided into two two-dimensional problems.[9] The mathematical solutions of these two-dimensional problems are added to calculate horizontal-well flow rate. Results described in Reference 9 show an excellent agreement between Equation 3–10 and the laboratory experiments. The laboratory experiments were electrical experiments based upon the similarity between Ohm's and Darcy's laws. The comparison of various equations, Equations 3–5 through 3–10, shows a small difference between the various equations by a term of $(h/L)\ln\pi$ in the denominator of the flow equations. However,

TABLE 3–1 RELATIONSHIP BETWEEN VARIOUS GEOMETRICAL FACTORS[9]

$\frac{L}{2r_{eh}}$	$\frac{L}{2a}$	$\frac{a}{r_{eh}}$
0.1	0.0998	1.002
0.2	0.198	1.010
0.3	0.293	1.024
0.4	0.384	1.042
0.5	0.470	1.064
0.6	0.549	1.093
0.7	0.620	1.129
0.8	0.683	1.171
0.9	0.739	1.218

the effect of this small difference on the calculations of production rate is normally minimal.

If the length of a horizontal well is significantly longer than the reservoir thickness, i.e., $L >> h$, then the second term in the denominator of Equation 3–5 is negligible and the solution reduces to

$$q_h = \frac{0.007078\, k_h h\, \Delta p/(\mu_o B_o)}{\ln\,(4r_{eh}/L)} \qquad (3\text{–}13)$$

Note Equation 3–13 is in U.S. field units. This can be rewritten as

$$q_h = \frac{0.007078\, k_h h\, \Delta p/(\mu_o B_o)}{\ln\,[r_{eh}/(L/4)]}. \qquad (3\text{–}14)$$

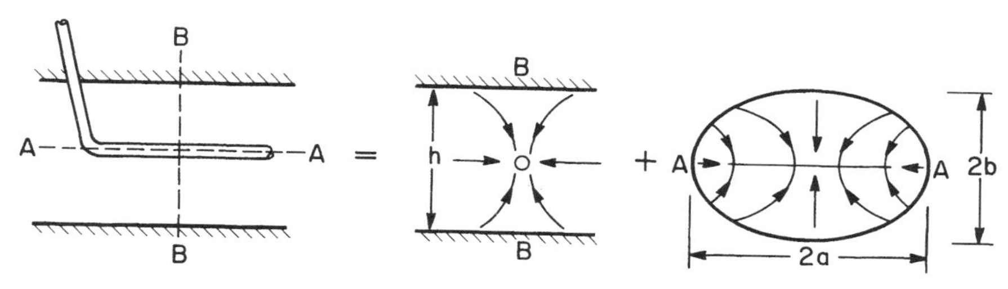

Figure 3–1 A Division of a 3-D Problem into Two, 2-D Problems.[9]

Thus, for a long horizontal well, the effective wellbore radius, $r'_w = L/4$, is the same as that for a fully penetrating infinite-conductivity vertical fracture. Similarly, Equations 3–6 through 3–10 would reduce to Equation 3–14 if well length $L >> h$ and also if well length L is small as compared to drainage radius r_{eh}. (In Equation 3–10 one would also have to assume that well length L is smaller than the half-length of the major axis of drainage ellipse a.) Thus, in a limiting case, at least for a single-phase flow, productivity of a horizontal well approaches that of a fully penetrating, infinite-conductivity vertical fracture.

EXAMPLE 3–1

A 1000-ft-long horizontal well is drilled in a reservoir with the following characteristics.

$$k_v = k_h = 75 \text{ md}, \quad \mu_o = 0.62 \text{ cp}$$
$$h = 160 \text{ ft}, \quad B_o = 1.34 \text{ RB/STB}$$
$$\phi = 3.8\%, \quad r_w = 0.365 \text{ ft}$$

Calculate the steady-state horizontal well productivity using different methods if a typical vertical well drains 40 acres.

Solution

As noted in Chapter 2, if a vertical spacing is 40 acres, then a 1000-ft-long horizontal well would drain about 80 acres. For a vertical well draining 40 acres, drainage radius r_{ev} for a circular drainage area is 745 ft. The productivity index for a vertical well can be calculated as

$$J_v = \frac{0.007078 \times 75 \times 160/(0.62 \times 1.34)}{\ln(745/0.365)}$$
$$= 102.23/\ln(2041)$$
$$= 13.4 \text{ STB/(day-psi)}$$
$$\approx 13 \text{ STB/(day-psi)}.$$

For a horizontal well draining 80 acres, the drainage radius of a circular drainage area is 1053 ft. Thus, $r_{eh} = 1053$ ft.

1. Borisov Method (Eq. 3–5)

$$J_h = q_h/\Delta p = \frac{0.007078\, k_h h}{\mu_o B_o\, [\ln(4r_e/L) + (h/L)\ln(h/(2\pi r_w))]}$$

$$= \frac{0.007078 \times 75 \times 160}{0.62 \times 1.34 \left[\ln\left(\frac{4 \times 1053}{1000}\right) + \left(\frac{160}{1000}\right)\ln\left(\frac{160}{2\pi \times 0.365}\right)\right]}$$

$$= 48.3 \approx 48 \text{ STB/(day-psi)}$$

2. Giger Method (Eq. 3–6)

$$J_h = \frac{0.007078\, k_h L/(\mu_o B_o)}{(L/h)\ln\left[\dfrac{1+\sqrt{1-[L/(2r_{eh})]^2}}{L/(2r_{eh})}\right] + \ln[h/(2\pi r_w)]}$$

$$= \frac{0.007078 \times 75 \times 1000/(0.62 \times 1.34)}{(1000/160)\ln\left[\dfrac{1+\sqrt{1-(1000/(2\times 1053))^2}}{1000/(2\times 1053)}\right] + \ln[160/(2\pi \times 0.365)]}$$

$= 49.8 \approx 50$ STB/(day-psi).

3. Joshi Method (Eqs. 3–10 and 3–11)

$$a = (L/2)\,[0.5 + \sqrt{0.25 + (2r_{eh}/L)^4}]^{0.5} \quad (3\text{–}11)$$

$$= (1000/2)\,[0.5 + \sqrt{0.25 + (2 \times 1053/1000)^4}]^{0.5}$$

$= 1114$ ft

$$J_h = \frac{0.007078 k_h h/(\mu_o B_o)}{\ln\left[\dfrac{a+\sqrt{a^2-(L/2)^2}}{L/2}\right] + (h/L)\ln[h/(2r_w)]}$$

$$= \frac{0.007078 \times 75 \times 160/(0.62 \times 1.34)}{\ln\left[\dfrac{1114+\sqrt{1114^2-(1000/2)^2}}{(1000/2)}\right] + (160/1000)\ln[160/(2\times 0.365)]}$$

$= 44.4 \approx 44$ STB/(day-psi).

The productivity ratios for an 80-acre spacing horizontal well and a 40-acre spacing vertical well by different methods are listed below.

J_h/J_v by Different Methods

METHODS	PRODUCTIVITY INDEX J_h, STB/(day/psi)	J_h/J_v	AREAL PRODUCTIVITY INDEX $= J_h$/acre STB/(day-psi-acre)
Borisov	48	3.7	0.60
Giger	50	3.8	0.63
Joshi	44	3.4	0.56

It is important to note that the above productivity index comparison assumes an unstimulated vertical well.

INFLUENCE OF RESERVOIR HEIGHT ON WELL PRODUCTIVITY

The influence of reservoir height on horizontal wells is quite significant. For a given length of a horizontal well, the incremental gain in reservoir contact area in a thin reservoir is much more than that in a thick reservoir. For example, assume drilling a 1000-ft-long horizontal well in two possible target zones (one zone with a thickness of 50 ft and the other zone with a thickness of 500 ft). The incremental gain in the contact area in a 50-ft-thick reservoir by drilling a 1000-ft-long horizontal well is about 20 times more than that with a vertical well. In contrast, in a 500-ft-thick reservoir, the incremental gain in contact area by drilling a 1000-ft-long horizontal well is only twofold. Thus, significantly more gain in contact area can be achieved in a thin reservoir than in a thick reservoir. It is important to note that the terms *thick* and *thin* are relative. One should look for incremental contact area rather than using a specific definition of thick and thin reservoirs. Additionally, it is also important to note that *thick reservoirs have more reserves than thin reservoirs.*

The influence of reservoir height on horizontal well productivity can be estimated using steady-state equations. Figure 3–2 shows the change in productivity of a horizontal well in a 160-acre drainage area under steady-

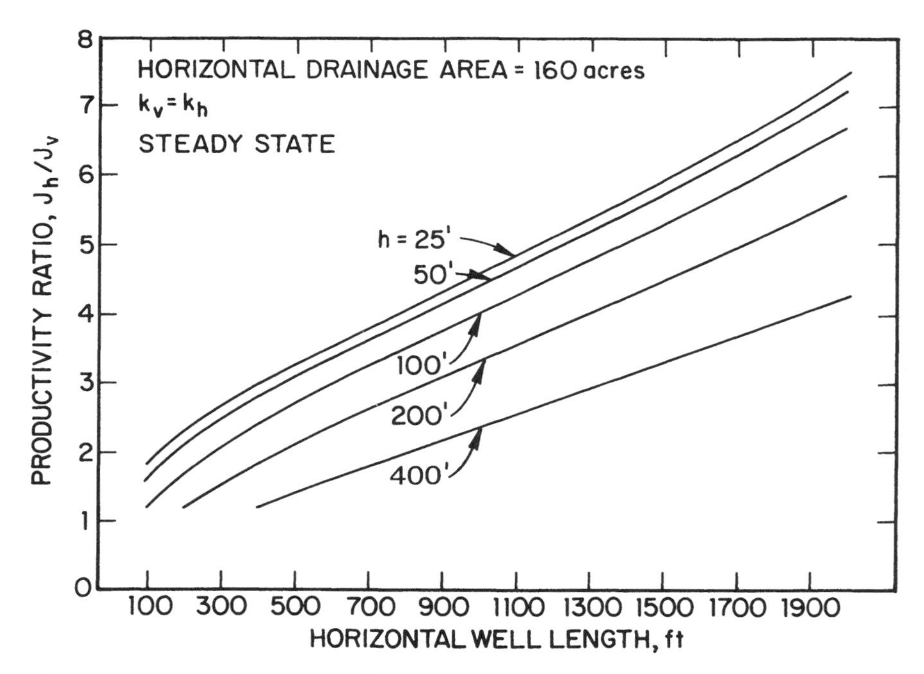

Figure 3–2 Productivity Ratio of Horizontal and Vertical Well versus Well Length for Different Reservoir Thickness.

state conditions. The results assume that the reservoir is isotropic ($k_h = k_v$). The top curve in Figure 3–2 is for a 25-ft-thick reservoir and the bottom curve is for a 400-ft-thick reservoir. As seen in the figure, the incremental gain in productivity is much higher in a thin reservoir than in a thick reservoir. Figure 3–3 shows the same results in terms of skin factors. As discussed earlier in Chapter 2, productivities can easily be converted to skin factors using Equations 2–5, 2–9 and 2–11.

EXAMPLE 3–2

Evaluate the productivity of a 1000-ft-long horizontal well in 25- and 400-ft-thick reservoirs with other reservoir parameters the same as those in Example 3–1. Compare ratios of productivity indices for horizontal and vertical wells, if vertical wells are spaced at 40 acres. What would be the productivity ratios if vertical wells are also spaced at 80 acres?

Solution

Equations 3–11 and 3–12 are used to calculate horizontal well productivity. As shown in Example 3–1, the half-length of the major axis of a drainage ellipse a is 1114 ft.

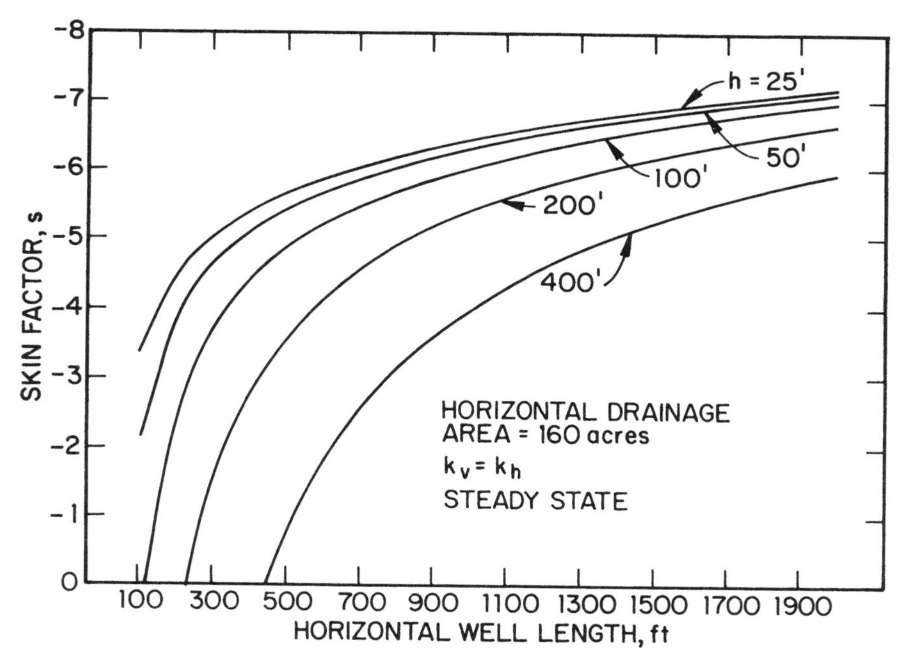

Figure 3–3 Skin Factor versus Well Length for Different Reservoir Thickness.

1. $h = 25$ ft

$$J_h = \frac{0.007078 \times 75 \times 25/(0.62 \times 1.34)}{\ln\left[\dfrac{1114 + \sqrt{(1114)^2 - (1000/2)^2}}{(1000/2)}\right] + (25/1000)\ln(25/(2 \times 0.365))}$$

$$= 15.97/(1.44 + 0.088) = 10.45 \approx 10 \text{ STB/(day-psi)}$$

Productivity of an unstimulated vertical well draining 40 acres is

$$J_v = \frac{0.007078 \times 75 \times 25/(0.62 \times 1.34)}{\ln(745/0.365)}$$

$$= 2.1 \text{ STB/(day-psi)}$$

where vertical well drainage radius is 745 ft. The productivity of an unstimulated vertical well draining 80 acres is

$$J_v = \frac{0.007078 \times 75 \times 25/(0.62 \times 1.34)}{\ln(1053/0.365)}$$

$$= 2.0 \text{ STB/(day-psi)}$$

where the vertical well drainage radius is 1053 ft. Thus, the productivity index of a vertical well is not strongly dependent upon drainage area. However, the productivity index of a vertical well does depend upon reservoir thickness.

2. $h = 400$ ft

$$J_h = \frac{0.007078 \times 75 \times 400/(0.62 \times 1.34)}{\ln\left[\dfrac{1114 + \sqrt{(1114)^2 - (1000/2)^2}}{(1000/2)}\right] + \dfrac{400}{1000}\ln\left[\dfrac{400}{2 \times 0.365}\right]}$$

$$= 255.58/(1.44 + 2.52) = 64.5 \approx 65 \text{ STB/(day-psi)}.$$

As noted above, for a vertical well, for a given drainage area, the productivity index is directly proportional to the payzone thickness. Therefore, a vertical well spaced at 40 acres in a 400-ft-thick reservoir will have a productivity index of

$$J_v = 2.1 \times 400/25 = 33.6 \text{ STB/(day-psi)}.$$

Similarly, a vertical well draining 80 acres in a 400-ft-thick reservoir will have a productivity index of

$$J_v = 2.0 \times 400/25 = 32 \text{ STB/(day-psi)}.$$

The comparison of horizontal well productivities for different payzone thicknesses is given below.

h, ft	J_h^+	J_v^*	J_v^{**}	J_h/J_v^*	J_h/J_v^{**}
25	10.5	2.1	2.0	5.00	5.25
400	64.5	33.6	32.0	1.92	2.02

+ horizontal well spacing = 80 acres
* vertical well spacing = 40 acres
** vertical well spacing = 80 acres.

INFLUENCE OF RESERVOIR ANISOTROPY

The influence of reservoir anisotropy has been dealt with quite extensively in the petroleum literature. If we have a reservoir with different horizontal and vertical permeabilities, then we can write the Laplace equation which represents steady-state flow as:

$$k_h \left(\frac{\partial^2 p}{dx^2}\right) + k_v \left(\frac{\partial^2 p}{dz^2}\right) = 0 \tag{3–15}$$

This can be rewritten as

$$\left(\frac{\partial^2 p}{\partial x^2}\right) + \left(\frac{\partial^2 p}{\partial z'^2}\right) = 0 \tag{3–16}$$

where

$$z' = z\sqrt{k_h/k_v} \tag{3–17}$$

and effective reservoir permeability, k_{eff}, is defined as

$$k_{eff} = \sqrt{k_v k_h} \tag{3–18}$$

Thus, the influence of reservoir anisotropy can be accounted for by modifying the reservoir thickness as

$$h' = h\sqrt{k_h/k_v}. \tag{3–19}$$

As noted earlier, the steady-state horizontal well solutions in Equations 3–5 through 3–10 represent a sum of two mathematical solutions, one representing horizontal flow, the other representing vertical flow. Thus, we can modify the vertical part of the steady-state equation to include the effect of reservoir anisotropy. Such a modification[9] is shown in Equations 3–20 and 3–21.

$$q_h = \frac{0.007078\, k_h h \Delta p/(\mu_o B_o)}{\ln\left[\dfrac{a + \sqrt{a^2 - (L/2)^2}}{L/2}\right] + (\beta h/L)\ln[\beta h/(2r_w)]} \tag{3–20}$$

$$q_h = \frac{0.007078\, k_h h \Delta p/(\mu_o B_o)}{\ln\left[\dfrac{a + \sqrt{a^2 - (L/2)^2}}{L/2}\right] + (\beta^2 h/L)\ln[h/(2r_w)]} \tag{3–21}$$

where $\beta = \sqrt{k_h/k_v}$.

Although Equation 3–21 is derived more rigorously than Equation 3–20, there is less than 14% difference in the productivity indices ($q_h/\Delta p$) calculated with these two equations for $L > 0.4\beta h$. Moreover, these productivity indices show less than 10% difference from the productivity index calculated using a mathematically rigorous pressure transient solution.[12] In general, Equation 3–21 gives a slightly higher productivity index than Equation 3–20. Although either Equation 3–20 or 3–21 could be used for engineering calculation purposes, Equation 3–20 is recommended for a conservative production forecast.

Recently, Renard and Dupuy[8] have presented an equation for an anisotropic reservoir. Their equation in U.S. oil field units is

$$J_h = \frac{0.007078\, k_h h}{\mu_o B_o}\left[\frac{1}{\cosh^{-1}(X) + (\beta h/L)\ln[h/(2\pi r'_w)]}\right] \tag{3–22}$$

where:

$$r'_w = \frac{1+\beta}{2\beta} r_w \tag{3–23}$$

where $X = 2a/L$ for an ellipsoidal drainage area, and a is defined by Equation 3–11.

EXAMPLE 3–3

A 2000-ft-long horizontal well is to be drilled in a reservoir with vertical permeability of about one-half of the horizontal permeability. The horizontal well is drilled on a 160-acre spacing. Other reservoir parameters are

$$k_h = 5 \text{ md}, \qquad h = 50 \text{ ft}$$
$$\mu_o = 0.3 \text{ cp}, \qquad B_o = 1.2 \text{ RB/STB}$$
$$r_w = 0.365 \text{ ft}$$

Calculate the horizontal well productivity by Joshi and Renard and Dupuy methods for $k_v/k_h = 0.5$. Also calculate horizontal well productivity by these two methods for $k_v/k_h = 0.1$.

Solution

1. $k_v/k_h = 0.5$
   a. **Joshi Method (Eqs. 3–20 and 3–11)**

$$\beta = \sqrt{k_h/k_v} = \sqrt{2} = 1.414$$
$$\beta h/L = 1.414 \times 50/2000 = 0.03536$$
$$\beta h/(2r_w) = 1.414 \times 50/(2 \times 0.365) = 96.86$$
$$r_{eh} = \sqrt{160 \times 43{,}560/\pi} = 1489 \text{ ft}$$
$$a = (2000/2)\,[0.5 + \sqrt{0.25 + (2 \times 1489/2000)^4}]^{0.5}$$
$$= 1665 \text{ ft}$$

$$J_h = \frac{0.007078 k_h h/(\mu_o B_o)}{\ln\left[\dfrac{a + \sqrt{a^2 - (L/2)^2}}{(L/2)}\right] + (\beta h/L)\ln[\beta h/(2r_w)]}$$

$$= \frac{0.007078 \times 5 \times 50/(0.3 \times 1.2)}{\ln\left[\dfrac{1665 + \sqrt{1665^2 - (2000/2)^2}}{(2000/2)}\right] + 0.03536 \ln(96.86)}$$

$$= \frac{4.915}{1.0974 + 0.1617}$$

$$= 3.9 \approx 4 \text{ STB/(day-psi)}$$

   b. **Renard and Dupuy Method (Eqs. 3–22 and 3–23)**

$$r'_w = \frac{1+\beta}{2\beta} r_w = \frac{1+1.414}{2 \times 1.414} \times 0.365 = 0.312 \text{ ft}$$

$$a = 1665 \text{ ft, and } L = 2000 \text{ ft}$$

$$X = 2a/L = 3330/2000 = 1.665$$

$$\beta h/L = 0.03536$$

$$J_h = \frac{0.007078\, k_h h/(\mu_o B_o)}{\cosh^{-1}(X) + (\beta h/L) \ln [h/(2\pi r'_w)]}$$

$$= \frac{0.007078 \times 5 \times 50/(0.3 \times 1.2)}{\cosh^{-1}(1.665) + 0.03536 \ln [50/(2\pi \times 0.312)]}$$

$$= \frac{4.915}{1.0974 + 0.1145}$$

$$= 4.06 \approx 4 \text{ STB/(day-psi)}$$

2. $k_v/k_h = 0.1$
   a. **Joshi Method**

$$\beta = \sqrt{k_h/k_v} = \sqrt{1/0.1} = 3.162$$

$$\beta h/L = 3.162 \times 50/2000 = 0.079$$

$$\beta h/(2r_w) = 3.162 \times 50/(2 \times 0.365) = 216.6$$

$$r_{eh} = 1489 \text{ ft}$$

$$a = 1665 \text{ ft}$$

$$J_h = \frac{0.007078 \times 5 \times 50/(0.3 \times 1.2)}{\ln\left[\dfrac{1665 + \sqrt{1665^2 - (2000/2)^2}}{(2000/2)}\right] + 0.079 \ln (216.6)}$$

$$= \frac{4.915}{1.0974 + 0.425}$$

$$= 3.2 \approx 3 \text{ STB/(day-psi)}$$

   b. **Renard and Dupuy Method**
   As noted above, $k_v/k_h = 0.1$ yields $\beta = 3.162$ and $\beta h/L = 0.079$. Additionally, as shown previously, $X = 1.665$. Thus,

$$r'_w = \frac{1+\beta}{2\beta} r_w = \frac{1+3.162}{2 \times 3.162} \times 0.365 = 0.24 \text{ ft}$$

$$J_h = \frac{0.007078 \times 5 \times 50/(0.3 \times 1.2)}{\cosh^{-1}(1.665) + 0.079 \ln [50/(2\pi \times 0.24)]}$$

$$= \frac{4.915}{1.0974 + 0.277}$$
$$= 3.6 \approx 4 \text{ STB/(day-psi)}$$

The results are summarized below:

Horizontal Well Productivity, J_h, STB/(day-psi)

METHOD	$k_v/k_h = 0.1$	$k_v/k_h = 0.5$
Joshi	3.2	3.9
Renard & Dupuy	3.6	4.1

Thus, both of the above methods give reasonably close productivity values.

If a 1000-ft-long horizontal well is drilled in a 50-ft-thick, isotropic reservoir ($k_h = k_v$), the incremental gain in the contact area for a horizontal well is about 20-fold. However, if the reservoir vertical permeability is 1/10th of the horizontal permeability, then the 50-ft reservoir acts as though it is 158 ft thick ($h' = h \times \beta = 50 \times \sqrt{1/0.1} = 158$ ft). Therefore, the incremental gain in the contact area by drilling a horizontal well would be only 6.3-fold (1000/158 = 6.3). This shows that reduction in vertical permeability has the same effect as drilling a horizontal well in a thicker reservoir and reducing incremental contact area.

Figure 3–4 shows the influence of reduced vertical permeability on horizontal well productivity under a steady-state condition in a 100-ft-thick reservoir. The top curve in the figure is for an isotropic reservoir where vertical permeability is equal to horizontal permeability. Similarly, results for k_v/k_h values of 0.5, 0.25, and 0.1 are included in Figure 3–4, which show a significant reduction in well productivity due to low vertical permeability. A review of Figure 3–4, Table 3–2 and Example 3–3 clearly indicate that good vertical permeability is essential for successful horizontal-well operations. If one has to drill a horizontal well in a low-vertical permeability reservoir, then it is essential to create reasonable vertical permeability artificially by fracturing a horizontal well. However, if one plans to increase vertical permeability by fracturing a horizontal well, then either medium- or long-radius drilling techniques will have to be used so that the small portions of a long well can be isolated for effective stimulation treatment. The zonal isolation in a long horizontal well can be obtained either by cementing and perforating the liner, or small portions of a solid liner can be isolated into several sections by using external casing packers (ECPS). In any other drilling technique, it would be difficult to selectively isolate a zone

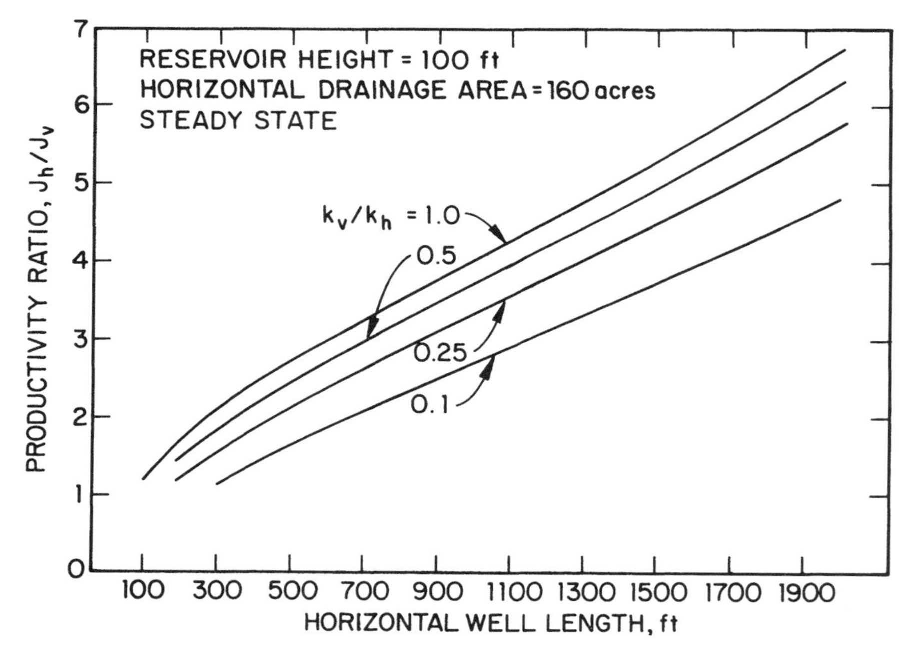

Figure 3–4 Effects of Vertical Permeability on Productivity Ratio of Horizontal and Vertical Wells.

for stimulating a horizontal well. Thus, preplanning of a horizontal well based upon geological, reservoir, and drilling considerations is essential.

ESTIMATION OF VERTICAL PERMEABILITY

In most reservoir engineering analyses, vertical permeability is normally considered one-tenth or even less than one-tenth of the horizontal permeability. By assuming a very low vertical permeability, the vertical component of the flow becomes negligible and flow to the vertical wellbore is essentially radial.

Experience with core analyses indicates that in most cases, for a given core, vertical and horizontal permeabilities are almost equal, unless a shale barrier or distinctly different layers are present. Occasionally it is difficult to obtain a complete core for the entire payzone. In these cases, core data must be used cautiously. Thus, in field operations, shale barriers and distinct layering may inhibit vertical permeability. However, in clean sands and fractured reservoirs vertical permeability may not be significantly different from the horizontal permeability. In naturally fractured reservoirs, the fracture orientation and intensity will dictate the ratio of vertical and horizontal permeabilities. However, even in the case of vertical fractures, the maximum

TABLE 3–2 INFLUENCE OF RESERVOIR ANISOTROPY ON HORIZONTAL WELL PRODUCTIVITY INCREASE AS COMPARED TO A VERTICAL WELL

	J_h/J_v		
HORIZONTAL LENGTH (ft)	$(k_v/k_h) = 0.1$	$(k_v/k_h) = 0.5$	$(k_v/k_h) = 1.0$
	RESERVOIR HEIGHT = 25 ft		
100	1.07	1.53	1.67
200	1.59	2.04	2.17
500	2.58	3.02	3.13
1000	3.86	4.33	4.44
1500	5.11	5.65	5.77
2000	6.48	7.13	7.28
	RESERVOIR HEIGHT = 50 ft		
100	—	1.14	1.34
200	1.09	1.66	1.87
500	1.99	2.66	2.87
1000	3.16	3.94	4.16
1500	4.27	5.21	5.46
2000	5.46	6.60	6.91
	RESERVOIR HEIGHT = 100 ft		
200	—	1.17	1.42
500	1.32	2.10	2.40
1000	2.24	3.29	3.65
1500	3.12	4.43	4.86
2000	4.04	5.66	6.19
	RESERVOIR HEIGHT = 200 ft		
500	—	1.42	1.76
1000	1.37	2.40	2.86
1500	1.96	3.32	3.90
2000	2.57	4.28	5.01
	RESERVOIR HEIGHT = 400 ft		
500	—	—	1.11
1000	—	1.50	1.93
1500	1.08	2.14	2.71
2000	1.43	2.79	3.52

* Horizontal well length is restricted to less than the drainage diameter (2979 ft) of a 160-acre spacing vertical well, and wellbore radius is $r_w = 0.365$ ft.

** Horizontal well length is assumed to be greater than the reservoir height.

ratio of k_v/k_h is one. This is because all fractures are three dimensional; i.e., they have length as well as width and height. This three-dimensional fracture will not only enhance horizontal but also vertical permeability, resulting in a maximum k_v/k_h ratio of one.

In areally anisotropic reservoirs, it is possible to have a higher vertical permeability than the effective horizontal permeability. Consider a reservoir with high permeability k_x in the x direction and a lower permeability k_y in the y direction. Then effective horizontal permeability is $\sqrt{k_x k_y}$, which is lower than the high horizontal permeability k_x in the x direction. In such cases the effective horizontal permeability can be lower than the vertical permeability k_v.

Vertical permeability can be estimated by the following three methods: 1) pressure transient test on a partially penetrating well 2) by knowing water breakthrough time in a reservoir with a bottom water zone (see Chapter 8 for further discussion) 3) by plotting depth against static pressure data to estimate communications between the zones and 4) reviewing core data. It is advisable to estimate statistical distribution of the vertical permeability core data. The statistical distribution of vertical permeability over depth, such as normal, log-normal, or harmonic, should be used to estimate the appropriate permeability averaging technique. The arithmetic averaging of the vertical permeability may not be the best averaging method to calculate average vertical permeability in all circumstances.

EFFECTIVE WELLBORE RADIUS OF A HORIZONTAL WELL

As discussed earlier in Chapter 2, one can calculate the effective wellbore radius of a horizontal well by converting productivity of a horizontal well into that of an equivalent vertical well. The effective wellbore radius is defined by

$$r'_w = r_w \exp(-s). \qquad (2\text{–}5)$$

To calculate the required vertical-wellbore diameter to produce oil at the same rate as that of a horizontal well, equal drainage volumes, $r_{eh} = r_{ev}$, and equal productivity indices, $(q/\Delta p)_h = (q/\Delta p)_v$ were assumed. This gives,

$$\left[\frac{2\pi k_h h/(\mu_o B_o)}{\ln(r_e/r'_w)}\right]_v = \left[\frac{2\pi k_h h/(\mu_o B_o)}{\ln\left[\dfrac{a+\sqrt{a^2-(L/2)^2}}{L/2}\right] + (h/L)\ln[h/(2r_w)]}\right]_h \qquad (3\text{–}24)$$

Solving Equation 3–24 for r_w' gives

$$r_w' = \frac{r_{eh}(L/2)}{a\,[1 + \sqrt{1 - [L/(2a)]^2}]\,[h/(2r_w)]^{h/L}} \tag{3–25}$$

where a can be obtained from either Equation 3–11 or Table 3–1. Equation 3–25 can be used to calculate effective wellbore radius r_w', and then Equation 2–5 can be used to calculate skin factor s.

If the reservoir is anisotropic, the effective wellbore radius is

$$r_w' = \frac{r_{eh}(L/2)}{a\,[1 + \sqrt{1 - [L/(2a)]^2}]\,[\beta h/(2r_w)]^{(\beta h/L)}}. \tag{3–26}$$

Van Der Vlis et al.[11] have also suggested an equation for the effective wellbore radius of a horizontal well drilled in an isotropic reservoir. Their equation is

$$r_w' = \frac{L}{4}\left[\sin\left(\frac{4r_w}{h} \times 90^\circ\right)\cos\left(\frac{\delta}{h} \times 180^\circ\right)\right]^{h/L} \tag{3–27}$$

δ = vertical distance between the well center and the reservoir mid-height

As shown in Chapter 2, the concept of effective wellbore radius can be extended to calculate the ratio of horizontal and vertical well productivity indices as shown below:

$$J_h/J_v = [\ln(r_{ev}/r_w)]/[\ln(r_{eh}/r_w')],\ \text{for } L > h\sqrt{k_h/k_v} \text{ and } (L/2) < 0.9r_{eh}. \tag{3–28}$$

It is important to note that the above productivity index comparison assumes an unstimulated vertical well. Because vertical well stimulation varies from region to region, only unstimulated vertical well productivities are used for general comparison. The productivity increases calculated from Equation 3–28 will have to be adjusted, depending on local experience with the vertical well stimulation treatments. Equation 3–28 is valid only for reservoirs operating above the bubble point. Nevertheless, in a solution gas-drive reservoir, the productivity index is a first derivative of an inflow-performance relationship curve. Therefore, in a solution gas-drive reservoir, Equation 3–28 gives a fair estimation of productivity improvements with horizontal wells.

EXAMPLE 3–4

A 2000-ft-long horizontal well is drilled on 160-acre spacing. The reservoir data are

$$k_h = 50 \text{ md}, \quad h = 100 \text{ ft}, \quad r_w = 0.3 \text{ ft}$$
$$\phi = 14\%, \quad \mu = 0.8 \text{ cp}, \quad B_o = 1.32 \text{ RB/STB}$$
$$k_v/k_h = 1.$$

Calculate the effective wellbore radius of the horizontal well.

Solution

Assuming a circular drainage area, the drainage radius is

$$\begin{aligned} r_{eh} &= \sqrt{160 \times 43{,}560/\pi} \\ &= 1489 \text{ ft} \end{aligned}$$

Using Equation 3–11, the half-length of the major axis for the drainage ellipse is calculated as

$$\begin{aligned} a &= (2000/2)\,[0.5 + \sqrt{0.25 + (2 \times 1489/2000)^4}]^{0.5} \\ &= 1665 \text{ ft} \end{aligned}$$

Using Equation 3–25, the effective wellbore radius for a horizontal well is given by

$$r'_w = \frac{r_{eh}\,(L/2)}{a[1 + \sqrt{1 - [L/(2a)]^2}]\,[h/(2r_w)]^{h/L}} \tag{3–25}$$

$$\begin{aligned} &= \frac{1489 \times (2000/2)}{1665[1 + \sqrt{1 - (2000/(2 \times 1665))^2}][100/(2 \times 0.3)]^{100/2000}} \\ &= 1{,}489{,}000/(1665 \times 1.80 \times 1.29) = 385.1 \text{ ft} \\ &\approx 385 \text{ ft} \end{aligned}$$

Skin factor

$$\begin{aligned} s &= -\ln(r'_w/r_w) \\ &= -\ln(385/0.3) \\ &= -7.2 \end{aligned}$$

The Van der Vlis et al. method, described in Equation 3–27, can also be used to calculate effective wellbore radius. In their method, if we assume that the well is drilled at the elevation center of the reservoir height, then in Equation 3–27, $\delta = 0$, and

$$r'_w = L/4\left[\sin\left(\frac{4r_w}{h} \times 90°\right)\right]^{h/L}$$

$$= \frac{2000}{4} [\sin(4 \times 0.3 \times 90°/100)]^{100/2000}$$
$$= 410 \text{ ft}$$
$$s = -\ln(r'_w/r_w)$$
$$= -\ln(410/0.3)$$
$$= -7.2$$

Based on the transient analysis of horizontal wells in an infinite reservoir, Ozkan, Raghavan & Joshi[12] report the following equations for effective wellbore radius of a horizontal well in a reservoir with two different permeabilities in the areal plane, namely k_x and k_y, and vertical permeability k_v.

$$r'_w = \frac{L/2}{\exp(1+\sigma)}\left[4\sin\left[\frac{\pi}{h}\left(z_w + \frac{r_w}{2}\sqrt{\frac{k_v}{k_y}}\right)\right]\sin\left(\frac{\pi}{2h}r_w\sqrt{\frac{k_v}{k_y}}\right)\right]^{m'} \qquad (3\text{–}29)$$

$$\text{for } [L/(2h)]\sqrt{k_v/k_h} \geq 1.15/(\sqrt{k_h/k_x} - |x_D|)$$

where $m' = (h/L)\sqrt{k_x/k_v}$, $k_h = \sqrt{k_x k_y}$, and $|x_D| = 0.732$ for infinite-conductivity wellbores and either 0 or 1 for uniform flux wellbores. It is important to note that in Equation 3–29, z_w represents vertical distance of a horizontal well from the bottom boundary of the payzone. For long horizontal wells, $m' \approx 0$, and hence r'_w approaches that of a vertically fractured well,

$$r'_w = \frac{L/2}{\exp(1+\sigma)} \qquad (3\text{–}30)$$

where $1 + \sigma = 1$ and $\ln 2$ for uniform-flux and infinite-conductivity fracture conditions, respectively. Equation 3–30 defines the effective wellbore radius for a vertically fractured well.

EXAMPLE 3–5

A 1000-ft-long horizontal well is drilled on a 320-acre well spacing in a 100-ft-thick reservoir. The permeabilities in the x, y, and z directions are 0.2 md, 5 md, and 2.5 md, respectively. (Effective horizontal permeability is 1 md.) Drilled wellbore radius is 0.3 ft.

Calculate the effective wellbore radius and the skin factor if the horizontal well is centrally located in the vertical plane. Also assume that the horizontal wellbore has infinite conductivity; i.e., there is no pressure drop in the wellbore.

Solution
$k_x = 0.2$ md, $k_y = 5$ md, and $k_v = 2.5$, hence, from Equation 3–29

$$r'_w = \frac{L/2}{\exp(1+\sigma)}\left[4\sin\left[\frac{\pi}{h}\left(z_w + \frac{r_w}{2}\sqrt{\frac{k_v}{k_y}}\right)\right]\right.$$
$$\left.\sin\left(\frac{\pi}{2h}r_w\sqrt{\frac{k_v}{k_y}}\right)\right]^{m'} \quad (3\text{–}29)$$

$$m' = (h/L)\sqrt{k_x/k_v} = (100/1000)\sqrt{0.2/2.5}$$
$$= 0.0283$$

$$r'_w = \frac{(1000/2)}{2}\left[4\sin\left[\frac{\pi}{100}\left(50 + \frac{0.3}{2}\sqrt{\frac{2.5}{5}}\right)\right]\right.$$
$$\left.\sin\left(\frac{\pi}{2}\times\frac{0.3}{100}\sqrt{\frac{2.5}{5}}\right)\right]^{0.0283}$$
$$= 250[4 \times 1 \times 0.00333]^{0.0283}$$
$$= 221 \text{ ft}$$
$$s = -\ln[r'_w/r_w] = -\ln[221/0.3]$$
$$= -6.6$$

PRODUCTIVITY OF SLANT WELLS

Figure 3–5 shows a schematic diagram of a slant well. Cinco, Miller, and Ramey[13] have presented a solution for a slanted well. They have also calculated skin factors for slanted wells and have shown that these skin factors depend only on the well geometry. Their equation to calculate the pseudo-skin factor due to a slant well is given below.

$$s_s = -(\alpha'/41)^{2.06} - (\alpha'/56)^{1.865}\log[h_D/100] \quad (3\text{–}31)$$
$$\text{for } t_D \geq t_{D1} \text{ and } \alpha' \leq 75°$$

where

$$h_D = (h/r_w)(\sqrt{k_h/k_v}) \quad (3\text{–}32)$$
$$\alpha' = \tan^{-1}[\sqrt{k_v/k_h}\tan\alpha] \quad (3\text{–}33)$$
$$t_D = 0.000264\, k_h t/(\phi\mu c_t r_w^2) \quad (3\text{–}34)$$

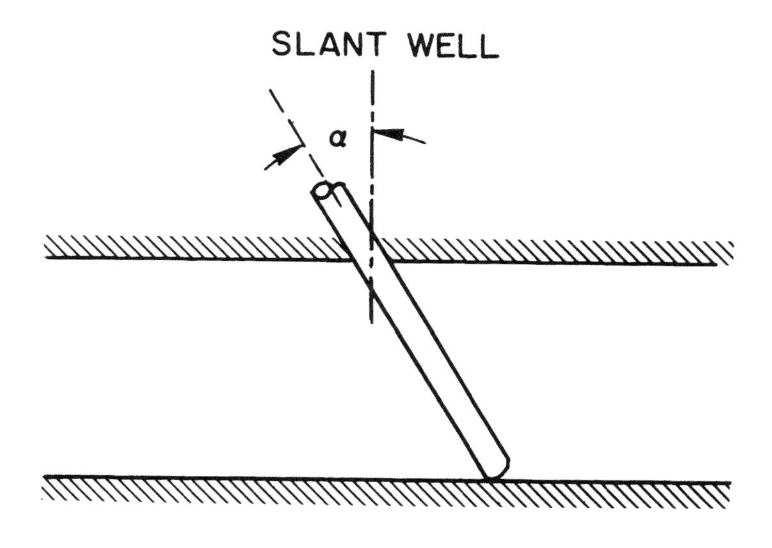

Figure 3–5 A Schematic of a Slant Well.[13]

and

$$t_{D1} = \max \begin{vmatrix} 70\, r_D^2 \\ (25/3)\,(r_D\cos\alpha + (h_D/2)\tan\alpha')^2 \\ (25/3)\,(r_D\cos\alpha - (h_D/2)\tan\alpha')^2 \end{vmatrix} \quad (3\text{–}35)$$

$$r_D = r/r_w \quad (3\text{–}36)$$

and α is slant angle. Note in Equation 3–34, t is in hours. The effective wellbore radius is given by

$$r'_w = r_w \exp(-s_s) \quad (3\text{–}37)$$

and the productivity index of a slant well can be compared to an unstimulated vertical well by using the following relationship:

$$J_s/J_v = \ln(r_e/r_w)/\ln(r_e/r'_w). \quad (3\text{–}38)$$

Figure 3–6 presents pseudo-skin factors of slant wells as a function h_D for different slant angles, α'.

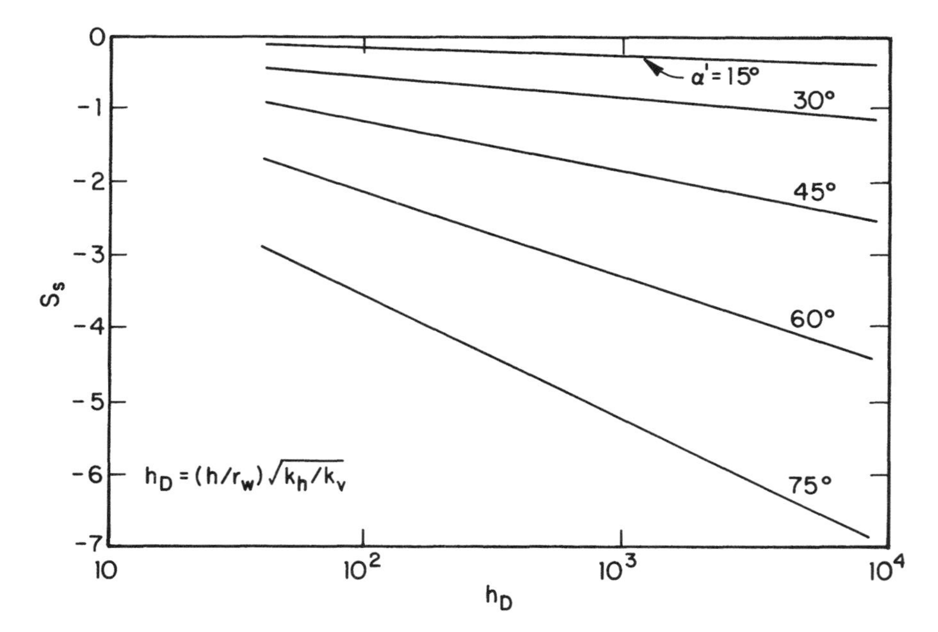

Figure 3–6 Slant Well Skin Factors.[13]

Van der Vlis et al.[11] have also presented an equation to calculate slant well productivity. Their equation, presented in terms of effective wellbore radius, is

$$r'_w = (L/4)\,[0.454 \sin (360° \, r_w/h)]^{h/L} \tag{3–39}$$

where

$$L = h/\cos\alpha. \tag{3–40}$$

Note that Equation 3–39 is obtained by replacing δ in Equation 3–27 with $0.35h$ and it only applies for $\alpha \geq 20°$. As noted earlier, Equation 3–38 can be used to calculate the productivity index. The results of effective wellbore radius and skin factors calculated from the Van der Vlis equation, Equation 3–39, as well as from the equation of Cinco et al., Equation 3–31, are in fairly good agreement with each other, and therefore either one of them could be used for calculation purposes (see Example 3–7).

It is important to note that Equation 3–31 can be used for anisotropic reservoirs, while Equation 3–39 is only for isotropic reservoirs. It is also important to note Equation 3–31 is applicable for wells with slant angles $\alpha \leq 75°$ (the maximum slant angle for which Equation 3–39 can be used is not specified). Typical results for an 80-acre drainage area in an isotropic reservoir are shown in Figure 3–7, and the results are tabulated in Table 3–3. All of these results assume that the well penetrates the entire reservoir height. It is interesting to note that the productivity of a slant well compared to the productivity of a vertical well (J_s/J_v) under steady-state conditions *increases as the reservoir thickness increases*. In contrast, J_h/J_v *decreases as the reservoir thickness increases* for a horizontal well. Thus, one can develop some guidelines to decide where horizontal wells and where slant wells are applicable. It is important to remember that in the case of slant wells, the length of the well is limited by reservoir height and slant angle of the well as shown by Equation 3–40.

EXAMPLE 3–6

It is proposed to drill a 60° slant well in a reservoir. The following reservoir parameters are known.

$$\text{well spacing} = 160 \text{ acres}, k_v/k_h = 1, r_w = 0.365 \text{ ft}$$

1. Calculate the productivity improvement of the slant well over a vertical well using the Cinco et al. correlation for a 100-ft-thick reservoir. Also, calculate the ratio J_s/J_v for reservoir thicknesses of 25 ft and 400 ft.

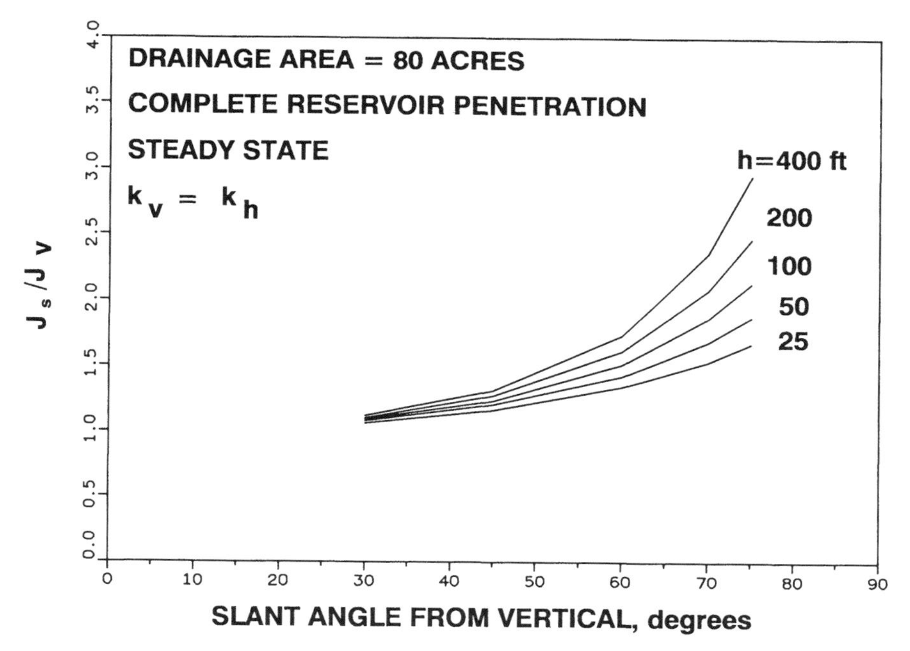

Figure 3–7 Productivity Ratio J_s/J_v as a Function of Slant Angle.[9]

TABLE 3–3 THE INFLUENCE OF SLANT ANGLE ON PRODUCTIVITY

$k_v/k_h = 1$

WELL ANGLE	SKIN FACTOR	J_s/J_v
	RESERVOIR HEIGHT = 25 ft	
30.0	−.47	1.06
45.0	−1.10	1.16
60.0	−2.00	1.34
70.0	−2.76	1.53
75.0	−3.19	1.67
	RESERVOIR HEIGHT = 50 ft	
WELL ANGLE	SKIN FACTOR	J_s/J_v
30.0	−.57	1.08
45.0	−1.30	1.20
60.0	−2.35	1.42
70.0	−3.22	1.68
75.0	−3.71	1.87
	RESERVOIR HEIGHT = 100 ft	
WELL ANGLE	SKIN FACTOR	J_s/J_v
30.0	−.66	1.09
45.0	−1.50	1.23
60.0	−2.69	1.51
70.0	−3.67	1.86
75.0	−4.22	2.13
	RESERVOIR HEIGHT = 200 ft	
WELL ANGLE	SKIN FACTOR	J_s/J_v
30.0	−.76	1.10
45.0	−1.70	1.27
60.0	−3.03	1.61
70.0	−4.13	2.08
75.0	−4.74	2.47
	RESERVOIR HEIGHT = 400 ft	
WELL ANGLE	SKIN FACTOR	J_s/J_v
30.0	−.85	1.12
45.0	−1.90	1.31
60.0	−3.37	1.73
70.0	−4.59	2.36
75.0	−5.26	2.95

* Complete reservoir penetration by a slant well in an 80-acre well spacing
* Steady state results

2. Calculate the productivity improvement ratio, J_s/J_v, for $k_v/k_h = 0.1$ and 0.5, if $h = 100$ ft.
3. What is the maximum length of a 70° slant well that can be drilled in a 300-ft-thick reservoir?

Solution

Using the Cinco et al. method (Equations 3–31 to 3–33)

$$h_D = \frac{h}{r_w}\sqrt{\frac{k_h}{k_v}} = \frac{100}{0.365}\sqrt{1} = 274$$

$$\alpha' = \tan^{-1}[\sqrt{1}\tan\alpha] = \alpha = 60°$$

Skin factor is estimated using Equation 3–31 as

$$\begin{aligned} s_s &= -(\alpha'/41)^{2.06} - (\alpha'/56)^{1.865}\log[h_D/100] \\ &= -(60/41)^{2.06} - (60/56)^{1.865}\log[274/100] \\ &= -2.69 \end{aligned}$$

Effective wellbore radius is calculated using Equation 3–37

$$\begin{aligned} r'_w &= r_w \exp(-s_s) = 0.365\exp(+2.69) \\ &= 5.38 \text{ ft} \approx 5.4 \text{ ft.} \end{aligned}$$

For a 160-acre drainage area,

$$r_e = \sqrt{160 \times 43{,}560/\pi} = 1489.5 \text{ ft}$$

and J_s/J_v is given by Equation 3–38 as

$$\begin{aligned} J_s/J_v &= \ln(r_e/r_w)/\ln(r_e/r'_w) \\ &= \ln(1489.5/0.365)/\ln(1489.5/5.37) \\ &= 1.48 \approx 1.5 \end{aligned}$$

The results are summarized below.

1. $k_v/k_h = 1$: J_s/J_v for Different h

***h* ft**	s_s	r'_w **ft**	J_s/J_v
25	−2.0	2.7	1.3
100	−2.7	5.4	1.5
400	−3.4	10.7	1.7

2. h = 100 ft: J_s/J_v for Different k_v/k_h

k_v/k_h	s_s	r'_w ft	J_s/J_v
0.1	−0.75	0.77	1.1
0.5	−2.0	2.8	1.3
1.0	−2.7	5.4	1.5

3. Maximum well length that can be drilled at 70° in a 300-ft-thick reservoir is

$$L = h/\cos(\alpha) = 300/\cos(70°) = 877 \text{ ft}$$

EXAMPLE 3–7

Calculate the productivity improvement of a slant well over a vertical well for a 60-acre well spacing for four different slant angles of 30°, 45°, 60° and 70°, using Cinco et al. and Van Der Vlis et al. correlations. Given: k_v/k_h = 1.0, r_w = 0.35 ft, and h = 600 ft.

Solution

The calculations are illustrated for a slant angle of 30°.

1. **Cinco et al. Method (Equations 3–31, 3–32, 3–33, 3–37 and 3–38)**
   For $k_v = k_h$ from Equation 3–33

$$\alpha' = \alpha = 30°$$

$$h_D = \frac{h}{r_w}\sqrt{\frac{k_h}{k_v}} = \frac{600}{0.35}\sqrt{1} = 1714.3$$

$$\begin{aligned} s_s &= -(\alpha'/41)^{2.06} - (\alpha'/56)^{1.865}\log(h_D/100) \\ &= -(30/41)^{2.06} - (30/56)^{1.865}\log(1714.3/100) \\ &= -0.91 \end{aligned}$$

$$\begin{aligned} r'_w &= r_w\exp(-s_s) = 0.35\exp(0.91) \\ &= 0.87 \text{ ft} \end{aligned}$$

$$r_e = \sqrt{60 \times 43{,}560/\pi} = 912 \text{ ft}$$

$$\begin{aligned} J_s/J_v &= \ln(r_e/r_w)/\ln(r_e/r'_w) \\ &= \ln(912/0.35)/\ln(912/0.87) \\ &= 7.865/6.955 \\ &= 1.13 \end{aligned}$$

2. **Van der Vlis et al. Method (Equations 3–39 and 3–40)**

$$L = h/\cos(\alpha) = 600/\cos(30°) = 693 \text{ ft}$$
$$r'_w = (L/4)[0.454 \sin(360° \, r_w/h)]^{h/L}$$
$$= (693/4)[0.454 \sin(360° \times 0.35/600)]^{600/693}$$
$$= 0.68 \text{ ft}$$
$$s_s = -\ln(0.68/0.35) = -0.66$$
$$J_s/J_v = \ln(912/0.35)/\ln(912/0.68) = 1.09$$

As shown below, the calculated results using these two methods are in good agreement with each other. Therefore, either equation could be used for predicting productivity improvements.

COMPARISON OF CINCO ET AL. AND VAN DER VLIS ET AL. METHODS

		Cinco et al.		Van der Vlis et al.	
α (deg)	***L*, ft**	s_s	J_s/J_v	r'_w **ft**	J_s/J_v
30	693	−0.91	1.14	0.68	1.09
45	849	−2.03	1.35	2.31	1.32
60	1200	−3.59	1.84	12.24	1.82
70	1754	−4.88	2.64	49.14	2.69

COMPARISON OF SLANT WELL AND HORIZONTAL WELL PRODUCTIVITIES

As noted earlier, horizontal wells are effective in thin reservoirs, while slant wells are highly effective in thick reservoirs. Therefore, one would like to find the optimum completion for a given reservoir thickness. One of the ways to do this is to assume a drilled well having a fixed length. This fixed drilled length could be either vertical, horizontal, or slant. Then one can compare productivities of horizontal, vertical, and slant wells with each other to determine the optimum completion method.

Figure 3–8 shows a typical comparison of slant wells and horizontal wells with vertical wells for a 100-ft-thick reservoir. The figure shows that a horizontal well will always do significantly better than any slant well in a 100-ft-thick reservoir, even for the small values of the permeability ratio, k_v/k_h. This clearly indicates that horizontal wells are preferred options in 100-ft-thick reservoirs.

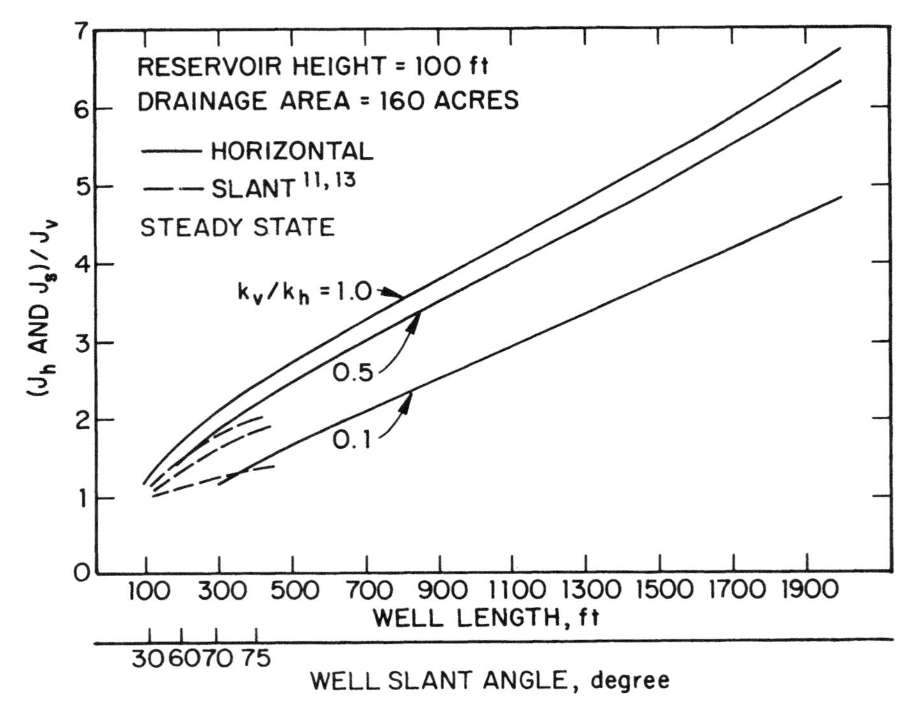

Figure 3–8 Comparison of Horizontal and Slant Well Productivities for a 100-ft-thick Reservoir.

Figure 3–9 shows the productivity comparisons in a 400-ft-thick reservoir. The figure shows that the productivities of horizontal wells are higher than the productivities of slant wells for isotropic reservoirs ($k_h = k_v$) and for reservoirs with $k_v/k_h = 0.5$. However, if vertical permeability is one-tenth of horizontal permeability, performance of the slant well is significantly better than the horizontal well. This indicates that in a thick reservoir, for horizontal wells to be effective, they will have to be drilled in high vertical permeability reservoirs. It also indicates that in thick, low vertical permeability reservoirs, stimulation may be necessary to enhance vertical permeability and improve project economics. This requires appropriate planning so that proper drilling techniques can be selected so as to facilitate effective stimulation of a horizontal well.

FORMATION DAMAGE IN HORIZONTAL WELLS

As noted in Chapter 2, the concept of skin factor was developed to account for loss in productivity due to near wellbore formation damage.[8,14–16]

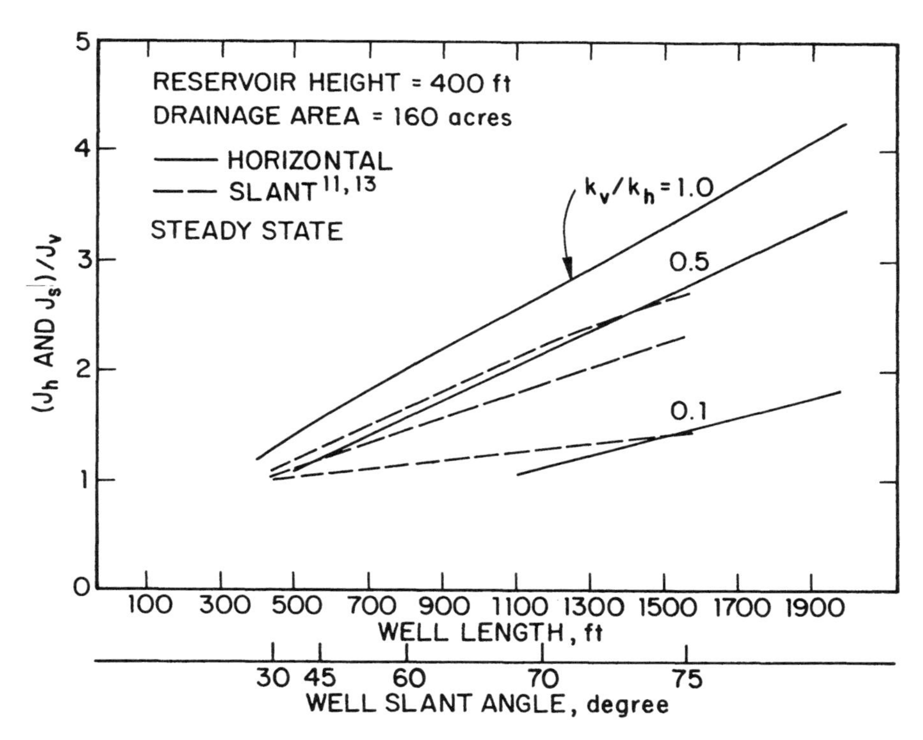

Figure 3–9 Comparison of Horizontal and Slant Well Productivities for a 400-ft-thick Reservoir.

The near wellbore damage causes an extra pressure drop near the wellbore resulting in loss of pressure drawdown. As described in Chapter 2, the pressure drop in the skin region is proportional to the flow rate per unit well length. (All skin pressure drop calculations assume a steady-state flow in the skin region.) As shown in Equations 2–3 and 2–4 in Chapter 2, for vertical wells, pressure drop due to positive skin factor $(\Delta p_{skin})_v$ is proportional to q_v/h. For horizontal wells, pressure drop due to positive skin factor $(\Delta p_{skin})_h$ is proportional to q_h/L.

Thus, because of lower flow rate per unit well length, long horizontal wells exhibit a smaller loss of well productivity due to drilling damage than a vertical well. The preceding statement is true, assuming that we have similar damage (or a skin factor) for horizontal and vertical wells (see Example 2–1). Nevertheless, in practice it takes a longer time to drill a horizontal well than a vertical well, resulting in exposing the producing formation to drilling fluid for a longer time period than in vertical drilling operations. Thus, for a similar set of mud conditions, horizontal wells may show more damage than a vertical well. (As noted in Section 2–3 of Chapter 2, in general, a high-permeability formation shows a relatively smaller damage than a low-permeability formation (see Equation 2–2)). Recently some methods have been devised to estimate the influence of formation damage on horizontal well productivity.

Sparlin and Hagen[16] have reported the following equations to calculate flow rate from a damaged horizontal well. As shown in Figure 3–10, if d represents thickness of the damaged zone around the horizontal well, then the average vertical permeability, $k_{avg\text{-}vert}$, and the average horizontal permeability, $k_{avg\text{-}horiz}$ are calculated as shown below.

$$k_{avg\text{-}vert} = \frac{k_s k \ln [h/(2r_w)]}{k \ln ((r_w + d)/r_w) + k_s \ln (h/(2r_w + 2d))} \quad (3\text{–}41)$$

$$k_{avg\text{-}horiz} = \frac{k_s k \ln (r_e/r_w)}{k \ln ((r_w + d)/r_w) + k_s \ln (r_e/(r_w + d))} \quad (3\text{–}42)$$

$$\frac{q_d}{q_h} = \frac{\ln (c) + (h/L) \ln [h/(2r_w)]}{(k/k_{avg\text{-}horiz}) \ln (c) + (k/k_{avg\text{-}vert})(h/L) \ln [h/(2r_w)]} \quad (3\text{–}43)$$

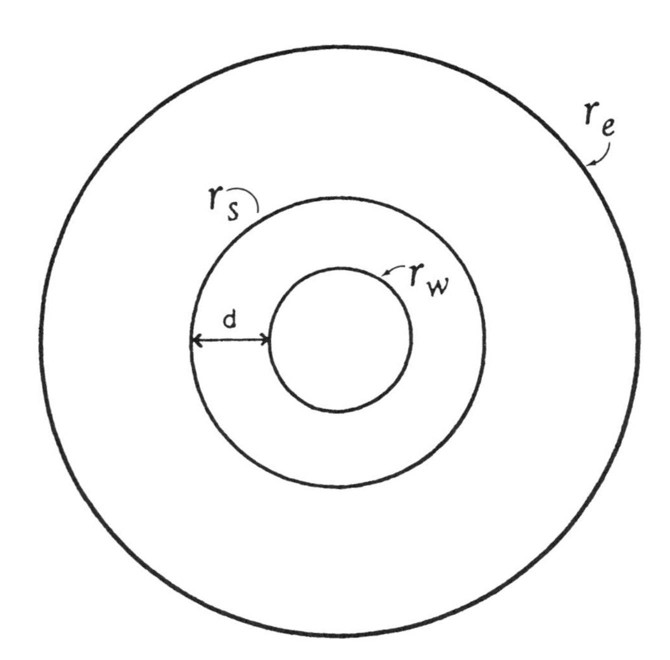

Figure 3–10 A Schematic View of a Skin Zone near Wellbore.

where

k_s = damage zone permeability
d = damage zone thickness
q_d = flow rate of a damaged horizontal well
q_h = flow rate of an undamaged horizontal well

$$c = [r_{eh} + (r_{eh}^2 - (L/2)^2)^{0.5}]/[L/2] \tag{3–44}$$

Note that Equations 3–41 through 3–43 are for isotropic reservoirs only; thus k simply represents reservoir permeability. Equation 3–43 represents a loss in production for a horizontal well due to near wellbore damage. Table 3–4 includes a set of calculations showing a drop in production of a 2000-ft-long horizontal well due to near wellbore damage. The table shows a significant drop in deliverability of a horizontal well due to near wellbore damage. Sparlin and Hagen have also reported calculations for estimating the production rate from a horizontal well when the formation collapses around a screen (or slotted liner). The set of calculations are similar to those reported in Equations 3–41 through 3–43; the only difference is that one

TABLE 3–4 COMPARISON OF q_d/q_h CALCULATIONS DESCRIBING THE EFFECT OF FORMATION DAMAGE ON HORIZONTAL WELL PRODUCTIVITY

h = 50 ft, k = 100 md, r_w = 0.33 ft, r_{eh} = 2106 ft, L = 2000 ft				
d, ft	k_s, md	$k_{avg\text{-}vert}$	$k_{avg\text{-}horiz}$	q_d/q_h
0.5	50	82.4	90.5	0.90
1	50	75.6	86.3	0.87
2	50	68.9	81.8	0.82
3	50	65.2	79.1	0.80
0.5	25	61.0	76.0	0.76
1	25	50.9	67.7	0.68
2	25	42.5	59.9	0.59
3	25	38.4	55.8	0.55
0.5	10	34.3	51.4	0.51
1	10	25.7	41.1	0.40
2	10	19.7	33.3	0.32
3	10	17.2	29.6	0.29

has to estimate equivalent damage-zone thickness d and equivalent damage-zone permeability k_s. Typical results are shown in Table 3–5.[17] It is important to note that in practice, it may be difficult to estimate a damage-zone thickness and the effective permeability. In general, one expects horizontal wells to exhibit a lesser sand control problem than a vertical well. This is because near the wellbore, fluid velocities in a horizontal well are smaller than those in a vertical well. Sparlin and Hagen's[16] study indicated that horizontal wells may have a problem of hole collapse in weakly consolidated formations. In such formations, use of a slotted liner or a screen completion is desirable. Also, prepacked screens with 10-darcy permeability sand would provide higher well productivities than those obtained with 20/40 or 40/60 U.S. mesh gravel pack.

Recently, Renard and Dupuy[8] presented a solution for productivity of a damaged horizontal well as

$$J_{h,d} = \frac{0.007078\, k_h h/(\mu_o B_o)}{\cosh^{-1} X + (\beta h/L)\ln[h/(2\pi r'_w)] + s_h} \tag{3–45}$$

where $x = 2a/L$

TABLE 3–5 COMPARISON OF q_d/q_h CALCULATIONS DESCRIBING THE EFFECT OF FORMATION DAMAGE COMBINED WITH THE EFFECT OF FORMATION COLLAPSE AROUND A SCREEN OR SLOTTED LINER ON HORIZONTAL WELL PRODUCTIVITY

$h = 50$ ft, $k = 100$ md, $r_w = 33$ ft
$r_s = 0.208$ ft, $r_{eh} = 2106$ ft, $L = 2000$ ft

d, ft	k_s, md	$k_{avg\text{-}vert}$	$k_{avg\text{-}horiz}$	q_d/q_h
0.5	50	77.6	87.0	0.86
1	50	72.1	83.3	0.82
2	50	66.5	79.2	0.78
3	50	63.3	76.9	0.76
0.5	25	53.6	69.0	0.67
1	25	46.3	62.4	0.61
2	25	39.8	56.0	0.54
3	25	36.5	52.6	0.51
0.5	10	27.8	42.5	0.41
1	10	22.3	35.6	0.34
2	10	18.1	29.8	0.28
3	10	16.1	27.0	0.26

where r'_w is defined by Equation 3–23 and s_h is the skin factor that represents near-wellbore damage. In addition, Renard and Dupuy assumed the following proportionality between horizontal and vertical well damage

$$s_h = (\beta h/L)\,[(k/k_s) - 1]\,\ln(r_s/r_w) = (\beta h/L)\,s_v \qquad (3\text{–}46)$$

where s_v is vertical well damage. The above equation tells us that as the horizontal well gets longer, effective formation damage s_h will become smaller and smaller. As noted earlier, in practice one may experience higher effective damage in a longer well than in a shorter well. This is because of a longer drilling time for long wells. The correlation assumed to describe the relationship between well length and the corresponding damage would have significant impact on calculated well productivity of a damaged horizontal well. The determination of this relationship is complex and is probably dependent upon drilling time and mud chemistry.

Combining Equations 3–45 and 3–22 one can show that

$$J_{h,d}/J_h = B'/(B' + s_v) \qquad (3\text{–}47)$$

where

$$B' = [L/(h\beta)]\cosh^{-1} X + \ln[h/(2\pi r'_w)] \qquad (3\text{–}48)$$

For a typical well length and drainage radius, Table 3–6 summarizes the results obtained using Equations 3–47 and 3–48. The table shows that as vertical permeability decreases, the productivity of a damaged horizontal well also decreases as compared to an undamaged horizontal well. The table also shows that formation damage can severely impair productivity of a

TABLE 3–6 COMPARISON OF $J_{v,d}/J_v$ AND $J_{h,d}/J_h$ FOR DIFFERENT VALUES OF s_v AND β

		$J_{h,d}/J_h$		
s_v	$J_{v,d}/J_v$	$\beta = 1$	$\beta = 1.41$	$\beta = 3.16$
1	0.90	0.98	0.98	0.96
5	0.64	0.92	0.90	0.81
10	0.47	0.86	0.82	0.69
20	0.30	0.75	0.69	0.52

$\beta = \sqrt{k_h/k_v}$

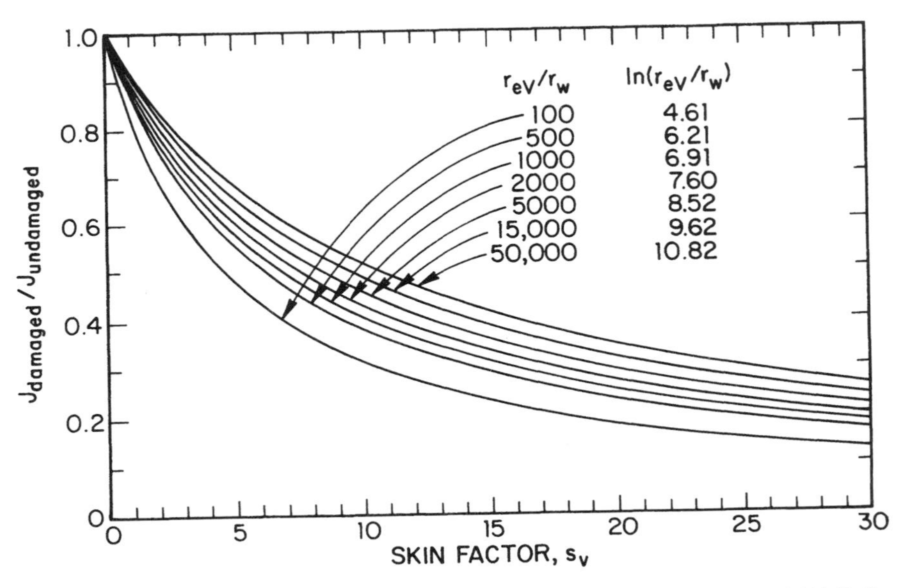

Figure 3–11 Productivity Ratio as a Function of Skin Factor for Vertical Wells.[8]

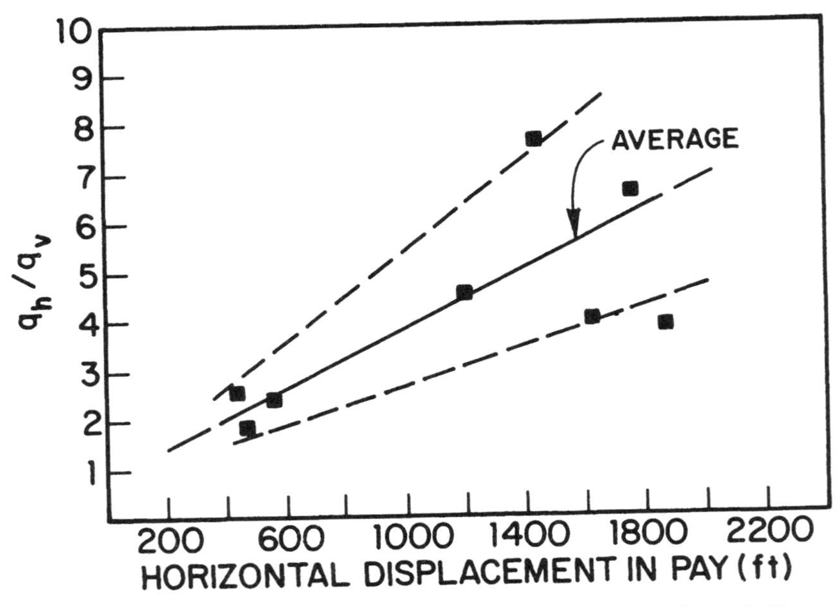

Figure 3–12 Flow Rate versus Well Length for Austin Chalk Wells.[18]

horizontal well. As noted above, in practice, one would expect a loss of productivity more than that indicated in Table 3–6, indicating severity of the formation damage. Figure 3–11 depicts the influence of formation damage on vertical well productivity. As noted earlier, Equation 3–41 can be used to estimate loss in productivity of a horizontal well due to formation damage.

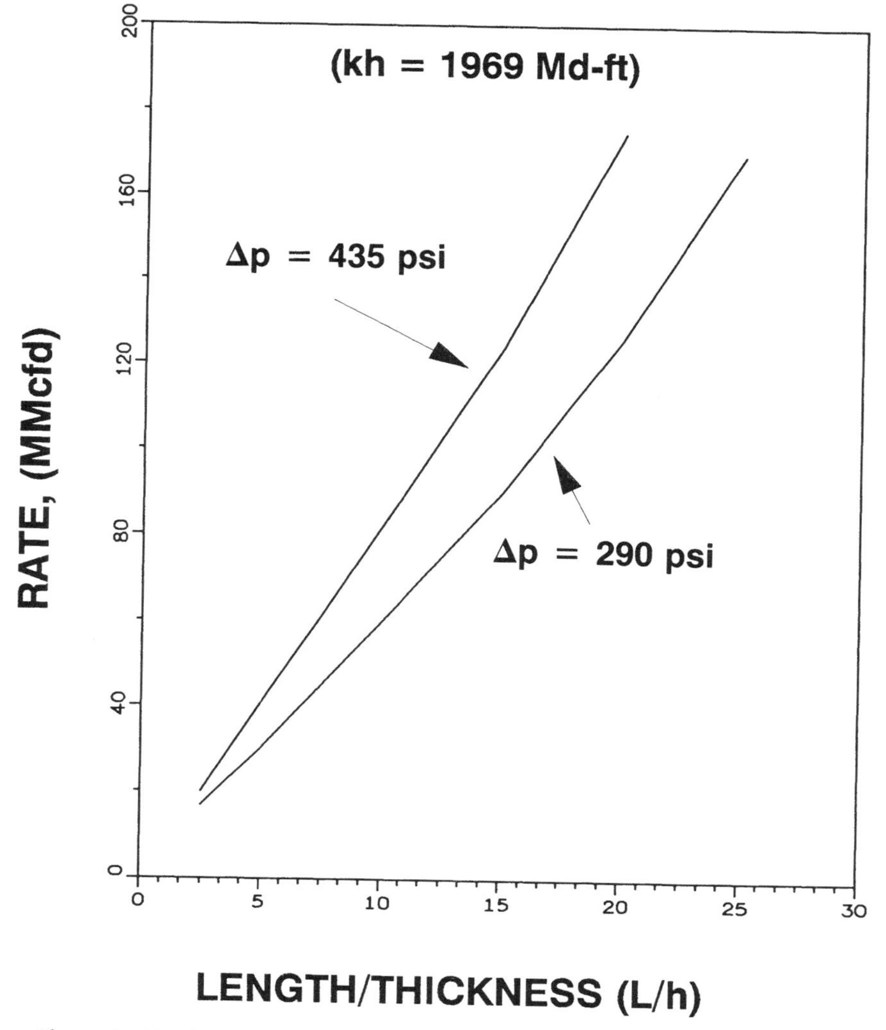

Figure 3–13 Flow Rate versus Ratio of Well Length to Reservoir Thickness for Zuidwal Field,[19] Δp represents pressure drawdown.

FIELD HISTORIES

The following two field histories along with Equations 3–5 through 3–10 demonstrate the importance of well length. It also tells us that long wells give us higher productivity than short wells. A few years ago, many wells were drilled with a 1000 ft long horizontal section. Recently, many wells were drilled with a 2000 ft long horizontal section. Now the trend seems to be drilling longer than 2000 ft horizontal sections.

1. **Austin Chalk Reservoir, Texas, U.S.A.**[18]

Since 1986, Amoco has drilled 10 horizontal wells in the Giddings Austin Chalk reservoir with considerable success. Their field experience showed a very strong correlation between horizontal well length and the oil production rates. As shown in Figure 3–12, for the same pressure drawdown, 500-ft- to 2200-ft-long horizontal wells produced at rates 2.5 to 7 times that of vertical wells.

The vertical wells were drilled at a much earlier time than the horizontal wells. In comparing horizontal and vertical production rates, the vertical well rates were reduced to account for reservoir pressure depletion over time.

As of June, 1990, over 50 horizontal wells have been drilled in the Austin chalk formation in the Pearsall field in West Texas. In general, well length varies from 1000 to 3000 ft.

2. **Zuidwal Field, Onshore the Netherlands**[19]

Zuidwal field in onshore the Netherlands is one of the first gas fields developed with horizontal wells. Since September 1987, three horizontal wells were drilled ranging from about 1800 to 2200 ft. Again, as shown in Figure 3–13, they observed higher flow rates for longer wells.

REFERENCES

1. Landrum, B. L. and Crawford, P. B.: "Effect of Drainhole Drilling on Production Capacity," *Petroleum Transactions, AIME,* Vol. 204, pp. 271–273, 1955.
2. Roemershauser, A. E. and Hawkins, M. F., Jr.: "The Effect of Slant Hole, Drainhole, and Lateral Hole Drilling on Well Productivity," *Journal of Petroleum Technology,* p. 11–14, February 1955.
3. Chierici, G. L., Ciucci, G. M., and Pizzi, G.: "A Systematic Study of Gas and Water Coning By Potentiometric Models," *Journal of Petroleum Technology,* pp. 923–929, August 1964.
4. Borisov, Ju. P.: "Oil Production Using Horizontal and Multiple Deviation Wells," Nedra, Moscow, 1964. Translated by J. Strauss, S. D. Joshi (ed.), Phillips Petroleum Co., the R & D Library Translation, Bartlesville, Oklahoma, 1984.

5. Merkulov, V. P.: "Le debit des puits devies et horizontaux," *Neft. Khoz.*, vol. 6, pp. 51–56, 1958.
6. Giger, F.: "Reduction du nombre de puits par l'utilisation de forages horizontaux," *Revue de l'Institut Francais du Petrole*, vol. 38, No 3, May–June 1983.
7. Giger, F. M., Reiss, L. H., and Jourdan, A. P.: "The Reservoir Engineering Aspect of Horizontal Drilling," paper SPE 13024 presented at the SPE 59th Annual Technical Conference and Exhibition, Houston, Texas, Sept. 16–19, 1984.
8. Renard, G. I. and Dupuy, J. M.: "Influence of Formation Damage on the Flow Efficiency of Horizontal Wells," paper SPE 19414, presented at the Formation Damage Control Symposium, Lafayette, Louisiana, Feb. 22–23, 1990.
9. Joshi, S. D.: "Augmentation of Well Productivity Using Slant and Horizontal Wells," *Journal of Petroleum Technology*, pp. 729–739, June 1988.
10. Joshi, S. D.: "A Review of Horizontal Well and Drainhole Technology," paper SPE 16868, presented at the 1987 Annual Technical Conference, Dallas, Texas. A revised version was presented at the SPE Rocky Mountain Regional Meeting, Casper, Wyoming, May 1988.
11. Van Der Vlis, A. C., Duns, H., and Luque, R. F.: "Increasing Well Productivity in Tight Chalk Reservoir," *Proc.*, vol. 3, pp. 71–78, 10th World Petroleum Congress, Bucharest, Romania, 1979.
12. Ozkan, E., Raghavan, R., and Joshi, S. D.: "Horizontal Well Pressure Analysis," *SPE Formation Evaluation*, pp. 567–575, December 1989.
13. Cinco, H., Miller, F. G., and Ramey, Jr., H. J.: "Unsteady-State Pressure Distribution Created by a Directionally Drilled Well," *Journal of Petroleum Technology*, pp. 1392–1402, November 1975.
14. Sparlin, D. D. and Hagen, Jr. R. W.: "Controlling Sand in a Horizontal Completion," *World Oil*, pp. 54–60, November 1988.
15. Mauduit, D.: "Determining the Productivity of Horizontal Completions," *World Oil*, pp. 55–61, December 1989.
16. Sparlin, D. D. and Hagen, Jr., R. W.: "Authors' Reply Determining the Productivity of Horizontal Completions," *World Oil*, pp. 56–61, December 1989.
17. Personal communication with D. D. Sparlin, April 17, 1990.
18. Sheikholeslami, B. A., Schlottman, B. W., Siedel, F. A., and Button, D. M.: "Drilling and Production Aspects of Horizontal Wells in the Austin Chalk," paper SPE 19825, presented at the 64th SPE Annual Technical Conference and Exhibition, San Antonio, Texas, Oct. 8–11, 1989.
19. Celier, G.C.M.R., Jouault, P., and de Montigny, O.A.M.C.: "Zuidwal: A Gas Field Development With Horizontal Wells," paper SPE 19826, presented at the 64th Annual Technical Conference and Exhibition of the Society of Petroleum Engineers, San Antonio, Texas, Oct. 8–11, 1989.

CHAPTER 4

Influence of Well Eccentricity

INTRODUCTION

For drilling a horizontal well, it is essential to decide on tolerance limits for well elevation. In other words, one has to decide how much deviation from a vertical elevation is tolerable. For small tolerance limits (± 5 ft), several measurements and surveys are required as the horizontal well is being drilled. Well drilling time is proportional to the number of directional surveys required. In the short-radius drilling technique, one may pull out of the hole with a drilling assembly and insert aluminum drill pipe for directional surveys. After surveying, aluminum drill pipe will have to be pulled out of the hole and drill pipe with flexible (wiggly) collars is reinserted to continue drilling. Thus, in the short-radius technique, surveying may involve several trips and can be expensive (note that by late 1990 to mid-1991, MWD (measurement while drilling) tools may be available for the short-radius drilling technique!).

For a medium-radius well, MWD tools are employed for directional control. Many MWD tools are actuated using pressure pulsing or some other activation technique. In general, the drilling rate slows down when MWD is activated. Thus, the number of surveys has a direct influence on drilling time and costs. Additionally, MWD tools for directional surveys are presently about 50 to 90 ft behind the bit. Thus, the exact bit location may not be known at all times. This results in slight inaccuracies in drilling and results in a need for estimating tolerance limits on drilling plans.

The type of reservoir determines the drilling elevation tolerance.

1. Reservoirs with closed top and bottom boundaries: In this case, bottom water and top gas are absent. Ideally, one would like to drill a well at the reservoir elevation center. A loss of productivity is expected when the well is not at the elevation center. As shown in this chapter, the loss in productivity is minimum for long wells. This is because a long horizontal well drilled in a thin reservoir acts as though it is a vertical fracture intersecting the entire reservoir height. A horizontal well, which acts as a fluid withdrawal conduit can be located anywhere in this vertical plane, with a minimum loss of productivity regardless of well location.
2. Reservoirs with water and/or gas coning: In these reservoirs, well location in the vertical plane is very important. A well location in the vertical plane, especially for a long well, would not cause a significant change in well productivity. However, the location of the well in the vertical plane would determine the breakthrough time of either gas or water or both, and subsequent changes in gas-oil ratio (GOR) and water-oil ratio (WOR). Thus, the well location in the vertical plane will affect the ultimate reserves producible from a well. A literature review reveals that horizontal wells have been more successful in reducing water coning than gas coning. Even for water coning, the successes are for oil zone thicknesses larger than 20 to 30 ft. In the case of gas coning, even when oil payzone thickness is above 50 ft, it has been difficult to minimize gas coning, even though some wells were drilled at the bottom of the oil payzone.

The above discussion demonstrates the importance of well tolerance limits on well performance. In this book, the well tolerance in a vertical plane is denoted as well eccentricity. In the following sections mathematical expressions are given to estimate the influence of well eccentricity on productivity of a horizontal well.[1–10] References 9 and 10 include results from very short drainholes producing in conjunction with the vertical portion of the well. These results, obtained using electrical analog experiments, are restricted to very short drainholes, and therefore, are not included in this text.

INFLUENCE OF WELL ECCENTRICITY

STEADY-STATE EQUATIONS

Figure 4–1 shows a schematic diagram of an off-centered horizontal well in a vertical plane. In the figure δ represents well eccentricity. The

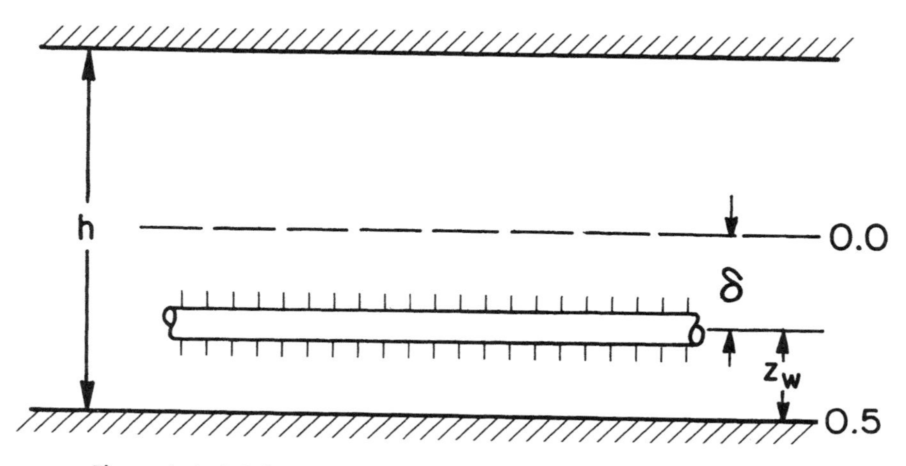

Figure 4–1 A Schematic View of an Off-Centered Horizontal Well.

influence of eccentricity on the well production rate is calculated using the following equation!

$$q_h = \frac{0.007078\, k_h\, h\, \Delta p/(\mu_o\, B_o)}{\ln\left[\dfrac{a + \sqrt{a^2 - (L/2)^2}}{L/2}\right] + (\beta h/L) \ln\left[\dfrac{(\beta h/2)^2 + \beta^2 \delta^2}{(\beta\, h\, r_w/2)}\right]} \quad (4\text{–}1)$$

for $L > \beta h$, $\delta < h/2$ and $L < 1.8\, r_{eh}$

q_h = oil flow rate, STB/day
h = reservoir height, ft
Δp = pressure drop, psi
k_h = horizontal permeability, md
μ_o = oil viscosity, cp
B_o = oil formation factor, RB/STB
r_w = wellbore radius, ft
L = horizontal well length, ft
δ = horizontal well eccentricity, ft

As previously defined, a can be calculated using Equation 3–11 and $\beta = \sqrt{k_h/k_v}$. Figure 4–2 compares the productivities of an off-centered horizontal well with that of a centered well for different well eccentricities. It can be seen that if the horizontal well is sufficiently long as compared to the reservoir height, the well can be located anywhere in the vertical plane without significant loss of productivity. In general, a horizontal well's performance is not significantly affected by eccentricity as long as the well is located *between* ± *25%* from the reservoir center. Strictly speaking, this is true for bounded reservoirs where top and bottom boundaries are closed (there is no bottom water and top gas).

EXAMPLE 4–1

A 1000-ft-long horizontal well is drilled 10 ft from the top of the payzone in a 50-ft-thick reservoir. Other reservoir parameters are

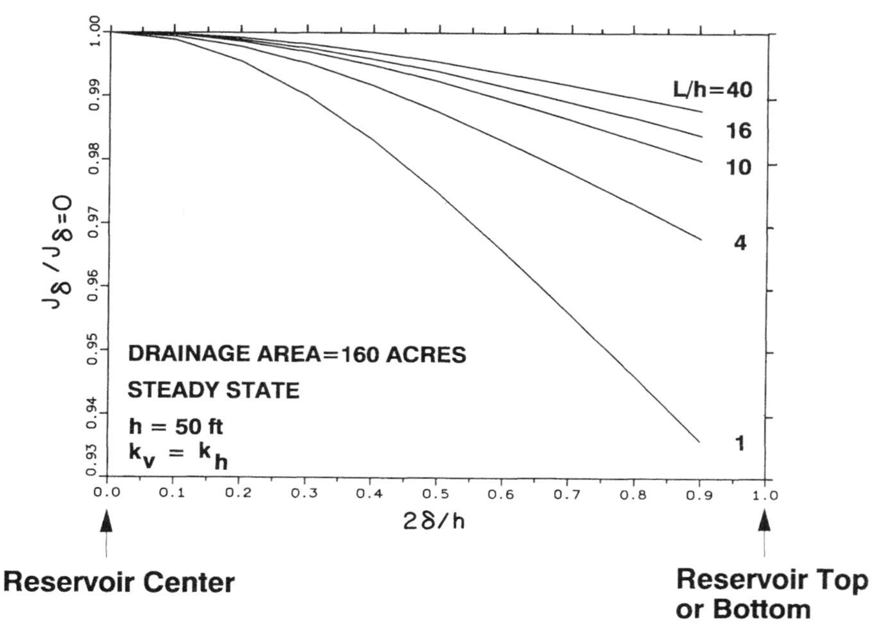

Figure 4–2 Influence of Horizontal Well Eccentricity on Productivity.

r_{eh} = 1490 ft (160-acre spacing) r_w = 0.33 ft
k = 0.5 md μ_o = 0.4 cp B_o = 1.35 RB/STB

1. Compare the horizontal well productivities for k_v/k_h values of 0.1, 0.5, and 1.0.
2. For k_v/k_h = 0.1, compare the horizontal well productivities if the horizontal well is located at distances of 10 ft and 20 ft from the top of the payzone.

Solution

The productivity of an off-centered horizontal well is given by

$$J_h = \frac{0.007078\, k_h\, h/(\mu_o\, B_o)}{\ln\left[\dfrac{a + \sqrt{a^2 - (L/2)^2}}{(L/2)}\right] + \dfrac{\beta h}{L} \ln\left[\dfrac{(\beta h/2)^2 + \beta^2\, \delta^2}{(\beta\, h\, r_w/2)}\right]}$$

where

$$\delta = \text{horizontal well eccentricity, ft}$$

1. The calculations are illustrated for $k_v/k_h = 0.1$ for a well located 10 ft from the top of the payzone.

$$\beta = \sqrt{k_h/k_v} = \sqrt{1/0.1} = 3.16$$
$$\delta = (50/2) - 10 = 15 \text{ ft}$$
$$a = 0.5\, L\, [0.5 + \sqrt{0.25 + (2\, r_{eh}/L)^4}]^{0.5}$$
$$= 0.5 \times 1{,}000\, [0.5 + \sqrt{0.25 + (2 \times 1490/1000)^4}]^{0.5}$$
$$= 1{,}532 \text{ ft}$$

$$J_h = \frac{0.007078 \times 0.5 \times 50/(0.4 \times 1.35)}{\ln\left[\dfrac{1532 + \sqrt{1532^2 - (1000/2)^2}}{(1000/2)}\right] + \left(\dfrac{3.16 \times 50}{1000}\right) \times \ln\left[\dfrac{(3.16 \times 50/2)^2 + 3.16^2 \times 15^2}{3.16 \times 50 \times 0.33/2}\right]}$$

$$J_h = 0.121 \text{ STB/(day-psi)}$$

2. For a well located 20 ft from the top of the payzone,

$$\delta = (50/2) - 20 \quad = 0.5$$
$$J_h = \frac{0.3277}{1.7854 + 0.87} = 0.123$$

The calculations for cases (1) and (2) are summarized below:

1. Influence of varying k_v/k_h: $\delta = 15$ ft

k_v/k_h	$\beta = \sqrt{k_h/k_v}$	J_h STB/(day-psi)
0.1	3.16	0.121
0.5	1.41	0.153
1.0	1.00	0.162

2. Influence of varying δ : $k_v/k_h = 0.1$, $\beta = 3.16$

δ ft	J_h STB/(day-psi)
15	0.121
5	0.123
0	0.124

This example demonstrates that if the horizontal well is sufficiently long as compared to the reservoir height, the well could be located anywhere in the vertical plane without significant loss of productivity.

The influence of eccentricity for a horizontal well can also be expressed in terms of pseudo-skin factors. By comparing late time-pressure responses of horizontal wells with that of a fully penetrating vertical fracture, Ozkan et al.[2] calculated pseudo-skin factors for horizontal wells in an infinite reservoir. Their equation gives pseudo-skin factors in terms of dimensionless length, L_D, dimensionless radius, r_{wD}, and dimensionless height, z_{wD}, which are defined below (also see Figure 4–1):

$$L_D = (L/(2h))\sqrt{k_v/k_h} \tag{4–2}$$

$$r_{wD} = r_w/(L/2) \tag{4–3}$$

$$z_{wD} = z_w/h \tag{4–4}$$

$$\log s' = A' - B'\log L_D + C'(\log L_D)^2 \tag{4–5}$$

for $0.15 \leq L_D \leq 100$ and $0.125 \leq z_{wD} \leq 0.5$, and for $0.1 \leq L_D \leq 25$ and $0.0625 \leq z_{wD} \leq 0.125$. The constants A', B', and C' are given as

$$A' = A_1 + A_2 \log r_{wD} + A_3 (\log r_{wD})^2 \tag{4–6}$$

$$B' = B_1 + B_2 \log r_{wD} + B_3 (\log r_{wD})^2 \tag{4–7}$$

$$C' = C_1 + C_2 \log r_{wD} + C_3 (\log r_{wD})^2 \tag{4–8}$$

for $10^{-4} \leq r_{wD} \leq 10^{-2}$. Correlation constants are given in Table 4–1. For $z_{wD} > 0.5$, replace z_{wD} by $1 - z_{wD}$ in all calculations.

In the above equation, the pseudo-skin factor term, s' indicates the deviation of a horizontal well from a fully penetrating infinite-conductivity fracture. This is defined as

$$s_h = s' - \ln(0.25\, L/r_w) \tag{4–9}$$

Thus, the smaller the value of pseudo-skin factor s' the closer will be a horizontal well's performance to that of an infinite-conductivity fracture. Figure 4–3 shows the variation of pseudo-skin factor for various values of z_{wD} (As noted above, for $z_{wD} > 0.5$, replace z_{wD} by $1 - z_{wD}$). As shown in the figure, pseudo-skin factors are negligibly small for large values of L_D, i.e., for $L >> 2h$ (if $k_v/k_h = 1$). This indicates that well location does not have any significant influence on the pseudo-skin factor, especially if $L_D > 5$.

TABLE 4–1 PSEUDO-SKIN FACTOR CORRELATION CONSTANTS[2] (Equations 4–6 through 4–8)

Well* Location ($z_{wD} = z_w/h$) →	0.5	0.25	0.125	0.0625
Constants				
A_1	−0.8761	−0.5475	0.02620	0.1027
A_2	−0.6829	−0.4778	−0.1599	−0.1369
A_3	−0.08058	−0.04881	−0.003549	−0.0007541
B_1	2.8521	2.4183	2.1247	1.8377
B_2	0.9297	0.6929	0.5502	0.4048
B_3	0.1243	0.09135	0.07284	0.05342
C_1	−1.1258	−1.1817	−1.7673	−1.4129
C_2	−0.4764	−0.5726	−0.9499	−0.7704
C_3	−0.05145	−0.07205	−0.1296	−0.1067

* For $z_{wD} > 0.5$ use values of $(1 - z_{wD})$ instead of z_{wD}.

RESERVOIRS WITH A TOP GAS CAP

For reservoirs with a gas cap, the steady-state equation for a well influenced by a constant pressure boundary can be written in terms of productivity index, J as[3]

$$J = \frac{q_h}{\bar{p} - p_{wf}} \tag{4–10}$$

where

$$\bar{p} - p_{wf} = \frac{162.2 q\, B_O\, \mu_O}{\sqrt{k_h k_v}\,(L/2)} \left[\log\left\{ \frac{8\, h\, \beta}{\pi r_w (1 + \beta)} \cot(\pi z_w/(2h)) \right\} + 0.4343 \left\{ s_m - \frac{(h - z_w)\,\beta}{L/2} \right\} \right] \tag{4–11}$$

where

$$\beta = \sqrt{k_h/k_v}$$

In this equation, z_w is the distance of a horizontal well from the reservoir bottom boundary and $\bar{p}$ is the average reservoir pressure. Equation 4–11 is

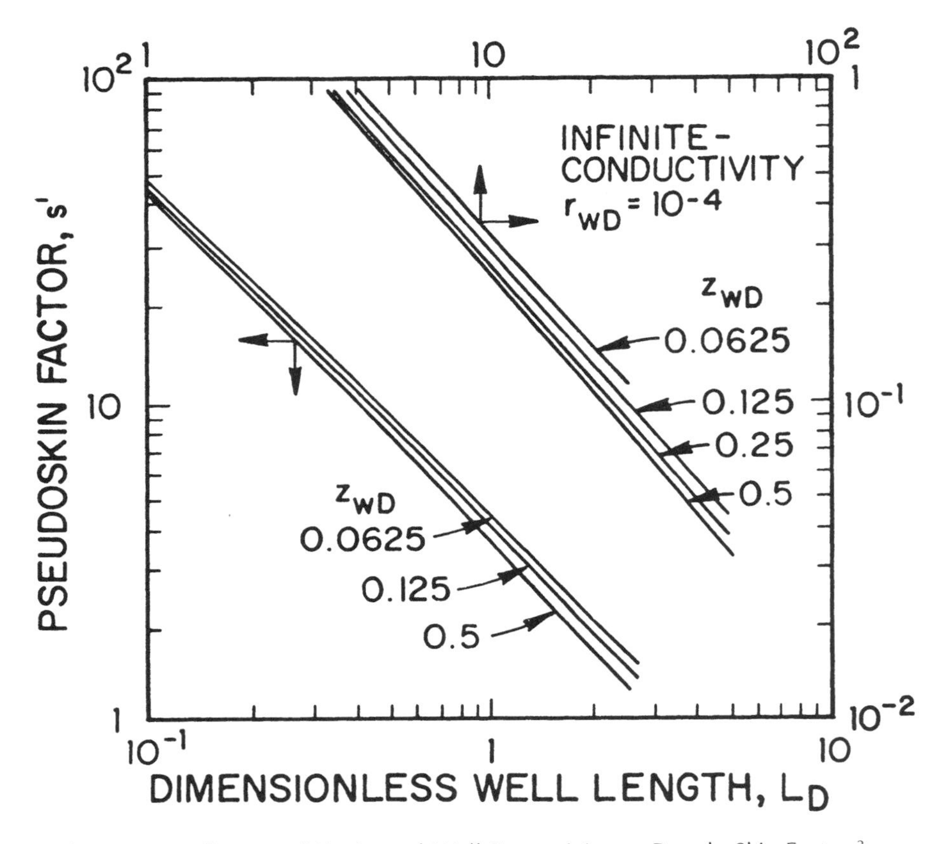

Figure 4–3 Influence of Horizontal Well Eccentricity on Pseudo-Skin Factor.[2]

only valid when $\sqrt{(k_h/k_v)}(2\ h/L) < 5$. It is important to note that in the above equation, the gas cap is represented as a constant pressure boundary condition. Additionally, the bottom boundary is assumed to be impermeable, i.e., a no-flow boundary.

In the above equation, s_m is a mechanical skin factor which can be evaluated from pressure transient testing or from the following equation:

$$s_m = \frac{\sqrt{k_h k_v}\ L/2}{374.4\ qB_o\ \mu_o}(p_i - p_{wfss}) - 2.303 \log\left[\frac{8h\beta}{\pi r_w(1 + \beta)}\cot(\pi z_w/(2\ h)) + \frac{(h - z_w)\ \beta}{L/2}\right] \quad (4\text{–}12)$$

In this equation, the subscript *ss* denotes steady state. Figure 4–4 shows typical calculations for the influence of well eccentricity in a reservoir with top gas and an impermeable bottom boundary.[3]

EXAMPLE 4–2

A 2000-ft-long horizontal well is drilled in a 50-ft-thick oil reservoir. The reservoir is overlain by a gas cap. Other reservoir parameters are:

$k_h = 2$ md $k_v = 1$ md
$\mu_o = 0.6$ cp $B_o = 1.2$ RB/STB
$r_w = 0.365$ ft Mechanical skin, $s_m = +2$

Compare the productivities for a horizontal well located 5, 10, 15, 20 and 25 ft from the reservoir bottom.

Solution

For a horizontal well located 5 ft from the reservoir bottom

$z_w = 5$ ft $\beta = \sqrt{k_h/k_v} = \sqrt{2/1} = 1.414$

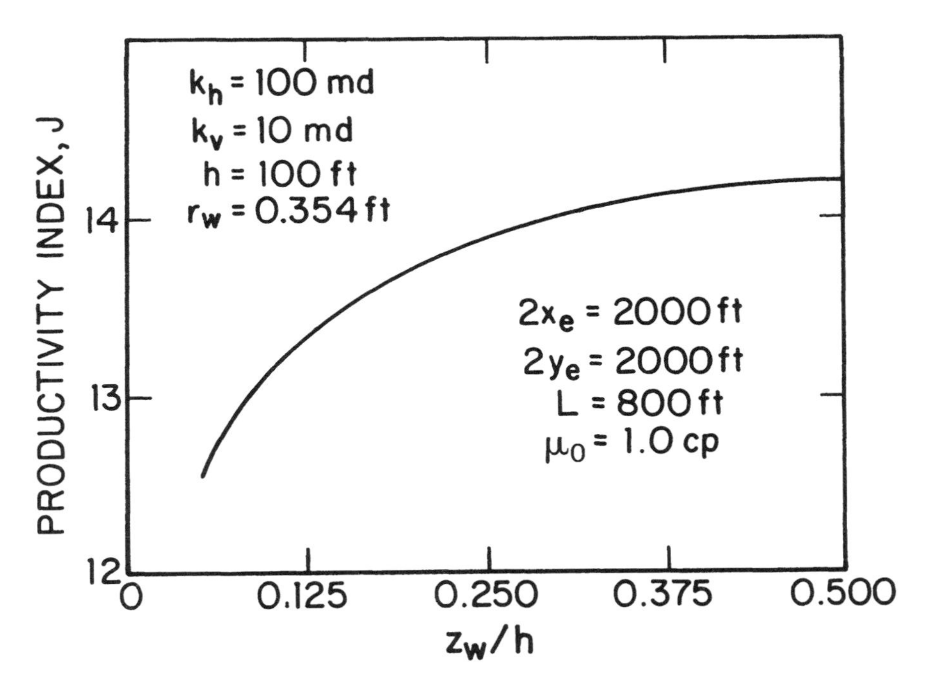

Figure 4–4 Influence of Horizontal Well Eccentricity on Productivity Index.[3]

Equations 4–10 and 4–11 can be used to calculate horizontal well productivity.

$$\bar{p} - p_{wf} = \frac{162.2q\, B_o\, \mu_o}{\sqrt{k_h k_v}\,(L/2)} \left[\log \left\{ \frac{8\, h\, \beta}{\pi r_w (1 + \beta)} \cot(\pi z_w/(2h)) \right\} + 0.4343 \left\{ s_m - \frac{(h - z_w)\, \beta}{L/2} \right\} \right] \tag{4–11}$$

$$\bar{p} - p_{wf} = \frac{162.2q_h \times 1.2 \times 0.6}{\sqrt{2 \times 1}\,(2000/2)} \log \left\{ \frac{8 \times 50 \times 1.414}{\pi \times 0.365\,(1 + 1.414)} \cot(\pi \times 5/(2 \times 50)) \right\} + 0.4343 \left\{ 2 - \frac{(50 - 5)\, 1.414}{2000/2} \right\} \Bigg]$$

$$= 0.08269 q_h\, [3.11 + 0.896]$$

$$= 0.331q$$

$$J_h = \frac{q_h}{p - p_{wf}} = \frac{1}{0.331} = 3.05 \text{ bbl/day-psi}$$

The results for different horizontal well locations are summarized below.

z_w ft	z_w/h	J_h bbl/day-psi
5	0.1	3.05
10	0.2	3.32
15	0.3	3.50
20	0.4	3.66
25	0.5	3.82

DRILLING SEVERAL WELLS

In practice, one can drill several horizontal wells like the spokes of a wheel from a single spudding location. When horizontal wells originate from a single point (Fig. 2–5), the following equation can be used to calculate total oil production:[5]

$$q_h = \frac{0.007078\, k_h h\, \Delta p/(\mu_o\, B_o)}{\ln\,[F\, r_e/L] + (h/nL)\ln\,[h/(2\pi r_w)]} \tag{4–13}$$

where n represents the number of spokes and F = 4, 2, 1.86, and 1.78 for n = 1, 2, 3, and 4, respectively. It is obvious from Equation 4–13 and Figure

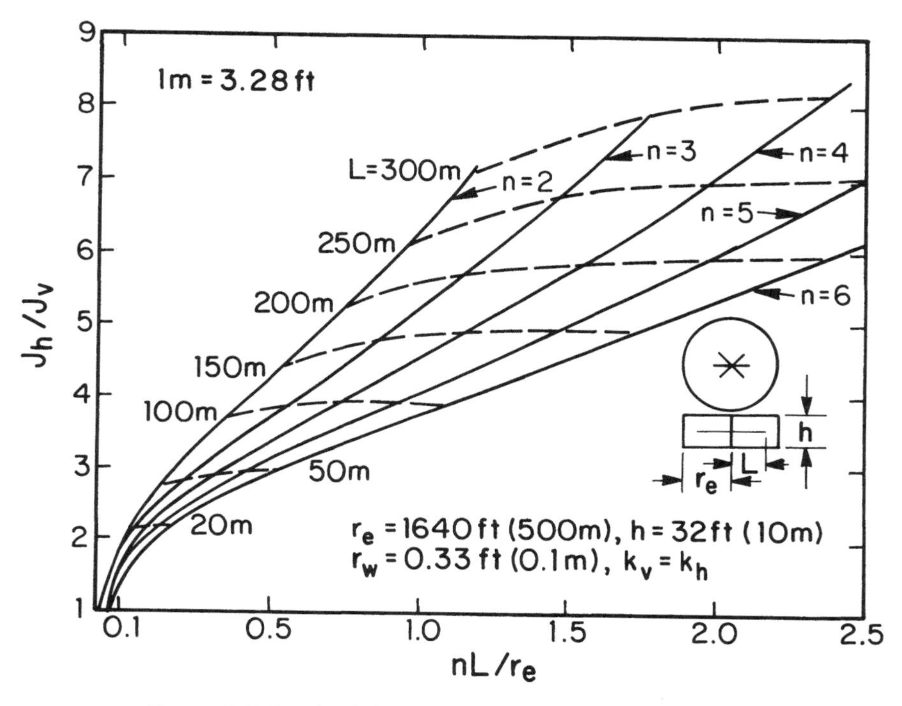

Figure 4–5 Productivity Ratio with Multiple Drainhole.[5,7]

4–5 that as the number of wells increases, the incremental productivity obtained by each additional well declines rapidly. In actual drilling operations, all horizontal wells will not originate from a single point as assumed in the above analysis. Depending upon the drilling method employed and its turning radius, there will be some distance between the spudding point at which the well enters the reservoir. This distance will give a smaller interference effect than that indicated by Equation 4–13. A single long well would probably give overall optimum recovery, considering the incremental recovery obtained from each additional well. Nevertheless, if one cannot drill a single long horizontal well from a given location, the next preferred option is to drill two diametrically opposite wells.

The pattern shown in Figure 2–5 with multiple wells at a given elevation is very suitable for an injection well in an EOR application. A large surface area provided by multiple drainholes will facilitate high injection rates. For many EOR projects, economic success is possible if high injection rates can be maintained. (In addition, in many EOR projects, such as those involving steam stimulation, gravity drainage can play an important role.[6]) However, many state regulatory bodies in the United States require operators to inject fluids below the formation parting pressure. (This is to prevent fracturing of a well, especially when the drinking water aquifer is close to the hydrocarbon formation. By injecting above the parting pressure, it is possible to create a fracture which may intersect the drinking water aquifers). From a production standpoint, it is imperative to maintain high injection rates. A horizontal well, especially with many laterals, provides such an option where we can inject fluids into the formation at high rates without fracturing the formations. Thus, from the regulatory standpoint, *horizontal wells have distinct advantages over conventional vertical well injectors.*

HORIZONTAL WELLS AT DIFFERENT ELEVATIONS

In some reservoirs, multiple drainholes have been drilled at different elevations. Figure 2–6 shows a schematic of horizontal wells at different elevations. If m represents the number of levels or elevations at which drainholes are drilled and if H represents reservoir thickness drained by each drainhole, then total reservoir height $h = Hm$. Equation 4–13 can be modified to calculate total flow rate from multiple drainholes as[5]

$$q_h = \frac{0.007078 k_h h \, \Delta p/(\mu_o \, B_o)}{\ln(r_e F/L) + [h/(Lmn)] \ln [h/(2\pi m r_w)]} \tag{4–14}$$

where F is calculated from values given below Equation 4–13.

EXAMPLE 4–3

Calculate the total oil production from four 250-ft-long horizontal wells spudded from the same location at the same depth given the following fluid and reservoir parameters.

$k_h = 10$ md $\quad \Delta p = 300$ psi
$h = 50$ ft $\quad r_e = 1000$ ft
$\mu_o = 1$ cp $\quad B_o = 1.1$ RB/STB
$L = 250$ ft $\quad r_w = 0.354$ ft

Solution

Using Equation 4–14 to calculate total oil production

$$q_h = \frac{0.007078 \, k_h h \, \Delta p/(\mu_o B_o)}{\ln(r_e F/L) + [h/(Lmn)] \ln[h/(2\pi m r_w)]}$$

It can be seen that for one elevation ($m = 1$), Equation 4–14 reduces to Equation 4–13, where $n = 4$ and therefore, $F = 1.78$

$$q_h = \frac{0.007078(10)(50)(300)/(1 \times 1.1)}{\ln[(1000)(1.78)/250] + 50/(250 \times 4)\ln[50/(2\pi \times 0.354)]}$$

$$q_h = 455.7 \text{ STB/day}$$

$$q_h \approx 456 \text{ STB/day}$$

In field operations, if the total horizontally drilled footage, i.e., product *Lmn*, is constant, then to produce the reservoir, one can use different combinations of well lengths, number of drainhole levels or layers, and number of drainholes at each level. It can be shown mathematically that if $L > h$, the maximum productivity in a given reservoir is obtained by drilling the single (or two diametrically opposite) longest possible horizontal well or drainhole.

In some thick reservoirs, more than one lateral are kicked off from a single vertical well. This results in multiple horizontal wells or drainholes stacked one above the other. Figures 4–6 and 4–7, respectively, show physical representations of single and multiple drainhole systems.[8] The wells are drilled at different elevations in a given layer as shown in Figure 4–8, a mathematical solution can be obtained for each drainhole of length L drilled in a reservoir of height z_e. The solutions are added to obtain production from

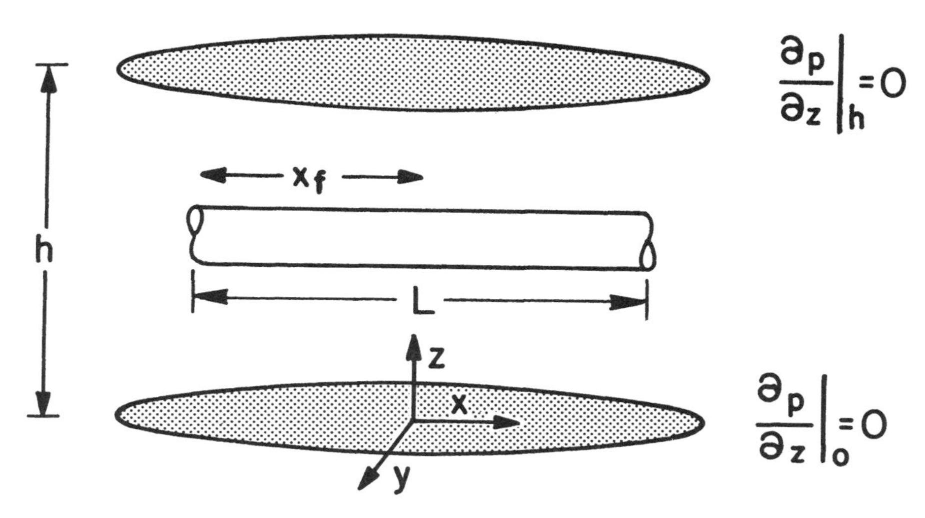

Figure 4–6 Physical Representation of a Single Drainhole.[8]

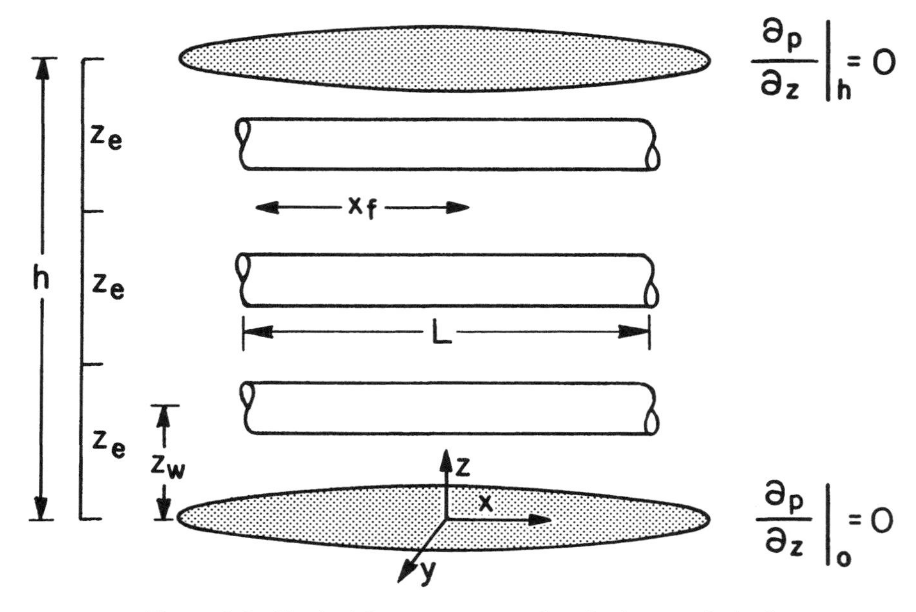

Figure 4–7 Physical Representation of Multiple Drainholes.[8]

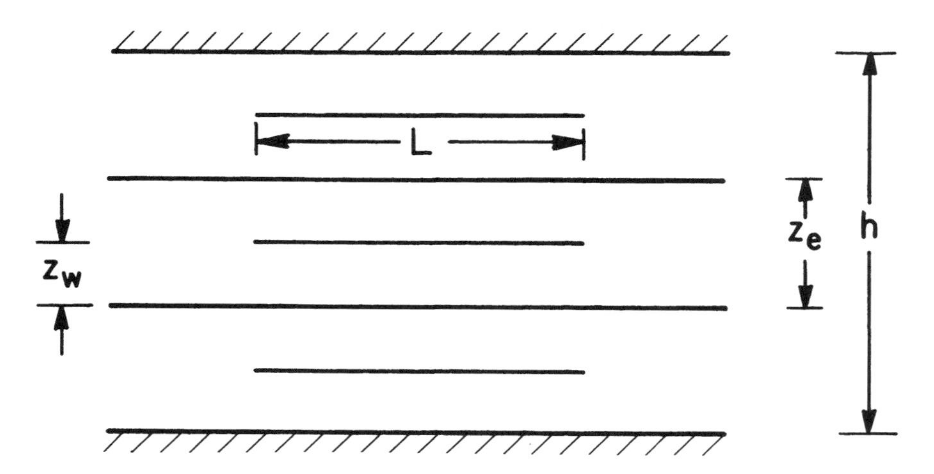

Figure 4–8 Mathematical Representation of Multiple Drainholes.[8]

multiple drainholes. This has the same effect as having a single long horizontal well with total well length equal to the sum of different lengths, drilled in a reservoir of total height, h.

For multiple drainholes drilled at different elevations, it is difficult to assess whether they are in the same zone due to difficulty in determining vertical communication. In these instances it is advisable to treat each well separately and add independent solutions to obtain total productivity. Similar to vertical wells, when drilling horizontal wells in different layers from a single vertical well, differential depletion due to permeability variations in different layers needs to be considered.

REFERENCES

1. Joshi, S. D.: "Augmentation of Well Productivity Using Slant and Horizontal Wells," *Journal of Petroleum Technology*, pp. 729–739, June 1988.
2. Ozkan, E., Raghavan, R. and Joshi, S. D.: "Horizontal Well Pressure Analysis," *SPE Formation Evaluation*, pp. 567–575, December 1989.
3. Kuchuk, F. J., Goode, P. A., Brice, B. W., Sherrard, D. W., and Thambynayagam, R. K. M.: "Pressure Transient Analysis and Inflow Performance for Horizontal Wells," paper SPE 18300, presented in the SPE 63rd Annual Technical Conference, Houston, Texas, Oct. 2–5, 1988.
4. Goode, P. A. and Thambynayagam, R. K. M.: "Pressure Drawdown and Buildup Analysis of Horizontal Wells in Anisotropic Media," *SPE Formation Evaluation*, pp. 683–697, December 1987.
5. Borisov, Ju P.: "Oil Production Using Horizontal and Multiple Deviation Wells," Nedra, Moscow, 1954. Translated into English by J. Strauss, edited by S. D. Joshi, Phillips Petroleum Co., the R & D Library Translation, Bartlesville, Oklahoma, 1984.
6. Dykstra, H. and Dickinson, W.: "Oil Recovery by Gravity Drainage into Horizontal Wells Compared with Recovery from Vertical Wells," Paper SPE 19827, presented at the 1989 SPE Annual Meeting and Exhibition, San Antonio, Texas, Oct. 8–11, 1989.
7. Reiss, L. H. and Giger, F.: "Purpose of Horizontal Drilling, Oil Field Considerations," *Petrole et Techniques*, No. 249, pp. 33–44, December 1982.
8. Clonts, M. D. and Ramey Jr., H. J.: "Pressure Transient Analysis for Wells with Horizontal Drainholes," paper SPE 15116, presented at the California Regional Meeting, Oakland, California, April 2–4, 1986.
9. Perrine, R. L.: "Well Productivity Increase from Drainholes as Measured by Model Studies," *Trans., AIME*, No. 204, pp. 30–34, 1955.
10. Roemershauser, A. E. and Hawkins, M. F., Jr.: "The Effect of Slant Hole, Drainhole and Lateral Hole Drilling on Well Productivity," *Journal of Petroleum Technology*, pp. 11–14, February 1955.

CHAPTER 5

Comparison of Horizontal and Fractured Vertical Wells

INTRODUCTION

This chapter includes a comparison of fractured vertical wells and horizontal wells. This comparison provides the estimated productivity gains that can be obtained either by stimulating vertical wells or by drilling horizontal wells. The chapter also includes a brief discussion about limitations, advantages, and gaps in technology of vertical well fracturing. Additionally, applications of horizontal wells in fractured reservoirs and recent field histories with stimulated horizontal wells are discussed.

VERTICAL WELL STIMULATION

Many vertical wells do experience damage while drilling. To establish a contact between the wellbore and the reservoir beyond the damage zone, the majority of the wells are normally acidized. In some cases, especially in the United States and Canada, wells are fractured using either propped or unpropped fractures. Technology for fracturing has been around for a long time. However, from a practical standpoint, the following issues are important.

1. During fracturing of a vertical well, one has no significant control over the fracture height. In general, high pumping rates result in tall fractures. Additionally, the fracture height strongly depends upon the strength of the cap and the base formations as compared to the strength of the reservoir rock. If the layers above and/or below the reservoir formation are weaker than the reservoir rock, then one experiences a large fracture height and limited reservoir penetration in a lateral direction for a given fracture treatment volume. An operator has minimum control over fracture height other than controlling injection rates to minimize fracture height, and therefore growth out of the payzone.
2. During fracturing, there is minimum control of fracture direction. The stimulated fractures are in a plane, which is perpendicular to the minimum principle stress. In naturally fractured reservoirs, the general observation is that artificially created fractures are parallel to the natural fracture orientation. In other words, if a vertical well is drilled in a naturally fractured reservoir and is fracture-stimulated, the induced fracture will be parallel to the natural fracture direction. This reduces the probability of the induced fracture intersecting several natural fractures and furthermore, results in low well productivity with limited drainage areas and possibly limits producible reserves.
3. In principle, it is desirable to obtain long infinite-conductivity fractures, i.e., fractures so conductive as compared to the formation that there is practically no pressure drop within the fracture itself. Figure 5–1 shows a schematic diagram of different types of horizontal wells and fractures. In practice, it is difficult to obtain long infinite-conductivity fractures. Additionally, the effect of proppant settling, crushing, and frac-fluid effects limit fracture conductivity.
4. In general, for reservoir engineering calculations, the effective fracture length is calculated either from a pressure buildup of a vertical well or from the production decline history. A well normally shows a smaller fracture length than what is estimated by fracture design calculations. In general, the fracture half-lengths calculated from a well test can be 30% to 50% shorter than the designed frac job length. This is because of the

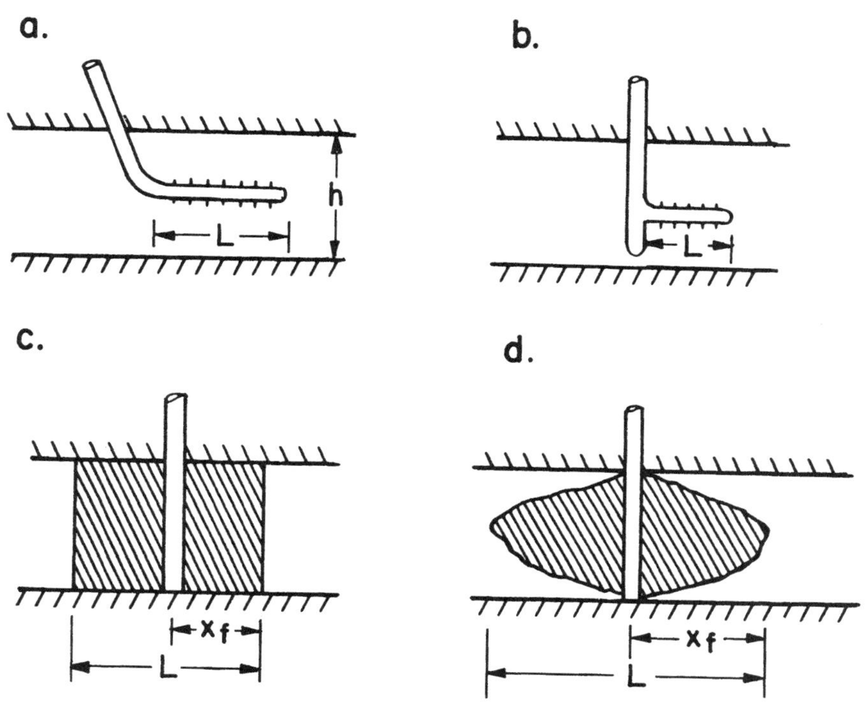

Figure 5–1 Schematic Diagrams of (a) Horizontal Well, (b) Drainhole, (c) Fully Penetrating Fracture, and (d) Realistic Fracture Geometry.

technological gap in fracture design as well as in interpretation techniques. Many times the major design gap is that the toughness of the rock, which is difficult to measure, is not included in the fracture design. Additionally, many other parameters such as rock strengths, fluid leak-off rates, and proppant strength at reservoir conditions are not known at every location. In general, the stimulated fractures have finite-conductivity, but infinite-conductivity or uniform-flux models are routinely used to calculate fracture half-length, resulting in shorter lengths calculated from a well test. If one assumes that the created fracture has the same length as the designed one, the fracture conductivity calculated from the well test is significantly smaller than the designed fracture conductivity. Additionally, the interpretation from the well test is based upon single-phase flow.

5. In some reservoirs, fracture conductivity decreases over time. This has been observed in laboratory experiments as well as in field measurements (see Figs. 5–2 and 5–3).[1,2] It is important to know that closure stress increases as the reservoir depletes, resulting in lower fracture conductivity.

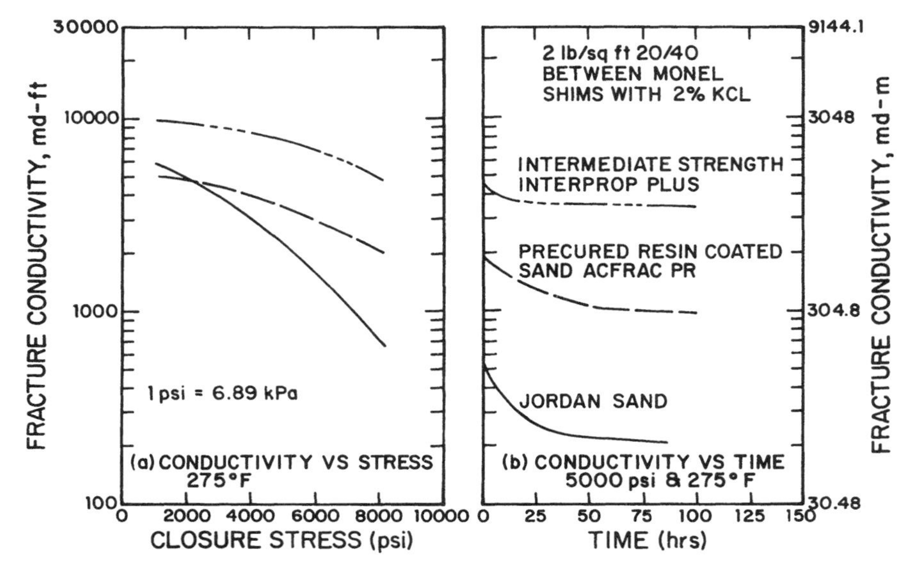

Figure 5–2 Short-term and Long-term Finite Conductivities of Proppants.[1]

It is also important to note that in some reservoirs, even after several years, initial fracture conductivity may not change. Thus, changes in fracture conductivity over time is dependent upon local conditions.

6. To achieve long fracture length normally requires pumping large fracture volumes. This means pumping large volumes of fluids into a formation. In many fracture jobs, after the fracture treatment, during the flowback period, very little frac fluid is recovered. In many reservoirs, especially in low permeability reservoirs, one may recover only 30% or even less of the injected frac fluids. The fluid which is left behind in the formation probably accumulates in the induced fracture and thereby reduces relative permeability to oil or gas flow.
7. In some reservoirs, serious sand flowback problems exist. One can use resin-coated sand to minimize the proppant flowback. However, in some cases, sand flowback problems persist even with resin-coated sands.
8. Fracture stimulation may not be an effective way to enhance well productivity in reservoirs with permeability over 10 millidarcies. This is because a fracture has finite conductivity, and therefore, there is a finite pressure drop within the fracture itself. Depending upon reservoir permeability and the proppant used, beyond a certain length pressure drops through fractures become comparable to pressure drops through the formation. Hence, beyond this point, any additional increase in fracture length will not yield any additional well productivity. Thus, for a given

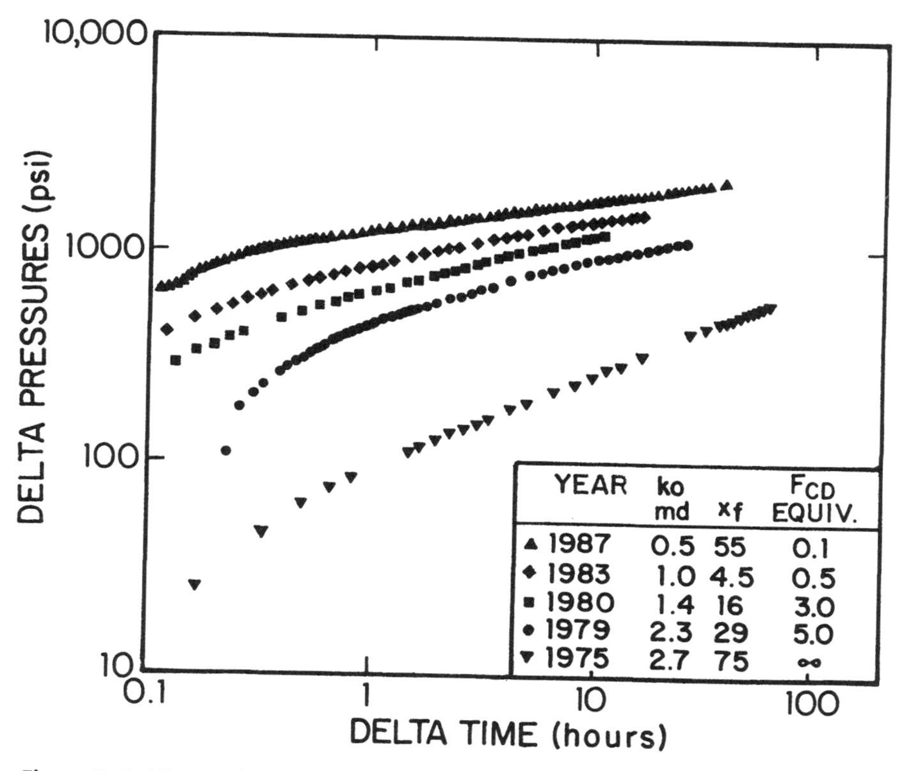

Figure 5–3 Historical Transient Data from a Well Showing a Decline in Induced Fracture Conductivity.[2]

reservoir permeability and given proppants, there is an optimum fracture length that one can create to optimize well productivity. (In some high permeability formations wells are fracture stimulated to overcome near wellbore skin damage.)

The above discussion illustrates practical problems with conventional fracture jobs. As noted earlier in Chapters 1 and 3, a horizontal well can be looked upon as a limiting case of an infinite-conductivity, fully penetrating fracture, with fracture height equal to the wellbore diameter. A horizontal well provides the means to mechanically induce a fracture in a desired direction for a desired length. In contrast, in conventional well stimulation, one has no control over induced fracture direction. As noted earlier, in naturally fractured reservoirs, the induced fracture may be parallel to the natural fracture direction, making it difficult to connect several fractures and drain them effectively. However, one can choose the horizontal-well direction to be perpendicular to the natural fractures and drain the reservoir effectively. Thus, horizontal wells have significant advantage over conventional vertically fractured wells. Whenever large fracture jobs are considered, a horizontal well should also be considered as a completion option. The major disadvantage of horizontal wells is their cost as compared to the fracturing treatment.

TYPES OF FRACTURES

From the reservoir standpoint, the artificial or induced fractures are subdivided into three different categories, namely: infinite-conductivity fractures, uniform-flux fractures, and finite-conductivity fractures. The following definitions of each assume that after a stimulation, a fracture extends into two diametrically opposite directions from a vertical well (see Fig. 5–1). Each wing of the fracture has length x_f, and the fracture is fully penetrating; i.e., it intersects the entire reservoir height.

INFINITE-CONDUCTIVITY FRACTURES

Infinite-conductivity fractures are those fractures where there is no pressure drop within the fracture itself. Thus, the entire fracture length is at the same pressure. This condition mathematically requires that maximum fluid entry rate into the fracture is at the fracture tips and minimum fluid entry is at the fracture center, i.e., at the vertical wellbore itself. The effective wellbore radius of an infinite-conductivity fracture which penetrates the entire reservoir height is defined as

$$r'_w = L/4 = x_f/2 \qquad \text{for}(x_f/x_e) \leq 0.3 \tag{5–1}$$

where

r'_w = effective wellbore radius, ft
L = total fracture length, ft
x_f = fracture half-length, ft
x_e = half-length of a side of a drainage area square, ft

UNIFORM FLUX FRACTURES

In this case, there is a finite pressure drop within the fracture itself. However, fluid entry rate along the entire fracture length is constant. This boundary condition requires a maximum pressure at the fracture tips with a minimum pressure at the fracture center, i.e., at the center of the vertical wellbore. Many water injection wells are known to have uniform-flux fractures. An effective wellbore radius of a uniform-flux fracture is defined as

$$r'_w = L/(2e) = x_f/e \quad (5\text{–}2)$$

where e is Napierian logarithmic constant with a value equal to 2.718.

FINITE-CONDUCTIVITY FRACTURES

Many induced fractures have finite-conductivity or flow capacity, unless of course reservoir permeability is very low. (An induced fracture in a low-permeability reservoir exhibits an infinite-conductivity.) As shown in Figure 5–2, the flow capacity of the proppant, expressed in md-ft, depends upon proppant type.[1] The figure shows typical laboratory data of fracture conductivity measurements. As shown in the figure, Jordan sand proppants exhibit conductivity of the order of 200 md-ft, while resin-coated sand exhibits conductivity of the order of 1000 md-ft. An interprop plus, or bauxite, exhibits conductivity of the order of 3000 md-ft. The dimensionless fracture conductivity F_{CD} is defined as[3]

$$F_{CD} = k_f b_f/(k\, x_f) \quad (5\text{–}3)$$

where

k_f = fracture conductivity, md
b_f = fracture width, ft
k = formation conductivity, md
x_f = fracture half-length, ft ($x_f = L/2$)
L = total fracture length, ft

Once the F_{CD} value is known, one can use either Figure 5–4 or 5–5 to calculate an effective wellbore radius of finite-conductivity fractures.[4,5] For F_{CD} values less than 0.1, the following equation can be used to calculate the effective wellbore radius:[6]

$$r'_w = 0.2807\, k_f b_f/k \quad (5\text{–}4)$$

EFFECTIVE WELLBORE RADIUS OF A FINITE-CONDUCTIVITY FRACTURE

Equation 5–3 requires four parameters to calculate dimensionless fracture conductivity F_{CD}. Most fracturing companies provide proppant selection guides with various charts showing fracture flow capacity $k_f b_f$ for various proppants. Figure 5–6 shows a typical chart provided by a fracturing company.[7,8] (Extensive discussion on fracturing can be found in Reference 9.) Thus, by selecting a proppant one can estimate the $k_f b_f$ value. To calculate the F_{CD} value one also needs to estimate fracture half-length x_f. As noted above, generally there is a significant difference between fracture half-length

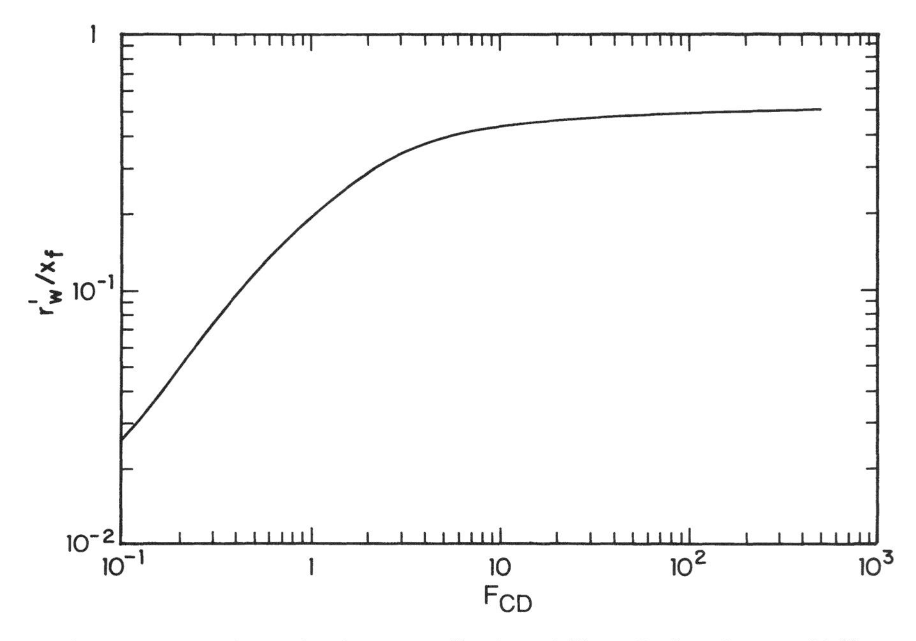

Figure 5–4 A Relationship between Effective Wellbore Radius, Fracture Half-Length and Dimensionless Fracture Conductivity.[4]

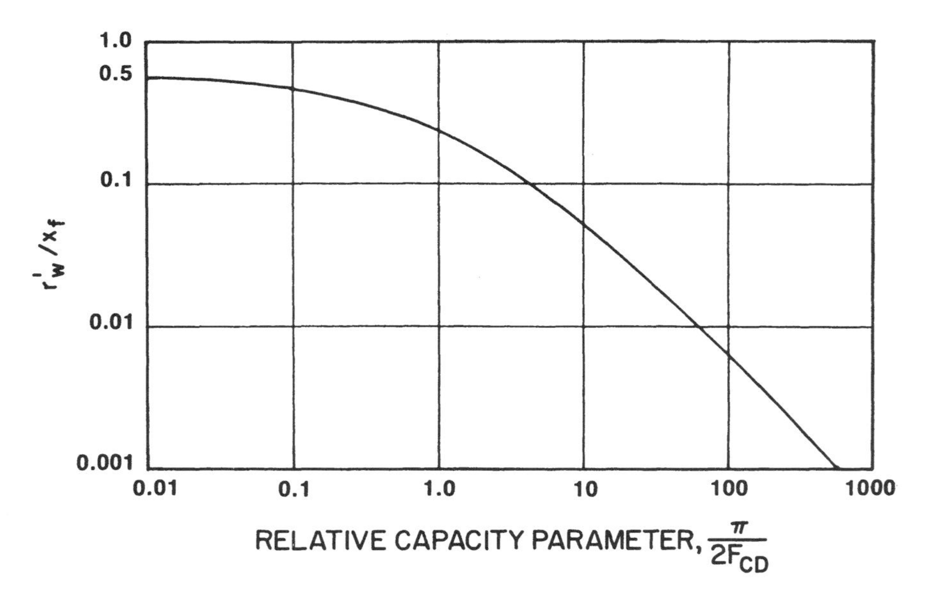

Figure 5–5 A Relationship between Effective Wellbore Radius, Fracture Half-Lengths and Dimensionless Fracture Conductivity.[5]

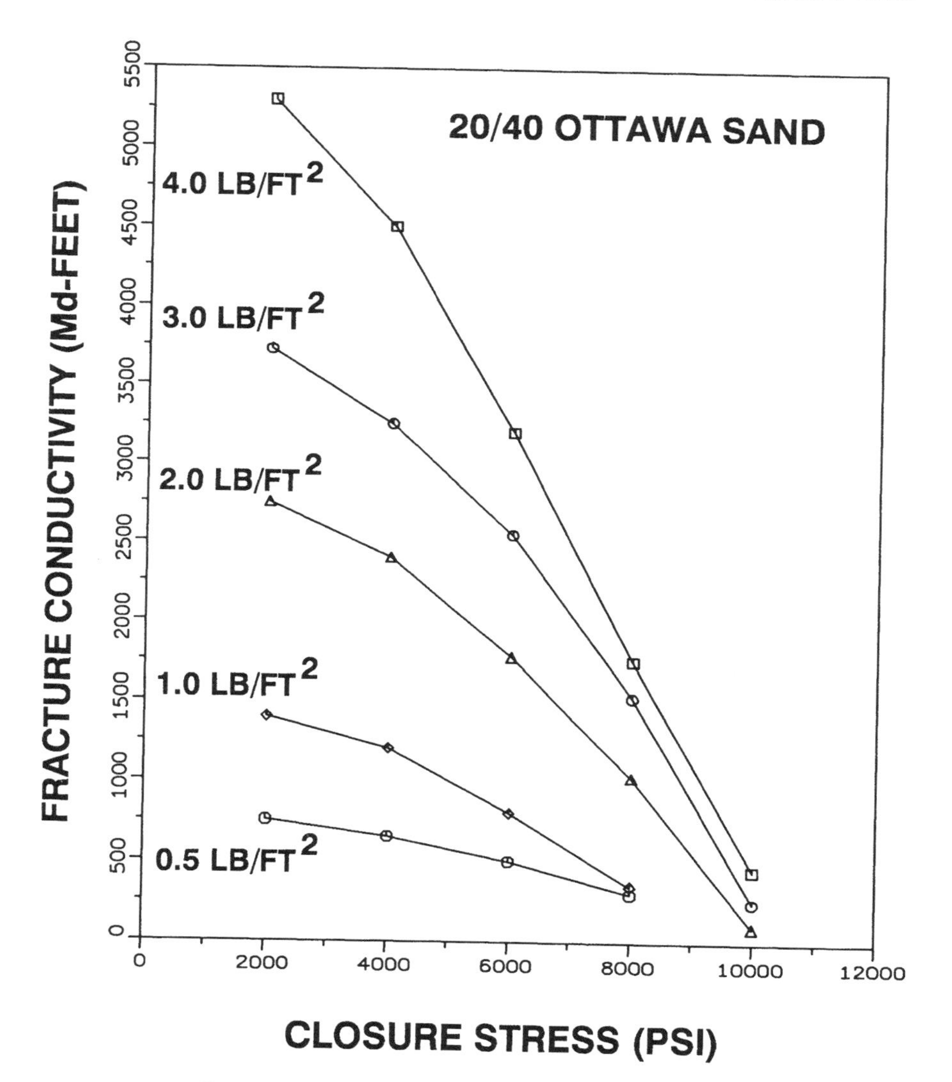

Figure 5–6 A Typical Fracture Conductivity Plot.[7,8]

estimated by fracture design and that obtained from a well test. An engineer needs to use local experience to establish a correlation between the designed and calculated fracture half-lengths. Once the effective fracture length is estimated, then for a given reservoir permeability k the dimensionless fracture conductivity F_{CD} can be estimated using Equation 5–3. For a fractured well, the following criteria can be used to estimate the effectiveness of a fracture treatment:

1. $F_{CD} < 10$ — poor fracture job
2. $10 < F_{CD} < 50$ — good to excellent fracture job
3. $F_{CD} > 50$ — excellent fracture job

Equation 5–3 and Figures 5–4 and 5–5 tell us that for a fixed proppant type and a fixed reservoir permeability, as the fracture half-length increases, F_{CD} value decreases and the ratio of r'_w/x_f decreases. (For infinite-conductivity fractures, as seen from Equation 5–1 and Figures 5–4 and 5–5, ratio of r'_w/x_f is 0.5.) Thus, beyond a certain length, an increase in the fracture length will not necessarily yield a significant gain in effective wellbore radius r'_w and therefore, no significant gain in well productivity. In other words, long fractures would definitely give higher productivity than short fractures. Nevertheless, the incremental gain in productivity diminishes as finite-conductivity fractures become longer and longer. This is because of the pressure drops within the fracture itself, which may be comparable to the reservoir pressure drop, resulting in a diminishing incremental gain in the productivity with length. The following example will amplify this point.

EXAMPLE 5–1

An oil reservoir with a 4 md permeability is to be developed at 80-acre spacing. The wells are to be stimulated using resin-coated sand as a proppant ($k_f b_f$ = 984 md-ft). Fracture designs indicate that jobs can be designed to create 200-, 400-, and 600-ft fracture half-lengths. Calculate steady-state productivity for each of the fracture designs and compare it with unstimulated vertical-well productivity.

Given:

$$h = 50 \text{ ft} \qquad \mu_o = 0.7 \text{ cp}$$
$$B_o = 1.1 \text{RB/STB}, \qquad r_w = 0.3 \text{ ft}$$
$$\phi = 19\%$$

Solution

1. Fracture half-length, x_f = 200 ft

$$F_{CD} = k_f b_f/(k\, x_f) = 984/(4 \times 200) = 1.23$$

From Figure 5–4, corresponding to F_{CD} = 1.23,

$$r'_w/x_f = 0.22$$
$$r'_w = 0.22 \times 200 = 44 \text{ ft}$$

For 80-acre spacing, $r_e = \sqrt{80 \times 43{,}560/\pi} = 1053$ ft

$$J_{frac} = \frac{0.007078\ kh/(\mu_0 B_0)}{\ln (r_e/r'_w)}$$
$$= \frac{0.007078 \times 4 \times 50/(0.7 \times 1.1)}{\ln (1053/44)}$$
$$= 0.58 \text{ STB/(day} - \text{psi)}$$
$$J_{frac}/J_v = \frac{\ln(r_e/r_w)}{\ln(r_e/r'_w)} = \frac{\ln(1053/0.3)}{\ln(1053/44)} = 2.57$$

2. Fracture half-length, $x_f = 400$ ft

$$F_{CD} = k_f b_f/(k\, x_f) = 984/(4 \times 400) = 0.615$$

From Figure 5–4,

$$r'_w/x_f = 0.14$$
$$r'_w = 0.14 \times 400 = 56 \text{ ft}$$
$$J_{frac} = \frac{0.007078\ kh/(\mu_0 B_0)}{\ln(r_e/r'_w)} = \frac{0.007078 \times 4 \times 50/(0.7 \times 1.1)}{\ln(1053/56)}$$
$$= 0.63 \text{ STB/(day} - \text{psi)}$$
$$J_{frac}/J_v = \frac{\ln(r_e/r_w)}{\ln(r_e/r'_w)} = \frac{\ln(1053/0.3)}{\ln(1053/56)} = 2.78$$

3. Fracture half-length, $x_f = 600$ ft

$$F_{CD} = k_f b_f/(k\, x_f) = 984/(4 \times 600) = 0.41$$

From Figure 5–4, $r'_w/x_f = 0.097$,

$$r'_w = 0.097 \times 600 = 58 \text{ ft}$$
$$J_{frac} = \frac{0.007078\ kh/(\mu_0 B_0)}{\ln(r_e/r'_w)} = \frac{0.007078 \times 4 \times 50/(0.7 \times 1.1)}{\ln(1053/58)}$$
$$= 0.634 \text{ STB/(day} - \text{psi)}$$
$$J_{frac}/J_v = \frac{\ln(r_e/r_w)}{\ln(r_e/r'_w)} = \frac{\ln(1053/0.3)}{\ln(1053/58)}$$
$$= 2.82$$

The above example shows that increasing fracture treatment size, i.e., increasing fracture half-length from 200 to 400 to 600 ft, gives steady-state productivity ratios of fractured and unfractured wells of 2.57, 2.78, and 2.82 respectively. This indicates diminishing gain in well productivity as fracture treatment size (or length) increases for a fixed proppant material.

In this example, F_{CD} values obtained, even for a 200-ft-long fracture, are low. In practice, one would design a fracture length shorter than 200 ft or use proppant with conductivity higher than 1000 md-ft.

EXAMPLE 5–2

It is proposed to develop an oil reservoir on 160-acre well spacing. Initial development wells show reservoir permeability varying from 0.1 to 1 md. (It is contemplated that in some "sweet spots" reservoir permeability can be as high as 10 md.)

Drainage radius, r_e = 1490 ft. (160-acre well spacing)

$h = 100$ ft $\qquad \phi = 0.11$

$\mu_o = 0.4$ cp $\qquad B_o = 1.2$ RB/STB

$r_w = 0.3$ ft

The wells need stimulation to enhance their productivity. Experience with well stimulation in similar fields shows that the calculated fracture half-length for the Jordan sand proppants is about half that of the fracture design. The fracture design using Jordan sand proppant ($k_f\, b_f = 230$ md-ft) gives a fracture half-length of 400 ft. Compare the productivity improvement obtained over a vertical well for reservoir permeabilities of 0.1 md, 1 md, and 10 md.

Solution

The fracture design length is 400 ft. Based on experience, the effective fracture half-length is 200 ft.

1. If formation permeability is 0.1 md then

$$F_{CD} = k_f b_f/(k\, x_f) = 230/(0.1 \times 200) = 11.5$$

From Figure 5–4, for $F_{CD} = 11.5$

$$r'_w/x_f = 0.44$$

$$r'_w = 0.44 \times 200 \text{ ft} = 88 \text{ ft}$$

$$J_{frac}/J_v = \frac{\ln(r_e/r_w)}{\ln(r_e/r'_w)} = \frac{\ln(1490/0.3)}{\ln(1490/88)} = 3$$

2. If formation permeability is 1.0 md then,

$$F_{CD} = k_f x_f/(k\, x_f) = 230/(1 \times 200) = 1.15$$

From Figure 5–4 for $F_{CD} = 1.15$

$$r'_w/x_f = 0.185$$
$$r'_w = 0.185 \times 200 = 37 \text{ ft}$$
$$J_{frac}/J_v = \frac{\ln(r_e/r_w)}{\ln(r_e/r'_w)} = \frac{\ln(1490/0.3)}{\ln(1490/37)} = 2.3$$

3. If formation permeability is 10 md, then $F_{CD} = 0.115$. From Figure 5–4 for $F_{CD} = 0.115$

$$r'_w/x_f = 0.028$$
$$r'_w = 0.028 \times 200 = 5.6 \text{ ft}$$
$$J_{frac}/J_v = 1.5$$

The above results clearly indicate that in high-permeability reservoirs, it is very difficult to get significant enhancement using fracture treatments.

Figure 5–7 shows transient performance of finite-conductivity fractures.[3] The figure shows that in late time, a finite-conductivity fracture with $F_{CD} \geq 10$ may behave as an infinite-conductivity fracture of the same length. In other words, most fractured vertical wells can be modeled either with infinite-conductivity or uniform-flux models. Thus for modeling purposes one has two options: (1) calculate F_{CD} and then use Figure 5–7 to predict well performance;[3] (2) calculate F_{CD} and using Figures 5–4, or 5–5,[4,5] calculate equivalent effective wellbore radius and an equivalent infinite-conductivity fracture length and use an infinite-conductivity model for future predictions. In short, one can adjust fracture conductivity (or flow capacity), fracture half-lengths, or both to estimate future production.

COMPARISON OF HORIZONTAL WELLS AND FINITE-CONDUCTIVITY FRACTURES

Figures 5–8 through 5–11 depict comparisons of horizontal-well productivities with productivities of fully penetrating fractures of the same length.[10] Thus, loss in productivity due to partial fracture penetration is ignored. The comparison is based upon a steady-state productivity index for horizontal wells and fractured vertical wells, i.e., using Equations 2–10 and 2–11 for comparison purposes. The comparisons are made for a 160-acre drainage area for the following variables:

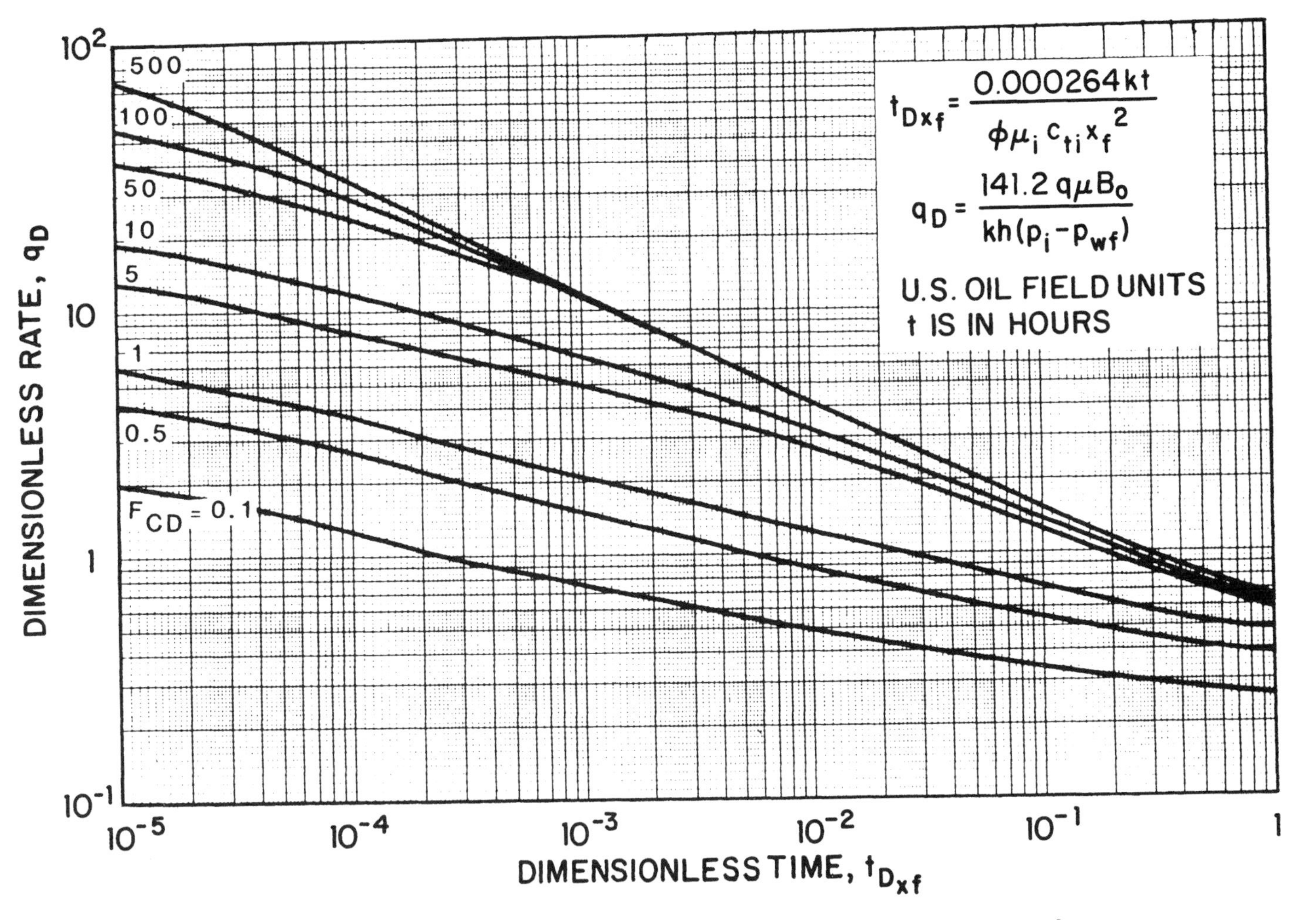

Figure 5–7 Transient Performance of Vertical Wells with Finite Conductivity Fractures.[3]

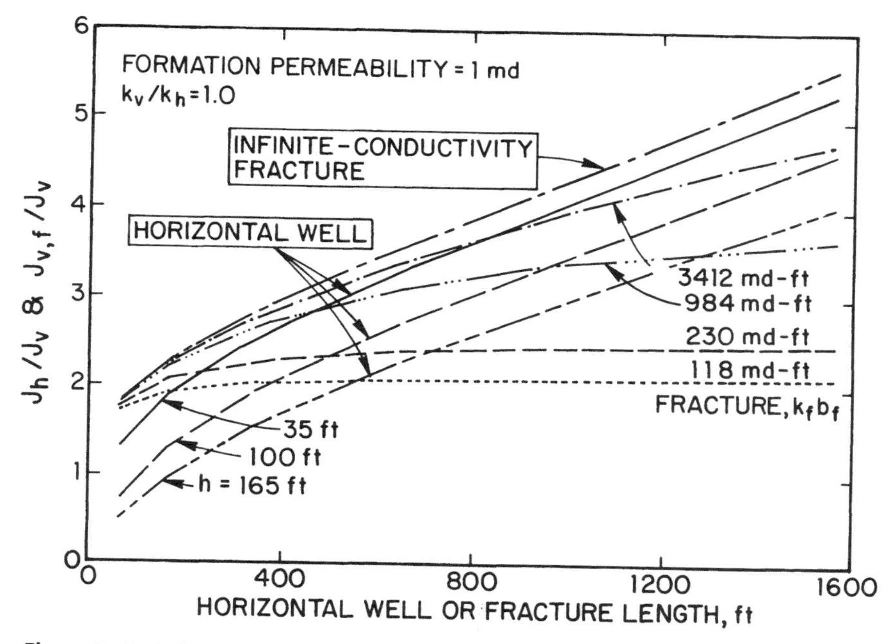

Figure 5–8 A Comparison of Productivities of Horizontal and Stimulated Vertical Wells.[10]

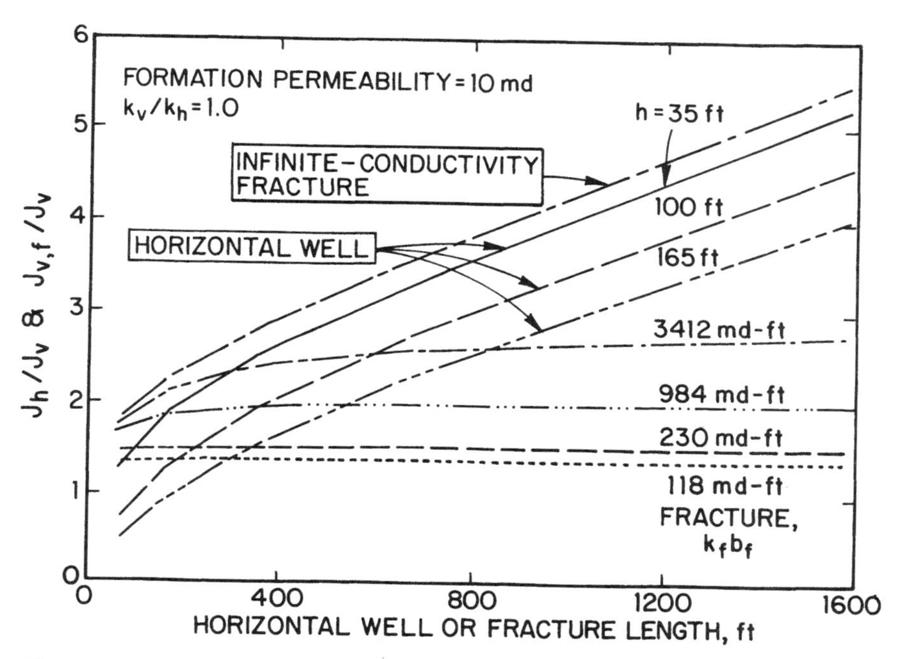

Figure 5–9 A Comparison of Productivities of Horizontal and Stimulated Vertical Wells.[10]

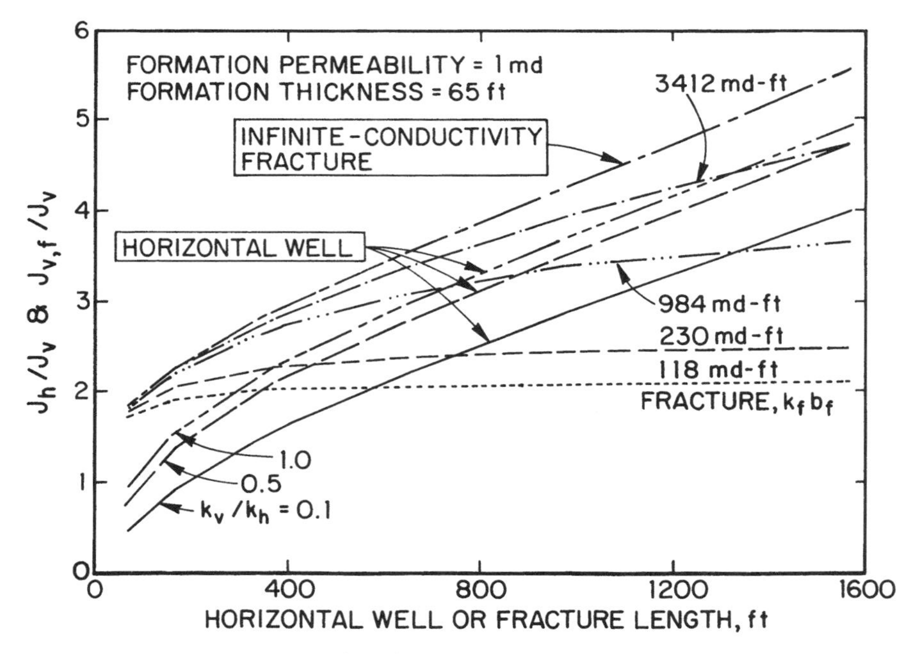

Figure 5–10 A Comparison of Productivities of Horizontal and Stimulated Vertical Wells.[10]

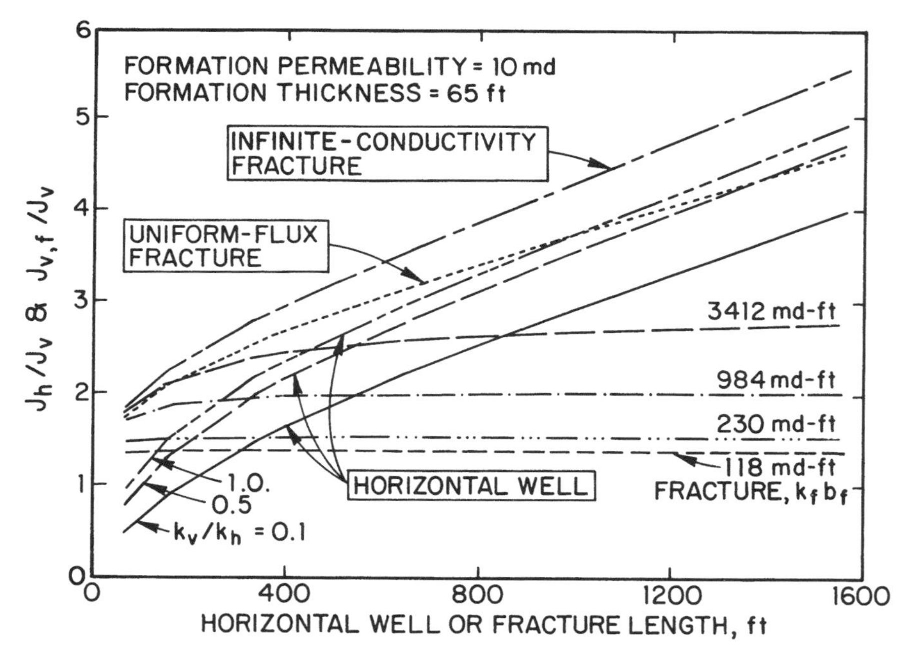

Figure 5–11 A Comparison of Productivities of Horizontal and Stimulated Vertical Wells.[10]

1. reservoir permeability: 1 and 10 md
2. horizontal well and fracture length: 100 to 1600 ft
3. reservoir height: 35, 100 and 165 ft
4. vertical to horizontal permeability ratio, k_v/k_h: 0.1, 0.5 and 1
5. fracture types: infinite conductivity, uniform flux, and finite conductivity with $k_f b_f$ = 118, 230, 984, and 3412 md-ft

The comparison of horizontal wells and fractured vertical wells is restricted to a length of 1600 ft. It is possible to drill more than 1600 ft horizontally. In contrast, most common fracture lengths obtained from well tests are about 200 ft (with an infinite conductivity). A massive hydraulic fracturing could give fracture half-lengths of about 800 ft and a total length of 1600 ft.

In the present comparison, fracture conductivities for various fracture jobs are represented as $k_f b_f$ = 118 md-ft for an acid treatment,[11] $k_f b_f$ = 230 md-ft for Jordan sand, $k_f b_f$ = 984 md-ft for precured resin-coated sand, and $k_f b_f$ = 3412 md-ft for Mullite (20/40 Intermediate strength Interprop™ plus).[1] For a given stimulation treatment, fracture half-length, and reservoir permeability, the dimensionless fracture conductivity (F_{CD}) was calculated using Equation 5–3. For F_{CD} values greater than 0.1, Figure 5–4 was used to calculate effective wellbore radius r'_w of the fracture treatment.

For low permeability reservoirs ($k \leq 0.1$ md), horizontal drilling and conventional fracturing can be very competitive. Of course, this assumes that one can obtain long fractures. (In general, it is difficult to obtain long fractures in a low-permeability formation.) Wherever large fracture extensions are difficult to achieve, long horizontal wells provide an alternative completion option.

Figures 5–9 and 5–11 tell us that in reservoirs with permeability greater than 10 md, conventional fracturing has a minimum effect in enhancing well productivity. As noted earlier, this is because of a pressure drop within the fracture itself, which is comparable to reservoir pressure drop. In contrast, with a horizontal well, one can achieve significant productivity improvement by drilling long wells, as long as vertical permeability is sufficiently high ($k_v/k_h \geq 0.1$). Figures 5–8 through 5–11 tell us the minimum length of a horizontal well required to obtain the same productivity as that obtained by conventional well stimulation.

It is important to note that in Figures 5–8 through 5–11 the fractures were assumed to be fully penetrating, and loss in fracture productivity due to partial fracture height was ignored.

Another method to compare horizontal wells with hydraulically fractured vertical wells is to calculate the effective wellbore radius of a fractured vertical well that is required to produce at the same rate as a horizontal well. This can be accomplished using the following expression:[12,13]

$$r'_w = \frac{r_{ev}\,(L/2)}{[a + \sqrt{a^2 - (L/2)^2}]\,[\beta h/(2r_w)]^{(\beta h/L)}} \qquad (5\text{–}5)$$

and also

$$r'_w = m\,x_f \qquad (5\text{–}6)$$

In the above equations, $m = 0.5$ for an infinite-conductivity vertical fracture. Additionally, r'_w is an effective wellbore radius of a fractured vertical well in ft, a is calculated using Equation 3–11, r_{ev} is drainage radius of a fractured vertical well in ft, and β is equal to $\sqrt{k_h/k_v}$. This equation shows that in low-permeability formations, the economic feasibility of a horizontal well strongly depends on the anisotropy, i.e. on the value of $\beta = \sqrt{k_h/k_v}$. For high β values, i.e., in reservoirs with low vertical permeability, short fracture lengths are required in a vertical well to match the productivity of a horizontal well. Figure 5–12 shows fracture half-length sizes that are required to match pro-

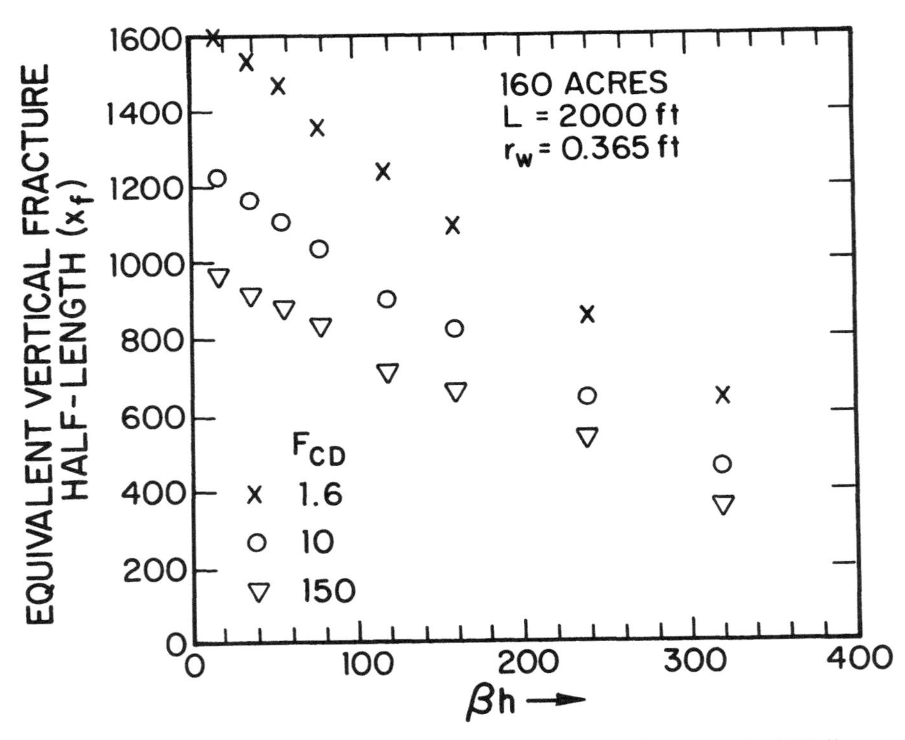

Figure 5–12 An Equivalent Fracture Half-Length Needed in a Vertical Well to Match Horizontal Well Productivity.[13]

ductivities of horizontal wells. The results are plotted for $F_{CD} = 1.6$, 10, and 150, respectively.[13] The figure clearly shows that in thin formations and in high vertical permeability formations, i.e., at small βh values, fractured vertical wells should have very long fracture lengths to match the productivity of horizontal wells.

EXAMPLE 5–3

It is proposed to drill a 2000-ft-long horizontal well to boost production from a low-permeability reservoir. Another proposed alternative is to hydraulically fracture a vertical well with Jordan sand ($k_f b_f = 230$ md-ft) as proppant. Calculate the equivalent fracture half-length in a vertical well needed to match the production from the horizontal well. Other reservoir parameters are

$$r_{ev} = 1490 \text{ ft (160-acre spacing)} \qquad r_w = 0.365 \text{ ft}$$
$$k = 0.05 \text{ md} \qquad h = 50 \text{ ft}$$
$$k_v/k_h = 0.1$$

Solution

$$\beta = \sqrt{k_h/k_v} = \sqrt{1/0.1} = 3.162$$
$$\beta h = 3.162 \times 50 \text{ ft} = 158.1 \text{ ft}$$

Using Equation 3–11, a is estimated as

$$a = 0.5L\left[0.5 + \sqrt{0.25 + (2\, r_{ev}/L)^4}\right]^{0.5}$$
$$= 0.5 \times 2000\left[0.5 + \sqrt{0.25 + (2 \times 1490/2000)^4}\right]^{0.5} = 1666 \text{ ft}$$

Using Equation 3–26, effective wellbore radius of a horizontal well, r'_w is estimated as

$$r'_w = \frac{r_{ev}(L/2)}{[a + \sqrt{a^2 - (L/2)^2}]\,[\beta h/(2r_w)]^{\beta h/L}}$$
$$= \frac{1490 \times (2000/2)}{[1666 + \sqrt{1666^2 - (2000/2)^2}]\,[158.1/(2 \times 0.365)]^{158.1/2000}}$$
$$= \frac{1{,}490{,}000}{2998.5 \times 1.53}$$
$$= 325 \text{ ft}$$

For Jordan sand $k_f b_f$ is 230 md-ft. Hence, F_{CD} can be calculated using Equation 5–3 as

$$F_{CD} = k_f b_f/(k\, x_f) = 230/(0.05\, x_f) = 4600/x_f$$

The value of x_f is estimated by trial and error. The procedure involves the following steps:

1. Assume x_f.
2. Calculate F_{CD}.
3. Using Figure 5–4, calculate r'_w/x_f.
4. Back calculate x_f for $r'_w = 325$ ft and compare with the assumed value.
5. Repeat the calculations until both the assumed and calculated values of x_f match.

These steps are illustrated below.

Assumed x_f, ft	F_{CD}	$\frac{r'_w}{x_f}$	Calculated x_f, ft
500	9.2	0.42	774
1000	4.6	0.37	878
800	5.75	0.35	822

Thus, with Jordan sand, the equivalent fracture half-length needed to match horizontal well production is about 800 ft. As noted in the table, the F_{CD} value for the fracture job is 5.75, which is a low value. It would be better to use resin-coated sand or other material of higher fracture flow capacity than Jordan sand. This would reduce the fracture half-length required to match horizontal well productivity. It may not be easy to obtain a fracture half-length of 800 ft. Of course, this depends upon local conditions.

VERTICAL FRACTURING SUMMARY

In summary, it is difficult to obtain more than 200- to 300-ft-long infinite-conductivity fractures. Obtaining an infinite conductivity, fully penetrating fracture is difficult due to proppant embedment, fines generation, proppant settling, and frac-fluid effects.[14–17] At least, in principle, long horizontal wells approximate infinite-conductivity fractures, and it is possible to drill 2000- to 3000-, even 5000-ft-long horizontal wells. Thus, horizontal wells can be used to enhance well productivity in low-permeability as well as in high-permeability reservoirs. The disadvantage of horizontal wells is their cost as compared to vertical-well stimulation treatments.

Finally, it is important to note here that although horizontal wells represent an infinite-conductivity fracture, in practice the drilled horizontal wells may need some sort of cleaning. This cleaning job may involve a simple swab job, a simple wash, a light acid cleanup, a matrix acidization job, or a limited acid-fracture job depending upon local production experience.

HORIZONTAL WELLS IN FRACTURED RESERVOIRS

Recently, several horizontal wells in the United States have been drilled in naturally fractured reservoirs such as Austin Chalk in Texas and Bakken Shale in North Dakota. The Austin Chalk is a highly fractured reservoir with many large fractures.[18,19] Bakken Shale is a low-permeability reservoir with permeability of about 0.2 md.[20,21] Many vertical wells drilled in these two reservoirs are not economical, probably because artificial fractures created from the vertical wells are parallel with the natural fractures. By drilling horizontal wells perpendicular to natural fractures, one can intersect several natural fractures and drain them effectively. Typical production improvement from a horizontal well in the Austin Chalk reservoir is shown in Figure 3–12.[18,19] One can see that as well length increases, so does the enhancement in productivity. Additionally, in Figure 3–12, one can connect two points above the average line. The straight line drawn through the above average points probably reflects a performance curve when a horizontal well intersects oil-filled fractures every 250 to 300 ft. On the other hand, if one draws a line through the bottom three points, it probably reflects the performance of a well which intersects an oil-filled fracture every 600 ft. The success of horizontal wells in the heavily drilled Austin Chalk reservoir indicates that due to reservoir heterogeneities, a large amount of hydrocarbon can be left behind, and it is possible to recover it using horizontal wells.

It might be interesting to note that similar to fractured vertical wells, horizontal wells show a higher kh value than an unstimulated vertical well. (The kh value can be estimated either by a well test analysis or by analysis of production decline curves.) Since h, the reservoir height, is fixed, the only parameter that could have increased is k, the reservoir permeability. In naturally fractured reservoirs, fractured vertical wells as well as horizontal wells intersect natural fractures, resulting in the enhancement of effective permeability. By drilling a long horizontal well, enhancement of permeability can be significantly greater than that obtained from a fractured vertical well. Thus, long horizontal wells drilled perpendicular to the natural fracture trend would improve well productivity significantly.

FRACTURED HORIZONTAL WELLS

Under the following circumstances one may consider stimulating horizontal wells:

1. In low-permeability formations to enhance drainage volume.
2. In formations with low vertical permeability, creating vertical fracture results in enhanced vertical permeability and thus, enhanced well productivity.

3. In a layered reservoir, by creating vertical fractures along the well length, one can connect different producing layers at different elevations.
4. In a cemented horizontal well, one can regain lost productivity of a horizontal well. Drilling and completion considerations may require cementing a horizontal well. By creating several fractures along the length of a cemented horizontal well one can achieve, at least, the same productivity as an open hole horizontal well.
5. In situations where the productivity obtained from a drilled horizontal well is not sufficient.

One of the main reasons for stimulating horizontal wells is to enhance vertical permeability. Ideally one would like to create fractures which are perpendicular to the horizontal wells in the vertical plane. The direction of the fracture orienting from the horizontal well is the same as that for vertical wells, i.e., parallel to the plane of minimum principle stress. Therefore, if a horizontal well is drilled along the low principle stress direction, then the stimulated fractures will be perpendicular to the horizontal well as shown in Figure 5–13.[22] However, if the horizontal well is also drilled along the maximum principle stress direction, then the stimulated fractures will be parallel to the horizontal well. This is shown in Figure 5–14.[22] Therefore, if the horizontal well is to be stimulated, it is important to consider the local stress directions. In an EOR application, artificial fractures perpendicular to

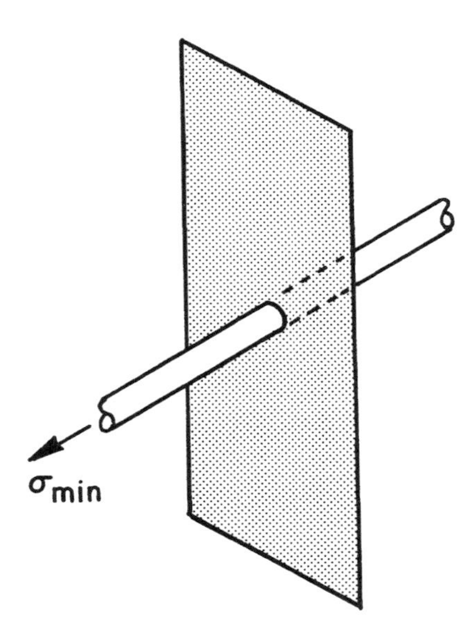

Figure 5–13 Fractured Horizontal Wells with Fractures Perpendicular to the Well.

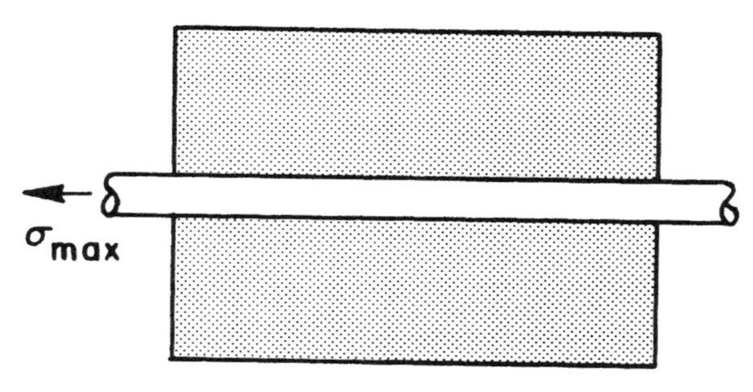

Figure 5–14 Fractured Horizontal Wells with Fractures Parallel to Well.

the horizontal wellbore would have an adverse effect on well productivity.

At present the local stress directions can be estimated using the following techniques:

1. *Microfracturing.* While drilling using mud, the formation is fractured as shown in Figure 5–15, and then the oriented core is taken. This microfracture indicates the direction of the induced fracture.[23]
2. *Strain relaxation.*[24,25] In this case, a well is cored under pressure, and this pressurized, oriented core is brought to the surface pressure and temperature conditions. Because of stress relaxation, the core will also relax, with maximum core relaxation, i.e., the maximum core expansion occurring along the direction of the maximum stress. In contrast, a minimum relaxation, and therefore, minimum core expansion, will occur along the direction of the least principle stress. This will identify maximum and minimum principle stress directions. (In practice, taking oriented cores can be expensive. Additionally, recovery of an oriented core is not always easy.)
3. *Caliper Logs.* In some cases, it is possible to identify stress directions by looking at caliper logs. In many cases, a drilled vertical wellbore may not be circular but rather highly elliptical because of the difference in the stresses in the horizontal plane. One expects a maximum borehole size along the minimum stress direction and a minimum borehole size along the maximum stress direction. To identify borehole shape, a tool with four arms will be very useful. The tool can be either a caliper tool, preferably with orientation, or a dipmeter.

Thus, the above three techniques can be used to locate the directions of the maximum and minimum principal stresses. A schematic of a long horizontal well with multiple fractures along the length is shown in Figure 5–16. The steady-state productivity increase for a fractured horizontal well is shown in Figure 5–17.[26,27] As shown in Figures 5–18 and 5–19 numerical simulation

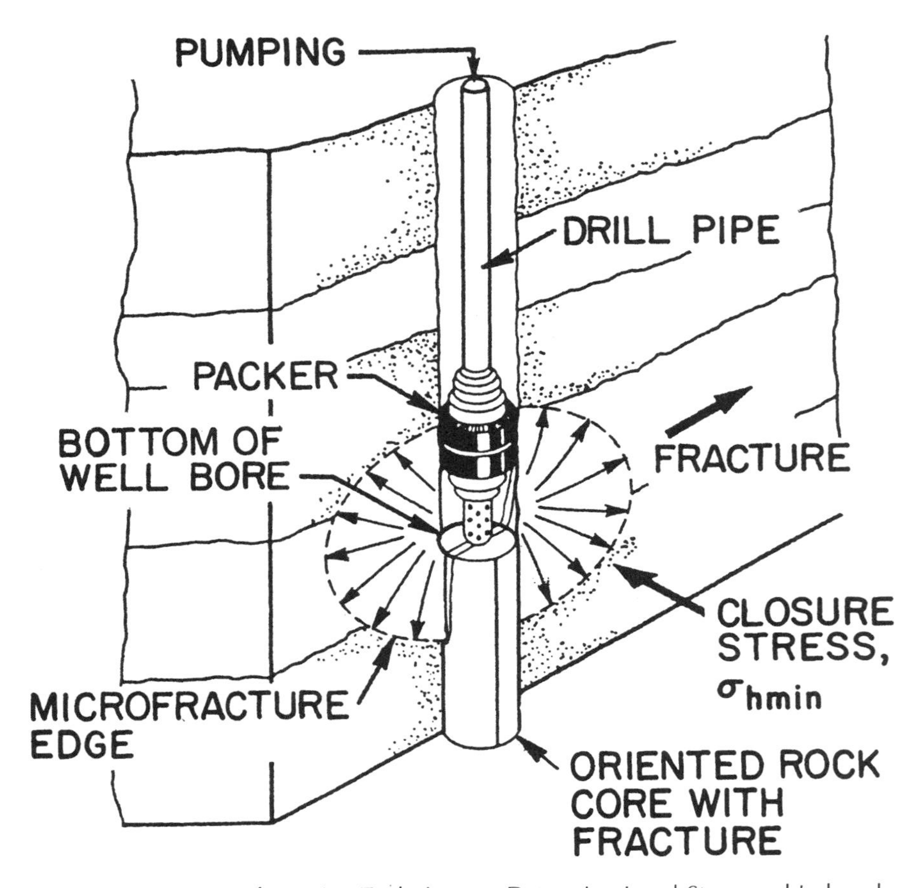

Figure 5–15 Microfracturing Technique to Determine Local Stress and Induced Fracture Direction.[23]

gives similar results.[28] As we increase the number of fractures along the length of a horizontal well, we can get higher and higher productivity. To address the problem of multiple fractures along the wellbore, it is easier to visualize each section of a fractured horizontal well as a fractured vertical well, but with some additional pressure drop. The additional pressure drop is caused by the fracture, which is perpendicular to the well. This is different from a vertical well where the fracture is along the well length. An additional pressure drop due to the fracture perpendicular to the well is accounted for as an additional skin factor as shown below.[13]

$$\Delta p_{skin} = (141.2 q B_0 \mu/(kh))(kh/(k_f b_f))[\ln(h/2r_w) - (\pi/2)] \quad (5\text{–}7)$$

$$(s_{ch})_c = (kh/(k_f b_f))\ [\ln\ (h/2r_w) - \pi/2] \quad (5\text{–}8)$$

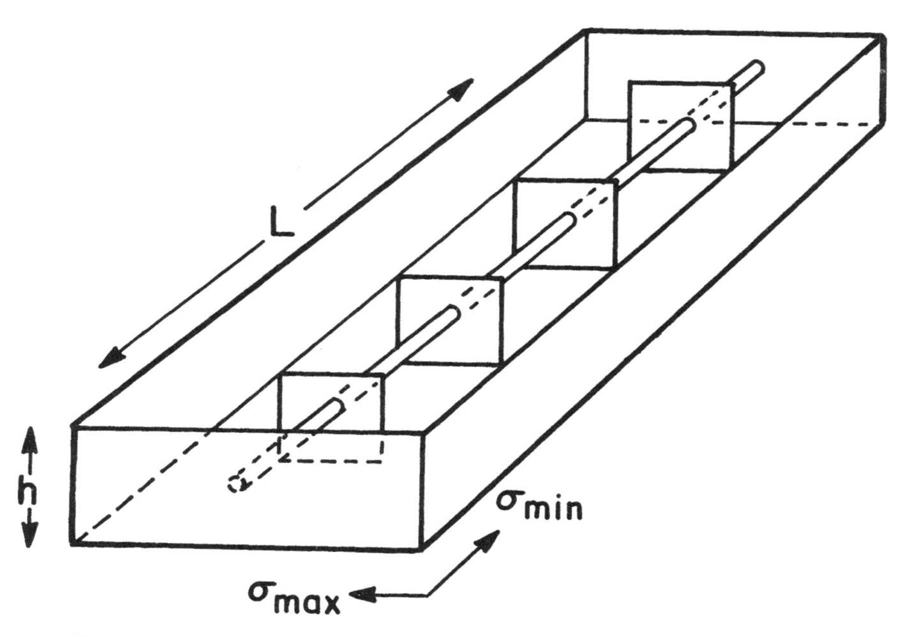

Figure 5–16 A Schematic of a Horizontal Well with Multiple Fractures.[13]

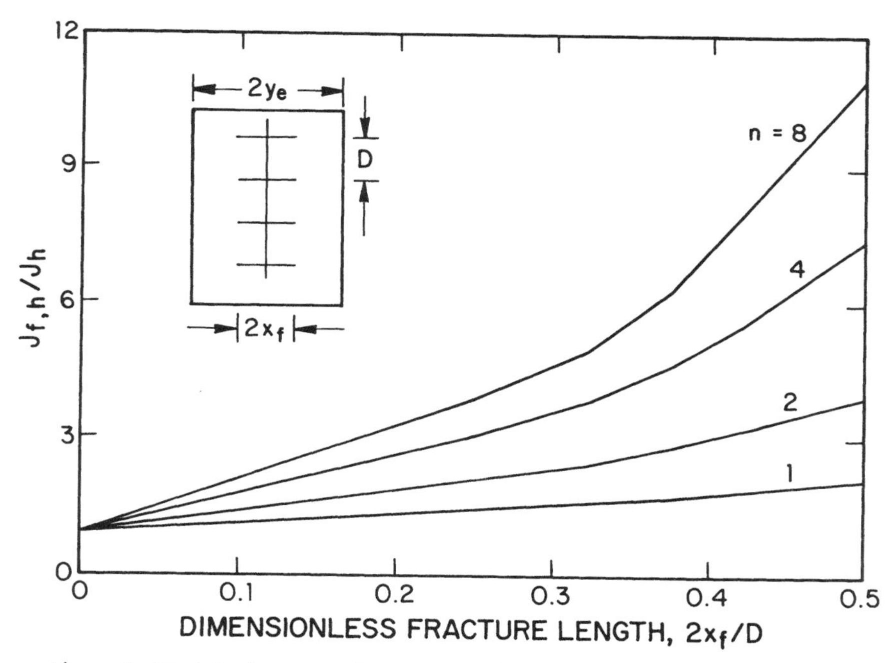

Figure 5–17 A Performance Comparison of Stimulated and Unstimulated Horizontal Wells.[27] (n represents the number of fractures perpendicular to the wellbore)

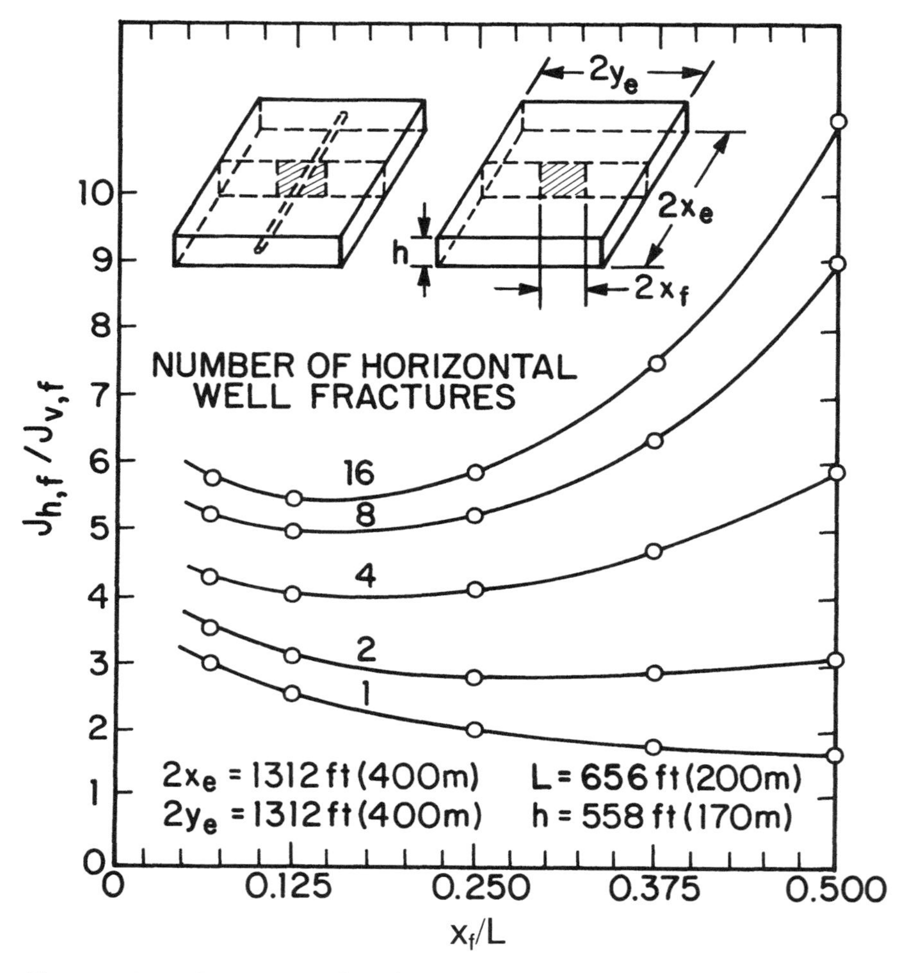

Figure 5–18 A Comparison of Productivities of Fractured Horizontal and Vertical Wells.[26]

$(s_{ch})_c$ = skin effect due to limited fracture/well contact
q = oil flow rate, STB/day
Δp = pressure drop, psi $\quad h$ = reservoir height, ft
k = permeability, md $\quad k_f$ = fracture permeability, md
r_w = wellbore radius, ft $\quad b_f$ = fracture width, ft

The factors which determine the optimum number of fractures along the well length are economics and the time it will take for the fractures to interfere

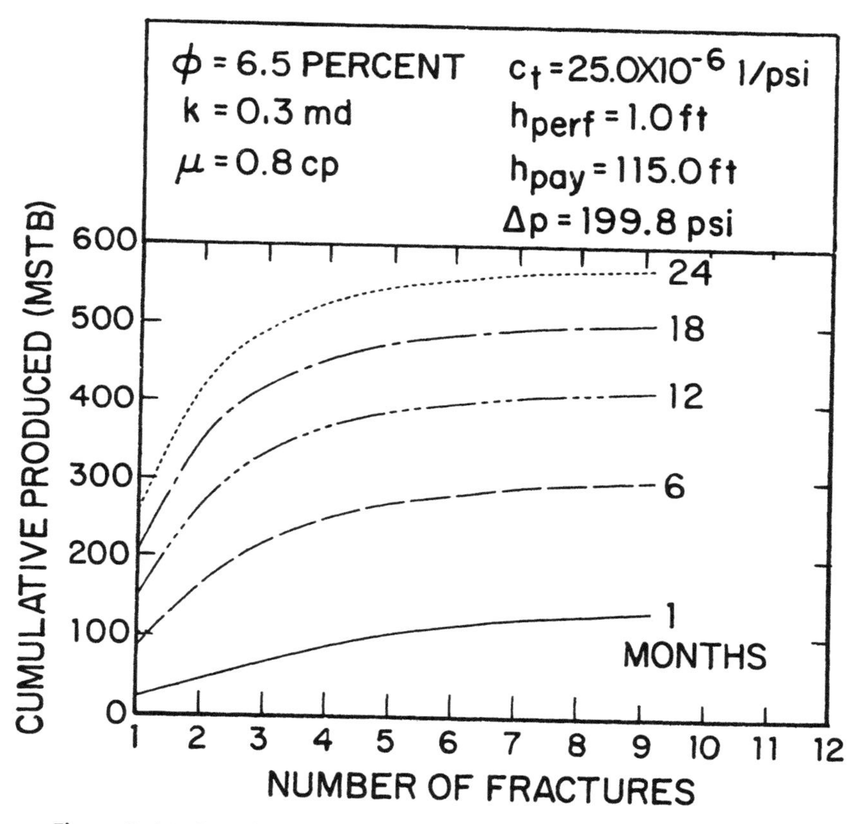

Figure 5–19 Cumulative Production from Fractured Horizontal Wells.[28]

with each other. When one fracture interferes with the other fracture, essentially they will be producing from the same zone, thereby reducing their effectiveness.

One can estimate the number of vertical fractures necessary to get the same productivity from a fractured horizontal well as that expected from an open-hole, uncased horizontal well.[13] Naturally, the purpose of the horizontal well is to increase the reservoir contact area. When the horizontal well is cemented, the reservoir contact enhancement is lost and therefore stimulation along the well length may be necessary to regain the reservoir contact. The number of infinite-conductivity fractures n_f which are required to give the same productivity as an uncased hole are given by the following equations: [13]

$$n = \frac{1 + \sqrt{1 + 4D}}{2} \tag{5-9}$$

$$D = \frac{L\,\pi}{4\,C\,x_f} \tag{5-10}$$

$$C = \ln\left[\frac{a + \sqrt{a^2 - (L/2)^2}}{L/2}\right] + \frac{\beta h}{L}\ln\frac{\beta h}{2r_w} \tag{5-11}$$

$$\beta = \sqrt{k_h/k_v} \tag{5-12}$$

a is calculated using Equation 3–11.

EXAMPLE 5–4

A 1000-ft-long horizontal well is drilled on an 80 acre spacing (r_{eh} = 1053 ft) in a 50-ft-thick low permeability (k ≈ 0.03 md) oil reservoir. It is proposed to induce multiple vertical fractures of 100 ft half-length along the length of the horizontal well. The horizontal well is cased and selectively perforated. Calculate the minimum number of infinite-conductivity vertical fractures needed to match the productivity of an (unfractured) open-hole horizontal well.

Given: $k_v/k_h = 0.1$ $r_w = 0.365$ ft

Solution

$$\beta = \sqrt{k_h/k_v} = \sqrt{1/0.1} = 3.162$$

$$\begin{aligned} a &= 0.5\,L\,[0.5 + \sqrt{0.25 + (2\,r_{eh}/L)^4}]^{0.5} \\ &= 0.5 \times 1000\,[0.5 + \sqrt{0.25 + (2 \times 1053/1000)^4}]^{0.5} \\ &= 1114 \text{ ft.} \end{aligned}$$

From Equation 5–11,

$$\begin{aligned} C &= \ln\left[\frac{a + \sqrt{a^2 - (L/2)^2}}{L/2}\right] + \frac{\beta h}{L}\ln\frac{\beta h}{2r_w} \\ &= \ln\left[\frac{1114 + \sqrt{1114^2 - (1000/2)^2}}{1000/2}\right] \\ &\quad + \left(\frac{3.162 \times 50}{1000}\right)\ln\left(\frac{3.162 \times 50}{2 \times 0.365}\right) \\ &= 1.44 + 0.85 \\ &= 2.29 \end{aligned}$$

From Equation 5–10,

$$D = \frac{L\,\pi}{4\,C\,x_f} = \frac{1000 \times \pi}{4 \times 2.29 \times 100} = 3.43$$

From Equation 5–9,

$$n = (1 + \sqrt{4D})/2 = (1 + \sqrt{1 + 4 \times 3.43})/2 = 2.4 \approx 3$$

Thus, at least three 100-ft-long infinite-conductivity fractures are needed for the productivity of a cemented horizontal well to be the same as that of an open-hole horizontal well. One can design more than two fractures to further improve the productivity. This additional productivity would depend upon the time it would take for fractures to interfere with each other.

FIELD HISTORIES

Recently, several horizontal wells have been fracture-stimulated. A gas well in West Virginia was completed using a cased liner with external casing packers and port holes.[29,30] Several fractures were created along the well length. The details of this well are included in Chapter 9.[30] In the Danish sector of the North Sea, 1000 to 3000 ft long, stimulated horizontal wells were drilled in the Dan field.[31] The low permeability chalk reservoir is underlain by bottom water and in some cases with top gas. In the area where horizontal wells are drilled, the gross reservoir thickness varies from 400 to 700 ft. Generally, conventional wells completed in Maastrichtian chalk are acid fractured only, while wells completed in relatively less permeable Danian formation are hydraulically fractured using proppant.

The first well, MFB–14 i.e., MFB-14B (1097 ft long), was planned to be drilled in Maastrichtian chalk and acid stimulated. However, the reservoir dip was one degree higher than estimated. This resulted in drilling two-thirds of the 1090 ft well length in the upper Danian chalk rather than in the lower, softer Maastrichtian chalk. Initial plans included acid stimulation only. However, since the well was drilled in the tighter chalk, acid stimulation of one zone indicated that the well would not produce without propping the fracture. Therefore, the initial completion plan was changed and the well was stimulated using propped fracture. This result increased the completion time from 28 to 40 days. This well's production stabilized at 1.8 times a conventional well, which is half of what was expected. The log revealed that 60% of the flow came from the first fracture, 20% from the second, and 20% from the remaining three fractures.

Based upon the above experiences of MFB–14, a 2500-ft-long MFB–15 and a 2600-ft-long MFB–13 were drilled in Maastrichtian chalk and were stimulated using acid fractures. The well schematics, production histories, and stimulation tool assemblies are shown in Figures 5–20 through 5–22. It is important to note that out of 41 producing wells in the Dan field, 25% of the production comes from three multifractured horizontal wells.[31] Additionally, the cost of a multifractured horizontal well is U.S. $12 million, while the cost of a conventional well is U.S. $6 million. The average cost

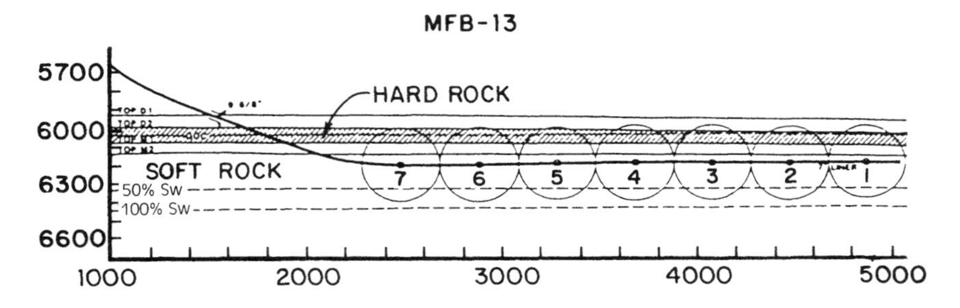

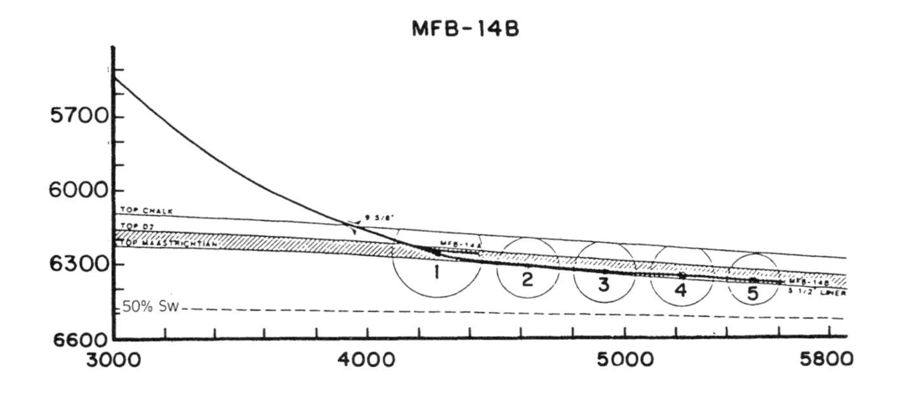

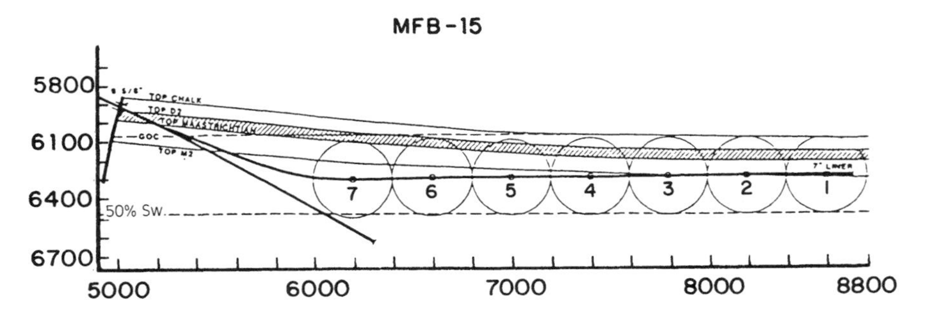

Figure 5–20 Horizontal Well Trajectories in the Dan Field, offshore Denmark.[31] (Units on x axis and y axis are in ft)

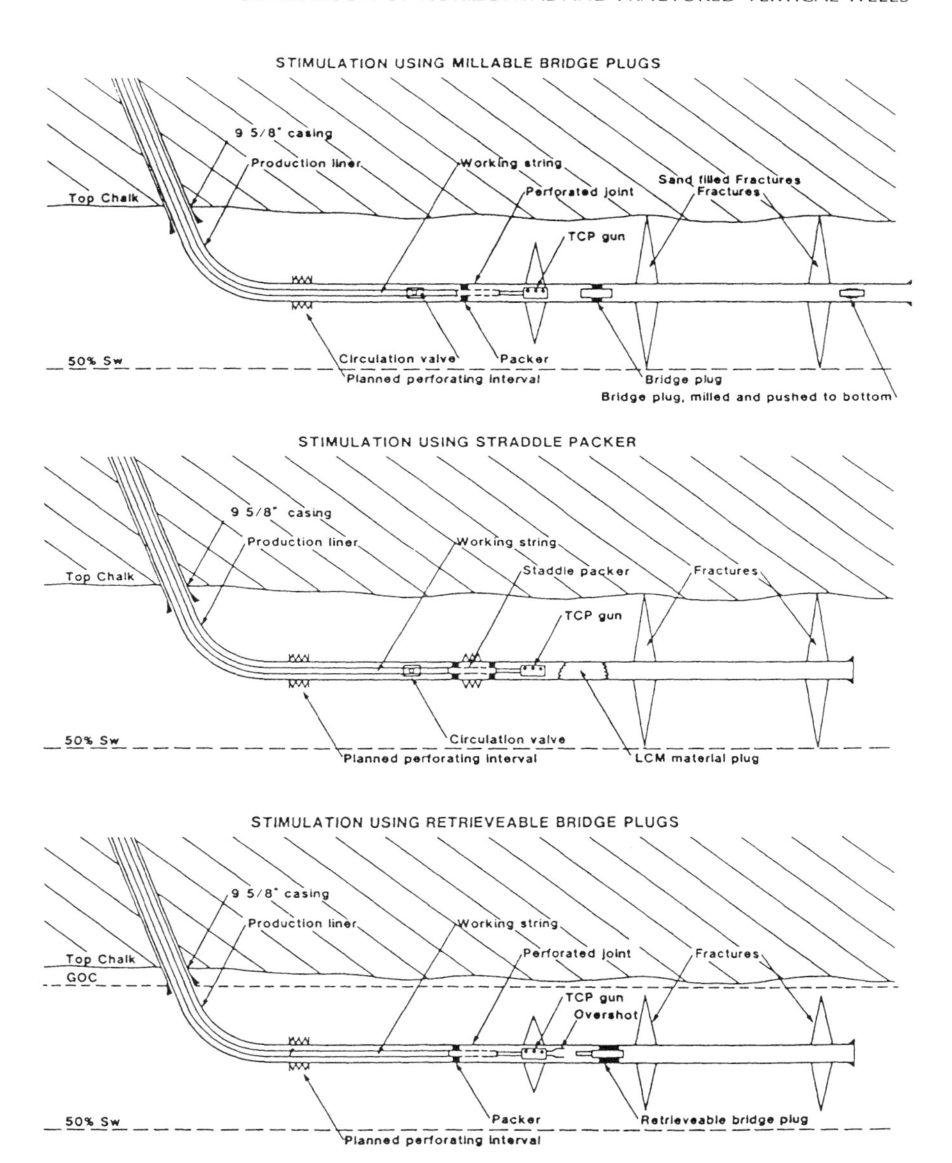

Figure 5–21 A Schematic of Horizontal Well Stimulation Assemblies used in the Dan Field, North Sea.[31]

of a well slot in the offshore Dan field is U.S. \$7.5 million. Thus the total cost of a horizontal well is U.S. \$19.5 million as compared to U.S. \$13.5 million for a conventional well. Therefore, if the horizontal well is producing at a rate 1.4 times greater than a conventional well, horizontal wells are economically justified.[31]

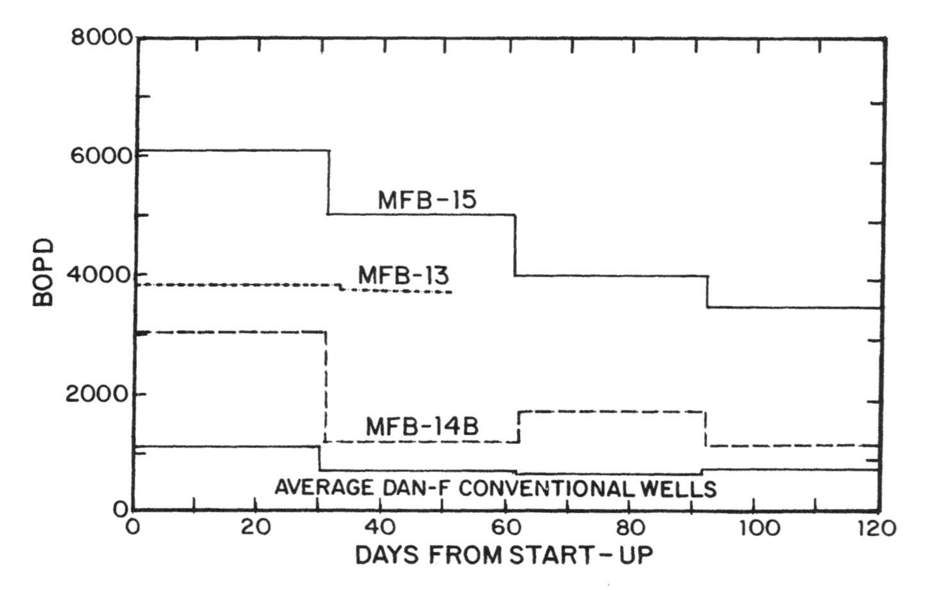

Figure 5–22 Production Rates from Fractured Horizontal Wells in the Dan Field, North Sea.[31]

Figure 5–23 shows schematic diagrams of a 1620-ft-long horizontal well drilled in the lower Spraberry formation in West Texas.[32] The well was completed partially (8320 ft to 9905 ft) by a cemented liner while part of the well (9905 ft to 9948 ft) was open hole. The end part was left open hole because of difficulties in inserting a $5\frac{1}{2}$-in. liner in an $8\frac{1}{2}$-in. horizontal hole. (Two $7\frac{1}{2}$-in., 4-bladed positive centralizers were installed at each joint of a liner portion in the build curve and horizontal section.) Using 20–40 Brady sand and resin-coated sand (in some cases, precured sand) at tail end, five fractures were created. A total of 1,408,000 pounds of sand was pumped into the well. The pressures at the end of the pad in each of the fractures were as follows:

Interval, ft (MD)	Pressure, psi
Open Hole	3167
9563–9567	1506
9063–9067	3296
8694–8698	1559
8320–8324	3140

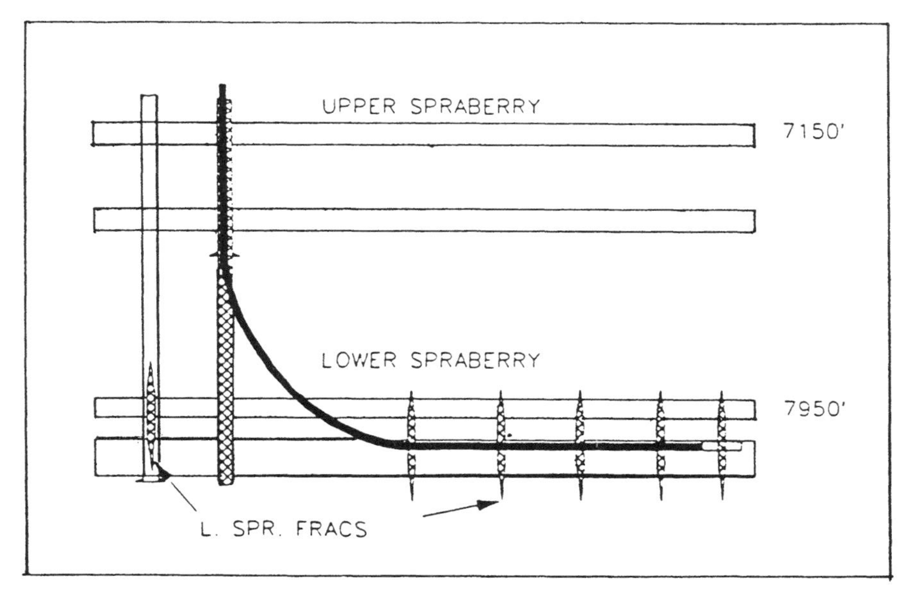

Figure 5–23 A Schematic Diagram of a Fractured Horizontal Well in the Spraberry Formation, Texas.[32]

The above variation in pressures indicates that five individual fractures were created. Additionally, the above data also indicate reservoir heterogeneity along a 1648-ft-long well in a 2.6 md reservoir. The initial well test showed a rate of 69 BOPD, 326 BWPD, and 64 MCFD. The low rates could be attributed to the well intersecting partially depleted zones.

PRACTICAL CONSIDERATIONS

An engineer has to make several practical decisions from the operation standpoint. Once a decision is made to fracture a horizontal well, an appropriate drilling and completion scheme should be chosen. At the present time, there is no effective way to fracture short-radius wells. Due to a sharp turning radius of 20 to 40 ft, cementing a short-radius well is not practical. In the short radius method a large open wellbore causes a significant fluid loss, resulting in an ineffective stimulation treatment. Of course, one can inject fluids above the fracture pressure and stimulate the well. This may or may not work since one would not know where the fluid went or where the fracture, or fractures, were initiated. Thus, it may be difficult to draw definite engineering conclusions and guidelines based upon open-hole stimulation. For a medium-radius well completed with a slotted liner, similar problems exist. From an engineering standpoint, selectively completed horizontal wells (i.e., wells completed with cemented and perforated liners or wells completed with a combination of liners and external casing packers) would be necessary for an effective, localized well stimulation treatment and for reliable interpretations of the test results. (Note that external casing packers may not hold pressure differentials that are required for propped fracture treatments.)

Another important set of decisions an engineer has to make is the number of fractures along the well length and fracture treatment sizes. Additionally, decisions need to be made about propped or unpropped fractures. Equations 5–9 through 5–11 can be used to decide on the number of fractures. It is important to note that these equations are for infinite-conductivity fractures, and cannot be used with very long fracture half-lengths (more than 150 ft). Additionally, it can be expensive to perforate a horizontal well. It is possible to reduce the perforation cost by minimizing the number of perforated intervals along the well length. Furthermore, cost can be reduced by using unpropped fractures. If proppants are used, a large amount of fluid injection may be necessary to pump in a large proppant volume. Once again an engineer has to evaluate local experience about percentage of frac-fluid return. In a horizontal well, the unreturned fluids will tend to separate in a lower fracture portion of a tall fracture. This may result in altering relative permeability for various fluids around the wellbore. Thus, to get a long fracture extension, one needs a large fracture volume. To pump a large fracture volume one needs to use a large amount of frac-fluid, and large frac-fluid pumping may have an adverse effect on well productivity. Therefore, one will have to optimize fracturing design by considering the above discussed issues.

SUMMARY

The technology to fracture a horizontal well is developing rapidly. The production results to date have been mixed. A successful application of this technique has been documented for the Dan Field in the North Sea, while not-so-successful applications are reported in the United States. In this author's view, stimulated horizontal wells are desirable in low-permeability reservoirs where formation damage and reservoir drainage in a reasonable time frame are serious problems.

REFERENCES

1. Penny, G. S.: "An Evaluation of the Effects of Environmental Conditions and Fracturing Fluids Upon the Long-Term Conductivity of Proppants," paper SPE 16900, presented at the SPE 62nd Annual Technical Conference and Exhibition, Dallas, Texas, Sept. 27–30, 1987.

2. Snow, S. E. and Brownlee, M. H.: "Practical and Theoretical Aspects of Well Testing in the Ekofisk Area Chalk Fields," paper SPE 19776, presented at the SPE 64th Annual Technical Conference and Exhibition, San Antonio, Texas, Oct. 8–11, 1989.
3. Agarwal, R. G., Carter, R. D., Pollock, C. B.: "Evaluation and Performance Prediction of Low-Permeability Gas Wells Stimulated by Massive Hydraulic Fracturing," *Journal of Petroleum Technology,* pp. 362–372, March 1979.
4. Cinco-Ley, H. and Samaniego, F.: "Transient Pressure Analysis for Fractured Wells," *Journal of Petroleum Technology,* pp. 1749–1766, September 1981.
5. Prats, M.: "Effect of Vertical Fractures on Reservoir Behavior-Incompressible Fluid Case," *SPE Journal,* pp. 105–118, June 1961.
6. Cinco-Ley, H., Ramey, Jr., H. J., Samaniego, F., Rodriguez, F.: "Behavior of Wells with Low-Conductivity Vertical Fractures," paper SPE 16766, presented at the SPE 62nd Annual Technical Conference and Exhibition, Dallas, Texas, Sept. 27–30, 1987.
7. *Halliburton Services—Proppants,* Halliburton Services, Duncan, Oklahoma.
8. *Proppant Selection Guide,* Dowell Schlumberger, Tulsa, Oklahoma.
9. Economides, M. J. and Nolte, K. G.: *Reservoir Stimulation,* Houston, Texas, 1978.
10. Joshi, S. D.: "Production Forecasting Methods for Horizontal Wells," paper SPE 17580, presented at the SPE International Meeting, Tianjin, China, Nov. 1–4, 1988.
11. Williams, B. B., Gridley, J. L., Schechter, R. S.: *Acidizing Fundamentals,* SPE Monograph No. 6, Dallas, Texas, 1979.
12. Joshi, S. D.: "Augmentation of Well Productivity Using Slant and Horizontal Wells," *Journal of Petroleum Technology,* pp. 729–739, June 1988.
13. Mukherjee, H. and Economides, M. J.: "A Parametric Comparison of Horizontal and Vertical Well Performance," paper SPE 18303, presented at the SPE 63rd Annual Technical Conference and Exhibition, Houston, Texas, Oct. 2–5, 1988.
14. Crooke, Jr., C. E.: "Effect of Fracturing Fluids on Fracture Conductivity," *Journal of Petroleum Technology,* pp. 1273–1282, October 1975.
15. Crooke, Jr., C. E.: "Conductivity of Fracture Proppants in Multiple Layers," *Journal of Petroleum Technology,* pp. 1101–1107, September 1973.
16. Fast, C. R., Holman, G. B., and Covlin, R. J.: "The Application of Massive Hydraulic Fracturing to the Tight Muddy 'J' Formation, Wattenberg Field, Colorado," *Journal of Petroleum Technology,* pp. 10–16, January 1977.
17. Novotny, E. J.: "Proppant Transport," paper SPE 6813, presented at the SPE Annual Meeting, Denver, Colorado, Oct. 9–12, 1977.
18. Sheikholeslami, B. A., Schlottman, B. W., Siedel, F. A., Button, D. M.: "Drilling and Production Aspects of Horizontal Wells in the Austin Chalk," paper SPE 19825, presented at the SPE 64th Annual Technical Conference and Exhibition, San Antonio, Texas, Oct. 8–11, 1989.
19. Moritis, G.: "Worldwide Horizontal Drilling Surges," *Oil & Gas Journal,* pp. 53–63, Feb. 27, 1989. Also see pp. 53–64, Feb 26, 1990.
20. "Horizontal Drilling Grows in Williston," *Oil & Gas Journal,* pp. 22–25, Nov. 6, 1989.
21. McCaslin, J.: "Bakken Drilling to Expand," *Oil & Gas Journal,* p. 129, March 26, 1990.
22. El Rabaa, W.: "Experimental Study of Hydraulic Fracture Geometry Initiated from Horizontal Wells," paper SPE 19720, presented at the SPE 64th Annual Technical Conference and Exhibition, San Antonio, Texas, Oct. 8–11, 1989.

23. Soliman, M., Rose, B., El Rabaa, W., Hunt, J. L.: "Planning Hydraulically Fractured Horizontal Completions," *World Oil,* pp. 54–58, September 1989.
24. Technical Progress Report, "In situ stress and Fracture Permeability: A co-operative DOE-Industry Research Program," *EOR Progress Review,* U.S. DOE, No. 59, June 30, 1989.
25. *Horizontal Wells,* Dowell Schlumberger, Tulsa, Oklahoma.
26. Giger, F. M.: "Horizontal Well Production Techniques in Heterogeneous Reservoirs," paper SPE 13710, presented at the Middle East Oil Technical Conference, Bahrain, March 11–14, 1985.
27. Karcher, B. J., Giger, F. M., and Combe, J.: "Some Practical Formulas to Predict Horizontal Well Behavior," paper SPE 15430, presented at the SPE 61st Annual Technical Conference and Exhibition, New Orleans, Louisiana, Oct. 5–8, 1986.
28. Austin, C. E., Rose, R. E. and Schuh, F. J.: "Simultaneous Multiple Entry Hydraulic Fracture Treatments of Horizontally Drilled Wells," paper SPE 18263, presented at the SPE 63rd Annual Technical Conference and Exhibition, Houston, Texas, Oct. 2–5, 1988.
29. Layne, A. W. and Siriwardane, H. J.: "Insights into Hydraulic Fracturing of a Horizontal Well in a Naturally Fractured Formation," paper SPE 18255, presented at the Annual Conference, Houston, Texas, Oct. 2–5, 1988.
30. Yost, A. B., II, Overbey, W. K., Salamy, S. P., Okoye, C. O., and Saradji, B. S.: "Devonian Shale Horizontal Well: Rationale for Wellsite Selection and Well Design," paper SPE/DOE 16410, presented at the Joint Symposium on Low Permeability Reservoirs, Denver, Colorado, May 18–19, 1987.
31. Anderson, S. A., Hansen, S. A., and Fjeldgaard, K.: "Horizontal Drilling and Completion: Denmark," paper SPE 18349, presented at the European Petroleum Conference, London, U.K., Oct. 16–19, 1988.
32. White, C. W., "Drilling & Completion of a Horizontal Lower Spraberry Well Including Multiple Hydraulic Fracture Treatments," SPE Paper 19721, presented at the 64th SPE Annual Technical Conference and Exhibition, San Antonio, Texas Oct. 8–11, 1989.

CHAPTER 6

Transient Well Testing

INTRODUCTION

Well test analysis of a horizontal well is complex and on many occasions difficult to interpret. Before discussing the analysis procedure, it is appropriate to state the goals of the well test analysis. In general, a well test analysis of a horizontal well is conducted to meet the following objectives:

1. To obtain reservoir properties,
2. To determine whether all the drilled length of a horizontal well is also a producing length, and
3. To estimate mechanical skin factor or drilling and completion related damage to a horizontal well. Based upon magnitude of the damage a decision regarding well stimulation can be made.

A horizontal well test is difficult to analyze because of the following reasons:

1. Most horizontal well mathematical models assume that horizontal wells are perfectly horizontal and are parallel to the top and bottom boundaries of the reservoir. In general, the drilled horizontal wellbores are rarely horizontal but rather snake-like with many variations in the vertical plane along the well length. This vertical variation of a well is dependent upon tolerance used in drilling the well. Some horizontal wells, as typically shown in Figure 6–1, illustrate that a severe "snake" effect can exist along the well length. Because of this, one portion of the well may see the top reservoir boundary earlier than the remaining portion of the well, while a certain portion of the well may see the bottom reservoir boundary earlier than the remaining portion of the well. All of these variations along the well length affect a pressure gauge inserted at the producing end of a horizontal well. However, the influence of these variations along the well length on the pressure response is not known. At present, well tests are carried out either by mounting a pressure gauge at the bottom of the producing end or at the well head. Recently, a few operators have installed permanent gauges at or near the bottom of the producing end.
2. If one wants to know whether the drilled length is also the producing length, the calculation is not straightforward, because horizontal wells exhibit negative skin factors, depending upon their lengths and reservoir properties (see Chapters 3 and 7). In practice, part of this negative skin factor is offset by the positive skin factors due to near wellbore damage and the effects of perforation density and the completion scheme. Except for the early flow period, the analysis of all other periods provides effective

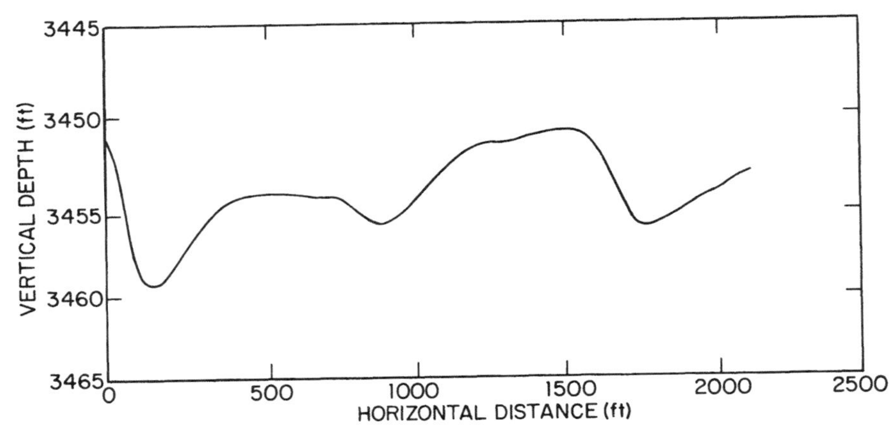

Figure 6–1 A Typical Horizontal Well Profile.

skin factor, which is a summation of the above noted positive and negative skin factors. It is difficult to separate them, unless, of course, one assumes that the producing length is the same as the drilled length. Moreover, if part of the horizontal wellbore intersects an unproductive interval, it will also appear as a positive skin factor. Thus, it may be difficult to distinguish the effect of intersecting nonproductive intervals from mechanical skin damage due to drilling and completion.

The only way to estimate the mechanical skin damage reliably is by analyzing an early radial flow in the vertical plane. Initially a horizontal well acts as though it is a vertical well drilled in a horizontal plane. Once the well is put on production, initially it develops a radial flow around it until the top or the bottom reservoir boundary affects the pressure response of the well. Normally, horizontal wells are drilled in thin reservoirs. Therefore, the initial flow period may last only for a few minutes to hours and it may be difficult to identify in the analysis of the well test data. Hence, it may be impossible to estimate mechanical skin damage. In addition, the wellbore storage effects may overwhelm the initial radial-flow period, especially if the wellbore is damaged. Thus, initial radial-flow period may not be detected resulting in an inability to reliably estimate mechanical skin damage of a horizontal well.

3. In general, if one conducts a test on a horizontal well where the productivity is less than expected or where one is trying to estimate whether a horizontal well is producing at its full potential, the interpretation of test results may be difficult. As noted above, the only way one can reliably estimate the mechanical skin damage is if, and only if, one can identify the initial radial-flow period. As discussed earlier, on many occasions it is difficult to identify the initial radial-flow period. (At least theoretically, it may be possible to estimate mechanical skin factor by analyzing initial linear flow. For example, as seen in Equations 6–11 through 6–14, by assuming drilled length is also a producing length one may be able to estimate mechanical skin damage.)

For a horizontal well showing a high negative skin factor in a well test, let's say on the order of -4 or -5, the stimulation decision is simple and one may decide not to stimulate the horizontal well. Similarly, for a horizontal well which shows skin factors on the order of $+5$, $+10$, or $+15$, the decision is fairly simple and one can decide to stimulate the horizontal well. The problem arises when the skin factors of a horizontal well are on the order of -2 to -1. Here, it tells us that part of the horizontal well is unproductive. The reasons for these unproductive intervals could be several: (1) Many horizontal wells, especially those drilled in naturally fractured zones, do show that due to reservoir heterogeneities, the entire well length may not be productive. (2) Occasionally, horizontal wells have been drilled out of the payzone due to drilling errors. Now the question remains whether the partially unproductive zone is due to mechanical skin damage caused by drilling and completion or whether a portion of a horizontal well intersects an unproductive reservoir formation. The producing intervals can be identified by production logging. The production logging in some instances is not only expensive but may not be useful if the well is completed as an open hole or with a slotted liner. With these completions, it is difficult to estimate the exact production length of a long horizontal well. For example, with a slotted liner the flow in the annulus between the slotted liner and the drilled hole makes it difficult to identify fluid entry into the wellbore. Thus, without proper completion, it will be difficult not only to identify nonproducing intervals along the well length but also to identify portions along the well length which need stimulation.

In this chapter the limitations and use of horizontal-well testing are outlined.

MATHEMATICAL SOLUTIONS AND THEIR PRACTICAL IMPLICATIONS

The available solutions to analyze horizontal-well performance during the transient state can be divided into three broad categories, namely: (1) uniform-flux solutions, (2) infinite-conductivity solutions (uniform-flux solution evaluated at equivalent pressure point), and (3) uniform flux with wellbore pressure averaging solutions.[1–10]

For a fractured vertical well a uniform-flux solution, which assumes uniform fluid entry along the fracture length, was developed by Gringerten, Ramey, and Raghavan.[11] In a vertical well, the fracture extends in two diametrically opposite directions from the well center. Therefore, to have uniform-fluid entry along the fracture length requires the lowest pressure at the well center with increasing pressure along the fracture length, and with the highest pressure at the fracture tips.

For a vertical well with an infinite-conductivity fracture, the wellbore pressure and the pressure along the length of the fracture is the same. Under this condition, maximum fluid enters the fracture near its tips with a minimum fluid entry at the fracture center, i.e., at the drilled vertical well. Reference 11 has shown that, for the uniform-flux fracture, the pressure response is given by

$$p_D = \frac{\sqrt{\pi}}{4} \int_0^{t_D^*} \left[\operatorname{erf} \frac{(1 + x_D)}{2\sqrt{\tau}} + \operatorname{erf} \frac{(1 - x_D)}{2\sqrt{\tau}} \right] \left[\exp\left(\frac{-y_D^2}{4\tau} \right) \right] \frac{d\tau}{\sqrt{\tau}} \qquad (6\text{–}1)$$

where

p_D = dimensionless pressure = $k_h h\,(p_i - p)/(141.2\, q\, B\mu)$ (6–2)

t_D^* = dimensionless time = $2.637 \times 10^{-4}\, k_h t/(\phi\, \mu\, c_t\, x_f^2)$ (6–3)

$y_D = y/x_f$

$x_D = x/x_f$ (6–4)

p_i = initial reservoir pressure, psi

p = present wellbore pressure, psi

q = flow rate, STB/day

B = formation volume factor, RB/STB

μ = oil viscosity, cp

t = time, hours

ϕ = porosity, fraction

c_t = total compressibility, 1/psi

x_f = half fracture length, ft

x = distance measured from the well center along the fracture, ft

y = distance perpendicular to the fracture, ft

The wellbore pressure is obtained by evaluating Equation 6–1 at $x_D = 0$ and $y_D = 0$. Reference 11 has also shown that the wellbore pressure response of infinite-conductivity fracture can be obtained by evaluating Equation 6–1 at $x_D = 0.732$ and $y_D = 0$ where uniform flux and infinite-conductivity solutions give the same pressure response. Thus Equation 6–1 can be used to calculate pressure responses of both uniform flux and infinite-conductivity fractures.

The same concept of uniform flux and infinite-conductivity was extended to horizontal wells.[1–3] References 1 to 3 have assumed that horizontal wells have virtually no pressure drop within the wellbore, resulting in an infinite-conductivity solution. The equation for a horizontal well in an infinite reservoir is (see Fig. 6–2).[1–3]

$$p_D(x_D, y_D, z_D, z_{wD}, L_D, t_D) = \frac{\sqrt{\pi}}{4}\sqrt{\frac{k_h}{k_y}}\int_0^{t_D}\left[\operatorname{erf}\frac{(\sqrt{k_h/k_x} + x_D)}{2\sqrt{\tau}} + \operatorname{erf}\frac{(\sqrt{k_h/k_x} - x_D)}{2\sqrt{\tau}}\right] \times [\exp(-y_D^2/4\tau)] \times \left[1 + 2\sum_{n=1}^{\infty}\exp(-n^2\pi^2 L_D^2\tau)\cos n\pi z_D \cos n\pi z_{wD}\right]\frac{d\tau}{\sqrt{\tau}}. \quad (6\text{–}5)$$

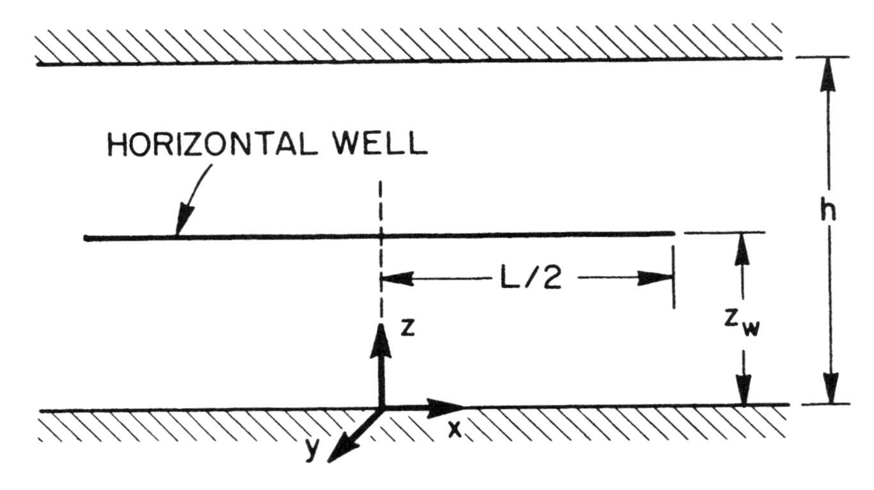

Figure 6–2 A Horizontal Well Model.

where

$$L_D = [L/(2h)]\sqrt{k_v/k_h} \tag{6–6}$$

$$x_D = (2x/L)\sqrt{k_h/k_x}$$

$$y_D = (2y/L)\sqrt{k_h/k_y}$$

$$z_D = z/h$$

$$z_{wD} = z_w/h$$

$$r_{wD} = (2r_w/L)\sqrt{k_h/k_y} \tag{6–7}$$

$$t_D = 0.001055 k_h t/(\phi\ \mu\ c_t\ L^2) \tag{6–8}$$

$$k_h = \sqrt{k_x k_y} \tag{6–9}$$

and

z_w = vertical distance measured from the bottom boundary of the payzone to the well
k_x = permeability in the x direction in the areal plane, md
k_y = permeability in the y direction in the areal plane, md
k_v = permeability in the z direction, md

It is important to note that

1. The above is a line source solution, i.e., a well with only one dimension of length.
2. For long wells, L_D is very large and the summation term in Equation 6–5 approaches zero. Thus, in the limiting case, horizontal well solution,

Equation 6–5, reduces to the fracture solution, Equation 6–1. (In addition to large L_D, one would have to assume an areally isotropic reservoir, i.e., $k_x = k_y$ to obtain the fracture solution given in Equation 6–1.)

3. By substituting $x = \pm L/2$ into Equation 6–5, one can obtain a uniform-flux solution for a horizontal well producing either from the right or left end of the wellbore.
4. By substituting $x_D = 0.732$ into Equation 6–5, one can obtain an infinite-conductivity solution.[1–3]

Reference 7 indicates that x_D value for the infinite-conductivity horizontal-well solution should depend upon L_D, dimensionless well length. For example, Figure 6–3 shows that for $L_D = 5$ the pressure response should be evaluated at $x_D = 0.7$, which is an intersection of uniform flux and infinite-conductivity solutions. Their pressure response results show that by fixing $x_D = 0.732$ instead of variable x_D, slightly different pressure response (maximum deviation of 13%, but generally less than 5%) is obtained in early time. (Note Reference 7 assumes an areally isotropic reservoir, i.e., $k_x = k_y = k_h$).

Babu and Odeh[9] have reported a uniform-flux solution for the pressure response of a horizontal well in a finite reservoir. Kuchuk et al.[5] have reported a solution by averaging pressure along the well length. Additionally, Reference 5 also includes a solution for a reservoir with gas cap by assuming that gas cap can be represented as a constant-pressure boundary.

From the practical standpoint, it is important to note that the majority of the available solutions are for reservoirs with closed top and bottom boundaries. In addition, depending upon type of wellbore assumption, i.e.,

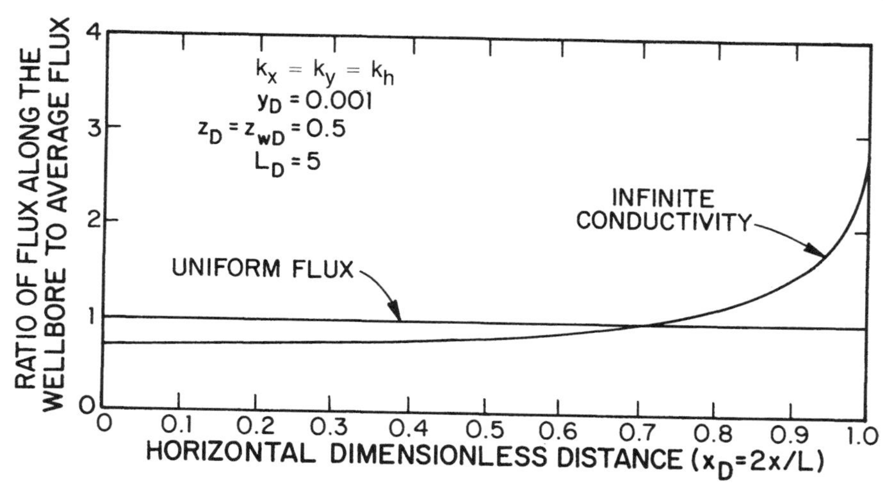

Figure 6–3 A Stabilized Flux Distribution along the Horizontal Well.[7]

uniform flux, infinite conductivity, etc., *one may obtain different time intervals which correspond to the beginning and the end of different flow regimes in a horizontal well.*

GENERALIZED FLOW REGIMES

As shown in Figure 6–4, in general, horizontal wells may exhibit four distinct flow regimes depending upon the well and reservoir geometries: (1) early time radial-flow period in a vertical plane, Figure 6–4(a), (2) early-time linear flow, Figure 6–4(b), (3) late time pseudo-radial flow in a horizontal plane, Figure 6–4(c) and, (4) late-time linear flow, Figure 6–4(d). Initially, as the well is put on production, a radial flow develops in the vertical plane perpendicular to the well. The well acts as though it is a vertical well turned sideways in a laterally infinite reservoir with thickness L. This flow period ends when the effect of the top or bottom boundary is felt or when flow across the well tip affects pressure response. If the well length is sufficiently long compared with reservoir thickness, a linear flow period may exist. In a sufficiently large reservoir, pseudo-radial flow will develop eventually. After pseudo-radial flow it is possible to develop late-time linear flow (see Fig. 6–4(d)).[4,9]

PRESSURE RESPONSE

As noted earlier, Figure 6–3 represents a horizontal well in an infinite reservoir with closed top and bottom boundaries. The horizontal well was assumed to be a line source (only a line, no dimensions) and the wellbore pressure is evaluated at a certain distance away from the well to represent finite wellbore.[1–3] Horizontal-well pressure response is evaluated based on two parameters: dimensionless well length L_D and dimensionless wellbore radius r_{wD}; these two variables are defined in Equations 6–6 and 6–7, respectively. Additionally, k_v is the vertical permeability and k_h is the equivalent horizontal permeability defined by

$$k_h = \sqrt{k_x k_y} \tag{6–9}$$

where k_x and k_y represent permeabilities in the x and y directions, respectively. Note that for an isotropic reservoir, $L_D = 0.5$ represents the solution for a horizontal well whose length is the same as the reservoir thickness. Most drilled horizontal wells have lengths longer than the reservoir thickness. So from a practical standpoint, the solutions for $L_D \geq 1$ are important. As noted

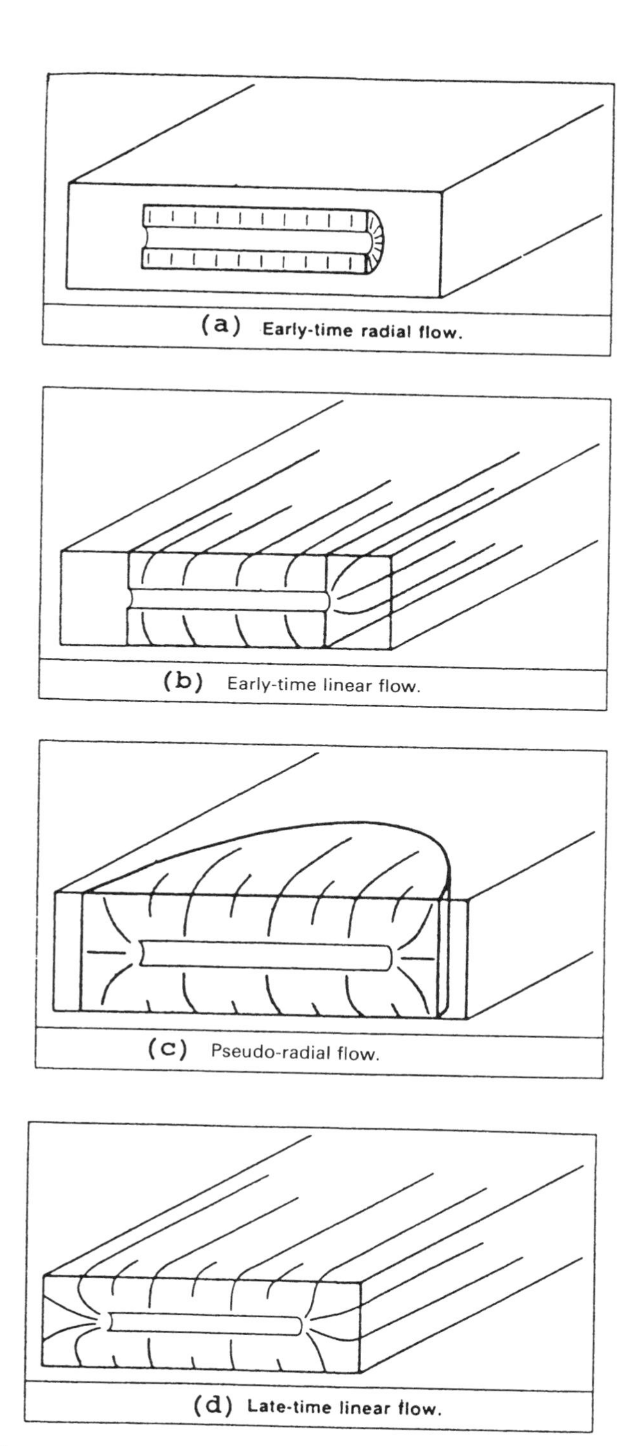

Figure 6–4 Four Possible Flow Regimes in Horizontal Wells.[4]

earlier, the dimensionless pressure and dimensionless time are defined, respectively, as

$$p_D = \frac{k_h h\,(p_i - p)}{141.2 q\, B\, \mu} \tag{6–2}$$

and

$$t_D = \frac{2.637 \times 10^{-4}\, k_h t}{\phi\, \mu\, c_t\, (L/2)^2} = \frac{0.001055\, k_h t}{\phi\, \mu\, c_t\, L^2} \tag{6–8}$$

For an infinite-conductivity wellbore, Figure 6–5 presents logarithmic plots of dimensionless pressure, p_{wD}, versus dimensionless time, t_D, for various values of dimensionless wellbore length, L_D. The solutions shown here are for a well located at the reservoir midheight ($z_{wD} = 0.5$), $r_{wD} = 10^{-4}$, and $k_x = k_y$. The bottom curve represents the pressure response of a vertical well with a fully penetrating, infinite-conductivity fracture. The time between dashed lines *AA* and *BB* represents transitional flow period from early-time radial flow (vertical radial flow) to pseudo-radial flow. (Note that the intermediate linear flow regime is not included in Figure 6–5.) As shown in Figure 6–5, once the pseudo-radial flow starts, the horizontal well solutions for $L_D \geq 10$ are practically the same as the vertically fractured well solution. Thus, during late times (i.e., when the well either reaches pseudo-radial flow

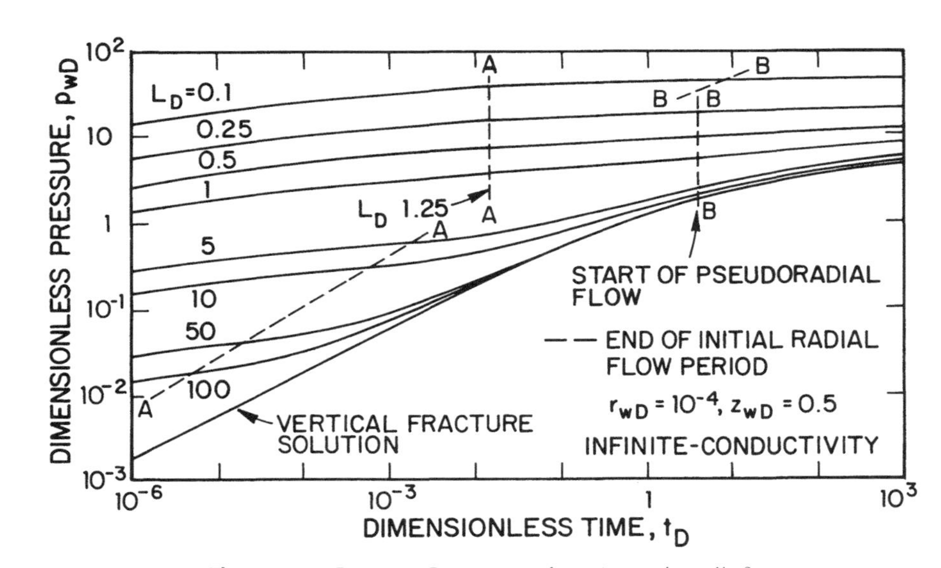

Figure 6–5 Pressure Response of Horizontal Wells.[2]

in an infinite-reservoir or a pseudo-steady state flow in a bounded reservoir), from a single-phase flow standpoint, the performance of horizontal wells is similar to that of a fully penetrating, infinite-conductivity fractured vertical well, where the fracture has the same length as the horizontal well. Figure 6–5 also shows that there is a significant difference between horizontal-well solutions and fractured-well solutions at early times before the onset of pseudo-radial flow. The early period is important in low-permeability reservoirs. As discussed in Chapter 2, in the case of oil wells with reservoir permeability greater than 10 md, especially when permeability is greater than 100 md, the transient flow only lasts for a few hours to a few days. In these cases, it does not matter whether $L_D = 5$, 10, or 50, because pseudo-steady state is reached rapidly and the productivity of a horizontal well will be very close to that of a fully-penetrating infinite-conductivity vertical fracture of the same length. However, in a low-permeability reservoir, the early part may last over a significant portion of the well's economic life. Hence, similar to a fractured vertical well, a horizontal well can also significantly increase productivity in low-permeability reservoirs.

In the theoretical sense, horizontal wells are similar to stimulated vertical wells with fully penetrating, infinite-conductivity fractures. Nevertheless, practically, horizontal wells have a significant advantage because one can drill 1000- to 4000-ft-long wells and in some instances even 5000-ft-long wells, while it is almost impossible to obtain such long infinite-conductivity fractures.

PRACTICAL DIFFICULTIES

Pressure responses shown in Figure 6–5 tell us that horizontal well pressure response depends upon dimensionless well length L_D and dimensionless wellbore radius r_{wD}. As shown in Equations 6–6 and 6–7, both L_D and r_{wD} depend upon well length. Thus, in order to obtain results from a type curve match, one has to assume that drilled wellbore length L is also a producing length. As discussed earlier, due to reservoir heterogeneities, and sometimes due to drilling errors, all the drilled length may not be producing length. Thus, changing L would change L_D and r_{wD}, resulting in generating different type curves for different r_{wD} values to match the pressure response. This difficulty can be overcome by defining dimensionless wellbore radius r_{wD} based upon reservoir height h instead of well length L.[2] This of course would require knowing average reservoir height along the well length.

DETAILED WELL TESTING FLOW REGIMES

The analytical solutions, including Equation 6–5, cannot be used directly for well testing purposes. (Note the horizontal well solution given in

Reference 4 has a different form.) Mathematical approximation techniques have to be used to simplify the analytical solutions such that pressure data can be analyzed in a practical way. Theoretical studies show that there are possibly four transient flow regimes depending on the well length relative to the reservoir thickness and well length relative to the drainage area.[1,3,4,5,9] Under certain circumstances, permeability, permeability anisotropy, and mechanical skin factors can be estimated by analyzing these transient flow pressure data. The flow regime details are outlined below.

EARLY-TIME RADIAL FLOW

The flow is radial in the vertical plane perpendicular to the well during this flow period. This is equivalent to a fully penetrating vertical well in an infinite reservoir of thickness L. The wellbore pressure response during this flow period is given by (or Equation 6.5 reduces to)

$$p_i - p_{wf} = \frac{162.6qB\mu}{\sqrt{k_v k_y}\,L}\left[\log\left(\frac{\sqrt{k_y k_v}\,t}{\phi\mu\,c_t\,r_w^2}\right) - 3.23 + 0.868s\right] \quad (6\text{–}10)$$

where s is the skin factor caused by damage or stimulation. If this skin factor has positive value, it is denoted as s_m, mechanical skin damage due to drilling and completion.

Equation 6–10 indicates that plotting wellbore pressure p_{wf} versus producing time t on semilog coordinates will exhibit a semilog straight line with slope given by

$$m_1 = \frac{162.6qB\mu}{\sqrt{k_v k_y}\,L}. \quad (6\text{–}11)$$

The equivalent permeability in a vertical plane around the wellbore can be calculated as

$$\sqrt{k_v k_y} = \frac{162.6qB\mu}{m_1\,L}. \quad (6\text{–}12)$$

Extrapolating the straight line to $t = 1$ hour, the following equation is obtained

$$p_i - p_{1,hr} = m_1\left[\log\left(\frac{\sqrt{k_v k_y}}{\phi\mu\,c_t\,r_w^2}\right) - 3.23 + 0.868\,s\right] \quad (6\text{–}13)$$

where p_i is initial reservoir pressure, and $p_{1,hr}$ is the pressure obtained by extrapolating the semilog straight line to $t = 1$ hour. Rearranging the above equation gives

$$s = 1.151 \left[\frac{p_i - p_{1,hr}}{m_1} - \log \left(\frac{\sqrt{k_y k_v}}{\phi \mu \, c_t \, r_w^2} \right) + 3.23 \right] \tag{6–14}$$

Equation 6–14 tells us that if early-time pressure response exhibits a semilog straight line, one can estimate skin factor s. In addition, using Equation 6–11 one can estimate parameter $L\sqrt{k_v k_y}$. If the reservoir is areal isotropic ($k_x = k_y = k_h$), then Lk_{eff} can be estimated where k_{eff} ($= \sqrt{k_h k_v}$) is the effective reservoir permeability. Thus, if the effective reservoir permeability is known, one can estimate producing well length L. Vice versa, if producing well length L is known (say by well logging), then one can estimate the effective reservoir permeability. As noted earlier and as shown below, this flow regime can be short and may be difficult to identify in field applications.

TIME TO END THE EARLY RADIAL FLOW

Physically, this flow period ends when the effect of the top or bottom boundary is felt or when the flow across the well tips affect pressure response. Reference 4 provides the following equation to estimate the time required to end the early-time radial flow:

$$t_{e1} = \frac{190 d_z^{2.095} \, r_w^{-0.095} \, \phi \mu \, c_t}{k_v} \tag{6–15}$$

where d_z is the distance of the well to the closest boundary (top or bottom) in ft, r_w is wellbore radius in ft, ϕ is porosity in fraction, μ is viscosity in cp, c_t is total compressibility in 1/psi, k_v is vertical permeability in md, and time t_{e1} is in hours. Odeh and Babu[9] showed that the duration of early time flow period in hours is given by the *minimum* of the following two terms:

$$t_{e1} = \frac{1800 \, d_z^2 \, \phi \mu \, c_t}{k_v} \tag{6–16}$$

or

$$t_{e1} = \frac{125 L^2 \phi \mu \, c_t}{k_y}. \tag{6–17}$$

Note that Equation 6–16 gives the time at which the closest boundary (top or bottom) starts affecting the pressure response while Equation 6–17 represents the time at which fluids start crossing the well tips. Additionally, depending upon dimensionless well length L_D, Figure 6–5 can also be used to estimate a value of time corresponding to the end of early radial flow period. It is noted with dashed line *AA* in Figure 6–5. Note that if the anisotropy ratio, k_h/k_v, is large, this early radial-flow period may not develop.

EXAMPLE 6–1

A 2100-ft-long well is completed in a 100-ft-thick zone with closed top and bottom boundaries. The estimated average horizontal permeability from several vertical-well tests is 0.7 md. The estimated vertical permeability from the core analysis of several vertical wells is 0.2 md. An 8½-in. diameter horizontal well drilling plan called for drilling the well at the central elevation of the reservoir with elevation tolerance ±20 ft. The horizontal wells were drilled using conventional water base mud, which is normally used to drill vertical wells. A reservoir engineer plans to conduct a buildup test with a downhole shut-in device. Estimate the time required to end initial radial flow.

Given: $\phi = 2\%$, $\mu = 0.3$ cp, $c_t = 2 \times 10^{-5}$ 1/psi.

Solution

Wellbore radius, $r_w = 8.5/(2 \times 12) = 0.354$ ft

Method I: Using Equation 6–15

$$t_{e1} = \frac{190 \times (50 - 20)^{2.095} (0.354)^{-0.095} \times 0.02 \times 0.3 \times 2 \times 10^{-5}}{0.2}$$

$$= 0.156 \text{ hrs} \approx 10 \text{ min.}$$

Method II: Minimum value of time duration calculated from Equations 6–16 and 6–17 is used to estimate the duration of the early radial flow.

$$t_{e1} = \frac{1800 \times (50 - 20)^2 \times 0.02 \times 0.3 \times 2 \times 10^{-5}}{0.2}$$

$$= 0.972 \text{ hrs} \approx 1 \text{ hr.}$$

$$t_{e1} = \frac{125 \times 2100^2 \times 0.02 \times 0.3 \times 2 \times 10^{-5}}{0.7}$$

$$= 94.5 \text{ hrs.}$$

The minimum of these two values is 1 hr. Thus, initial radial flow period will end in one hour.

Method III: For this well, dimensionless length L_D and dimensionless wellbore radius r_{wD} can be obtained using Equations 6–6 and 6–7, respectively.

$$L_D = [L/(2h)]\sqrt{k_v/k_h} \tag{6–6}$$

$$= \frac{2100}{2 \times 100}\sqrt{\frac{0.2}{0.7}} = 5.6$$

If we assume an areally isotropic reservoir, i.e., $k_x = k_y$, Equation 6–7 reduces to

$$r_{wD} = r_w/(L/2)$$
$$= 0.354/(2100/2) = 3.4 \times 10^{-4}$$

Note Figure 6–5 is for $r_{wD} = 10^{-4}$. Assuming it can be used for $r_{wD} = 3.4 \times 10^{-4}$, then, for $L_D = 5.6$, the time to end initial radial flow period is given by the dotted line *A-A* as $t_D = 1.5 \times 10^{-3}$. Substituting this value of t_D in Equation 6–8

$$t_D = 1.5 \times 10^{-3} = \frac{0.001055\, k_h t_{e1}}{\phi\, \mu\, c_t\, L^2}$$

and

$$t_{e1} = \frac{1.5 \times 10^{-3} \times 0.02 \times 0.3 \times 2 \times 10^{-5} \times 2100^2}{0.001055 \times 0.7}$$
$$= 1.07 \text{ hrs.}$$

Thus, depending upon the method used, the initial flow period would last between 10 minutes to 1 hour. If the wellbore is highly damaged, then the wellbore storage effect may wipe out this initial radial flow and make it difficult to estimate skin damage. In a damaged well, if one installs a surface readout device, it will be impossible to see initial radial flow due to a large wellbore volume in the horizontal, curved, and vertical sections of the wellbore. The reservoir engineers will have to use downhole shut-in and pressure measurement devices to enhance the chances of measuring this early radial-flow regime.

EXAMPLE 6–2

A 1900-ft-long horizontal well is drilled in an offshore reservoir. The current reservoir pressure is about 3400 psi. Well tests with the surrounding vertical wells show an average $k_h h$ of 4290 md-ft. The log data indicate reservoir height to be 110 ft, giving effective horizontal permeability of 39 md. Other given parameters are:

$$\mu = 0.8 \text{ cp}, \qquad c_t = 6.1 \times 10^{-6} \text{ 1/psi},$$
$$r_w = 0.35 \text{ ft}, \qquad \phi = 16\%.$$

The vertical-well logs show shale barriers, however, there is great uncertainty about effective vertical permeability. It is contemplated to run a well test on the horizontal well. Estimate the time required for a buildup test to reach the end of early radial flow.

Solution

In order to calculate the time corresponding to the end of the early radial flow period, one needs to have an estimate of vertical permeability which is not available in this case. Let us assume $k_v = 0.1 k_h = 3.9$ md, and the well is at the central elevation of the reservoir with elevation tolerance ± 20 ft.

Method I: Equation 6–15 gives

$$t_{e1} = \frac{190 \times (55 - 20)^{2.095} \times 0.35^{-0.095} \times 0.16 \times 0.8 \times 6.1 \times 10^{-6}}{3.9}$$
$$= 0.072 \text{ hrs} = 4.3 \text{ minutes}$$

Method II: Equations 6–16 and 6–17 give

$$t_{e1} = \frac{1800 d_z^2 \, \phi \, \mu \, c_t}{k_v} \tag{6–16}$$
$$= \frac{1800 \times (55 - 20)^2 \times 0.16 \times 0.8 \times 6.1 \times 10^{-6}}{3.9}$$
$$= 0.44 \text{ hrs.} \approx 26 \text{ minutes}$$

and

$$t_{e1} = \frac{125 \, L^2 \, \phi \, \mu \, c_t}{k_y} \tag{6–17}$$
$$= \frac{125 \times 1900^2 \times 0.16 \times 0.8 \times 6.1 \times 10^{-6}}{39}$$
$$= 9 \text{ hrs.} = 540 \text{ minutes}$$

The minimum of t_{e1} values obtained from Equations 6–16 and 6–17 is 26 minutes, which denotes the duration of the early radial flow period.

Method III:

$$L_D = \frac{L}{2h}\sqrt{k_v/k_h}$$
$$= \frac{1900}{2 \times 110}\sqrt{\frac{3.9}{39}} = 2.73 \qquad (6\text{–}6)$$

and assuming areally isotropic reservoir, i.e., $k_x = k_y$, Equation 6–7 reduces to

$$r_{wD} = r_w/(L/2)$$
$$= 0.35/(1900/2) = 3.68 \times 10^{-4}$$

Assume Figure 6–5 can be used for $r_{wD} = 3.68 \times 10^{-4}$, then $t_D = 5 \times 10^{-3}$ and

$$t_{e1} = \frac{5 \times 10^{-3} \times 0.16 \times 0.8 \times 6.1 \times 10^{-6} \times 1900^2}{0.001055 \times 39}$$
$$= 0.34 \text{ hrs.} \approx 21 \text{ minutes}$$

Thus, if $k_v = 3.9$ md, then early-time flow may last only for 4 to 26 minutes. Changing vertical permeability would also change the duration of the early radial flow.

Note that Equation 6–15 generally gives a lower value to end the early-time radial flow than the other two methods.

EARLY-TIME LINEAR FLOW

If the horizontal well is long enough compared to the formation thickness, a period of linear flow may develop once the pressure transient reaches the upper and lower boundaries. Pressure response during this flow period is given by

$$p_i - p_{wf} = \frac{8.128qB\mu}{Lh}\sqrt{\frac{t}{\phi\mu\, c_t\, k_y}} + \frac{141.2qB\mu}{L\sqrt{k_y k_v}}(s_z + s) \qquad (6\text{–}18)$$

where s_z is the pseudo-skin factor caused by partial penetration in the vertical direction, and is given by[9]

$$s_z = \ln\left(\frac{h}{r_w}\right) + 0.25\ln\left(\frac{k_y}{k_v}\right) - \ln\left(\sin\frac{180°z_w}{h}\right) - 1.838 \qquad (6\text{–}19)$$

(Reference 4 gives a different expression for pseudo-skin factor s_z.) Calculation of s_z from Equation 6–19 requires knowledge of permeability ratio k_y/k_v and the vertical location of the well z_w.

Equation 6–18 indicates that plotting $\Delta p = p_i - p_{wf}$ versus $\sqrt{t}$ will exhibit a straight line with slope given by

$$m_2 = \frac{8.128qB}{Lh}\sqrt{\frac{\mu}{\phi\, c_t\, k_y}}. \quad (6\text{–}20)$$

A product of producing well length square L^2 and permeability k_y can be obtained from the slope

$$L^2\, k_y = \left(\frac{8.128qB}{h\, m_2}\right)^2 \frac{\mu}{\phi\, c_t}. \quad (6\text{–}21)$$

If one assumes that drilled length is also a producing length, permeability k_y can be calculated. Additionally, extrapolating the straight line to $\sqrt{t} = 0$ gives

$$\Delta p\Big|_{t=0} = \frac{141.2qB\mu}{L\sqrt{k_y k_v}}(s_z + s). \quad (6\text{–}22)$$

If pseudo-skin factor s_z (Eq. 6–19) is known, then mechanical skin factor s can be estimated, assuming that the drilled length L is also a producing length.

As mentioned earlier, this flow regime exists only if the well is long enough compared with formation thickness. This flow period approximately ends at[4]

$$t_{e2} = \frac{20.8\, \phi\mu\, c_t\, L^2}{k_x}. \quad (6\text{–}23)$$

If the time given by Equation 6–23 is smaller than that given by Equation 6–15, the well is not sufficiently long compared with formation thickness and linear flow does not develop.[4] Reference 9 also gives criteria for the start and the end of linear flow as

$$t_{s2} = \frac{1800D_z^2\, \phi\mu\, c_t}{k_v} \quad (6\text{–}24)$$

and

$$t_{e2} = \frac{160\ \phi\ \mu\ c_t\ L^2}{k_x}, \tag{6–25}$$

where D_z $(=h - d_z)$ is the maximum distance between the well and the z-boundaries, i.e., the top or the bottom boundary. If $t_{s2} \geq t_{e2}$, this flow period will not occur. In many well tests, this flow period may be very short, making its identification difficult. Note that t_{e2} from Equation 6–25 is 7.7 times greater than that from Equation 6–23.

EXAMPLE 6–3

Calculate the time to start and the time to end early-time linear flow for the well described in Example 6–1, assuming $k_x = k_y = 0.7$ md.

Solution

Since the drilling plan has ±20 ft tolerance, the maximum distance of a well from either the top or the bottom boundary is 70 ft. Substituting this value into Equation 6–24, time to start the linear flow is

$$t_{s2} = \frac{1800\ D_z^2\ \phi\ \mu\ c_t}{k_v} \tag{6–24}$$

$$= \frac{1800 \times 70^2 \times 0.02 \times 0.3 \times 2 \times 10^{-5}}{0.2}$$

$$= 5.3 \text{ hrs.}$$

End of the linear-flow period is

$$t_{e2} = \frac{160\ \phi\ \mu\ c_t\ L^2}{k_x} \tag{6–25}$$

$$t_{e2} = \frac{160 \times 0.02 \times 0.3 \times 2 \times 10^{-5} \times 2100^2}{0.7}$$

$$= 121 \text{ hrs.}$$

Similarly, using Equation 6–23, a different estimate of time to end linear flow is

$$t_{e2} = \frac{20.8 \times 0.02 \times 0.3 \times 2 \times 10^{-5} \times 2100^2}{0.7}$$

$$= 15.7 \text{ hrs.}$$

Thus, this flow period will end in about 15 to 120 hours. This indicates that the current well is sufficiently long compared to the reservoir height. Therefore, it is possible to analyze pressure data of this flow period. Another important point to note is that the time to end early-time linear-flow period is proportional to the well length squared. In the above calculation, drilled length is assumed to be the producing length. However, if part of the well is damaged and unproductive or if part of the well is drilled in an unproductive zone, the effective producing length will be smaller than 2100 ft. For example, if 50% of the drilled length is unproductive, then the producing length is 1050 ft and the early linear-flow period would end between 14 and 30 hours. Thus, plotting test data on Δp against $\sqrt{t}$ plot, one may be able to approximately calculate the producing length and compare it with drilled length. This can be accomplished by comparing the time to end early linear-flow period from the theoretical calculations and field data.

EXAMPLE 6–4

Calculate the times corresponding to the beginning and the end of early-time linear-flow period for the well considered in Example 6–2.

Solution

Since the drilled hole has tolerance of ±20 ft, the maximum distance between the well and the top and bottom boundaries is 75 ft. Using Equation 6–24, the time to start the linear flow period is

$$t_{s2} = \frac{1800\, D_z^2\, \phi\, \mu\, c_t}{k_v} \tag{6–24}$$

$$= \frac{1800 \times 75^2 \times 0.16 \times 0.8 \times 6.1 \times 10^{-6}}{3.9}$$

$$= 2.03 \text{ hrs.}$$

and Equation 6–25 gives the time to end the linear flow period as

$$t_{e2} = \frac{160\, \phi\, \mu\, c_t\, L^2}{k_x} \tag{6–25}$$

$$= \frac{160 \times 0.16 \times 0.8 \times 6.1 \times 10^{-6} \times 1900^2}{39}$$

$$= 11.6 \text{ hrs.}$$

Equation 6–23 gives $t_{e2} = 1.51$ hrs.

PSEUDO-RADIAL FLOW PERIOD

If the well length is sufficiently short as compared to the reservoir size, pseudo-radial flow will develop at late times. This flow period begins at[4]

$$t_{s3} = \frac{1230 L^2\, \phi \mu\, c_t}{k_x} \tag{6–26}$$

Reference 9 suggested the following equation to predict the beginning of pseudo-radial flow

$$t_{s3} = \frac{1480 L^2 \phi \mu c_t}{k_x}. \tag{6–27}$$

This flow period ends when the pressure transient reaches one of the outer boundaries. Figure 6–5 can also be used to estimate the beginning of pseudo-radial flow.[2] (Note that the difference in the coefficients of Equations 6–26 and 6–27 is caused by the fact that different wellbore boundary conditions are used. Additionally, different mathematical convergence criterion might have been used to derive these two different equations.)

Pressure response during pseudo-radial flow period is given by[4]

$$p_i - p_{wf} = \frac{162.6 q B \mu}{\sqrt{k_x k_y}\, h}\left[\log\left(\frac{k_x t}{\phi \mu c_t L^2}\right) - 2.023\right] + \frac{141.2 q B \mu}{L\sqrt{k_y k_v}}(s_z + s) \tag{6–28}$$

Reference 9 presented an equation similar to Equation 6–28 with 2.023 replaced by 1.76. Equation 6–28 indicates that plotting p_{wf} versus t on semi-log coordinates will exhibit a semilog straight line of slope m_3, where

$$m_3 = \frac{162.6 q B \mu}{\sqrt{k_x k_y}\, h}. \tag{6–29}$$

The equivalent horizontal permeability $\sqrt{k_x k_y}$ can be obtained as

$$\sqrt{k_x k_y} = \frac{162.6 q B \mu}{m_3 h}. \tag{6–30}$$

Skin factor can also be obtained by

$$s = \frac{1.151 L}{h}\sqrt{\frac{k_v}{k_x}}\left[\frac{p_i - p_{1hr}}{m_3} - \log\left(\frac{k_x}{\phi \mu c_t L^2}\right) + 2.023\right] - s_z \tag{6–31}$$

where p_{1hr} is obtained by extrapolating the pseudo-radial flow semilog straight line to $t = 1$ hour. Pseudo-skin factor s_z is given by Equation 6–19.

If the top or bottom boundary is maintained at constant pressure (with, say, gas cap) then the pseudo-radial flow period will not develop.[5] Instead, one will have steady-state flow at the late time.

EXAMPLE 6–5

For reservoir and well data given in Example 6–1, calculate the time required to start a pseudo-radial flow.

Solution

Method I: From Equation 6–26

$$\begin{aligned} t_{s3} &= 1230\ \phi\mu c_t L^2/k_x \quad (6\text{–}26) \\ &= 1230 \times 0.02 \times 0.3 \times 2 \times 10^{-5} \times 2100^2/0.7 \\ &= 929.9 \text{ hrs} = 38.7 \text{ days.} \end{aligned}$$

Method II: From Equation 6–27

$$\begin{aligned} t_{s3} &= 1480\ \phi\mu c_t L^2/k_x \quad (6\text{–}27) \\ &= 1480 \times 0.02 \times 0.3 \times 2 \times 10^{-5} \times 2100^2/0.7 \\ &= 1118.9 \text{ hrs} = 46.6 \text{ days.} \end{aligned}$$

Method III: From Figure 6–5, time to start pseudo-radial flow (dashed line *B–B*) is $t_D = 3$, thus

$$\begin{aligned} t_{s3} &= \frac{\phi\mu c_t L^2 t_D}{0.001055 k_h} \\ &= \frac{0.02 \times 0.3 \times 2 \times 10^{-5} \times 2100^2 \times 3}{0.001055 \times 0.7} \\ &= 2150 \text{ hrs} = 89.6 \text{ days} \approx 90 \text{ days} \end{aligned}$$

Thus it will take about 39 to 90 days to reach pseudo-radial flow. It may be economically difficult to shut in a well for such a long time. In this case one will have to obtain the necessary information from an early radial or linear-flow period.

EXAMPLE 6–6

Calculate the time corresponding to the start of pseudo-radial flow for the data given in Example 6–2.

Solution

Method I: One can use Equation 6–26 to calculate the start of the pseudo-radial flow.

$$\begin{aligned} t_{s3} &= 1230\ \phi\mu c_t L^2/k_x \quad (6\text{–}26) \\ &= 1230 \times 0.16 \times 0.8 \times 6.1 \times 10^{-6} \times 1900^2/39 \\ &= 88.9 \text{ hrs} = 3.7 \text{ days} \end{aligned}$$

Method II: Equation 6–27 can also be used to calculate the start of the pseudo-radial flow.

$$\begin{aligned} t_{s3} &= 1480\, \phi\mu c_t L^2/k_x \\ &= 1480 \times 0.16 \times 0.8 \times 6.1 \times 10^{-6} \times 1900^2/39 \\ &= 107 \text{ hrs} = 4.46 \approx 5 \text{ days} \end{aligned}$$

Method III: In Figure 6–5, dashed line B–B at $t_D = 3$ indicates the beginning of pseudo-radial flow.

$$\begin{aligned} t_{s3} &= \frac{\phi\mu c_t L^2 t_D}{0.001055k} \\ &= \frac{0.16 \times 0.8 \times 6.1 \times 10^{-6} \times 1900^2 \times 3}{0.001055 \times 39} \\ &= 205.5 \text{ hrs} = 8.6 \text{ days} \end{aligned}$$

The well needs to be shut in for four to nine days to reach the pseudo-radial flow period.

LATE-TIME LINEAR FLOW

For reservoirs of finite width, there may exist a second linear-flow period. This flow period occurs when the pressure transient reaches the lateral extremities and the flow in this direction has become pseudo-steady state. Pressure response in this flow period is given by[4]

$$p_i - p_{wf} = \frac{8.128qB}{2x_e h}\sqrt{\frac{\mu t}{k_y \phi c_t}} + \frac{141.2qB\mu}{L\sqrt{k_y k_v}}(s_x + s_z + s) \qquad (6\text{–}32)$$

where $2x_e$ is the width of reservoir, s_z is the pseudo-skin factor due to partial penetration in a vertical direction (Equation 6–19), and s_x is the pseudo-skin factor due to partial penetration in the x direction. Reference 4 gives an expression for s_x.

PRESSURE DERIVATIVES

Well test analysis based on pressure derivatives can be used to improve the chances of obtaining a unique answer. The normalized pressure derivative is defined as

$$p'_{wD} = \frac{dp_{wD}}{d\ln t_D} = t_D\frac{dp_{wD}}{dt_D} = t\frac{dp_{wD}}{dt} \qquad (6\text{–}33)$$

In Equation 6–33 the first equality defines the logarithmic derivative of p_{wD}, the second equality follows from the chain rule, and the last equality follows by cancelling out the terms used in defining dimensionless time. During the pseudo-radial flow period in an infinite-reservoir, the pressure derivative is equal to ½, i.e.,

$$p'_{wD} = \frac{1}{2} \tag{6–34}$$

and thus,

$$\frac{p_{wD}}{2p'_{wD}} = p_{wD}. \tag{6–35}$$

We also have

$$\frac{p_{wD}}{2p'_{wD}} = \frac{\Delta p}{2\Delta p'} \tag{6–36}$$

where $\Delta p = p_i - p_{wf}$. Equation 6–36 indicates that the vertical scale on a plot of $\Delta p/(2\Delta p')$ versus t is identical to the vertical scale on a plot of $p_{wD}/2p'_{wD}$ versus t_D and thus, when using type curves based on the derivative group on the left side of Equation 6–36, type-curve matching of the derivative data will require only moving field data in the horizontal direction. This helps to obtain a unique match with type curves.[2,12,13] (References 12 and 13 used this concept to develop a new set of wellbore storage and skin problem type curves for vertical wells.) Figure 6–6 presents the type curves for infinite-conductivity horizontal wells.[2,14] The dashed curves represent the pressure responses and the solid curves represent the pressure derivative group (left side of Equation 6–35). At early times, the pressure/pressure derivative solutions show that the influence of dimensionless well length L_D is negligible. As time increases, these solutions diverge and ultimately merge with the appropriate p_{wD} curves after the onset of pseudo-radial flow. Thus, Figure 6–6 can also be used to identify the appropriate semilog straight lines.

Pressure derivative can also be used to determine the time at which pseudo-steady state flow begins for a bounded reservoir. Pressure response during pseudo-steady state flow is given by

$$p_{wD} = 2\pi t_{DA} + ln\left(\frac{4A}{e^{\gamma}C_A r_w^2}\right) + s \tag{6–37}$$

where s represents total skin factor.

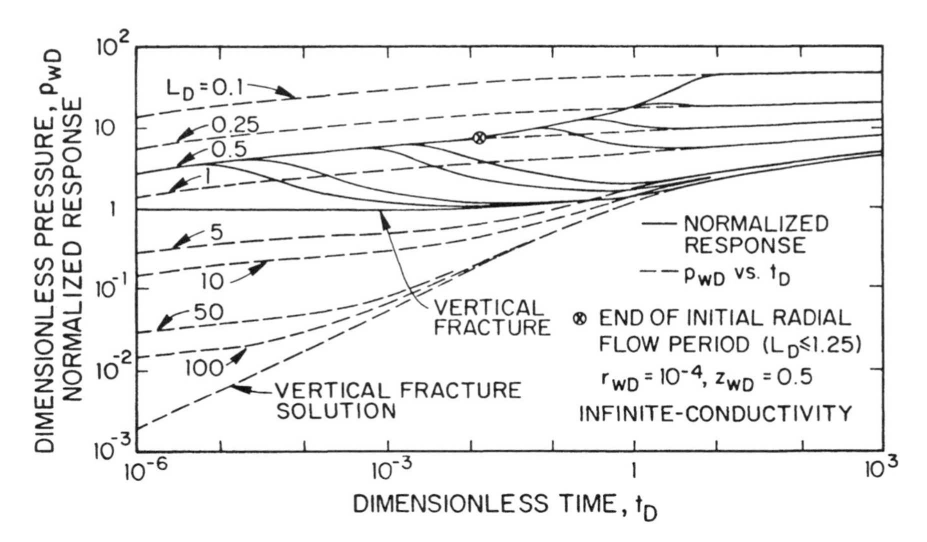

Figure 6–6 Horizontal Well Pressure and Normalized Pressure Derivative.[2]

Differentiating Equation 6–37 with respect to t_{DA} gives

$$\frac{dp_{wD}}{dt_{DA}} = 2\pi \qquad (6\text{–}38)$$

Equation 6–38 indicates that once the slope of the Cartesian plot of the p_{wD} versus t_{DA} becomes 2π, pseudo-steady state flow begins. Many investigators use Equation 6–38 to determine the beginning of pseudo-steady state flow. As noted in Chapters 2 and 7, different investigators use different tolerance limits to evaluate whether the value is close enough to 2π and then determine the beginning of pseudo-steady state. The different tolerance limits would give different answers to estimate the time to start pseudo-steady state flow.

WELLBORE STORAGE EFFECTS

So far we have only considered the cases where the wellbore storage effects are negligible. Unfortunately, reservoir fluids are compressible and the wellbore has a finite volume. Thus, during the production phase the wellbore unloads fluids and during the shut-in phase, it stores fluid. This phenomenon is called wellbore storage. Wellbore storage can occur in a variety of ways, due to compression, change in liquid levels, fluid cooling or heating, or phase segregation. Initially, upon starting a test, all production is from the wellbore, and thus pressure response is completely controlled by storage effects. Data obtained during this flow period cannot be interpreted to calculate formation flow capacity or skin factor. As time increases, the sandface rate increases and the wellbore unloading rate decreases. This corresponds to a transition from wellbore storage dominated flow to radial flow in the reservoir. Eventually, the wellbore unloading rate goes to zero and the sandface rate becomes equal to the surface rate.

Let C represent volume of fluid (bbls) unloaded per unit change in pressure within the wellbore, then

$$q_{wb}B = -C dp_{wf}/dt \tag{6–39}$$

where C is in bbl/psi and t is in days. If unloading occurs due to compression then $C = V_w c_w$, where V_w is the wellbore volume and c_w is the mean compressibility of the fluid in the wellbore. If unloading is due to changing liquid level then $C = V_u/\gamma$, where V_u is the wellbore volume per unit length in RB/ft and γ is average specific gravity of wellbore fluids.

The sandface rate is given by:

$$q_{sf} = \frac{kh}{141.2B\mu}\left[r\frac{dp}{dr}\right]_{r = r_w} \tag{6–40}$$

The surface rate q is

$$q = q_{sf} - (24C/B)(dp_{wf}/dt) \tag{6–41}$$

where t is in hours. Using dimensionless variables, Equation 6–41 can be written as

$$q_{sf} = q\left[1 - C_D\frac{dp_{wD}}{dt_D}\right] \tag{6–42}$$

where

$$C_D = \frac{5.615C}{2\pi\phi h c_t r_w^2} \tag{6–43}$$

The general range of C_D for vertical wells: $10^2 \leq C_D \leq 10^5$

gas well, C_D: 10^2 to 10^3
liquid compression, C_D: 5×10^3 to 5×10^4
liquid level change, C_D: 10^4 to 10^5

For horizontal wells, the wellbore storage dominated flow period could be long, especially if the wellbore is severely damaged. The wellbore storage dominated flow can be even longer for severely damaged wells with surface shut-in devices.

At the present time, for horizontal wells, methods are not available to estimate the time required to end the wellbore storage flow regime precisely. This information is required to evaluate the feasibility of conducting a test. However, for infinite-conductivity fractures, dimensionless wellbore storage has been defined as[15]

$$C_{Df} = 5.615C/(2\pi\phi hc_t x_f^2) \tag{6–44}$$

where x_f is the fracture half-length. This concept for fractured vertical wells was extended to horizontal wells as[3]

$$C_{Dh} = 5.615C/(2\pi\phi hc_t(L/2)^2) \tag{6–45}$$

For horizontal wells, the value of C in Equation 6–45 may be large, resulting in a large wellbore storage dominated flow. In this author's experience and also as noted in Reference 3 the wellbore storage may last beyond the early radial flow period. Until more precise techniques are available for horizontal wells, one can use wellbore storage type curves for infinite-conductivity, fractured vertical wells to estimate the end of wellbore storage and to roughly analyze the pressure data. Such curves are shown in Figure 6–7, where t_D^* is defined by Eq. (6–3).[15]

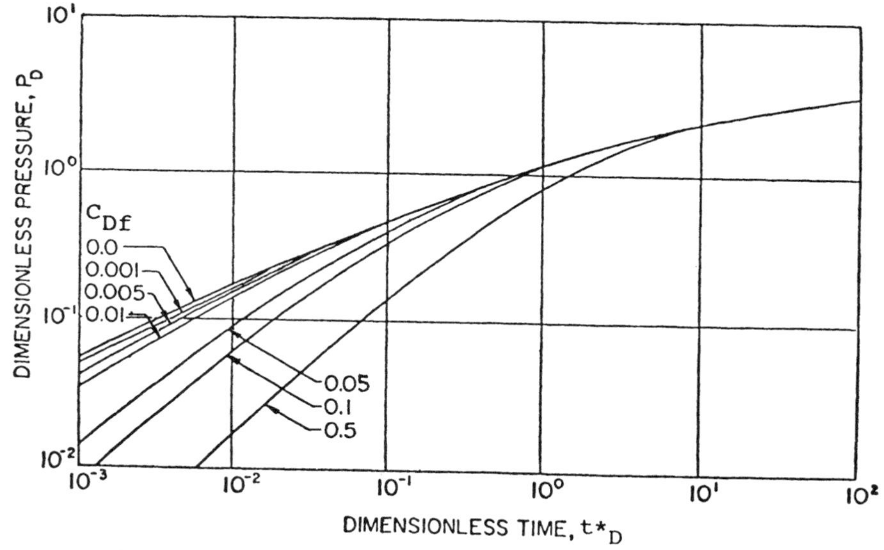

Figure 6–7 Type Curves for Infinite-Conductivity Vertically-Fractured Well with Wellbore Storage.[15]

It is difficult to analyze wellbore storage dominated well test data. To minimize wellbore storage effects, it is preferred that horizontal well tests are carried out with a bottom hole shut-in device. Otherwise a long wellbore-dominated flow may exist. A wellbore storage dominated flow shows a unit slope line (45°) on a log-log plot of pressure drop versus time; see Figure 6–8. This is normally used to identify the wellbore storage dominated flow. However, a derivative plot will help to ascertain the end of wellbore storage.

Daviau et al.[3] studied the effects of wellbore storage and skin factors on the pressure responses of horizontal wells. They showed that the first semilog straight line which represents radial flow in a vertical plane almost always disappears because of wellbore storage effects. The semilog analysis of this straight line would be possible only for a highly negative skin factor

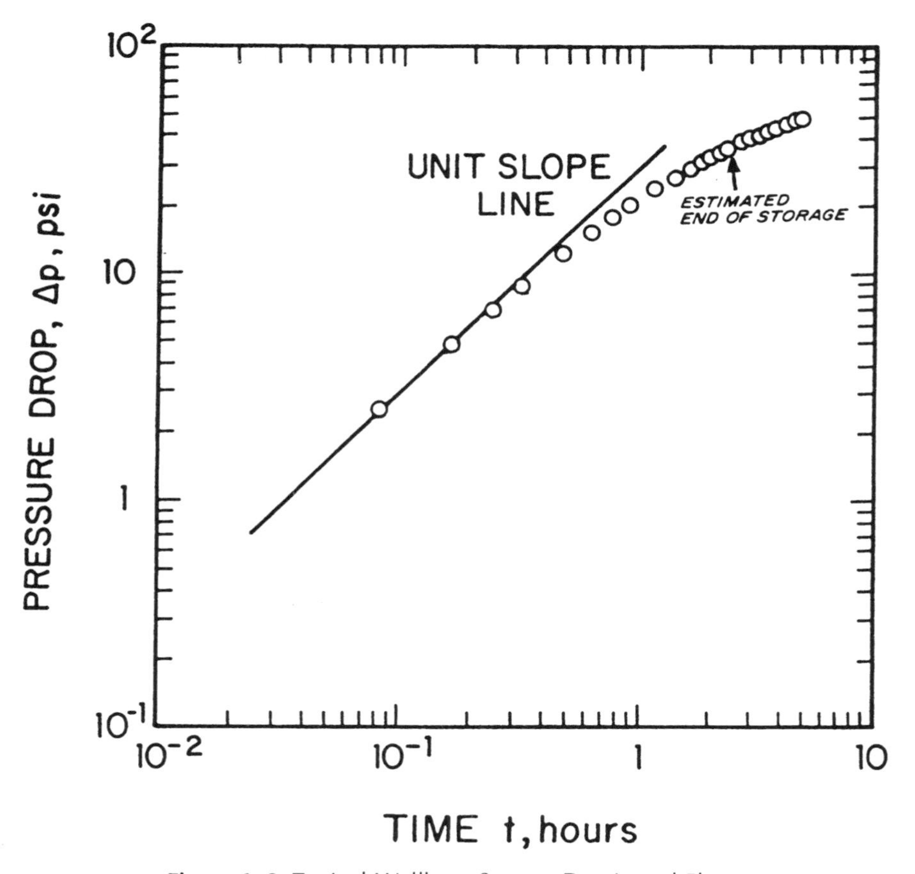

Figure 6–8 Typical Wellbore Storage Dominated Flow.

and small wellbore storage. (It should be mentioned that the first semilog straight line is very difficult to identify in practice regardless of wellbore storage effects.) Reference 3 shows that generally, the pseudo-radial flow is never disturbed by wellbore storage effects.

PRACTICAL CONSIDERATIONS

Well test analysis of horizontal wells is considerably more complicated than that of conventional vertical wells due to the complex geometry and possible existence of four flow regimes in contrast to the existence of essentially one radial-flow period for unfractured vertical wells. The possible problems associated with the skin factor make the analysis more difficult. Unlike conventional vertical wells, damage along the horizontal wellbore is not uniform; the wellbore damage close to the vertical section is more severe. Skin factor is defined as dimensionless pressure drop across skin zone; i.e.,

$$s = \frac{kh\Delta p_{skin}}{141.2qB\mu} \tag{6–46}$$

where Δp_{skin} is the additional pressure drop caused by damage. As mentioned earlier, that damage is unlikely to be uniform along the horizontal well; thus, Δp_{skin} will be a function of position (or varies along the wellbore). The well test analysis provides an average value for the entire well.

A horizontal well might penetrate an unproducing interval, or a section of wellbore may be so damaged that fluid flow is completely blocked. In this case, the horizontal well is only partially penetrated, and the skin factor obtained via well test analysis reflects both mechanical skin and pseudo-skin. The question is how bad the damage is. This needs to be answered before one designs a stimulation project.

Due to the nature of horizontal wells, it usually takes a considerably long time to reach pseudo-radial flow. In some finite reservoirs, pseudo-radial flow may never develop; i.e., reservoir boundaries start to affect pressure response before pseudo-radial flow begins to develop. Thus, the flow capacity and skin factor cannot be obtained by analyzing pseudo-radial flow pressure data. Even for a horizontal well in a large drainage area, it may be economically difficult to conduct a test long enough to reach pseudo-radial flow. Consider the following example.

EXAMPLE 6–7

Initial pressure, p_i = 7000 psi
Permeability, k = 0.75 md
System compressibility, c_t = 1.56×10^{-5} psi^{-1}

Porosity, ϕ = 2%
Oil viscosity, μ = 0.28 cp
Well length, L = 2750 ft
Reservoir height, h = 75 ft
Wellbore radius, r_w = 0.33 ft

The reservoir under consideration is infinite in extent and isotropic. Design a buildup test such that pseudo-radial flow can be reached and analyzed.

Solution

Using Equations 6–26 and 6–27 one can calculate the time at which pseudo-radial flow begins.

Method I: Use Equation 6–26

$$t_{s3} = \frac{1230L^2\phi\mu c_t}{k} = \frac{1230 \times 2750^2 \times 0.02 \times 0.28 \times 1.56 \times 10^{-5}}{0.75}$$
$$= 1083.4 \text{ hours} = 45 \text{ days.}$$

Method II: Use Equation 6–27

$$t_{s3} = \frac{1480L^2\phi\mu c_t}{k} = \frac{1480 \times 2750^2 \times 0.02 \times 0.28 \times 1.56 \times 10^{-5}}{0.75}$$
$$= 1303.7 \text{ hours} = 54 \text{ days.}$$

Method III:

$$\begin{aligned} L_D &= L/(2h)\sqrt{k_v/k_h} \qquad (6\text{–}6)\\ &= 2750/(2 \times 75) \times 1\\ &= 18.3\\ r_{wD} &= r_w/(L/2)\\ &= 0.33/(2750/2)\\ &= 2.4 \times 10^{-4} \end{aligned}$$

Assuming Figure 6–5 can be used for $r_{wD} = 2.4 \times 10^{-4}$, then $t_D = 3$ and

$$\begin{aligned} t_{s3} &= \frac{\phi\mu c_t L^2 t_D}{0.001055k}\\ &= \frac{0.02 \times 0.28 \times 1.56 \times 10^{-5} \times 2750^2 \times 3}{0.001055 \times 0.75}\\ &= 2505 \text{ hrs} = 104 \text{ days.} \end{aligned}$$

The well needs to be shut-in for at least 45 days in order to reach pseudo-radial flow.

EXAMPLE 6–8

Consider a horizontal well in an isotropic ($k_v = k_x = k_y = k_h = k$) and finite reservoir. The following parameters are given:[16]

Rotary Kelly Bushing, ft	78
Perforations, ft (MD)	3771–4599 and 4678–5381
Total Perforated Length, ft	1531
Pump Depth, ft (MD)/TVD (SS)	1965 / 1817
Gauge Depth, ft (MD)/TVD (SS)	2182 / 1980
Top of Perforation, ft (MD)/TVD (SS)	3771 / 2583
Bottom Hole Temperature, °F	155
Wellbore Radius, r_w, ft	.265
API Gravity	18.5
Formation Volume factor, B, RB/STB	1.078
Oil Viscosity, μ, cp	31.0
Porosity, ϕ	33%
System Compressibility, c_t, psi^{-1}	10^{-5}
Reservoir Thickness, h, ft	45
Average Oil Gradient, psi/ft	.377
Average Water Gradient, psi/ft	.440

Production Schedule

First Shut-in Period	5.47 hours	Stimulating well
First Pumping Period	5 minutes	Using submersible Pump
Second Shut-in Period	9.25 hours	Take out plug at 1963 (MD) and BHP gauge to 2182 (MD)
Second Pumping Period	8.08 hours	Using submersible Pump
Final Shut-in Period	12.08 hours	Buildup Test

After stimulation, the well pumped 811 STB in 8.08 hours and then was shut-in for a 12.08 hour buildup test. The measured pressure data are given in Table 6–1. The well volume between the pump suction and bottom perforation is

$$V = \pi r_w^2 L' = \pi\,(0.265\ \text{ft})^2\,(5381\ \text{ft} - 1965\ \text{ft})$$

$$= 753.6\ \text{ft}^3 \times \frac{1\ \text{bbl}}{5.615\ \text{ft}^3} = 134.2\ \text{barrels.}$$

TABLE 6–1 PRESSURE BUILDUP DATA FOR EXAMPLE 6–8

Δt (mins)	t_e (mins)	$\frac{t + \Delta t}{\Delta t}$	$\log\left(\frac{t + \Delta t}{\Delta t}\right)$	p_{ws} (psi)
0.0				808.49
5.0	4.95	98	1.991	810.98
10.0	9.80	49.5	1.695	813.25
15.0	14.55	33.33	1.523	815.33
20.0	19.21	25.25	1.402	817.25
30.0	28.25	17.17	1.235	820.66
40.0	36.95	13.13	1.118	823.60
50.0	45.33	10.70	1.029	826.17
60.0	53.39	9.08	0.958	828.42
80.0	68.67	7.06	0.849	832.20
100.0	82.91	5.85	0.767	835.25
120.0	96.20	5.04	0.703	837.75
140.0	108.64	4.46	0.650	839.85
160.0	120.31	4.03	0.605	841.63
180.0	131.28	3.69	0.568	843.16
200.0	141.61	3.43	0.535	844.49
240.0	160.55	3.02	0.480	846.69
280.0	177.52	2.73	0.437	848.43
320.0	192.80	2.52	0.401	849.85
360.0	206.63	2.35	0.371	851.02
420.0	222.76	2.15	0.333	852.45
480.0	241.24	1.99	0.303	853.58
540.0	255.51	1.90	0.278	854.50
600.0	268.20	1.81	0.257	855.26
660.0	279.56	1.73	0.239	855.91
725.0	290.60	1.67	0.222	856.50

The flow rate prior to shut-in is

$$q = \left(\frac{811 \text{ STB}}{8.08 \text{ hours}}\right)\left(\frac{24 \text{ hrs}}{1 \text{ day}}\right) = 2409 \text{ STB/D}$$

Based on the pressure buildup data in Table 6–1 calculate permeability and skin factors.

Solution

Horner time ratio $(t + \Delta t)/\Delta t$ and equivalent time t_e are calculated and listed in Table 6–1.[17,18] Figure 6–9 presents a semilog plot of buildup pressure, p_{ws}, versus Horner time ratio. From Figure 6–9 there appears to be a well-defined semilog straight line with slope $m = 38.5$ psi/cycle. Extrapolating this semilog straight line to $\Delta t = 1$ hour gives $p_{ws} = 828$ psia. This semilog straight line might correspond to early-time radial flow or pseudoradial flow.

1. Let us assume that the semilog straight line shown on Figure 6–9 corresponds to early-time radial flow around the wellbore. From Equation 6–12 for isotropic formation the permeability is

$$k = \frac{162.6 \times 2409 \times 31 \times 1.078}{38.5 \times 1531} = 222.1 \text{ md.}$$

Now we can use this permeability value to estimate the time corresponding to the end of early-time radial-flow period.

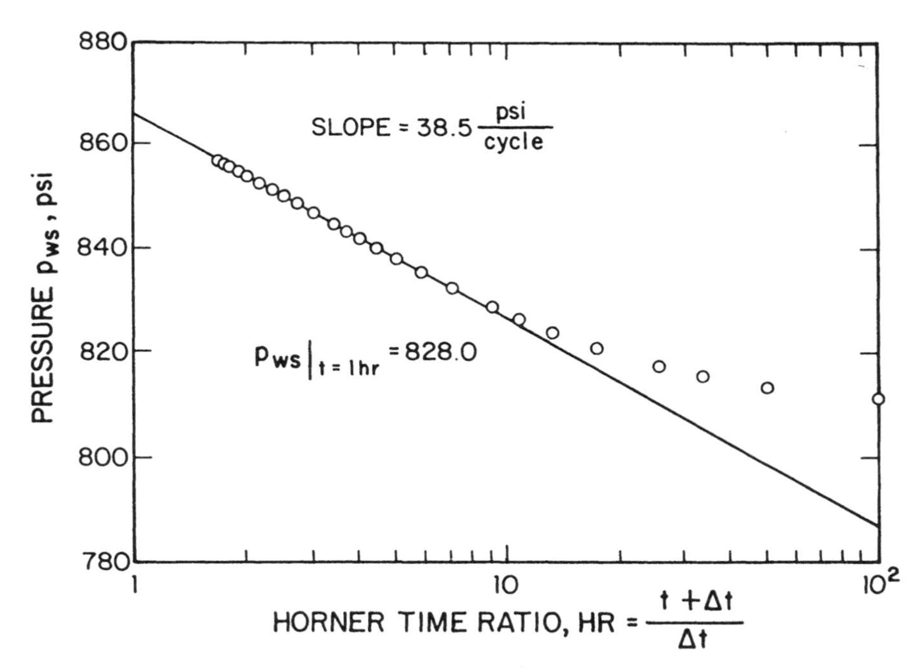

Figure 6–9 A Semilog Plot of Pressure versus Horner Time Ratio.

Method I:

$$t_{e1} = \frac{190\, d_z^{2.905}\, r_w^{-0.095}\, \phi\mu c_t}{k_v} \qquad (6\text{–}15)$$

$$= \frac{190 \times (22.5)^{2.095} \times 0.265^{-0.095} \times 0.33 \times 31 \times 10^{-5}}{222.1}$$

$$= 0.068 \text{ hours}$$

Method II:

$$t_{e1} = \frac{1800\, d_z^2\, \phi\mu c_t}{k_v} \qquad (6\text{–}16)$$

$$= \frac{1800 \times 22.5^2 \times 0.33 \times 31 \times 10^{-5}}{222.1}$$

$$= 0.42 \text{ hours}$$

and

$$t_{e1} = \frac{125\, L^2\, \phi\mu c_t}{k_y} \qquad (6\text{–}17)$$

$$= \frac{125 \times 1531^2 \times 0.33 \times 31 \times 10^{-5}}{222.1}$$

$$= 134 \text{ hrs} = 5.6 \text{ days.}$$

The minimum is 0.42 hours, which represents time to end early-time radial flow. (Note we assumed that the well is in the middle of formation thickness.)

The semilog straight line shown on Figure 6–9 starts at $t = 1.33$ hours. Obviously, it does not correspond to the early-time radial flow for an isotropic reservoir.

2. Let us assume the semilog straight line corresponds to pseudo-radial flow. From Equation 6–30:

$$k = \frac{162.6 \times 2409 \times 31 \times 1.078}{38.5 \times 45} = 7555.5 \text{ md}$$

The beginning time of pseudo-radial flow:
Method I:

$$t_{s3} = \frac{1230L^2\,\phi\mu c_t}{k_x} \tag{6–26}$$

$$= \frac{1230 \times 1531^2 \times 0.33 \times 31 \times 10^{-5}}{7555} = 39.04 \text{ hrs}$$

Method II:

$$t_{s3} = \frac{1480L^2\,\phi\mu c_t}{k_y} \tag{6–27}$$

$$= \frac{1480 \times 1531^2 \times 0.33 \times 31 \times 10^{-5}}{7555}$$

$$= 46.97 \text{ hrs}$$

Actual buildup time is $\Delta t = 12.08$ hours. The inconsistency indicates that the semilog straight line shown on Figure 6–9 does not represent the pseudo-radial flow.

Next we plotted $\Delta p\ (= p_{ws} - p_{wf,s})$ and its derivative versus equivalent time t_e on log-log scale; see Figure 6–10. It is clear that there are severe wellbore storage effects. Even at the end of the buildup period, the wellbore storage effects are still not negligible. Figure 6–10 shows that the straight line shown on Figure 6–9 does not represent radial flow. The results of Figure 6–10 also emphasize that one should first plot pressure data on log-log scale to detect wellbore storage effects when analyzing well testing data. Additionally, derivatives can be used to estimate the end of the wellbore storage effects. Since data given in Example 6–8 is within the wellbore storage dominated region, it cannot be analyzed.

SUMMARY

In conclusion, practical considerations may prevent us from conducting pressure tests and the complexity may limit the information we can get. Thus, it is important for an engineer to determine the objectives of the well test and make some preliminary calculations about various flow periods and their durations. Based upon test durations, an engineer can then estimate the shut-in time required to obtain various flow periods, information obtainable from these flow periods, and costs for conducting these tests. Then one can establish a reasonable basis to decide on conducting a well test. In

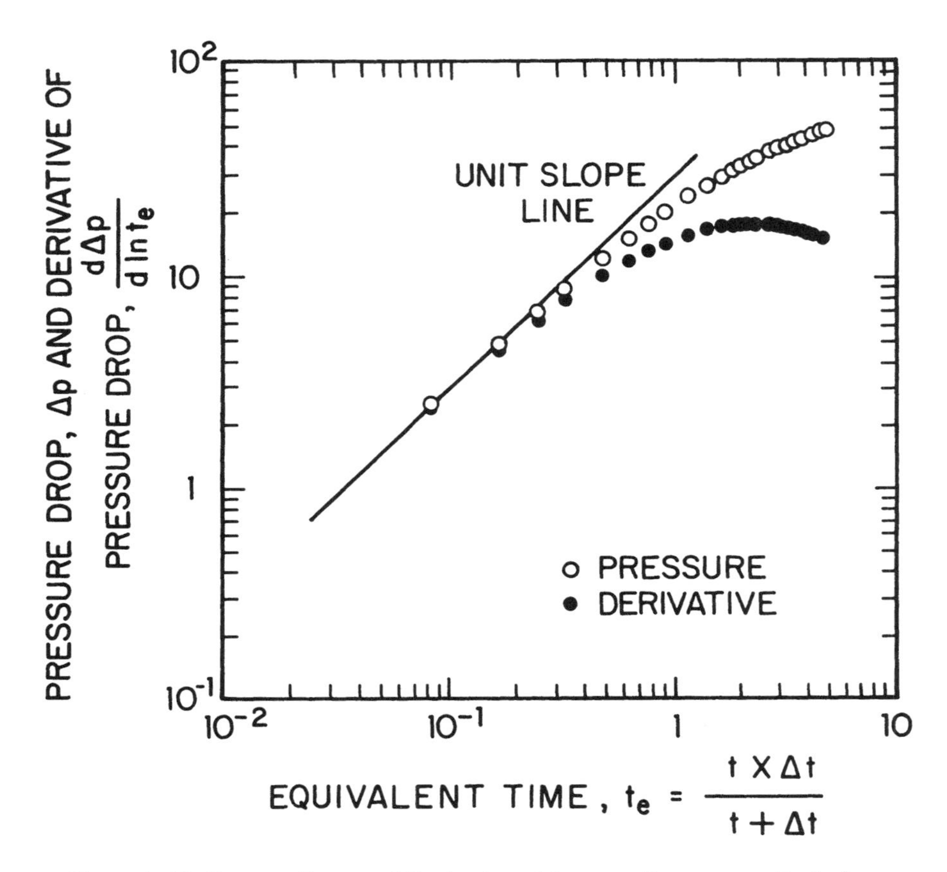

Figure 6–10 Pressure Drop and Derivative of Pressure Drop versus Equivalent Time.

some instances, it may be practical to initially conduct a drawdown test, although it may be difficult to execute, to estimate a range of various reservoir parameters. The build-up test can then be undertaken based upon evaluation of the initial drawdown testing. In some instances, one may have to use a reservoir simulator to match pressure buildup tests of horizontal wells.[19,20]

REFERENCES

1. Clonts, M. D. and Ramey, H. J., Jr.: "Pressure Transient Analysis for Wells with Horizontal Drainholes," paper SPE 15116, presented at the SPE California Regional Meeting, Oakland, California, April 2–4, 1986.

2. Ozkan, E., Raghavan, R. and Joshi, S. D.: "Horizontal-Well Pressure Analysis," *SPE Formation Evaluation,* pp. 567–575, December 1989.
3. Daviau, F., Mouronval, G., Bourdarot, G. and Curutchet, P.: "Pressure Analysis for Horizontal Wells," *SPE Formation Evaluation,* pp. 716–724, December 1988.
4. Goode, P. A. and Thambynayagam, R. K. M.: "Pressure Drawdown and Buildup Analysis for Horizontal Wells in Anisotropic Media," *SPE Formation Evaluation,* pp. 683–697, December 1987.
5. Kuchuk, F. J., Goode, P. A., Wilkinson, D. J. and Thambynayagam, R. K. M.: "Pressure Transient Behavior of Horizontal Wells With and Without Gas Cap or Aquifer," paper SPE 17413, presented at the SPE California Regional Meeting, Long Beach, California, March 23–25, 1988.
6. Kuchuk, F. J., Goode, P. A., Brice, B. W. and Sherrard, D. W.: "Pressure Transient Analysis and Inflow Performance for Horizontal Wells," paper SPE 18300, presented at the 63rd SPE Annual Conference, Houston, Texas, Oct. 2–5, 1988.
7. Rosa, A. J. and Carvalho, R. S.: "A Mathematical Model for Pressure Evaluation in an Infinite Conductivity Horizontal Well," *SPE Formation Evaluation,* pp. 559–566, Dec. 1989.
8. Carvalho, R. De S., and Rosa, A. J.: "Transient Pressure Behavior for Horizontal Wells in Naturally Fractured Reservoir," paper SPE 18302, presented at the 63rd SPE Annual Conference, Houston, Texas, Oct. 2–5, 1988.
9. Odeh, A. S. and Babu, D. K.: "Transient Flow Behavior of Horizontal Wells, Pressure Drawdown and Buildup Analysis," *SPE Formation Evaluation,* pp. 7–15, March 1990.
10. Aguilera, R. and Ng., M. C.: "Transient Pressure Analysis of Horizontal Wells in Anisotropic Naturally Fractured Reservoirs," paper SPE 19002, presented at the SPE Joint Rocky Mountain Regional/Low Permeability Reservoirs Symposium and Exhibition, Denver, Colorado, March 6–8, 1989.
11. Gringarten, A. C., Ramey, H. J. Jr. and Raghavan, R.: "Unsteady-State Pressure Distributions Created by a Well with Single Infinite-Conductivity Vertical Fracture," Society of Petroleum Engineers Journal, pp. 300–347, August 1974.
12. Onur, M. and Reynolds, A. C.: "A New Approach for Constructing Derivative Type Curves," *SPE Formation Evaluation,* pp. 197–206, March 1988.
13. Onur, M., Yeh, N. S. and Reynolds, A. C.: "New Applications of the Pressure Derivative in Well Test Analysis," *SPE Formation Evaluation,* pp. 429–437, September 1989.
14. Ozkan, E.: "Performance of Horizontal Wells," Ph.D. Dissertation, The University of Tulsa, Tulsa, Oklahoma, 1988.
15. Ramey, Jr., H. J. and Gringarten, A. C.: "Effect of High Volume Vertical Fracture on Geothermal Steam Well Behavior," Second United Nations Symposium on the Use and Development of Geothermal Energy, San Francisco, California, May 20–29, 1975.
16. Mead, H. N.: "Using Finite System Buildup Analysis To Investigate Fractured, Vugular, Stimulated, and Horizontal Wells," paper SPE 16379, presented at the SPE California Regional Meeting, Ventura, California, April 8–10, 1987.
17. Horner, D. R.: "Pressure Buildup in Wells," Proc. Third World Petroleum Congress, The Hague, Sec. II, pp. 503–521, 1951.
18. Agarwal, R. G.: "A New Method to Account for Producing Time Effects When Drawdown Type Curves Are Used to Analyze Pressure Buildup and Other Test Data," paper SPE 9289, presented at the SPE Annual Technical Conference and Exhibition, Dallas, Texas, Sept. 21–24, 1980.

19. Bourdarot, G. and Daviau, F.: "Vertical Permeability: Field Cases," paper SPE 19777, presented at SPE Annual Technical Conference and Exhibition, San Antonio, Texas, Oct. 8–11, 1989.
20. Broman, W. H., Stagg, T. O., and Rosenzweig, J. J.: "Horizontal Well Performance Evaluation at Prudhoe Bay," paper CIM/SPE 90–124, presented at the 41st CIM Annual Technical Meeting, Calgary, Alberta, Canada, June 10–13, 1990.

CHAPTER 7

Pseudo-Steady State Flow

INTRODUCTION

Pseudo-steady state begins when the pressure disturbance created by the producing well is felt at the boundary of the well drainage area. In other words, when the fluid mass situated at the drainage boundary starts moving towards the producing well, pseudo-steady state begins. This pseudo-steady state is also described as semi-steady state or depletion state. The name *depletion state* is probably the most appropriate, because it tells us that the reservoir has reached a point where the pressure at all the reservoir boundaries and also the average reservoir pressure will decrease over time as more and more fluid is withdrawn from the reservoir.

As noted in Chapter 2 and shown in Figure 2–12, for a circular drainage boundary, the fluid from the drainage boundary would start moving to the wellbore at a given instant in time. Similarly for a square drainage area, fluid at the boundary would start moving toward the producing well at one instant in time. In practice, a square drainage boundary is important, because most oil and gas fields are developed on square areas using 10- to 640-acre well spacings. If areal anisotropy exists; i.e., horizontal permeability in one direction is different than that in another direction, the resulting drainage area will be rectangular rather than square. In a given time, a well could drain more distance along the high-permeability direction than along the low-permeability direction. In many naturally fractured reservoirs, the permeability along fracture trends is higher than the permeability perpendicular to the fracture trend. This results in a longer drainage length along the fracture trend than in the direction perpendicular to the fracture trend. In such situations, it may be advisable to review conventional uniform well spacing. (See Chapter 2 for further discussion.)

In reservoirs with uniform horizontal permeability, if the wells are drilled in a rectangular area, then the well pressure disturbance would reach the closest boundary first (see Fig. 2–12). This is the end of the infinite acting period and the beginning of the transition period. When the well pressure disturbance reaches all the boundaries, only then pseudo-steady state begins. The times required to start pseudo-steady state for vertical and horizontal wells are discussed in Chapter 2.

MATERIAL BALANCE AND MATHEMATICAL FORMULATION

As noted above, pseudo-steady state assumes closed reservoir boundaries; i.e., there is no flow across the boundaries. The pseudo-steady state solutions are useful to predict well performance in the depletion stage of the reservoir. Moreover, in most reservoirs it is difficult to measure pressure at the drainage boundary; however, an average reservoir pressure can be estimated by a pressure buildup or drawdown test. Hence, pseudo-steady state solutions, based upon the average reservoir pressure, are useful in determining well performance.

Assuming a slightly compressible fluid, one can write a material balance equation by equating the reservoir voidage due to oil production with expansion of the remaining oil as

$$N_pB_O = V_pc_t(p_i - \bar{p})/5.615 \qquad (7\text{–}1)$$

$$= Ah\phi c_t(p_i - \bar{p})/5.615 \qquad (7\text{–}2)$$

where

V_p = reservoir pore volume, ft^3
A = drainage area, ft^2
c_t = total compressibility, 1/psi
N_p = cumulative oil produced, STB

$\bar{p}$ = average reservoir pressure, psi
p_i = initial reservoir pressure, psi
B_0 = oil formation volume factor, RB/STB
ϕ = porosity, fraction
h = pay zone thickness, ft

If the well is produced at a constant rate, q STB/day, for a time period, t days, then cumulative oil produced, N_p, is:

$$N_p = qt \tag{7–3}$$

Substituting Equation 7–3 into Equation 7–2 results in the following expression.

$$\bar{p} = p_i - \left(\frac{5.615qB_0}{Ah\phi c_t}\right) t \tag{7–4}$$

Equation 7–4 tells us that average reservoir pressure would decline over time as fluid is withdrawn from the reservoir.

GENERALIZED PSEUDO-STEADY STATE EQUATION FOR VERTICAL WELLS

In general, during pseudo-steady state flow of an ideal fluid (liquid) in a closed circular drainage area, the reservoir pressure at a distance r from the well center is described as[1,2]

$$p(r) = p_{wf} + \left(\frac{141.2q\mu_0 B_0}{kh}\right) [\ln(r/r_w) - 0.5\,(r/r_e)^2] \tag{7–5}$$

where

k = reservoir permeability, md
h = pay zone thickness, ft
p_{wf} = bottomhole pressure, psia
μ_0 = oil viscosity, cp
B_0 = oil formation factor, RB/STB
r_w = wellbore radius, ft
r_e = drainage radius, ft
q = production rate, STB/day

Substituting $p = p_e$, pressure at the drainage boundary at $r = r_e$ and rearranging gives the familiar equation for flow rate of a vertical well located centrally in the drainage plane.

$$q = \frac{kh(p_e - p_{wf})}{141.2\ \mu_o B_o[\ln(r_e/r_w) - 0.5]} \tag{7–6}$$

The generalized pseudo-steady state equation, based upon average reservoir pressure $\bar{p}$ for a vertical well, located at any place in a drainage area A can be expressed as:[1,2]

$$q = \frac{kh(\bar{p} - p_{wf})/(141.2\ \mu_o B_o)}{\ln[\sqrt{2.2458A/(C_A r_w^2)}] + s + s_m + Dq} \tag{7–7}$$

In the above equation, s_m represents mechanical skin factor due to drilling and completion related well damage. The skin factor, s, is an arithmetic addition of skin factors due to partial well penetration, perforations, and fracture and acid stimulation. Additionally, Dq represents near wellbore turbulence, and C_A represents shape factor. The discussion in this chapter is restricted to oil wells. Chapter 9 includes pseudo-steady state equations for gas wells.

SHAPE FACTORS FOR VERTICAL WELLS

SHAPE FACTOR DEFINITION

During pseudo-steady state flow for a vertical well located in a bounded reservoir, the dimensionless wellbore pressure, p_{wD}, varies linearly with dimensionless time, t_{DA}, and is expressed as[3]

$$p_{wD} = 2\pi t_{DA} + \tfrac{1}{2}\ln(A/r_w^2) + \tfrac{1}{2}\ln(2.2458/C_A) \tag{7–8}$$

where

C_A = shape factor, dimensionless.

Equation 7–8 is an extension of the transient flow equation in a pseudo-steady state region. Similar to the transient-flow well testing solution, the derivation includes an assumption of a single-phase flow. Additionally, the well is assumed to be producing a slightly compressible fluid at a constant rate. In this equation, C_A is a geometric factor, which depends upon drainage shape and the well location. Thus C_A accounts for the influence of well location within the drainage plane on well productivity.

The shape factor values, C_A, for various vertical well locations are listed in Table 7–1. As noted in Table 7–1, for a well located at the center of a

TABLE 7–1 SHAPE FACTOR DEPENDENT SKIN FACTORS, s_{CA}, FOR VERTICAL WELLS[4]

Geometry	C_A	s_{CA}*	t_{DApss}
	31.62	0.000	0.1
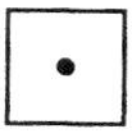	30.88	0.012	0.1
	31.60	0.000	0.1
	27.6	0.068	0.2
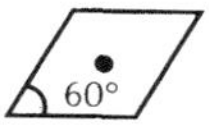 60°	27.1	0.077	0.2
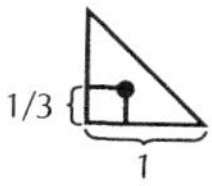 1/3, 1	21.9	0.184	0.4
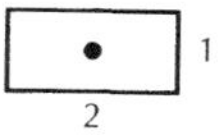 1, 2	21.84	0.185	0.3
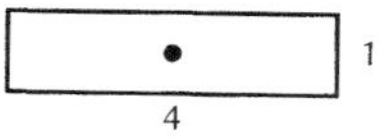 1, 4	5.379	0.886	0.8
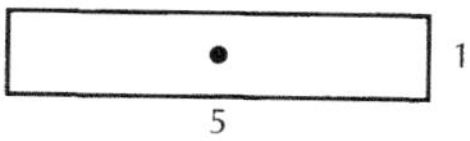 1, 5	2.361	1.298	1.0
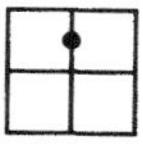	12.98	0.445	0.7
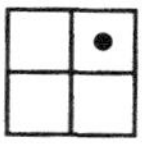	4.513	0.973	0.6
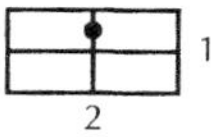 1, 2	10.84	0.535	0.4
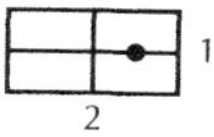 1, 2	4.514	0.973	1.5
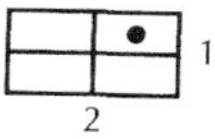 1, 2	2.077	1.362	1.7
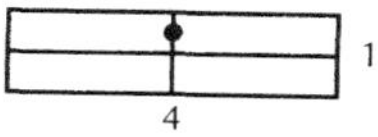 1, 4	2.690	1.232	0.8
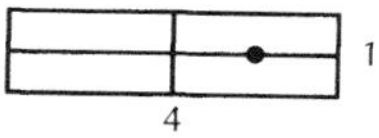 1, 4	0.232	2.458	4.0
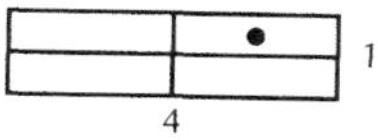 1, 4	0.115	2.806	4.0

* s_{CA} calculated using a vertical well at the center of a circle as a reference case.

circular drainage area, C_A value is 31.62. Additionally, a circular drainage area, A, can be represented as πr_e^2. Substituting for C_A and A in Equation 7–7, and assuming $s = 0$, $s_m = 0$ and $D = 0$, one obtains the classical single-phase pseudo-steady state equation for oil flow as

$$q = \frac{kh(\bar{p} - p_{wf})/(141.2\ \mu_O B_O)}{\ln(r_e/r_w) - 0.75} \tag{7–9}$$

It is important to note that Equation 7–9 is for a vertical well located centrally in a circular drainage area. For any other well location, an appropriate C_A value should be substituted in Equation 7–7 from Table 7–1 to obtain an expression for a flow rate. For example: For a well located at the center of a square with each side being $2x_e$ in length, the value of C_A is 30.88. Additionally, one can calculate an equivalent drainage radius r_e' of a square area

$$r_e' = \sqrt{A/\pi} \tag{7–10}$$

where A represents drainage area in ft^2 and r_e' is equivalent drainage radius in ft. Substituting $C_A = 30.88$ and r_e' in Equation 7–7, the flow equation for a well located at the center of a square is

$$q = \frac{kh(\bar{p} - p_{wf})/(141.2\ \mu_O B_O)}{\ln(r_e'/r_w) - 0.738} \tag{7–11}$$

SHAPE FACTORS EXPRESSED AS SKIN FACTORS

Another method to write pseudo-steady state equations is to express shape factors in terms of equivalent skin factors.[4] By choosing shape factor, $C_A = 31.62$, for a well at the center of a circular drainage area as a reference shape factor, for any other well location, a shape-related skin factor s_{CA} is expressed as[4]

$$s_{CA} = \ln[\sqrt{C_{A,ref}/C_A}] = \ln[\sqrt{31.62/C_A}] \tag{7–12}$$

Multiplying and dividing by the $C_{A,ref}$ term in Equation 7–7 results in the following expression:

$$q = \frac{kh(\bar{p} - p_{wf})/(141.2\ \mu_O B_O)}{\ln\left[\sqrt{\dfrac{2.2458A}{C_A \times r_w^2}}\sqrt{\dfrac{C_{A,ref}}{C_{A,ref}}}\right] + s + s_m + Dq} \tag{7–13}$$

The above equation can be rewritten for the circular reference area, with $A = \pi r_e^2$ and $C_{A,ref} = 31.62$, as

$$q = \frac{kh(\bar{p} - p_{wf})/(141.2\ \mu_0 B_0)}{\ln(r_e/r_w) - 0.75 + s_{CA} + s + s_m + Dq} \qquad (7\text{–}14)$$

where s_{CA} is shape-related skin factor. A list of s_{CA} values for vertical wells is given in Table 7–1. Obviously, for a well located at the center of a circle, $s_{CA} = 0$. Similarly, one can rewrite the above equations by choosing a well located at the center of a square as a reference shape factor, i.e., by choosing $C_{A,ref} = 30.88$ as a reference shape factor. Thus, an equation based upon square drainage area as a reference area is

$$q = \frac{kh(\bar{p} - p_{wf})/(141.2\ \mu_0 B_0)}{\ln(r_e'/r_w) - 0.738 + s_{CA} + s + s_m + Dq} \qquad (7\text{–}15)$$

Equations 7–14 and 7–15 tell us that to maximize productivity, a desirable value is $s_{CA} = 0$, i.e., a well located at the center of a circle or square is desired. For any other vertical well location, s_{CA} has a positive value, resulting in a loss of well productivity. Thus the concept of shape factors allows us to optimize well location in a given drainage area.

It is important to note that the above discussion was restricted to unstimulated (non-fractured) vertical wells. Additionally, the above discussion implicitly assumes that the horizontal permeability is uniform in the areal drainage plane.

SHAPE FACTORS FOR FRACTURED VERTICAL WELLS

Depending upon well location within a drainage plane, one can estimate shape factors in a similar manner for fractured vertical wells. Figure 7–1 shows a schematic of a fractured vertical well in a drainage area of dimensions, $2x_e$ and $2y_e$. In fractured vertical wells, the shape factor depends upon:

1. drainage area shape, i.e., ratio of $2x_e/(2y_e)$,
2. fracture penetration ratio, x_f/x_e,
3. fracture flow condition, i.e., uniform-flux or infinite-conductivity fracture.

In this section, only two types of fractures are considered. The uniform-flux fracture assumes a constant flow rate per unit fracture length along the entire fracture length. This condition requires a finite pressure drop from the tip of

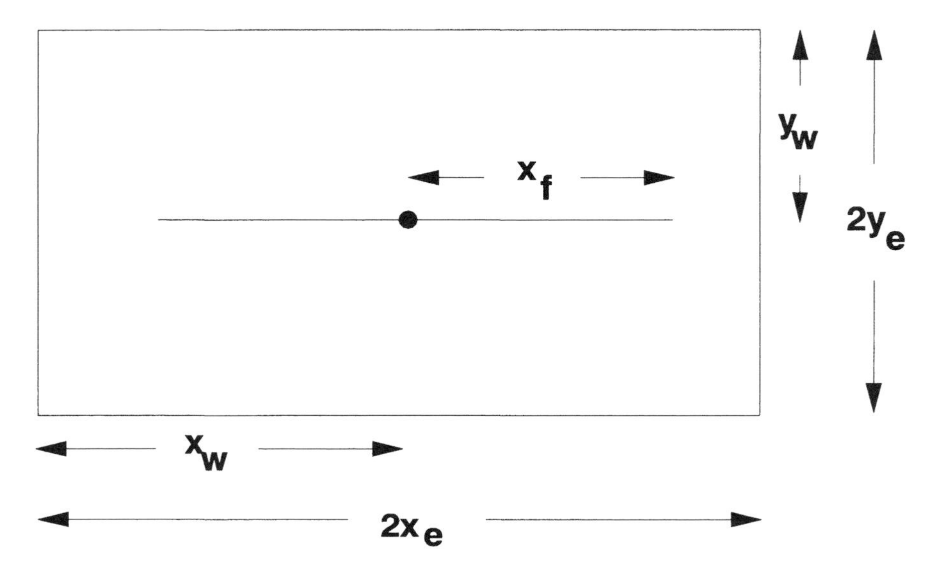

Figure 7–1 An Areal View of a Fractured Vertical Well.

the fracture to the fracture center. In contrast, in an infinite-conductivity fracture, the pressure along the fracture length is constant, but the rate of fluid entry into the fracture along its length varies. The fluid entry rate into a fracture near the tip is highest, while, at the fracture center, it is the lowest. In this chapter all discussion is restricted to infinite-conductivity fractures, unless noted otherwise.

The effective wellbore radius of an infinite-conductivity fully penetrating vertical fracture is

$$r'_w = x_f/2 \qquad \text{for } x_f/x_e < 0.4 \tag{7–16}$$

where

x_f = fracture half-length, ft
x_e = half the side of the square drainage area, ft.

Also, for a square drainage area, area is

$$A = (2x_e)^2 = 4x_e^2 \tag{7–17}$$

Substituting Equation 7–16 and Equation 7–17 in Equation 7–8, the dimensionless pressure during pseudo-steady state for a fractured vertical well in a square drainage area is

$$p_D = 2\pi t_{DA} + \frac{1}{2}\ln\left(\frac{x_e}{x_f}\right)^2 + \frac{1}{2}\ln\left(\frac{2.2458}{C_f}\right) + \frac{1}{2}\ln(16) \qquad (7\text{–}18)$$

C_f = shape factor for a fractured vertical well.

Shape factors C_f for different fracture penetrations in a square drainage area are listed in Table 7–2.[3] Shape factors for fractured wells draining rectangular areas are listed in Table 7–3.[6] Moreover, shape factors for off-centered fractured wells in rectangular drainage areas are given in Table 7–4.[6] For $x_f/\sqrt{A}$ = 0.01 to 1, Figure 7–2 shows a graphic correlation for shape factor C_f'.[5] To convert this shape factor C_f' to C_f the following equation can be used:

$$C_f = 0.25\, C_f' \qquad (7\text{–}19)$$

It is important to note that the definition of shape factor for a fractured vertical well, C_f, is different from that of C_A, the shape factor for an unfractured vertical well. (This is primarily because of the changes in parameters involved when converting unstimulated vertical well Equation 7–8 to fractured well Equation 7–18.) Caution must be exercised while using three different definitions (C_A, C_f', and C_f) of shape factors for fractured vertical wells. The following three methods can be used to calculate production rate, depending upon the shape factor definition. *The three different methods give identical answers.*

TABLE 7–2 SHAPE FACTORS, C_f, FOR FRACTURED VERTICAL WELLS IN A SQUARE DRAINAGE AREA[3]

x_f/x_e	Shape Factor, C_f
0.1	2.6541
0.2	2.0348
0.3	1.9986
0.5	1.6620
0.7	1.3127
1.0	0.7887

TABLE 7–3 SHAPE FACTORS, C_f, FOR FRACTURED VERTICAL WELLS LOCATED CENTRALLY IN THE RECTANGULAR DRAINAGE AREA[6]

C_f	x_e/y_e					
x_f/x_e	1	2	3	5	10	20
0.1	2.020	1.4100	0.751	0.2110	0.0026	0.000005
0.3	1.820	1.3611	0.836	0.2860	0.0205	0.000140
0.5	1.600	1.2890	0.924	0.6050	0.1179	0.010550
0.7	1.320	1.1100	0.880	0.5960	0.3000	0.122600
1.0	0.791	0.6662	0.528	0.3640	0.2010	0.106300

Method 1

The following equation can be used to calculate oil flow rate from a fractured vertical well during pseudo-steady state:

$$q = \frac{kh(\bar{p} - p_{wf})/(141.2\ \mu_o B_o)}{\ln[\sqrt{2.2458A/(C_A r_w^2)}] + s + s_m + Dq} \tag{7–7}$$

The C_A value of Equation 7–7 for fractured vertical wells depends upon the ratio x_f/x_e:

1. For $x_f/x_e > 0.1$, C_A is calculated as

$$C_A = 16\ C_f \tag{7–20}$$

where C_f values are listed in Tables 7–2, 7–3 and 7–4.

2. For $x_f/x_e < 0.1$, calculation of C_A involves three steps:
   a. first calculate C_f' from Figure 7–2;
   b. calculate C_f using Equation 7–19;
   c. calculate C_A using Equation 7–20.

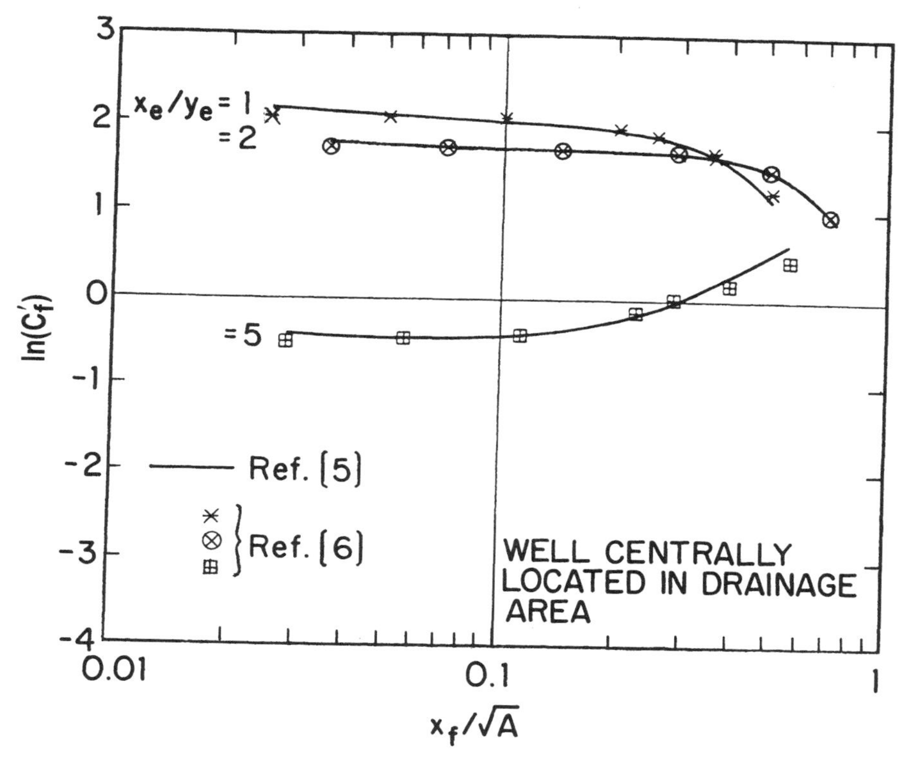

Figure 7–2 Shape Factors for Fractured Vertical Wells for Different Fractured Penetrations.[5]

Method 2

Similar to vertical wells, the generalized pseudo-steady state equation for a fractured vertical well is

$$q = \frac{kh(\bar{p} - p_{wf})/(141.2\ \mu_0 B_0)}{\ln(r'_e/r_w) - 0.738 + s_{CA,f} + s_f + s_m + Dq - c'} \qquad (7\text{–}21)$$

where

s_f = skin factor for an infinite-conductivity fully penetrating vertical fracture of half-length, x_f,

$$s_f = -\ln(r'_w/r_w) = -\ln[(x_f/2)/r_w] \qquad (7\text{–}22)$$

s_m = mechanical skin damage due to drilling and completion.

TABLE 7–4 SHAPE FACTORS, C_f, FOR OFF-CENTERED FRACTURED VERTICAL WELLS[6]

Influence of y_w/y_e*

	x_f/x_e	y_w/y_e = 0.25	0.5	1.0
x_e/y_e = 1				
	0.1	0.2240	0.8522	2.0200
	0.3	0.2365	0.7880	1.8220
	0.5	0.2401	0.7165	1.6040
	0.7	0.2004	0.5278	1.3170
	1.0	0.1351	0.3606	0.7909
x_e/y_e = 2				
	0.1	0.2272	0.7140	1.4100
	0.3	0.3355	0.7700	1.3610
	0.5	0.4325	0.8120	1.2890
	0.7	0.4431	0.7460	1.1105
	1.0	0.2754	0.4499	0.6660
x_e/y_e = 5				
	0.1	0.0375	0.09185	0.2110
	0.3	0.1271	0.20320	0.2864
	0.5	0.2758	0.38110	0.4841
	0.7	0.3851	0.49400	0.5960
	1.0	0.2557	0.31120	0.3642

Influence of x_w/x_e*

	x_f/x_e	x_w/x_e = 0.5	0.75	1.0
x_e/y_e = 1				
	0.1	0.9694	1.7440	2.0200
	0.3	1.1260	1.7800	1.8200
	0.5	1.2708	1.7800	1.6000
x_e/y_e = 2				
	0.1	0.3679	1.0680	1.4098
	0.3	0.5630	1.2980	1.3611
	0.5	0.8451	1.5470	1.2890
x_e/y_e = 5				
	0.1	0.0058	0.0828	0.2110
	0.3	0.0317	0.2540	0.2864
	0.5	0.1690	0.7634	0.6050

* x_w and y_w represent the distance of the fracture center from the nearest y and x boundary, respectively (see Figure 7–1).

$s_{CA,f}$ = shape related pseudo-skin factor, referenced to the center of a square for fractured vertical wells

$$= \ln [\sqrt{30.88/C_f}] \quad (7\text{–}22a)$$

(C_f is obtained from Tables 7–2 through 7–4.)

c' = shape factor conversion constant

$$= (1/2) \ln 16 = 1.386$$

In Equation 7–21, the constant c' ($c' = 1.386$) accounts for the difference in the definitions of shape factors, C_A (for unfractured vertical wells) and C_f (fractured vertical wells). This constant is applicable to all drainage patterns.

Method 3

Instead of calculating shape factors, one can adjust effective wellbore radius of a fractured vertical well to account for both fracture length as well as shape factor.[3] Figure 7–3 shows a plot of effective wellbore radius for vertical wells with uniform flux and infinite-conductivity fractures for different fracture penetrations.[7] The effective wellbore radius, r_w', calculated from Figure 7–3 can be directly substituted in place of r_w in Equation 7–9 to calculate oil flow rate in fractured vertical wells, where the vertical well is located centrally in the drainage area.

$$q = \frac{kh\,(\bar{p} - p_{wf})/(141.2\,\mu_o B_o)}{\ln (r_e/r_w') - 0.75} \quad (7\text{–}9)$$

The results of Reference 7 are for a square drainage boundary. These results can also be extended to rectangular drainage boundaries for varying $2x_e/(2y_e)$ ratios by replacing (x_f/x_e) with $(2x_f/\sqrt{A})$ on the x-axis or Figure 7–3.[8]

SHAPE FACTORS OF HORIZONTAL WELLS

For horizontal wells, in addition to the side boundaries of the areal drainage plane, the top and bottom reservoir boundaries also influence well productivity. Thus, a horizontal well shape factor depends upon:

1. drainage area shape, i.e., ratio $2x_e/(2y_e)$
2. well penetration, $L/(2x_e)$
3. dimensionless well length, $L_D = (L/2h)\sqrt{k_v/k_h}$

When a horizontal well is sufficiently long, i.e., $L_D > 10$, the influence of top and bottom boundaries becomes small and performance of a horizontal well approaches that of a fully penetrating infinite-conductivity fracture. Therefore, as shown in Figure 7–4, for a long horizontal well, its shape factor also approaches the shape factor of a fully penetrating infinite-conductivity fracture. Similar to fractured vertical wells, the dimensionless pressure drop

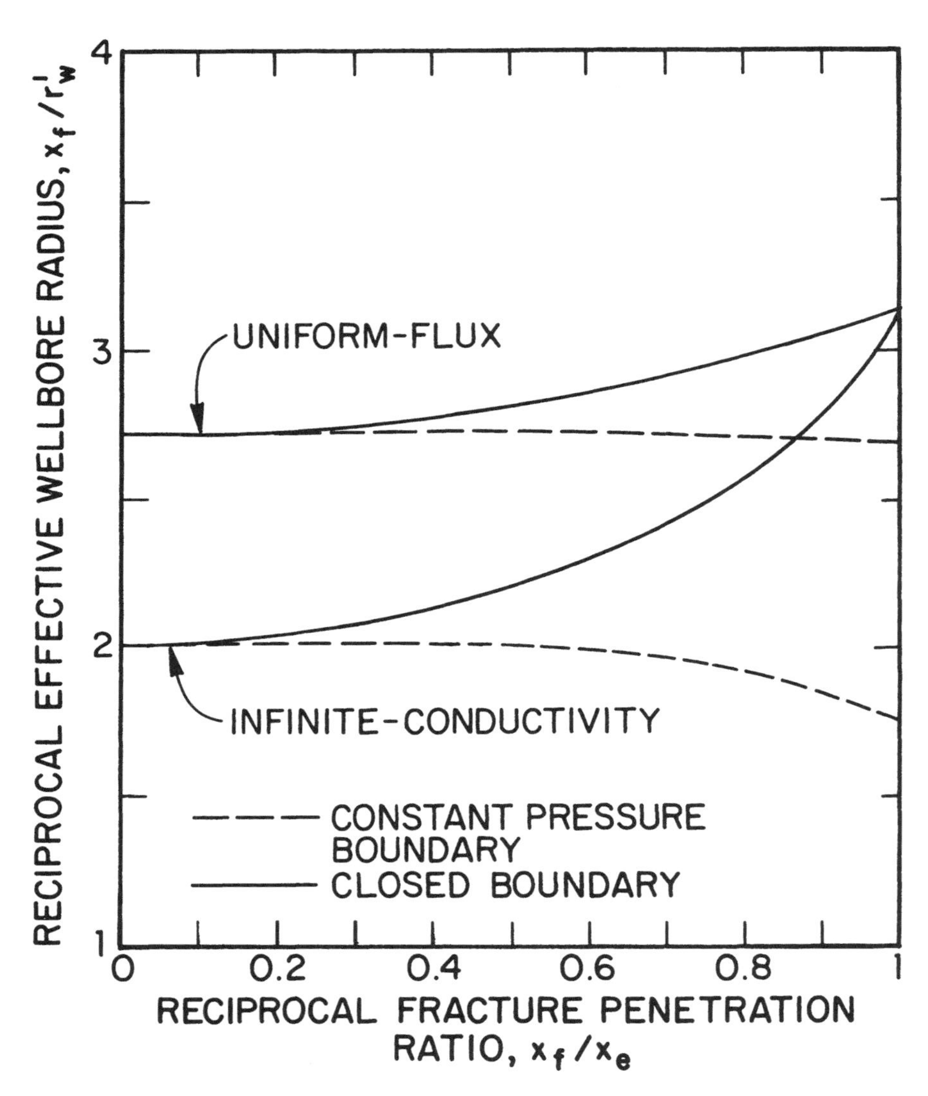

Figure 7–3 Effective Wellbore Radius for Fractured Vertical Wells for Different Fracture Penetrations.[7]

during pseudo-steady state for a horizontal well in a bounded reservoir is given as[6]

$$p_D = 2\pi t_{DA} + \frac{1}{2}\ln\left[\frac{A}{4\,(L/2)^2}\right] + \frac{1}{2}\ln\left[\frac{2.2458}{C_{A,h}}\right] + \frac{1}{2}\ln(16) \tag{7–23}$$

$C_{A,h}$ = shape factor for horizontal wells

L = length of horizontal well, ft

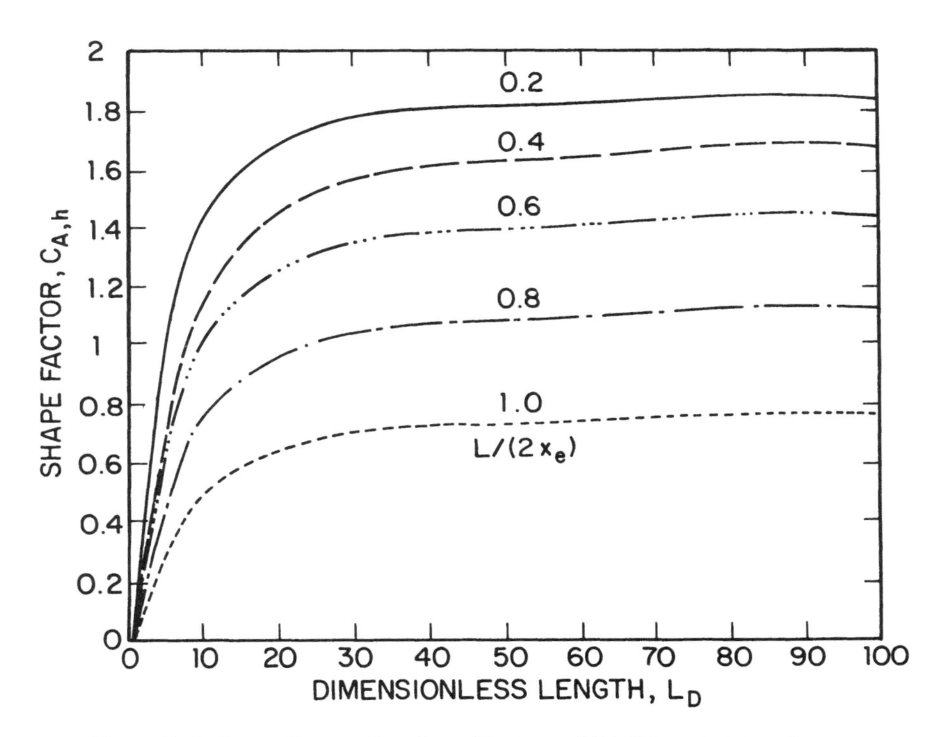

Figure 7–4 Shape Factor, $C_{A,h}$, for a Horizontal Well Located in a Square Drainage Area for Different Dimensionless Length.[6]

Shape-related skin factors for horizontal wells, $s_{CA,h}$ (based upon a square reference area), are tabulated in Table 7–5 and are plotted in Figures 7–5 through 7–7.[6,9]

HORIZONTAL WELL PSEUDO-STEADY STATE PRODUCTIVITY CALCULATIONS

Three methods are available to calculate pseudo-steady productivities of horizontal wells for single-phase flow.[6,10,11] In all these methods, the reservoir is assumed to be bounded in all directions and the horizontal well is located arbitrarily within a rectangular bounded drainage area. Figure 7–8 shows a schematic of a horizontal well drilled in a bounded reservoir. The difference between the three methods is in their mathematical solution methods and the boundary conditions used. For example, Method I assumes a horizontal well as an infinite-conductivity well.[6] Method II assumes a uniform-flux boundary condition.[10] Method III uses an approximate infinite-conductivity solution where the constant wellbore pressure is estimated by

TABLE 7–5 SHAPE RELATED SKIN FACTORS, $s_{CA,h}$, FOR HORIZONTAL WELLS FOR VARIOUS WELL PENETRATIONS AND DIFFERENT RECTANGULAR DRAINAGE AREAS[6]

	$L/(2x_e)$				
L_D	0.2	0.4	0.6	0.8	1.0
(1) $x_e/y_e = 1$					
1	3.772	4.439	4.557	4.819	5.250
2	2.321	2.732	2.927	3.141	3.354
3	1.983	2.240	2.437	2.626	2.832
5	1.724	1.891	1.948	2.125	2.356
10	1.536	1.644	1.703	1.851	2.061
20	1.452	1.526	1.598	1.733	1.930
50	1.420	1.471	1.546	1.672	1.863
100	1.412	1.458	1.533	1.656	1.845
(2) $x_e/y_e = 2$					
1	4.425	4.578	5.025	5.420	5.860
2	2.840	3.010	3.130	3.260	3.460
3	2.380	2.450	2.610	2.730	2.940
5	1.982	2.020	2.150	2.310	2.545
10	1.740	1.763	1.850	1.983	2.198
20	1.635	1.651	1.720	1.839	2.040
50	1.584	1.596	1.650	1.762	1.959
100	1.572	1.582	1.632	1.740	1.935
(3) $x_e/y_e = 5$					
1	5.500	5.270	5.110	5.140	5.440
2	3.960	3.720	3.540	3.650	3.780
3	3.440	3.190	3.020	3.020	3.250
5	2.942	2.667	2.554	2.493	2.758
10	2.629	2.343	2.189	2.155	2.399
20	2.491	2.196	2.022	2.044	2.236
50	2.420	2.120	1.934	1.925	2.150
100	2.408	2.100	1.909	1.903	2.126

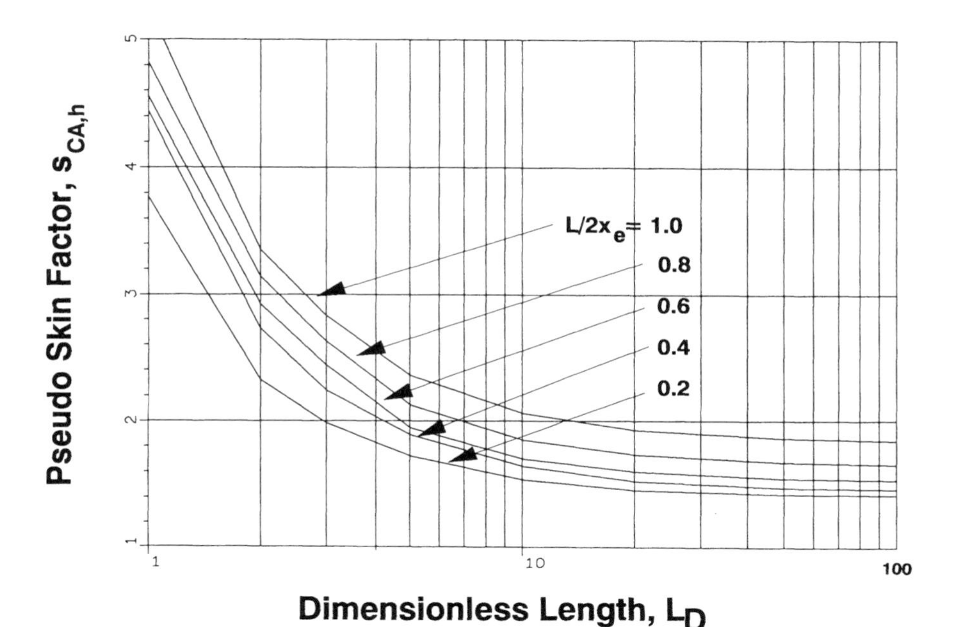

Figure 7–5 Shape Related Skin Factor, $s_{CA,h}$, for a Horizontal Well in a Square Drainage Area ($x_e/y_e = 1$).[6]

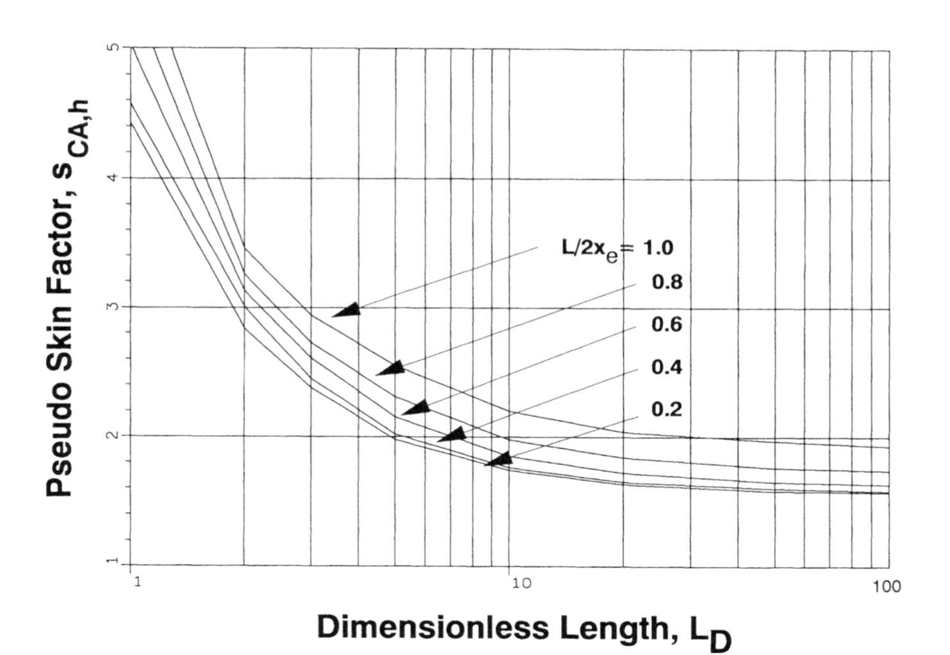

Figure 7–6 Shape Related Skin Factor, $s_{CA,h}$, for a Horizontal Well Located in a Rectangular Drainage Area ($x_e/y_e = 2$).[6]

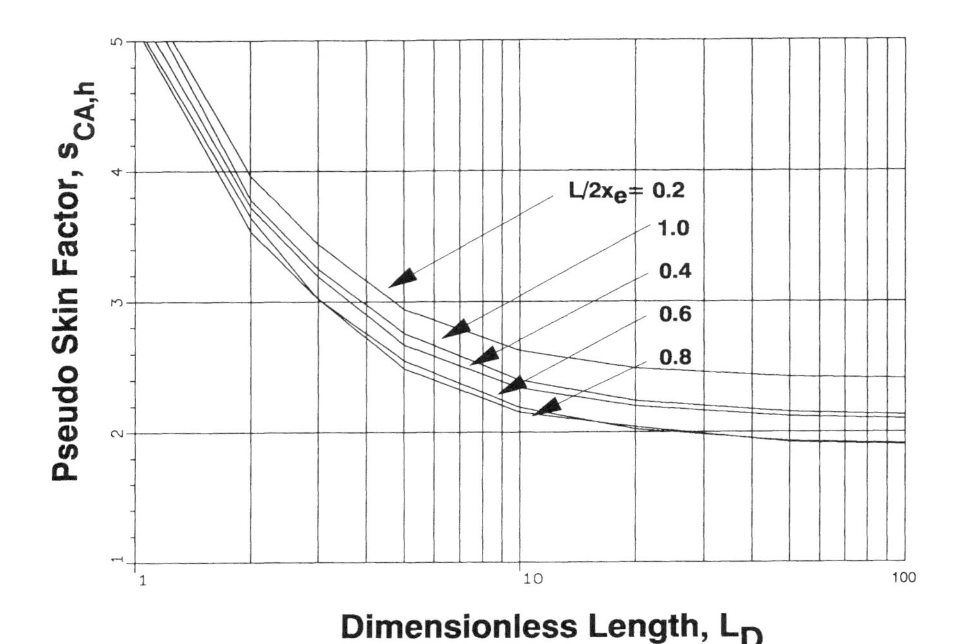

Figure 7–7 Shape Related Skin Factor, $s_{CA,h}$, for a Horizontal Well Located in a Rectangular Drainage Area ($x_e/y_e = 5$).[6]

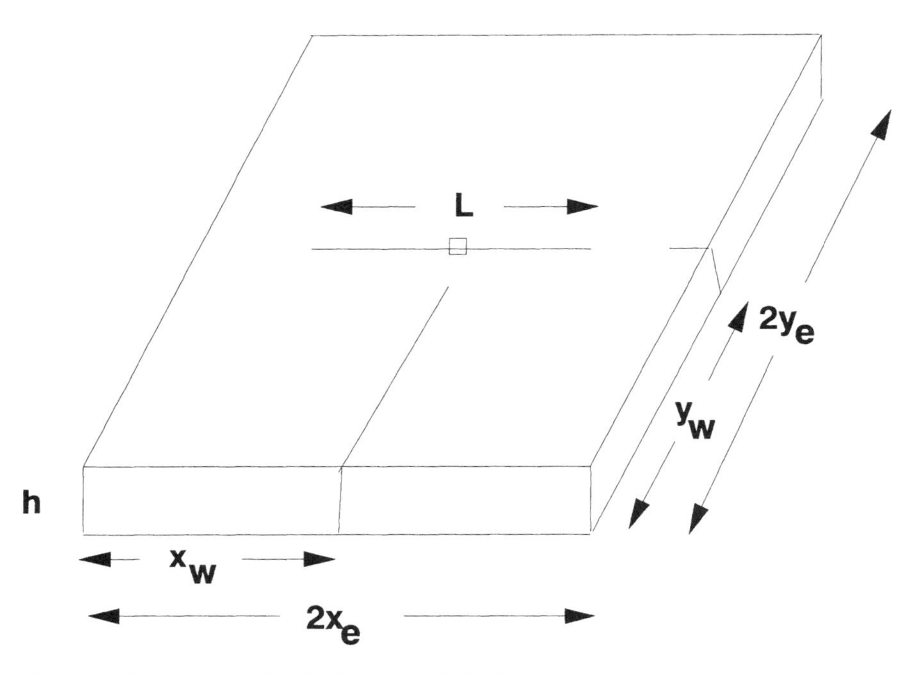

Figure 7–8 A Schematic of a Horizontal Well Located in a Rectangular Drainage Volume.

averaging pressure values of the uniform-flux solution along the wellbore length.[11] Because of the well boundary condition assumptions, Method III generally gives the highest flow rate and Method II gives the lowest rate of the three methods. However, the difference in the calculated flow rates using different methods is normally very small (±5%).

Method I

For rectangular drainage areas with $2x_e/(2y_e) = 1$ to 20, Mutalik et al.[6] reported the shape factors and the corresponding equivalent skin factors $s_{CA,h}$ for horizontal wells located at various positions within the drainage volume. The skin factors $s_{CA,h}$ for centrally located wells within drainage area with ratios of sides, $2x_e/(2y_e) = 1$, 2 and 5 are plotted in Figures 7–5 through 7–7 and are summarized in Table 7–5. The following equation can be used to calculate the productivity of a horizontal well:

$$J_h = \frac{q}{\bar{p}_R - p_{wf}} = \frac{0.007078\, kh/(\mu_o B_o)}{\ln\left(\frac{r_e'}{r_w}\right) - A' + s_f + s_m + s_{CA,h} - c' + Dq} \tag{7–24}$$

where

$$r_e' = \sqrt{A/\pi}, \text{ ft} \tag{7–10}$$

and

D = turbulence coefficient, 1/BOPD for oil and 1/MSCFD for gas
s_m = mechanical skin factor, dimensionless
s_f = skin factor of an infinite-conductivity, fully penetrating fracture of length, L
$s_f = -\ln[L/(4r_w)]$ (7–25)
$s_{CA,h}$ = shape-related skin factor
c' = shape factor conversion constant = 1.386

Note: s_m here is calculated as $s_m = s\,(h/L)\sqrt{kh/kv}$ where s is the skin factor obtained from Eq. 16–14).

2. Determine the value of $s_{CA,h}$ from Figures 7–5 through 7–7.
3. Knowing s_f and $s_{CA,h}$, the productivity can then be calculated using Equation 7–24.

Method II

In this method, a horizontal well problem is looked upon as a problem similar to that for a partially penetrating vertical well. If this partially penetrating vertical well problem is turned sideways, it results in a horizontal well problem. Babu and Odeh[10] derived the following equation for horizontal well pseudo-steady state productivity:

$$J_h = \frac{0.007078\,(2x_e)\,\sqrt{k_y k_v}/(\mu_o B_o)}{\ln(\sqrt{A_1}/r_w) + \ln C_H - 0.75 + s_R} \qquad (7\text{–}26)$$

where k_y is the horizontal permeability in the direction perpendicular to the wellbore. The value s_R accounts for the skin factor due to partial penetration of the horizontal well in the areal plane. $s_R = 0$ when $L = 2x_e$. C_H is the shape factor. A_1 is horizontal well drainage area in the vertical plane ($A_1 = 2y_e h$). The values $2x_e$ and $2y_e$ are reservoir dimensions shown in Figure 7–8. Equation 7–26 includes no formation damage, but it can be included as an additive term in the denominator.

Calculation of $\ln C_H$

$$\ln C_H = 6.28\left(\frac{2y_e}{h}\right)\sqrt{\frac{k_v}{k_y}}\left[\frac{1}{3} - \left(\frac{y_w}{2y_e}\right) + \left(\frac{y_w}{2y_e}\right)^2\right] - \ln\left[\sin\left(180^\circ \frac{z_w}{h}\right)\right] - 0.5\ln\left[\left(\frac{2y_e}{h}\right)\sqrt{\frac{k_v}{k_y}}\right] - 1.088 \qquad (7\text{–}27)$$

where z_w is the vertical distance between the horizontal well and the bottom boundary and as noted in Figure 7–8, y_w denotes the distance from the horizontal well to the closest boundary in the y direction.

Calculation of s_R

As stated previously, $s_R = 0$ when $L = 2x_e$. If $L < 2x_e$, then the value of partial penetration skin factor s_R depends upon the following two conditions:

Case 1: $2y_e/\sqrt{k_y} \geq 1.5x_e/\sqrt{k_x} >> 0.75h/\sqrt{k_v}$

Case 2: $2x_e/\sqrt{k_x} \geq 2.66y_e/\sqrt{k_y} >> 1.33h/\sqrt{k_v}$

Case 1

Here, $$s_R = PXYZ + PXY' \tag{7–28}$$

The *PXYZ* Component

$$PXYZ = \left[\frac{2x_e}{L} - 1\right]\left[\ln\left(\frac{h}{r_w}\right) + 0.25\ln\left(\frac{k_y}{k_v}\right) - \ln\left(\sin\frac{180^\circ z_w}{h}\right) - 1.84\right] \tag{7–29}$$

The *PXY'* Component

$$PXY' = \left(\frac{2(2x_e)^2}{L\,h}\sqrt{\frac{k_v}{k_x}}\right)\left[f(x) + 0.5\{f(y_1) - f(y_2)\}\right] \tag{7–30}$$

where f represents a function. The terms in parenthesis after f are their arguments defined as

$$x = \frac{L}{4x_e},\quad y_1 = \frac{4x_w + L}{4x_e},\ \text{and}\ y_2 = \frac{4x_w - L}{4x_e}$$

where x_w is the distance from the horizontal well mid-point to the closest boundary in the x direction (see Figure 7–8). Additionally, pressure computations are made at the mid-point along the well length, and function $f(x)$ is defined as

$$f(x) = -x\,[0.145 + \ln(x) - 0.137(x)^2] \tag{7–31}$$

The evaluation of $f(y_1)$ and $f(y_2)$ depends upon their arguments, $[(4x_w + L)/(4x_e)]$ and $[(4x_w - L)/(4x_e)]$, respectively. If the argument, $(y_1 \text{ or } y_2) \leq 1$, Equation 7–31 is used by replacing x with y_1 or y_2. On the other hand, if $(y_1 \text{ or } y_2)$ is > 1, then, the following equation can be used.

$$f(y) = (2 - y)\,[0.145 + \ln(2 - y) - 0.137(2 - y)^2] \tag{7–32}$$

where $$y = y_1 \text{ or } y_2 \tag{7–33}$$

Case 2

Here, $$s_R = PXYZ + PY + PXY \tag{7–34}$$

The definition of the three components in Equation 7–34 are given below.

The *PXYZ* Component

This is calculated using Equation 7–29.

The *PY* Component

$$PY = 6.28\,\frac{(2x_e)^2}{2y_e h}\,\frac{\sqrt{k_y k_v}}{k_x}\left[\left\{\frac{1}{3} - \left(\frac{x_w}{2x_e}\right) + \left(\frac{x_w}{2x_e}\right)^2\right\} + \frac{L}{48x_e}\left(\frac{L}{2x_e} - 3\right)\right] \qquad (7\text{–}35)$$

where x_w is the mid-point co-ordinate of the well.

The *PXY* Component

$$PXY = \left(\frac{2x_e}{L} - 1\right)\frac{6.28\,(2y_e)}{h}\sqrt{\frac{k_v}{k_y}}\left[\frac{1}{3} - \left(\frac{y_w}{2y_e}\right) + \left(\frac{y_w}{2y_e}\right)^2\right] \qquad (7\text{–}36)$$

$$\text{for } [\mathrm{Min}\,\{y_w, (2y_e - y_w)\} \geq 0.5y_e]$$

Eq. (7–36) is an approximation of the rigorous solution given below

$$PXY = \left(\frac{2x_e}{L} - 1\right)\frac{6.28\,(2y_e)}{h}\sqrt{\frac{k_v}{k_y}}\left[\frac{1}{3} - \left(\frac{y_w}{2y_e}\right) + \left(\frac{y_w}{2y_e}\right)^2\right] - \left[\frac{4(2x_e)(2y_e)}{\pi L h}\right]\sqrt{\frac{k_v}{k_y}}\sum_{1}^{3}\left[\frac{1}{n^2}\cos^2\frac{n\pi y_w}{2y_e}\exp\left(-\frac{n\pi L}{4y_e}\sqrt{\frac{k_y}{k_x}}\right)\right] \qquad (7\text{–}36a)$$

Although Equation 7–36a gives more accurate results, Equation 7–36 is an adequate approximation for many field applications.

Method III

Kuchuk et al.[11] used an approximate infinite-conductivity solution, where the constant wellbore pressure is obtained by averaging pressure values of the uniform-flux solution along the well length. Their productivity equation is expressed as

$$J_h = \frac{k_h h/(70.6\,\mu_o)}{F + (h/0.5L)\sqrt{k_h/k_v}\;s_x} \qquad (7\text{–}37)$$

F is a dimensionless function and depends upon $y_w/(2y_e)$, $x_w/(2x_e)$, $L/(4x_e)$ and $(y_e/x_e)\sqrt{k_x/k_y}$. Typical values of the function F are listed in Table 7–6. The value s_x is calculated using the following equation.

$$s_x = \ln\left[\left(\frac{\pi r_w}{h}\right)\left(1 + \sqrt{\frac{k_v}{k_h}}\right)\sin\left(\frac{\pi z_w}{h}\right)\right] - \sqrt{\frac{k_h}{k_v}}\left(\frac{2h}{L}\right)\left[\frac{1}{3} - \left(\frac{z_w}{h}\right) + \left(\frac{z_w}{h}\right)^2\right] \tag{7–38}$$

It is important to note that Equation 7–37 does not have B_0, i.e., the formation volume factor term. Hence, to obtain productivity for surface conditions, the B_0 term must be added in the denominator of Equation 7–37.

EXAMPLE 7–1

A horizontal well drilled in an oil reservoir has the following parameters

Area = 160 acres,	r_w = 0.365 ft	
h = 50 ft,	k_v/k_h = 0.1	
μ_0 = 0.5 cp,	$k_h = k_x = k_y$ = 1 md	
B_0 = 1.2 RB/STB,	s_m = 0,	$D = 0$.
z_w = 25 ft		

Calculate the pseudo-steady state productivity of a 2000-ft-long horizontal well by three different methods.

Solution

Method I

$$r_e' = \sqrt{\frac{A}{\pi}} = \sqrt{\frac{(160 \times 43560)}{\pi}} = 1489 \text{ ft}$$

1. **Calculation of s_f**

$$s_f = -\ln\left[\frac{L}{(4r_w)}\right] = -\ln\left[\frac{2000}{(4 \times 0.365)}\right] = -7.22$$

2. **Calculation of s_{CA}**

$$L_D = \frac{L}{2h}\sqrt{\frac{k_v}{k_h}} = \left(\frac{2000}{2 \times 50}\right)\sqrt{0.1} = 6.32$$

TABLE 7–6 VALUES OF DIMENSIONLESS FUNCTION, F, FOR CALCULATION OF PRODUCTIVITY OF HORIZONTAL WELLS (METHOD III)[11]

	$y_w/(2y_e) = 0.50, x_w/(2x_e) = 0.50$				
	$L/(4x_e)$				
$\frac{y_e}{x_e}\sqrt{\frac{k_x}{k_y}}$	**0.1**	**0.2**	**0.3**	**0.4**	**0.5**
0.25	3.80	2.11	1.09	0.48	0.26
0.50	3.25	1.87	1.12	0.69	0.52
1.00	3.62	2.30	1.60	1.21	1.05
2.00	4.66	3.34	2.65	2.25	2.09
4.00	6.75	5.44	4.74	4.35	4.19

	$y_w/(2y_e) = 0.25, x_w/(2x_e) = 0.50$				
	0.1	**0.2**	**0.3**	**0.4**	**0.5**
0.25	4.33	2.48	1.36	0.70	0.46
0.50	3.89	2.42	1.58	1.10	0.92
1.00	4.47	3.13	2.41	2.00	1.83
2.00	6.23	4.91	4.22	3.83	3.67
4.00	9.90	8.58	7.88	7.49	7.33

	$y_w/(2y_e) = 0.25, x_w/(2x_e) = 0.25$				
	0.05	**0.1**	**0.15**	**0.2**	**0.25**
0.25	9.08	7.48	6.43	5.65	5.05
0.50	6.97	5.56	4.71	4.12	3.71
1.00	6.91	5.54	4.76	4.24	3.90
2.00	8.38	7.02	6.26	5.76	5.44
4.00	11.97	10.61	9.85	9.36	9.04

	$y_w/(2y_e) = 0.50, x_w/(2x_e) = 0.25$				
	0.05	**0.1**	**0.15**	**0.2**	**0.25**
0.25	8.44	6.94	5.98	5.26	4.70
0.50	6.21	4.83	4.02	3.47	3.08
1.00	5.86	4.50	3.73	3.23	2.90
2.00	6.73	5.38	4.62	4.12	3.81
4.00	8.82	7.46	6.71	6.21	5.89

For a square drainage shape, $2x_e = 2y_e$.

$$2x_e = \sqrt{160 \times 43560} = 2640 \text{ ft}$$

$$\frac{L}{2x_e} = \frac{2000}{2640} = 0.757$$

For $L_D = 6.32$, $L/(2x_e) = 0.757$, from Figure 7–5, the shape-related pseudo-skin factor is $s_{CA,h} = 2.0$. Using Equation 7–24, with $s_m = 0$, $D = 0$

$$J_h = \frac{0.007078\, kh/(\mu_0 B_0)}{\ln (r_e'/r_w) - A' + s_{CA,h} + s_f - c'}$$

$$= \frac{0.007078 \times 1 \times 50/(0.5 \times 1.2)}{\ln (1489/0.365)] - 0.738 + 2.0 - 7.22 - 1.386}$$

$$= 0.61 \text{ STB/(day-psi)}$$

Method II

$$k_v/k_y = k_v/k_h = 0.1$$

1. **Calculation of $\ln C_H$:**

$$y_w/(2y_e) = 1320/2640 = 0.5$$

$$(z_w/h) = 25/50 = 0.5$$

Using Equation 7–27

$$\ln C_H = \left(6.28 \times \frac{2640}{50}\sqrt{0.1}\right)\left[\frac{1}{3} - 0.5 + 0.25\right]$$

$$- \ln [\sin (180° \times 0.5)] - 0.5 \ln \left[\left(\frac{2640}{50}\right)\sqrt{0.1}\right]$$

$$- 1.088 = 6.24$$

2. **Calculation of s_R:**

$$k_y = k_x = 1 \text{ md and } k_v = 0.1 \text{ md}$$

$$\frac{2y_e}{\sqrt{k_y}} = \frac{2640}{\sqrt{1}} = 2640$$

$$\frac{1.5x_e}{\sqrt{k_x}} = 1.5 \times \frac{1320}{\sqrt{1}} = 1980$$

$$\frac{0.75h}{\sqrt{k_v}} = 0.75 \times \frac{50}{\sqrt{0.1}} = 118.58$$

Hence, $$\frac{2y_e}{\sqrt{k_y}} > \frac{1.5x_e}{\sqrt{k_x}} > \frac{0.75h}{\sqrt{k_v}}$$

Therefore, one should use Case 1 where $s_R = PXYZ + PXY'$.

***PXYZ* Component**

Using Equation 7–29

$$PXYZ = \left[\frac{2640}{2000} - 1\right]\left[\ln\left(\frac{50}{0.365}\right) + 0.25\ln(10) - \ln[\sin(180° \times 0.5)] - 1.84\right] = 1.17$$

***PXY'* Component**

The first step is to calculate the three functions listed in Equation 7–30.

$$x = L/(4x_e) = 2000/(2 \times 2640) = 0.3788$$

Using Equation 7–31, the first function $f(x)$ is calculated as

$$\begin{aligned} f(x) &= -0.3788\,[0.145 + \ln(0.3788) - 0.137\,(0.3788)^2] \\ &= 0.32 \end{aligned}$$

Next, determine the value of the functions y_1 and y_2.

$$y_1 = \frac{4x_w + L}{4x_e} = \frac{4 \times 1320 + 2000}{2640 \times 2} = 1.3788$$

Since the argument y_1 is > 1, the function $f(y_1)$ is calculated using Equation 7–32.

$$\begin{aligned} 2 - y_1 &= 2 - 1.3788 \\ &= 0.62 \\ f(y_1) &= (0.62)\,[0.145 + \ln(0.62) - 0.137\,(0.62)^2] = -0.239 \end{aligned}$$

Next, the argument y_2 is calculated as

$$y_2 = \frac{4x_w - L}{4x_e} = \frac{4 \times 1320 - 2000}{2 \times 2640} = 0.62$$

Since $y_2 < 1.0$, Equation 7–31 is used to calculate $f(y_2)$ by replacing x with y_2. Thus,

$$f(y_2) = -(0.62)[0.145 + \ln(0.62) - 0.137\,(0.62)^2] = +0.239$$

Using Equation 7–30,

$$PXY' = \frac{2 \times 2640^2}{2000 \times 50}\sqrt{0.1}\,[0.32 + 0.5\,(-0.239 - 0.239)]$$
$$= 3.57$$

Using Equation 7–28,

$$s_R = 1.17 + 3.57 = 4.74$$

3. **Calculation of J_h**

Using Equation 7–26

$$J_h = \frac{0.007078 \times 2640\sqrt{1 \times 0.1}/(0.5 \times 1.2)}{\ln[(\sqrt{2640 \times 50})/0.365] + 6.24 - 0.75 + 4.74}$$
$$= 0.58 \text{ STB/(day-psi)}$$

Method III

1. **Calculation of F:**

The function F is a function of the following parameters.

$$\frac{y_w}{2y_e} = \frac{1320}{2640} = 0.5$$

$$\frac{L}{4x_e} = \frac{2000}{5280} = 0.3788$$

$$\frac{x_w}{2x_e} = \frac{1320}{2640} = 0.5$$

$$\frac{y_e}{x_e}\sqrt{\frac{k_x}{k_y}} = 1 \times \sqrt{1} = 1.0$$

From Table 7–6, for these sets of parameters, $F = 1.29$

2. **Calculation of s_x:**

$$\frac{z_w}{h} = \frac{25}{50} = 0.5$$

$$k_y/k_x = 1$$

$$k_h/k_v = 1/0.1 = 10$$

$$\frac{h}{0.5L} = \frac{50}{1000} = 0.05$$

Using Equations 7–37 and 7–38

$$s_x = -\ln\left[\left(\frac{\pi \times 0.365}{50}\right)(1 + \sqrt{0.1})\sin(180° \times 0.5)\right]$$
$$-\sqrt{10}\,(0.05)\,[0.333 - 0.5 + 0.25]$$
$$= 3.487$$

$$J_h = \frac{1 \times 50/(70.6 \times 0.5 \times 1.2)}{1.29 + (0.05)\sqrt{10} \times 3.487}$$
$$= 0.64 \text{ STB/(day-psi)}$$

A COMPARISON OF HORIZONTAL WELL PRODUCTIVITIES BY THE THREE METHODS FOR DIFFERENT VALUES OF k_v/k_h FOR EXAMPLE 7–1.

	J_h, (STB/(day-psi)), Horizontal Well Productivity		
Method	$k_v/k_h = 0.1$	$k_v/k_h = 0.5$	$k_v/k_h = 1$
Method I	0.61	0.75	0.80
Method II	0.58	0.69	0.74
Method III	0.64	0.78	0.82

EXAMPLE 7–2

A 2500-ft-long horizontal well is drilled on a 320-acre spacing. The reservoir is 50 ft thick. The permeabilities in the x, y, and z directions are 200, 200, and 100 md, respectively, and the wellbore radius = 0.25 ft. Additionally, μ_o = 0.7 cp and B_o = 1.25 RB/STB. Calculate the pseudo-steady productivity of the horizontal well by three different methods. Assume $s_m = 0$, $D = 0$.

Solution

Method I

$$r'_e = \sqrt{A/\pi} = \sqrt{320 \times 43560/\pi}$$
$$= 2106 \text{ ft}$$

Calculation of s_f

$$s_f = -\ln\left(\frac{L}{4r_w}\right)$$
$$= -\ln\left(\frac{2500}{4 \times 0.25}\right)$$
$$= -7.824$$

Calculation of $s_{CA,h}$

$$L_D = (L/(2h))\sqrt{k_v/k_h} = (2500/(2 \times 50))\sqrt{100/200} = 17.7$$
$$2x_e = 3733 \text{ ft (assuming a square drainage area)}$$
$$L/(2x_e) = 2500/3733 = 0.67$$

From Figure 7–5, in a square drainage area, corresponding to $L_D = 17.7$ and $L/(2x_e) = 0.67$

$$s_{CA,h} = 1.75$$

Calculation of Horizontal Well Productivity using Equation 7–24

$$J_h = \frac{0.007078kh/(\mu_0 B_0)}{\ln(r'_e/r_w) - 0.738 + s_f + s_{CA,h} - c'}$$
$$= \frac{0.007078 \times 200 \times 50/(0.7 \times 1.25)}{\ln(2106/0.25) - 0.738 + (-7.82) + 1.75 - 1.386}$$
$$= 96 \text{ STB/(day-psi)}$$

Method II

$$k_v/k_h = 100/200 = 0.5$$

1. **Calculation of $\ln C_H$:**

$$y_w/(2y_e) = 1867/3733 = 0.5$$
$$\frac{z_w}{h} = \frac{25}{50} = 0.5$$

Using Equation 7–27

$$\ln C_H = \left(6.28 \times \frac{3733}{50}\sqrt{0.5}\right)\left[\frac{1}{3} - 0.5 + 0.25\right] - \ln\left[\sin\left(180° \times 0.5\right)\right] - 0.5\ln\left[\left(\frac{3733}{50}\right)\sqrt{0.5}\right] - 1.088$$

$$= 24.56$$

2. **Calculation of s_R:**

$$k_y = k_x = 200 \text{ md and } k_v = 100 \text{ md}$$

$$\frac{2y_e}{\sqrt{k_y}} = \frac{3733}{\sqrt{200}} = 264$$

$$\frac{1.5x_e}{\sqrt{k_x}} = 0.75 \times \frac{3733}{\sqrt{200}} = 198$$

$$\frac{0.75h}{\sqrt{k_v}} = 0.75 \times \frac{50}{\sqrt{100}} = 3.75$$

Hence,

$$\frac{2y_e}{\sqrt{k_y}} > \frac{1.5x_e}{\sqrt{k_x}} > \frac{0.75h}{\sqrt{k_v}}$$

Therefore, one should use Case 1 where $s_R = PXYZ + PXY'$.

PXYZ Component

Using Equation 7–29

$$PXYZ = \left[\frac{3733}{2500} - 1\right]\left[\ln\left(\frac{50}{0.365}\right) + 0.25\ln(2) - \ln\left[\sin(180° \times 0.5)\right] - 1.84\right] = 1.6$$

PXY' Component

The first step is to calculate the three functions listed in Equation 7–30.

$$x = L/(4x_e) = 2500/(2 \times 3733) = 0.335$$

Using Equation 7–31, the first function $f(x)$ is calculated as

$$f(x) = -0.335\left[0.145 + \ln(0.335) - 0.137(0.335)^2\right]$$
$$= 0.323$$

Next, determine the value of the functions y_1 and y_2.

$$y_1 = \frac{4x_w + L}{4x_e} = \frac{4 \times 1867 + 2500}{3733 \times 2} = 1.335$$

Since the argument y_1 is > 1, the function $f(y_1)$ is calculated using Equation 7–32.

$$\begin{aligned} 2 - y_1 &= 2 - 1.335 \\ &= 0.665 \\ f(y_1) &= (0.665)\,[0.145 + \ln(0.665) - 0.137\,(0.665)^2] = -0.215 \end{aligned}$$

Next, the argument y_2 is calculated as

$$\begin{aligned} y_2 = \frac{4x_w - L}{4x_e} &= \frac{4 \times 1867 - 2500}{2 \times 3733} \\ &= 0.665 \end{aligned}$$

Since $y_2 < 1.0$, Equation 7.28 is used to calculate $f(y_2)$ by replacing x with y_2. Thus,

$$f(y_2) = -(0.665)[0.145 + \ln(0.665) - 0.137\,(0.665)^2] = +0.215$$

Using Equation 7–30,

$$\begin{aligned} PXY' &= \left(\frac{2 \times 3733^2}{2500 \times 50}\right) \sqrt{0.5}\,[0.323 + 0.5\,(-0.215 - 0.215)] \\ &= 17 \end{aligned}$$

Using Equation 7–28,

$$s_R = 1.6 + 17 = 18.6$$

3. **Calculation of J_h**

Using Equation 7–26

$$\begin{aligned} J_h &= \frac{0.007078 \times 3733\,\sqrt{200 \times 100}/(0.7 \times 1.25)}{\ln[(\sqrt{3733 \times 50})/0.25] + 24.56 - 0.75 + 18.6} \\ &= 86 \text{ STB/(day-psi)} \end{aligned}$$

Method III

1. **Calculation of F:**

The function F is a function of the following parameters.

$$\frac{y_w}{2y_e} = \frac{1867}{3733} = 0.5$$

$$\frac{L}{4x_e} = \frac{2500}{(2 \times 3733)} = 0.335$$

$$\frac{x_w}{2x_e} = \frac{1867}{3733} = 0.5$$

$$\frac{y_e}{x_e}\sqrt{\frac{k_x}{k_y}} = 1 \times \sqrt{1} = 1.0$$

From Table 7–6, F is calculated using linear interpolation as

$$F = 1.47$$

2. **Calculation of s_x:**

$$\frac{z_w}{h} = \frac{25}{50} = 0.5$$

$$k_y/k_x = 1$$

$$k_h/k_v = 200/100 = 2$$

$$\frac{h}{0.5L} = \frac{50}{1250} = 0.04$$

Using Equation 7–37 and 7–38

$$s_x = -\ln\left[\left(\frac{\pi \times 0.25}{50}\right)(1 + \sqrt{0.5})\sin(180° \times 0.5)\right]$$
$$-\sqrt{2}\,(0.04)\,[0.333 - 0.5 + 0.25)$$
$$= 3.61$$

$$J_h = \frac{200 \times 50/(70.6 \times 0.7 \times 1.25)}{1.47 + (0.04)\sqrt{2} \times 3.61}$$
$$= 96.7 \text{ STB/(day-psi)}$$

The following table shows a comparison of the horizontal well productivities obtained by the three methods.

Methods	J_h, STB/(day-psi)
1. Method I	96
2. Method II	86
3. Method III	97

INFLOW PERFORMANCE OF PARTIALLY OPEN HORIZONTAL WELLS

From a practical standpoint, inflow performance of a partially open horizontal well is very important. This is especially important in cemented horizontal wells, where only parts of the well are perforated.

If one perforates a horizontal well similar to a vertical well, with one to eight shots per foot length, perforation can be expensive. Additionally, from a practical standpoint, one may have to leave gaps in the perforations, where packers would be set. Thus, an engineer would have to evaluate the influence of well productivity loss due to the limited zone opening. Engineering intuition tells us that these open sections should be spaced uniformly along the well length. (Of course, this assumes a uniform-permeability reservoir.) At the present time, only one solution is available to calculate productivity of a partially open horizontal well. The mathematical solution is restricted to a horizontal well of length L drilled in a rectangular drainage area with $2x_e/(2y_e) = 2$.[12] The tabular results are available for the following three distributions of 20% open-flow length and for a fully open horizontal well.

Case 1 Five open intervals, each 4% of the well length, which are uniformly distributed along the well.

Case 2 Two open intervals, each being 10% of the well length and placed at each end of the well.

Case 3 Single open interval at the well center.

Case 4 Fully open horizontal well.

Productivity of a partially open horizontal well is[12]

$$J_h = \frac{0.007078\, k_h h/(\mu_o B_o)}{p_{wD} + s_m^*} \tag{7–39}$$

where

$$k_h = \sqrt{k_y k_x} = \text{effective horizontal permeability, md}$$
$$p_{wD} = \text{Dimensionless Inflow Pressure} \tag{7–40}$$
$$s_m^* = \text{skin factor}$$

As noted above, p_{wD} values are listed only when 20% of the well length is open to flow. For other open flow length fractions Figure 7–9 can be used.

Calculation of p_{wD}

p_{wD} depends upon the ratio, $L/(4x_e)$ and h_D^*, the dimensionless height defined as

$$h_D^* = \frac{h}{2x_e}\sqrt{\frac{k_h}{k_v}} \tag{7–41}$$

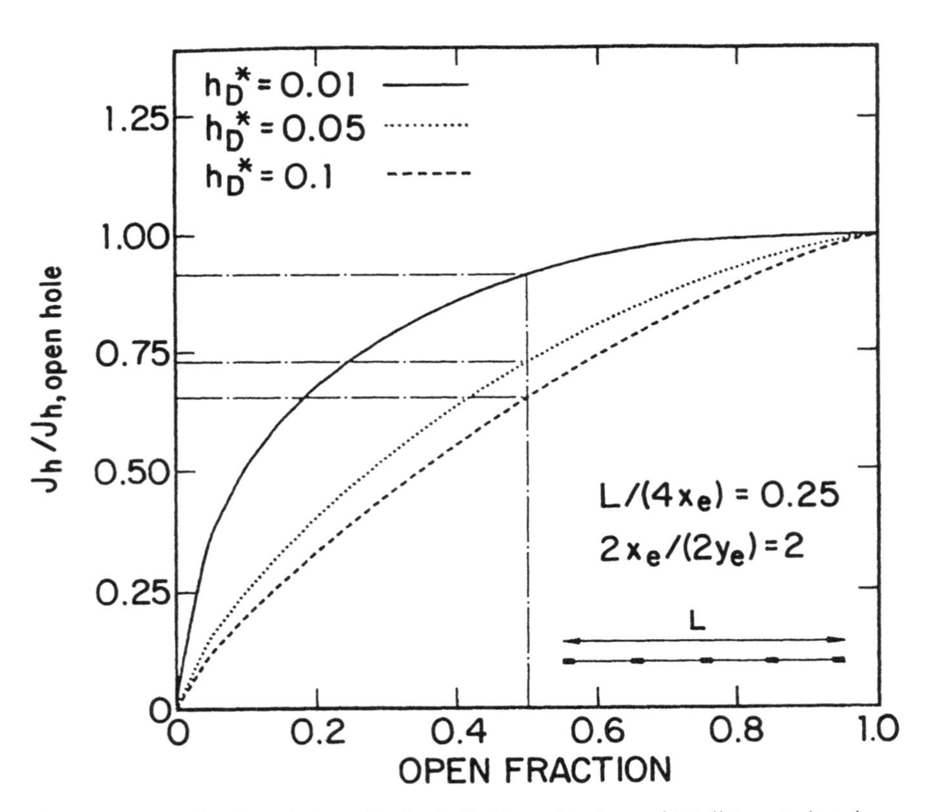

Figure 7–9 Productivity Index of a Partially Open Horizontal Well Located at the Formation Mid-height.[12]

For specific values of h_D^* and $L/(4x_e)$, p_{wD} values are listed in Table 7–7.

Calculation of s_m^*

The skin factor s_m^* is related to the van Everdingen definition of mechanical skin factor as

$$s_m^* = \frac{h}{L_p}\sqrt{\frac{k_h}{k_v}}\, s_m \tag{7–42}$$

where

s_m = mechanical skin factor due to formation damage or stimulation
L_p = total open length (perforated) of the well.

EXAMPLE 7–3

It is proposed to drill a 2000-ft-long horizontal well on a 320-acre well spacing. Assume that the drainage area is rectangular and $2x_e/(2y_e) = 2$. The horizontal well is fully cemented and is selectively perforated such that 20% of the drilled length is open to flow. Other reservoir parameters are:

$k_h = 10$ md, $k_v/k_h = 1.0$
$h = 50$ ft, $s_m = +2$
$\mu_0 = 0.4$ cp, $B_0 = 1.25$ RB/STB

Three completion schemes are being considered.

Case 1 Five open intervals, each 80 ft long, uniformly distributed along the well.
Case 2 Two open intervals, each 200 ft long and placed at each end of the well.
Case 3 Single open interval, 400 ft long, placed at the center of the well.
Case 4 Fully open horizontal well.

Estimate the horizontal well productivity for the three completion schemes and also for the fully open horizontal well.

Solution

The productivity of a partially open horizontal well is given by

$$J_h = \frac{0.007078\, k_h h/(\mu_0 B_0)}{p_{wD} + s_m^*}$$

TABLE 7–7 VALUES OF DIMENSIONLESS INFLOW PRESSURE, p_{wD}, FOR DIFFERENT h_D^* AND $L/(4x_e)$, FOR CALCULATION OF PRODUCTIVITY OF PARTIALLY OPEN HORIZONTAL WELLS LOCATED AT FORMATION MID-HEIGHT[12]

	$L/(4x_e)$	$h_D^* = 0.01$ p_{wD}	$h_D^* = 0.1$ p_{wD}	h_D^*	$L/(4x_e) = 0.1$ p_{wD}	$L/(4x_e) = 0.25$ p_{wD}
Open Hole*	0.025	3.583	12.739	0.001	1.627	0.726
Case 1		5.310	30.927		1.772	0.880
Case 2		6.027	38.478		2.094	1.261
Case 3		7.344	44.907		3.250	2.324
	0.050	2.605	7.344	0.003	1.650	0.729
		3.694	19.189		1.880	0.895
		4.134	22.917		2.204	1.276
		5.331	26.267		3.360	2.339
	0.075	2.104	5.300	0.005	1.679	0.736
		2.917	14.254		2.019	0.928
		3.293	16.695		2.349	1.310
		4.467	19.091		3.507	2.373
	0.100	1.769	4.179	0.007	1.713	0.751
		2.428	11.447		2.176	1.002
		2.781	13.249		2.513	1.385
		3.947	15.204		3.673	2.448
	0.125	1.517	3.452	0.010	1.769	0.782
		2.080	9.602		2.428	1.155
		2.424	11.029		2.781	1.541
		3.583	12.739		3.947	2.606
	0.150	1.316	2.932	0.030	2.210	0.848
		1.813	8.283		4.317	1.464
		2.156	9.468		4.884	1.861
		3.306	11.023		6.132	2.929
	0.175	1.150	2.537	0.050	2.739	0.980
		1.600	7.285		6.315	2.069
		1.946	8.306		7.198	2.500
		3.085	9.754		8.599	3.579
	0.200	1.009	2.224	0.070	3.296	1.242
		1.425	6.501		8.351	3.217
		1.780	7.408		0.590	3.746
		2.900	8.772		11.191	4.862
	0.225	0.888	1.969	0.100	4.179	1.756
		1.279	5.866		11.447	5.342
		1.647	6.696		13.249	6.119
		2.743	7.987		15.204	7.344
	0.250	0.782	1.756			
		1.155	5.342			
		1.541	6.119			
		2.606	7.344			

* Open hole represents Case 4 in the text

1. **Calculation of p_{wD}**

Since drainage area is rectangular and $2x_e/(2y_e) = 2$.

$$(2x_e)(2y_e) = 320 \times 43560$$

Also, $2x_e = 4y_e$
Hence, $(4y_e)(2y_e) = 320 \times 43560$

$$y_e = 1320 \text{ ft}$$
$$x_e = 2640 \text{ ft}$$
$$L = \text{Horizontal well length} = 2000 \text{ ft}$$
$$L/(4x_e) = 2000/10560 = 0.189$$

$$h_D^* = \frac{h}{2x_e}\sqrt{\frac{k_h}{k_v}} = \frac{50}{5280}\sqrt{1} = 0.01$$

Calculate p_{wD} from Table 7–7, for $L/(4x_e) \approx 0.19$ and $h_D^* = 0.01$.

	p_{WD}
1. 5 open intervals	1.502
2. 2 open intervals at the ends	1.853
3. 1 open interval at the center	2.981
4. Fully open horizontal well	1.071

2. **Calculation of s_m^* (from Eq. 7–42)**

$$L_p = (0.2 \times 2000) = 400 \text{ ft}$$

$$s_m^* = \frac{h}{L_p}\sqrt{\frac{k_h}{k_v}}\, s_m$$
$$= \frac{50}{400}\sqrt{1} \times 2$$
$$= 0.25$$

3. **Calculation of Productivity Index, J_h**

$$J_h = \frac{0.007078 \times k_h h/(\mu_o B_o)}{p_{WD} + s_m^*}$$
$$= \frac{0.007078 \times 10 \times 50/(0.4 \times 1.25)}{p_{WD} + 0.25}$$

The results are summarized below.

	J_h, STB/(day-psi)
1. 5 open intervals evenly spaced	4.0
2. 2 open intervals at the ends	3.4
3. 1 open interval at the center	2.2
4. Fully open horizontal well	5.4

INFLOW PERFORMANCE RELATIONSHIP (IPR) FOR HORIZONTAL WELLS IN SOLUTION GAS-DRIVE RESERVOIRS

For solution gas-drive reservoirs, Vogel[13] and Fetkovich[14] have reported inflow performance relationships for vertical wells.

Vogel Equation[13]

$$\frac{q_0}{q_{0,\max}} = \left[1 - 0.2\left(\frac{p_{wf}}{\bar{p}}\right) - 0.8\left(\frac{p_{wf}}{\bar{p}}\right)^2\right] \qquad (7\text{–}43)$$

Fetkovich Equation[14]

$$\frac{q_0}{q_{0,\max}} = \left[1 - \left(\frac{p_{wf}}{\bar{p}}\right)^2\right]^n \qquad (7\text{–}44)$$

where

q_0 = oil flow rate, STB/day
$\bar{p}$ = shut-in pressure or average reservoir pressure, psia
p_{wf} = flowing bottomhole pressure, psia
$q_{0,\max}$ = maximum flow rate for 100% drawdown, STB/day
n = exponent, dimensionless

Exponent n in the Fetkovich Equation is normally equal to one. For n = 1.24, Equations 7–43 and 7–44 yield identical results.

To estimate horizontal well production rates in a solution gas-drive reservoir, Bendakhlia and Aziz[15] developed the following equation.

$$\frac{q_0}{q_{0,\max}} = \left[1 - V\left(\frac{p_{wf}}{\bar{p}}\right) - (1 - V)\left(\frac{p_{wf}}{\bar{p}}\right)^2\right]^n \quad (7\text{–}45)$$

V = variable parameter, dimensionless

The above equation was developed using the results of numerical simulation. As shown in Figure 7–10, the parameters V and n were correlated as a function of recovery factor. The results show that horizontal wells exhibit a high initial oil productivity that decreases very rapidly with cumulative production. The high initial oil productivity is due to the large contact area provided by the horizontal well. Moreover, initially the bounded reservoir was assumed to be at the bubble point, i.e., the free-gas phase is absent. Therefore, the total reservoir permeability contributes to oil flow. As oil is produced, the average reservoir pressure in a bounded reservoir decreases, a gas phase is generated and the well starts producing both oil and gas. This multiphase production reduces effective reservoir permeability to oil, resulting in a drop of oil productivity of a horizontal well.

EXAMPLE 7–4

A 2000-ft-long horizontal well is drilled in an oil reservoir producing by a solution gas-drive mechanism. At a bottom hole pressure of 2000 psia, the well produced 400 STB/day. The reservoir pressure was 2500 psia and the recovery factor was 4%. If the bubble point pressure is 2500 psia: (a) calculate the maximum oil flow rate, $q_{0,\max}$; (b) calculate the oil rate for p_{wf} = 1500 psia; and (c) construct the IPR curve.

Solution

The IPR equation for a horizontal well in a solution gas-drive reservoir is

$$\frac{q_0}{q_{0,\max}} = \left[1 - V\left(\frac{p_{wf}}{\bar{p}}\right) - (1 - V)\left(\frac{p_{wf}}{\bar{p}}\right)^2\right]^n \quad (7\text{–}45)$$

1. **Calculation of Maximum Oil Rate, $q_{0,\max}$.**

The first step is to calculate the value of $q_{0,\max}$. It is known that q = 400 STB/day at p_{wf} = 2000 psia. Also, $\bar{p}$ = 2500 psia and recovery factor = 4%. From Figure 7–10, at a recovery factor of 4%, V = 0.1 and n = 1. Substituting these values in Equation 7–45,

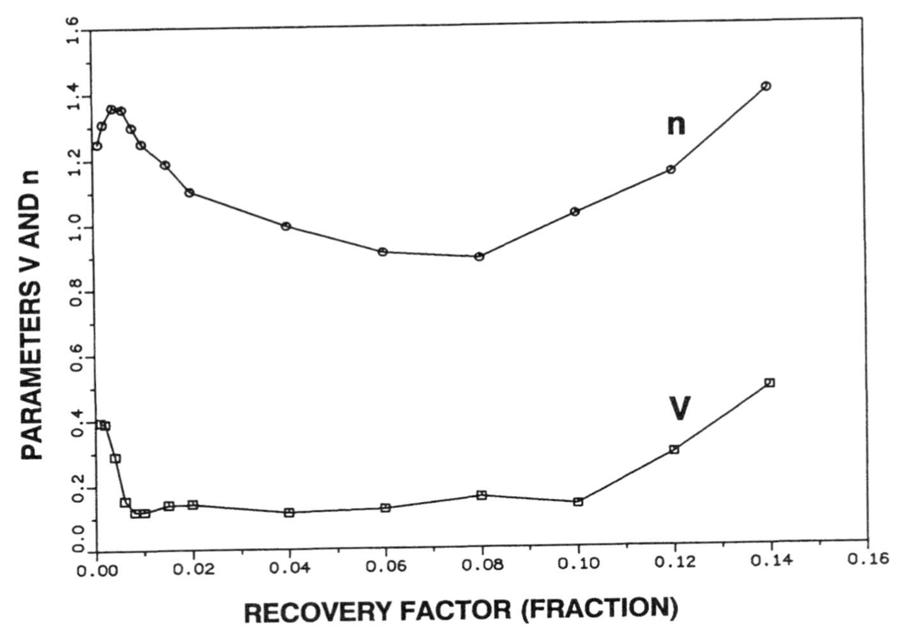

Figure 7–10 Correlation for Parameters, *V* and *n*, as a Function of Recovery Factor.[15]

$$q_{o,\max} = \frac{400}{\left[1 - 0.1\left(\frac{p_{wf}}{\bar{p}}\right) - (1 - 0.1)\left(\frac{p_{wf}}{\bar{p}}\right)^{2}\right]^{1}}$$

$$p_{wf}/\bar{p} = 2000/2500 = 0.8$$

$$q_{o,\max} = \frac{400}{1 - 0.1 \times 0.8 - (1 - 0.1) \times (0.8)^2}$$

$$= 1163 \text{ STB/day}$$

2. **Calculation of Oil Rate for p_{wf} = 1500 psia.**

$$p_{wf}/\bar{p} = 1500/2500 = 0.6$$

$$\frac{q_o}{q_{o,\max}} = 1 - V \times \left(\frac{p_{wf}}{\bar{p}}\right) - (1 - V) \times \left(\frac{p_{wf}}{\bar{p}}\right)^2$$

$$= 1 - 0.1 \times 0.6 - (1 - 0.1) \times (0.6)^2$$

$$= 0.616$$

$$q_o = q_{o,\max} \times 0.616 = 0.616 \times 1163$$

$$= 716 \text{ STB/day}$$

3. **IPR Curve.**

The same procedure is applied to calculate the oil flow rate at other bottomhole flowing pressures. The results are summarized in the following table and are plotted in Figure 7–11.

Bottomhole Pressure, p_{wf} psia	Oil Rate q_o STB/day
0	1163
500	1098
1000	949
1500	716
2000	400
2500	0

Similarly, IPRs can be generated at different recovery factors by using suitable parameters V and n obtained from Figure 7–10.

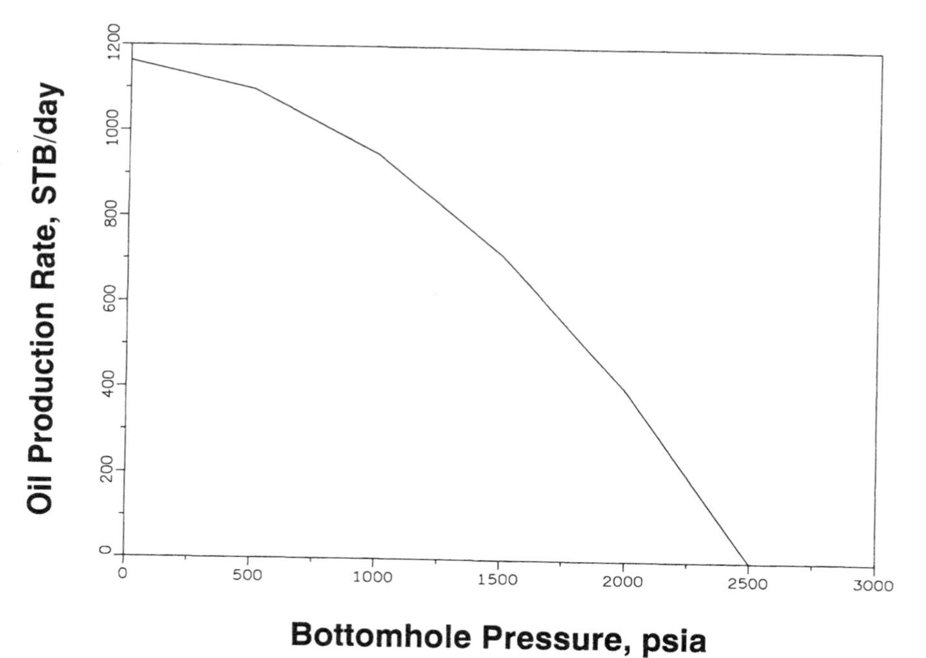

Figure 7–11 Inflow Performance Plot of a Horizontal Well for the Data Given in Example 7–4.

PREDICTING HORIZONTAL WELL PERFORMANCE IN SOLUTION GAS-DRIVE RESERVOIRS

Plahn et al.[16] have obtained numerical results for predicting horizontal well performance in solution gas drive reservoirs. They assume that the horizontal well is completed along the bottom of the productive interval, and the reservoir is initially at the bubble point. They presented type curves for forecasting oil production from horizontal wells producing at their maximum rate (with constant bottomhole pressure) from homogeneous isotropic solution gas-drive reservoirs. These type curves are shown in Figures 7–12 and 7–13. Based upon numerical results for square drainage areas, they have estimated upper, lower and average results. Their average graphical correlations have been correlated and are listed below:[17]

For 0 to 5% Critical Gas Saturation

$$y = 1.0271 - 0.41104\, x_1 - 0.1078\, x_1^2 \quad \text{for } x_1 \leq -2.2078 \tag{7–46}$$

$$y = 1.5526 + 0.06502\, x_1 \quad \text{for } x_1 \geq -2.2078 \tag{7–47}$$

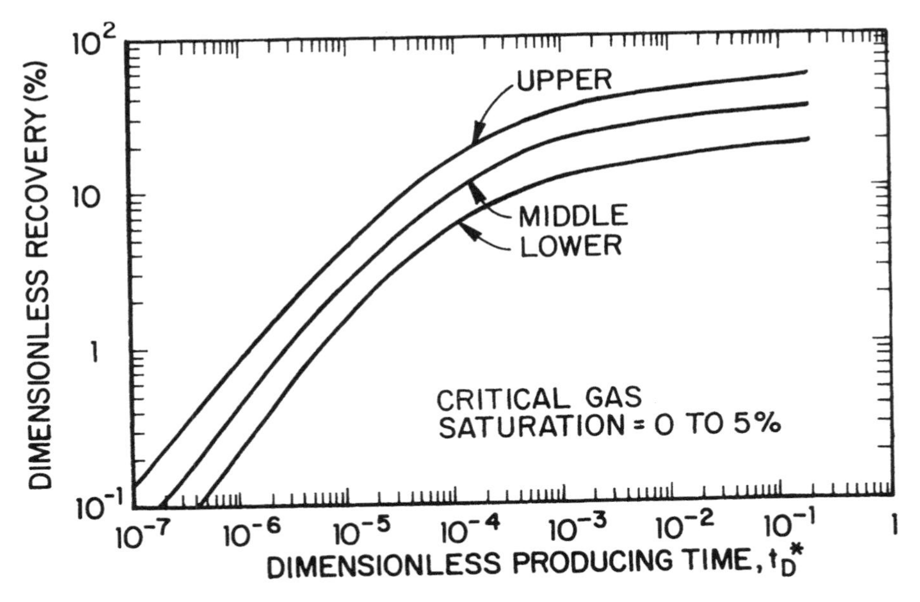

Figure 7–12 A Type Curve for a Horizontal Well in a Solution Gas-Drive Reservoir.[16]

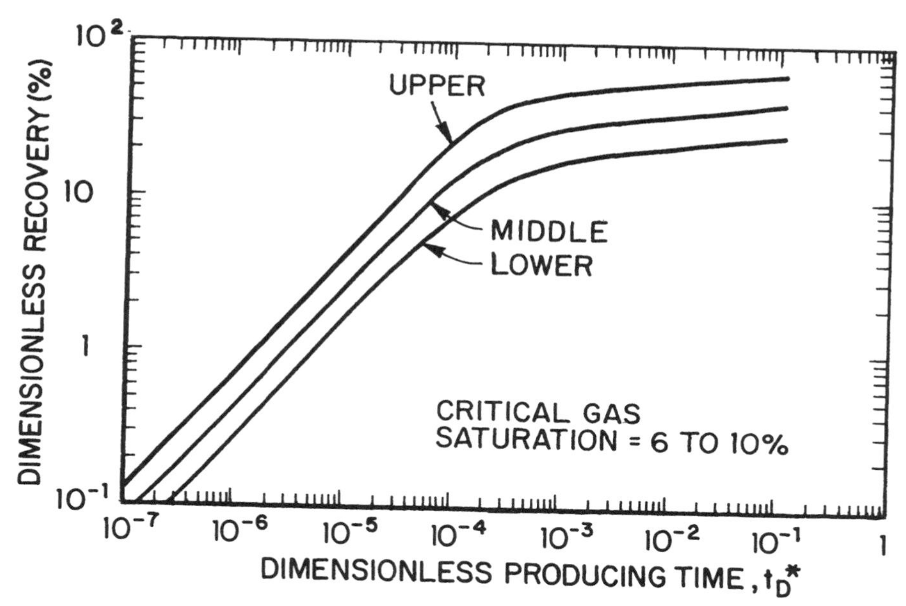

Figure 7–13 A Type Curve for a Horizontal Well in a Solution Gas-Drive Reservoir.[16]

For 6 to 10% Critical Gas Saturation

$$y = 1.2504 - 0.3903\, x_1 - 0.1097\, x_1^2 \quad \text{for } x_1 \leq -1.9469 \tag{7–48}$$

$$y = 1.6663 + 0.03701\, x_1 \quad \text{for } x_1 \geq -1.9469 \tag{7–49}$$

where

$$y = \log N_D, \text{ and } x_1 = \log t_D^* \tag{7–50}$$

Dimensionless producing time in U.S. oil field units is

$$t_D^* = \frac{0.00633\, k\, k_{roi}\, r_w\, L\, p_i\, t}{8\, \phi\, \mu_{oi}\, h\, x_e^3} \tag{7–51}$$

μ_{oi}, k_{roi} and p_i are evaluated at time zero and t is time in days. Dimensionless recovery N_D is

$$N_D = \frac{N_p}{N_m} \times 100 = \left(\frac{\text{cumulative oil produced at time } t}{\text{original moveable oil in place}}\right) \times 100 \tag{7–52}$$

The original moveable oil in place is defined as

$$N_m = \frac{(2x_e)^2 \, h\phi \, (S_{oi} - S_{or})}{5.615 \, B_{oi}} \tag{7–53}$$

S_{oi} = initial oil saturation, fraction
S_{or} = residual oil saturation, fraction
B_{oi} = oil formation volume factor at initial pressure, p_i, RB/STB
x_e = half length of side of square drainage area, ft.

It is important to note that the type curves in Figures 7–12 and 7–13 are developed assuming that the reservoir is isotropic and is initially at the bubble point. Additionally, it was assumed that the well is producing at a maximum possible rate with a constant bottomhole pressure. Equation 7–51, which defines dimensionless time, is based upon initial pressure, p_i, and does not include the effect of well flowing pressure, p_{wf}. Thus, when the influence of actual minimum flowing bottomhole pressure is included, the actual flow rate in a solution gas-drive reservoir using horizontal wells is less than that predicted using the type curves. The type curve represents the upper limit of the flow rate that can be produced using horizontal wells.

Reference 16 also included a study of influence of vertical permeability on maximum production rates. As shown in Figure 7–14, a decrease in vertical permeability reduces the oil production rate.

EXAMPLE 7–5

A 2000-ft-long horizontal well is drilled in 160-acre spacing in a solution gas drive reservoir.

Given:

B_{oi}	= 1.25 RB/STB,	p_i	= 1500 psig
h	= 60 ft,	r_w	= 0.33 ft
k	= 50 md,	S_{gc}	= 0.08
k_{roi}	= 0.60,	S_{oi}	= 0.65
$2x_e$	= 2640 ft (160 acres),	S_{or}	= 0.25
L	= 2000 ft,	μ_{oi}	= 1.8 cp
ϕ	= 0.20,	API	= 33°
γ_g	= 0.75,	BHT	= 200°F
z	= 0.90		

Using the Plahn et al. method, calculate the oil production rates over time.

Solution

Assuming a square drainage area, for 160-acre spacing, $2x_e$ = 2640 ft.

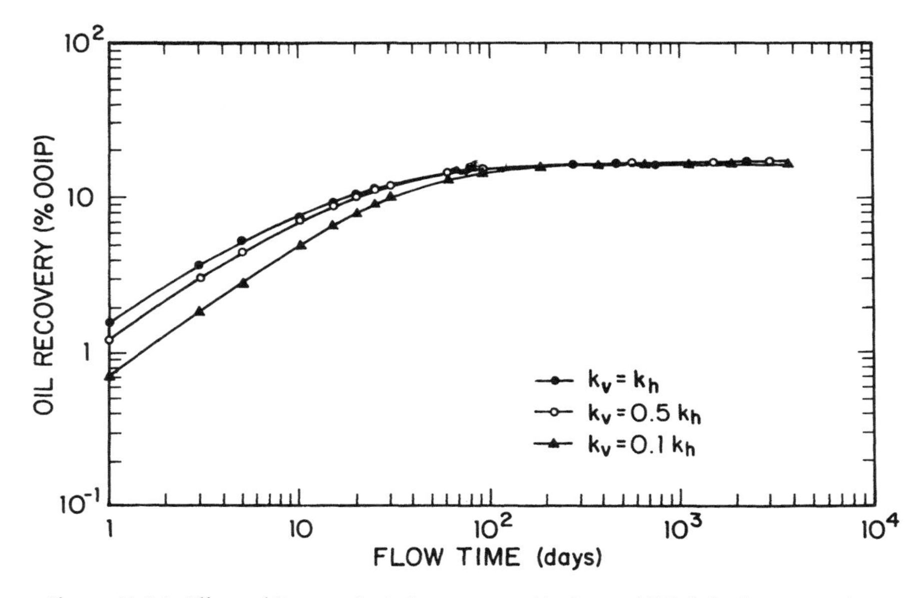

Figure 7–14 Effect of Reservoir Anisotropy on Horizontal Well Performance in a Solution Gas-Drive Reservoir.[16]

1. **Calculation of original movable oil in place, N_m**

$$N_m = \frac{(2x_e)^2\, h\, \phi\, (S_{oi} - S_{or})}{5.615\, B_{oi}} \tag{7–53}$$

$$= \frac{2640^2 \times 60 \times 0.2 \times (0.65 - 0.25)}{5.615 \times 1.25}$$

$$= 4.766 \times 10^6 \text{ STB}$$
$$= 4.77 \text{ MM STB}$$

2. **Dimensionless Time**

$$t_D^* = \frac{0.00633\, k\, k_{roi}\, r_w\, L\, p_i\, t}{\phi\, \mu\, h\, (2x_e)^3} \tag{7–51}$$

$$= \frac{0.00633 \times 50 \times 0.6 \times 0.33 \times 2000 \times 1500 \times t}{0.2 \times 1.8 \times 60 \times (2640)^3}$$

$$= (4.73 \times 10^{-7})\, t$$

3. **The calculations are illustrated for t = 1 day.**

$$t_D^* = 4.73 \times 10^{-7} \times 1 = 4.73 \times 10^{-7}$$

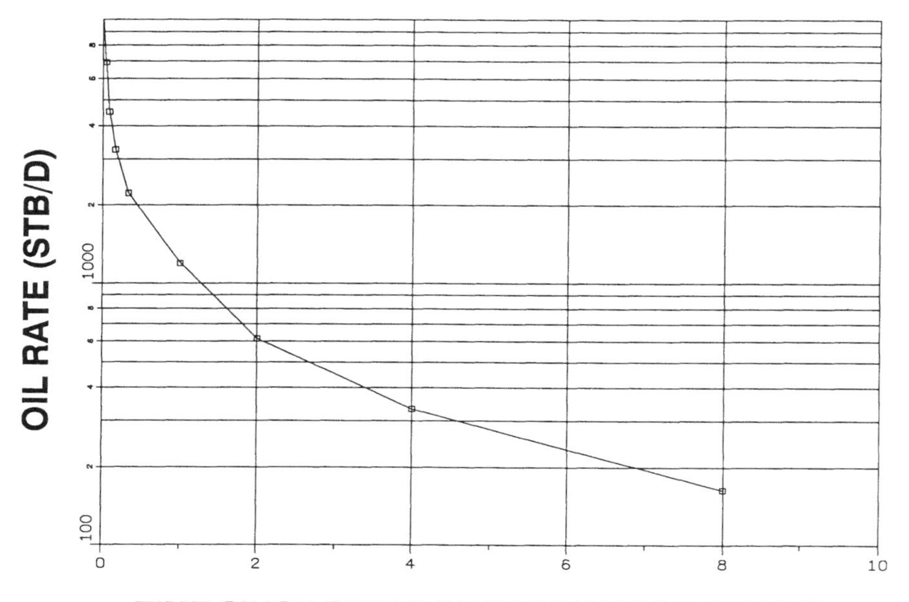

Figure 7–15 A Variation of Oil Rate over Time for the Data Given in Example 7–5.

4. **Calculation of N_D from Correlation**

$$x_1 = \log t_D^* = \log (4.73 \times 10^{-7}) = -6.325$$

Since critical gas saturation is 8% and $x_1 = -6.325 < -1.9469$, Equation 7–48 is used to calculate y.

$$y = 1.2504 - 0.3903x_1 - 0.1097x_1^2 \tag{7–48}$$

$$= 1.2504 - 0.3903 \times (-6.325) - 0.1097 (-6.325)^2$$

$$= -0.67$$

$$y = \log N_D$$

$$N_D = 10^{-y} = 10^{-0.67} = 0.214$$

5. **Oil Produced**

$$N_p = \frac{N_D N_m}{100}$$

$$= \frac{0.214 \times 4.77 \times 10^6}{100}$$

$$= 10197 \text{ STB}$$

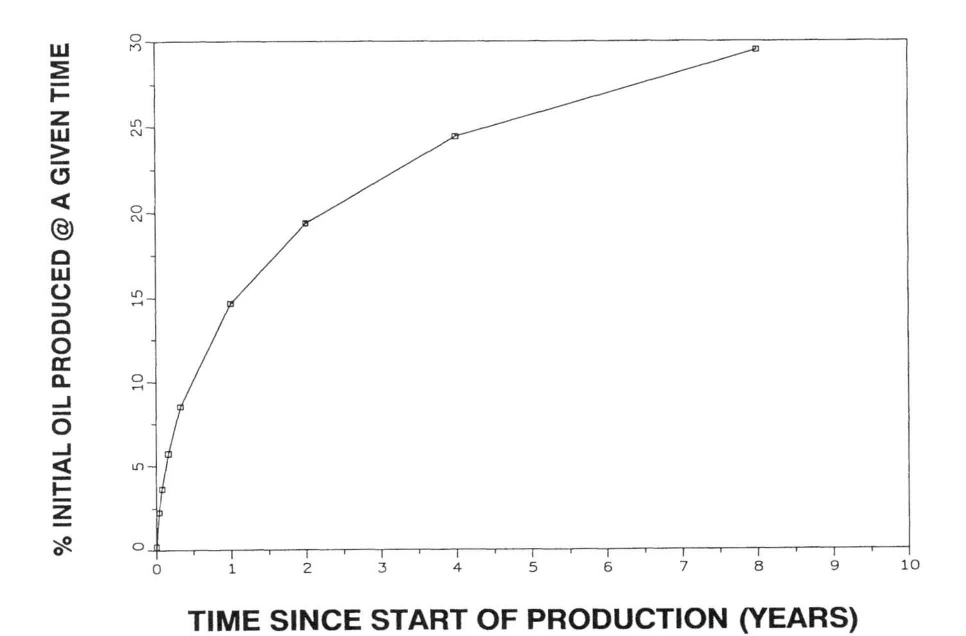

Figure 7–16 A Variation in Percentage of Initial Oil in Place Produced over Time for the Data Given in Example 7–5.

The production forecast calculations are summarized in the following table. The results are plotted in Figures 7–15 and 7–16.

t days	Δt days	t_D	$N_D\%$	N_p STB	ΔN_p STB	$q_o = \Delta N_p/\Delta t$ STB/day
1	1	4.73E-07	0.214	10197	10197	10197
15	14	7.09E-06	2.247	107108	96910	6922
30	15	1.42E-05	3.667	174768	67660	4511
60	30	2.84E-05	5.715	272408	97641	3255
120	60	5.68E-05	8.510	405600	133192	2220
365	245	1.73E-04	14.649	698208	292608	1194
730	365	3.45E-04	19.359	922726	224518	615
1460	730	6.91E-04	24.439	1164874	242148	332
2920	1460	1.38E-03	29.472	1404763	239890	164

REFERENCES

1. Dake, L. P.: *Fundamentals of Reservoir Engineering,* Elsevier Scientific Publishing, Co., New York, 1978.
2. Golan, M. and Whitson, C. H.: *Well Performance,* International Human Resources Corporation, Boston, Massachusetts, 1985.
3. Earlougher, Jr., R. C.: "Advances in Well Test Analysis," Monograph Vol. 5 of the Henry L. Doherty Series in Society of Petroleum Engineers of AIME, 1977.
4. Fetkovich, M. J. and Vienot, M. E.: "Shape Factors, C_A, Expressed as a Skin, s_{CA}," *Journal of Petroleum Technology,* pp. 321–322, February 1985.
5. Gringarten, A. C.: "Reservoir Limit Testing for Fractured Wells," paper SPE 7452, presented at the SPE 53rd Annual Fall Technical Conference and Exhibition, Houston, Texas, Oct. 1–3, 1978.
6. Mutalik, P. N., Godbole, S. P., and Joshi, S. D.: "Effect of Drainage Area Shapes on Horizontal Well Productivity," paper SPE 18301, presented in the SPE 63rd Annual Technical Conference, Houston, Texas, Oct. 2–5, 1988.
7. Gringarten, A. C., Ramey, Jr., H. J., and Raghavan, R.: "Unsteady-State Pressure Distribution Created by a Well with a Single Infinite-Conductivity Vertical Fracture," *Society of Petroleum Engineers Journal,* pp. 347–360, August, 1974.
8. Khan, A.: "Pressure Behavior of a Vertically Fractured Well Located at the Center of a Rectangular Drainage Region," M.S. Thesis, The University of Tulsa, Tulsa, Oklahoma, 1978.
9. Ozkan, E.: "Performance of Horizontal Wells," Ph.D. Dissertation, The University of Tulsa, Tulsa, Oklahoma, 1988.
10. Babu, D. K. and Odeh, A. S.: "Productivity of a Horizontal Well," *SPE Reservoir Engineering,* pp. 417–421, November 1989.
10a. Babu, D. K.: Personal communication, Nov. 1990.
11. Kuchuk, F. J., Goode, P. A., Brice, B. W., Sherrard, D. W., and Thambynayagam, R. K. M.: "Pressure Transient Analysis and Inflow Performance for Horizontal Wells," paper SPE 18300, presented in the SPE 63rd Annual Technical Conference and Exhibition, Houston, Texas, Oct. 2–5, 1988.
12. Goode, P. A. and Wilkinson, D. J.: "Inflow Performance of Partially Open Horizontal Wells," paper SPE 19341, presented in the SPE Eastern Regional Meeting, Morgantown, West Virginia, Oct. 24–27, 1989.
13. Vogel, J. H.: "Inflow Performance Relationships for Solution-Gas Drive Wells," *Journal of Petroleum Technology,* pp. 83–92, January 1968.
14. Fetkovich, M. J.: "The Isochronal Testing of Oil Wells," paper SPE 4529, presented at the 48th Annual Fall Meeting, Las Vegas, Nevada, Sept. 30–Oct. 3, 1973. (SPE Reprints series No. 14, 265).
15. Bendakhlia, H. and Aziz, K.: "Inflow Performance Relationship for Solution Gas Drive Horizontal Wells," paper SPE 19823, presented at the SPE 64th Annual Technical Conference, San Antonio, Texas, Oct. 8–11, 1989.
16. Plahn, S. V., Startzman, R. A. and Wattenbarger, R. A.: "A Method for Predicting Horizontal Well Performance in Solution-Gas Drive Reservoirs," paper SPE 16201, presented at the Production Operation Symposium, Oklahoma City, Oklahoma, March 8–10, 1987.
17. Joshi, S. D.: "Production Forecasting Methods for Horizontal Wells," paper SPE 17580, presented at the SPE International Meeting, Tianjin, China, Nov. 1–4, 1988.

CHAPTER 8

Water and Gas Coning in Vertical and Horizontal Wells

INTRODUCTION

Water and gas coning is a serious problem in many oil field applications. The production of coned water or gas can reduce oil production significantly. Therefore, it is important to minimize or at least delay coning. In reservoirs with bottom water, vertical wells are normally completed in the top section of the pay zone to minimize or delay water coning. Of course, this assumes that there is no gas cap. Similarly, in a reservoir with a gas cap, if there is no bottom water, a vertical well is perforated as low as possible so that the perforated interval is as far away from the gas cap as possible. If an oil reservoir has both, gas cap as well as bottom water, then the vertical well is normally perforated either near the center of the oil zone thickness or below the center, toward the water zone. This is because coning tendencies are inversely proportional to the density difference and are directly proportional to the viscosity. The density difference between gas and oil is normally larger than the density difference between water and oil. Hence, gas has less tendency to cone than water. However, gas viscosity is much lower than the water viscosity, and therefore, for the same pressure drawdown in a given reservoir, the gas flow rate will be higher than the water flow rate. Thus, density and viscosity differences between water and gas tend to balance each other. Therefore, to minimize gas as well as water coning, a preferred perforated interval is at the center of the oil pay zone. From the practical standpoint, however, many wells are perforated closer to water-oil contact than to the gas-oil contact.

One of the main reasons for coning is pressure drawdown. As shown in Figure 8–1, a vertical well exhibits a large pressure drawdown near the wellbore. This large pressure drawdown in the vicinity of the wellbore causes coning. Inversely, the low pressure drawdown exhibits minimum coning tendency. To achieve a given production rate, one has to impose a larger

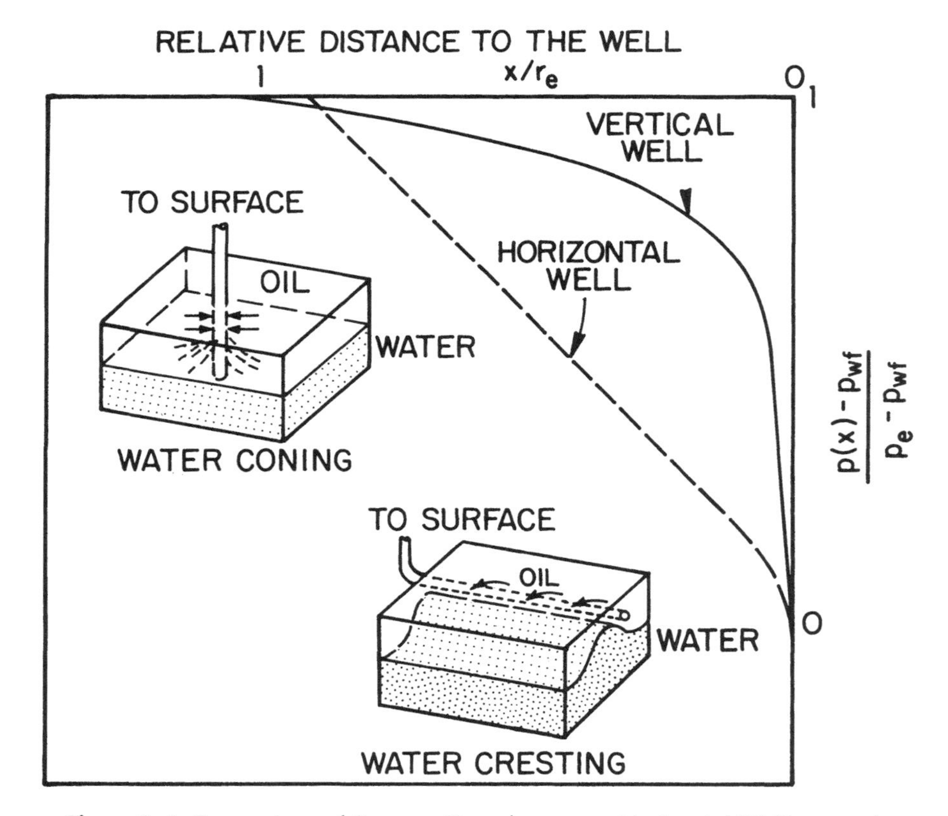

Figure 8–1 Comparison of Pressure Drawdown near Horizontal Wellbore and Vertical Wellbore.

pressure drawdown in a low-permeability reservoir than in a high-permeability reservoir. Thus, high-permeability reservoirs show less tendency for coning than low-permeability reservoirs. Similarly, for a given pressure drawdown, the magnitude of pressure drop in the near wellbore region is smaller in a high-permeability reservoir than in a low-permeability reservoir. Thus, due to minimum pressure drawdown in the near wellbore region, high-permeability reservoirs exhibit minimum coning tendencies. In general, reservoirs with a permeability of one darcy and above should exhibit minimum coning problems, unless, of course, they are very thin (oil rim) reservoirs.

In naturally fractured reservoirs, especially those with vertical fractures, one can have severe coning in spite of high reservoir permeability. This is because bottom water and top gas travel through high-permeability (vertical) fractures. This is especially true in fractured reservoirs with low matrix permeability, and large matrix blocks where water imbibition in the matrix is very slow. There are several fractured limestone and reef reservoirs, where coning problems are severe due to high vertical fracturing. Here again, the only way to reduce coning is to minimize pressure drawdown.

It is clear from the preceding discussion that coning can be reduced or minimized by minimizing pressure drawdown. However, this poses a practical difficulty. Oil production rates are proportional to the drawdown and by minimizing drawdown, one may avoid coning; but this would also result in reducing the oil production rate. In other words, rate in barrels of oil per day per foot length of a well would be small, and in many instances operationally impractical. By drilling a long horizontal well, one can achieve minimum pressure drawdown. Production per unit well length may still be small, but because of a long well length, high oil production rates can be obtained. Thus, horizontal wells provide a production option whereby pressure drawdown can be minimized, coning tendencies can be minimized, and high oil production rates can be sustained.

Horizontal wells can mitigate coning only in cases where reduction of pressure drawdown would minimize coning. In the extreme case, in a very high-permeability reservoir using horizontal wells, one can achieve pressure drawdown equal to the gravity head difference between oil and water or oil and gas zones.

This chapter includes discussion of oil and gas coning using horizontal wells. Nevertheless, to understand horizontal well behavior, it is important to understand vertical well coning behavior. Therefore, this chapter includes discussion on both vertical and horizontal well coning behaviors.

CRITICAL RATE DEFINITION

Since the early days, several experiments and mathematical analyses were conducted to solve coning problems. One of the basic conclusions of many analyses was if oil is produced at a sufficiently low rate (or if pressure drawdown in a vertical well is reduced), coning of water and gas can be avoided, and only oil is produced. This low rate is called the *critical rate*. Thus, the critical rate is defined as the maximum rate at which oil is produced without production of gas or water.

VERTICAL WELL CRITICAL RATE CORRELATIONS

Several vertical well critical rate correlations are available in the literature, some of which are summarized in Table 8–1.[1–5] It is important to note that these correlations are valid for a continuous oil pay zone with oil-water contact or gas-oil contact or both. These correlations show that the critical rate depends upon effective oil permeability, oil viscosity, density difference between oil and water or oil and gas, well penetration ratio h_p/h, and vertical permeability k_v.

TABLE 8–1 VERTICAL WELL CRITICAL RATE CORRELATIONS

Craft and Hawkins Method[1]

$$q_o = \frac{0.007078\, k_o\, h\, (p'_{ws} - p_{wf})}{\mu_o B_o \ln(r_e/r_w)} \times PR \tag{8–1–1}$$

$$PR = b'\left[1 + 7\sqrt{\frac{r_w}{2\,b'h}}\cos(b' \times 90°)\right] \tag{8–1–2}$$

where

q_o = critical rate (maximum oil rate without coning), STB/day
PR = productivity ratio
p'_{ws} = static well pressure corrected to the middle of the producing interval, psi
p_{wf} = flowing well pressure at the middle of the producing interval, psi
b' = penetration ratio, h_p/h
h_p = thickness of perforated interval, ft
h = oil column thickness, ft
μ_o = oil viscosity, *cp*
B_o = oil formation volume factor, RB/STB

TABLE 8–1 *(Continued)*

Meyer, Gardner, and Pirson Method[2,3]

Gas Coning

$$q_o = 0.001535 \frac{\rho_o - \rho_g}{\ln(r_e/r_w)} \left(\frac{k_o}{\mu_o B_o}\right) [h^2 - (h - h_p)^2] \qquad (8\text{–}1\text{–}3)$$

where

q_o = critical rate (maximum oil rate without gas coning), STB/day
ρ_o = oil density, gm/cc
ρ_g = gas density, gm/cc
r_e = drainage radius, ft
r_w = wellbore radius, ft
k_o = effective oil permeability, md
μ_o = oil viscosity, *cp*
B_o = oil formation volume factor, RB/STB
h_p = perforated interval, ft
h = oil column thickness, ft

Water Coning

$$q_o = 0.001535 \frac{\rho_w - \rho_o}{\ln(r_e/r_w)} \frac{k_o}{\mu_o B_o} (h^2 - h_p^2) \qquad (8\text{–}1\text{–}4)$$

where

ρ_w = water density, gm/cc

Simultaneous Gas and Water Coning

$$q_o = 0.001535 \frac{k_o}{\mu_o B_o} \frac{(h^2 - h_p^2)}{\ln(r_e/r_w)} \times \left[(\rho_w - \rho_o)\left(\frac{\rho_o - \rho_g}{\rho_w - \rho_g}\right)^2 + (\rho_o - \rho_g)\left(1 - \frac{\rho_o - \rho_g}{\rho_w - \rho_g}\right)^2 \right] \qquad (8\text{–}1\text{–}5)$$

Chaperon Method[4]

$$q_o = \frac{3.486 \times 10^{-5}}{B_o} \frac{k_h h^2}{\mu_o} [\Delta\rho] q_c^* \qquad (8\text{–}1\text{–}6)$$

TABLE 8–1 *(Continued)*

where

q_o = critical oil rate, m^3/hr
k_h = horizontal permeability, md
k_v = vertical permeability, md
h = oil column thickness, m
μ_o = viscosity, cp
$\Delta\rho$ = $\rho_w - \rho_o$, density difference, gm/cc
ρ_w = water density, gm/cc
ρ_o = oil density, gm/cc
B_o = formation volume factor, Res m^3/St m^3

In the U.S. oil field units, Equation 8–1–6 can be rewritten as:

$$q_o = \frac{4.888 \times 10^{-4}}{B_o} \frac{k_h h^2}{\mu_o} [\Delta\rho]\, q_c^* \qquad (8\text{–}1\text{–}7)$$

where

k_h is in md
h is in ft
μ_o is in cp
$\Delta\rho$ is in gm/cc
q_o is in STB/day

A relationship to calculate q_c^* is tabulated below.[4]

α''	q_c^*
4	1.2133
13	0.8962
40	0.7676

A correlation of tabulation is given in Equation 8–1–8.

$$q_c^* = 0.7311 + (1.9434/\alpha'') \qquad (8\text{–}1\text{–}8)$$

where

$\alpha'' = (r_e/h)\sqrt{k_v/k_h}$
r_e = drainage radius, ft
k_v = vertical permeability, md

TABLE 8–1 *(Continued)*

Schols Method[5]

$$q_o = \frac{(\rho_w - \rho_o)\, k_o\, (h^2 - h_p^2)}{2049\, \mu_o\, B_o} \times \left[0.432 + \frac{\pi}{\ln(r_e/r_w)}\right] [h/r_e]^{0.14} \quad (8\text{–}1\text{–}9)$$

where

q_o = critical rate, STB/day
ρ_w = water density, gm/cc
ρ_o = oil density, gm/cc
h = oil column thickness, ft
h_p = perforated interval, ft
B_o = formation volume factor, RB/STB
μ_o = oil viscosity, cp
r_e = drainage radius, ft
r_w = wellbore radius, ft

Hoyland, Papatzacos and Skjaeveland Method[5a]

Isotropic Reservoirs

$$q_o = \frac{k_o\, (\rho_w - \rho_o)}{173.35\, B_o \mu_o} \left[1 - \left(\frac{h_p}{h}\right)^2\right]^{1.325} h^{2.238}\, [\ln(r_e)]^{-1.990} \quad (8\text{–}1\text{–}10)$$

where

q_o = critical oil flow rate, STB/day
ρ_w = water density, gm/cc
ρ_o = oil density, gm/cc
B_o = oil formation volume factor, RB/STB
μ_o = oil viscosity, cp
k_o = effective oil permeability, md
h_p = perforated interval, ft
h = oil column thickness, ft
r_e = drainage radius, ft

Anisotropic Reservoirs

For an anisotropic reservoir, the correlations are given by the following figure where dimensionless critical rate q_{oD} is plotted versus dimensionless radius r_{eD}. Dimensionless critical rate q_{oD} and dimensionless radius are defined, respectively, by

$$q_{oD} = \frac{651.4\, \mu_o B_o q_o}{h^2\, (\rho_w - \rho_o)\, k_h} \quad (8\text{–}1\text{–}11)$$

TABLE 8–1 *(Continued)*

and

$$r_{eD} = \frac{r_e}{h}\sqrt{\frac{k_v}{k_h}} \qquad (8\text{–}1\text{–}12)$$

where

q_{oD} = dimensionless critical oil rate
q_o = critical oil flow rate, STB/day
ρ_w = water density, gm/cc
ρ_o = oil density, gm/cc
B_o = oil formation volume factor, RB/STB
μ_o = oil viscosity, cp
k_h = horizontal permeability, md
k_v = vertical permeability, md
h_p = perforated interval, ft
h = oil column thickness, ft
r_{eD} = dimensionless radius
r_e = drainage radius, ft

The figure can be used to obtain dimensionless critical rate q_{oD}, then Equation 8–1–11 can be used to calculate critical oil rate q_o.

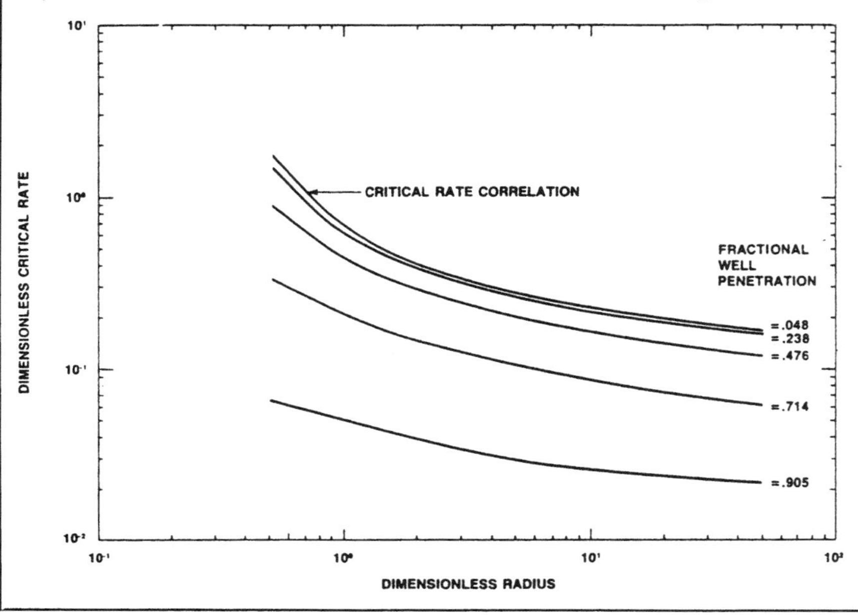

The first important parameter that affects coning is oil viscosity, μ_o. In heavy oil reservoirs, coning problems can be more severe than in light oil reservoirs. However, heavy oil reservoirs generally exhibit high permeabilities of the order of darcies. Therefore, the k_o/μ_o, or mobility ratio, exhibited by heavy oil may be comparable to k_o/μ_o exhibited by a light oil in a low-permeability reservoir. There is a range of k_o/μ_o over which reduction in pressure drawdown helps in reducing coning. The density difference, $\Delta\rho = \rho_w - \rho_o$, may vary from a low of about 0.05 gm/cc for heavy oil to a high of about 0.3 gm/cc for light oil (for oils with gravity less than 10° API, oil density is higher than fresh water density). In general, the higher the $\Delta\rho$, the higher the critical rate and the lower the coning tendency. The next parameter is well penetration ratio, $b' = h_p/h$. The lower the value of the perforated interval, h_p, the lower the penetration ratio and the higher the critical rate. (This, of course, assumes that the well is perforated at the top, in reservoirs with bottom water, or the well is perforated at the bottom, in reservoirs with a gas cap.) In some correlations, such as Chaperon's,[4] the perforated interval is assumed to be so small as compared to the reservoir height that for all practical purposes penetration ratio, $b' \approx 0.0$, i.e., $h_p/h \approx 0.0$ (see Table 8–1). Another important parameter is the vertical permeability. If the vertical permeability is zero, then there is no flow in the vertical direction and there is no coning problem. In the limiting cases, if vertical permeability is the same as the horizontal permeability, one would expect to have maximum coning tendencies. None of the correlations in Table 8–1, except Chaperon's[4] and Hoyland et al.[5a] include vertical permeability as a correlating parameter.[1–5] Moreover, even in Chaperon's correlation, Equation 8–1–6 in Table 8–1, one can normally assume $q_c^* = 1$, which essentially indicates minimum influence of vertical permeability on the critical production rate. In low vertical permeability reservoirs, q_c^* can be as high as 1.21, i.e., 21% increase in the critical rate due to low vertical permeability. In contrast, a high vertical permeability would give q_c^* as low as 0.76, giving a 24% drop in the critical rate. Thus, variation in critical rate is only ±20% depending upon the vertical permeability. The correlations in Table 8–1 tend to indicate that the critical rate is not strongly dependent upon the value of vertical permeability, as long as vertical permeability has a finite value, and well penetration in the reservoir is very small. Similarly, the anisotropic correlation of Hoyland et al.[5a] in Table 8–1 shows minimal influence of vertical permeability for $r_{eD} = (r_e/h)\sqrt{k_v/k_h}$ values between 7 and 50, especially when well penetration or the perforated interval, is very small as compared to the reservoir thickness.

EXAMPLE 8–1

A vertical well is drilled in an oil reservoir underlain by bottom water. The following reservoir data is given:

Density difference ($\rho_w - \rho_o$)	= 0.48 gm/cc
Permeability, k_o	= 200 md
Oil column thickness, h	= 80 ft
Perforated interval, h_p	= 8 ft
Oil viscosity, μ_o	= 0.4 cp
Oil formation volume factor, B_o	= 1.32 RB/STB
Drainage area	= 80 acres
Drainage radius, r_e	= 1053 ft
Wellbore radius, r_w	= 0.25 ft
Pressure differential between static and Flowing well pressures, $(p'_{ws} - p_{wf})$	= 80 psi

Calculate the critical oil production rate by the following five methods: (1) Craft and Hawkins,[1] (2) Meyer, Gardner, [2] and Pirson,[3] (3) Chaperon,[4] (4) Schols,[5] (5) Hoyland et al.[5a]

Solution

Critical coning equations listed in Table 8–1 are used to calculate critical production rates.

1. **Craft and Hawkins Method**

Well penetration ratio, b'

$$b' = h_p/h = 8/80 = 0.1$$

Productivity Ratio using Equation (8–1–2) from Table 8–1

$$PR = b'[1 + 7\sqrt{\frac{r_w}{2b'h}} \cos(b' \times 90°)] \tag{8–1–2}$$

$$= 0.1\ [1 + 7\sqrt{\frac{0.25}{2 \times 0.1 \times 80}} \cos(0.1 \times 90°)]$$

$$= 0.186$$

Maximum oil rate without coning using Equation 8–1–1 from Table 8–1

$$q_o = \frac{0.007078\, k_o\, h\, (p'_{ws} - p_{wf})}{\mu_o B_o \ln(r_e/r_w)} \times PR \tag{8–1–1}$$

$$= \frac{0.007078 \times 200 \times 80 \times 80}{0.4 \times 1.32 \times \ln(1053/0.25)} \times 0.186$$

$$= 383.3 \approx 383 \text{ STB/day}$$

2. **Meyer, Gardner and Pirson Method**

Maximum oil rate without coning using Equation 8–1–4 from Table 8–1

$$q_o = 0.001535 \frac{(\rho_w - \rho_o)}{\ln(r_e/r_w)} \frac{k_o}{\mu_o B_o} (h^2 - h_p^2) \quad (8\text{–}1\text{–}4)$$

$$= 0.001535 \times \frac{0.48}{\ln(1053/0.25)} \times \frac{200}{0.4 \times 1.32} \times (80^2 - 8^2)$$

$$= 211.9 \approx 212 \text{ STB/day}$$

3. **Chaperon's Method**

Assume $k_v = k_h = k_o = 200$ md

$$\alpha'' = (r_e/h)\sqrt{k_v/k_h} = (1053/80)\sqrt{1} = 13.2$$

$$q_c^* = 0.7311 + (1.9434/\alpha'') \quad (8\text{–}1\text{–}8)$$

$$= 0.7311 + (1.9434/13.2)$$

$$= 0.878$$

Using Equation 8–1–7 from Table 8–1 critical rate can be calculated as

$$q_o = \frac{4.888 \times 10^{-4}}{B_o} \frac{k_h h^2}{\mu_o} (\Delta\rho)\, q_c^* \quad (8\text{–}1\text{–}7)$$

$$= \frac{4.888 \times 10^{-4}}{1.32} \frac{200 \times (72)^2}{0.4} \times 0.48 \times 0.878$$

$$= 404.5 \approx 405 \text{ STB/day}$$

It is important to note that the Chaperon method includes h, oil column height, assuming that the perforated interval is very small. For convenience, in this problem, h, the oil pay zone height, is considered to be the distance between the bottom of the perforation and oil water contact.

4. **Schols' Method**[5]

$$q_o = \frac{(\rho_w - \rho_o)\, k_o\, (h^2 - h_p^2)}{2049\, \mu_o B_o} \times \left[0.432 + \frac{\pi}{\ln(r_e/r_w)}\right] [h/r_e]^{0.14} \quad (8\text{–}1\text{–}9)$$

$$= \frac{0.48 \times 200 \times (80^2 - 8^2)}{2049 \times 0.4 \times 1.32} \times \left[0.432 + \frac{\pi}{\ln(1053/0.25)}\right] \left[\frac{80}{1053}\right]^{0.14}$$

$$= 316.9 \approx 317 \text{ STB/day}$$

5. **Hoyland et al.'s Method**[5a]

$$q_o = \frac{k_o(\rho_w - \rho_o)}{173.35\ B_o\mu_o}\left[1 - \left(\frac{h_p}{h}\right)^2\right]^{1.325} h^{2.238}\ [\ln(r_e)]^{-1.990} \qquad (8\text{–}1\text{–}10)$$

$$= \frac{200 \times 0.48}{173.35 \times 1.32 \times 0.4}\left[1 - \left(\frac{8}{80}\right)^2\right]^{1.325} 80^{2.238}\ [\ln(1053)]^{-1.990}$$

$$= 395.67 \text{ STB/day} \approx 396 \text{ STB/day}$$

A comparison of critical oil rates calculated by the five methods is given below.

CRITICAL OIL RATES BY DIFFERENT METHODS

Method	Critical Oil Rate STB/day
Craft and Hawkins	383
Meyer, Gardner & Pirson	212
Chaperon	405
Schols	317
Hoyland, et. al.	396

CRITICAL RATE BY PRODUCTION TESTING

Example 8–1 shows that different theoretical correlations give different answers. Additionally, some correlations give too small a value to use the critical rate in field operations. Therefore, an engineer has to choose the correlation to be used in field applications.

In this author's view, there is no right or wrong critical rate correlation. If production testing data is available, one can easily choose an appropriate correlation that fits the field data. For example, in a given field, a few wells can be selected. Each well that is producing water or gas can be choked down. In practice, this can be done by (1) reducing choke size, (2) reducing gas injection in a gas lift operation, or (3) reducing pump rate or flow capacity for a pumping well. Choking the well down increases bottomhole producing pressure. The rise in bottomhole pressure reduces reservoir drawdown. This in turn results in a reduction of total fluid production from a well. Once the change is made, the well can be allowed to stabilize and a water cut can be measured at the new rate. A plot of flowing tubing pressure against water and oil rate can be made. The production test can be conducted at several chosen settings, i.e., at several flowing tubing pressures (FTP). A plot of oil rate and water rate against FTP can be used to estimate critical production rate (see Fig. 8–2). As shown in the figure, if one increases FTP, at a certain FTP setting, the water rate may drop significantly, indicating the critical rate. This critical rate can be compared with various theoretical correlations to determine the appropriate one. Additionally, this critical rate can be used to estimate reservoir properties, assuming applicability of a certain theoretical correlation. It is important to note that the critical rate obtained from the production testing refers to the oil column thickness at that point in time. In many low porosity reservoirs ($\phi < 5\%$), determination of oil column thickness from well logs is difficult. A well buildup test may be used to determine oil-water or gas-oil contact, assuming that the contact represents a constant pressure boundary. In some occasions, interpretation of the buildup test is also difficult. In these cases, production testing may provide another method to determine oil column thickness.

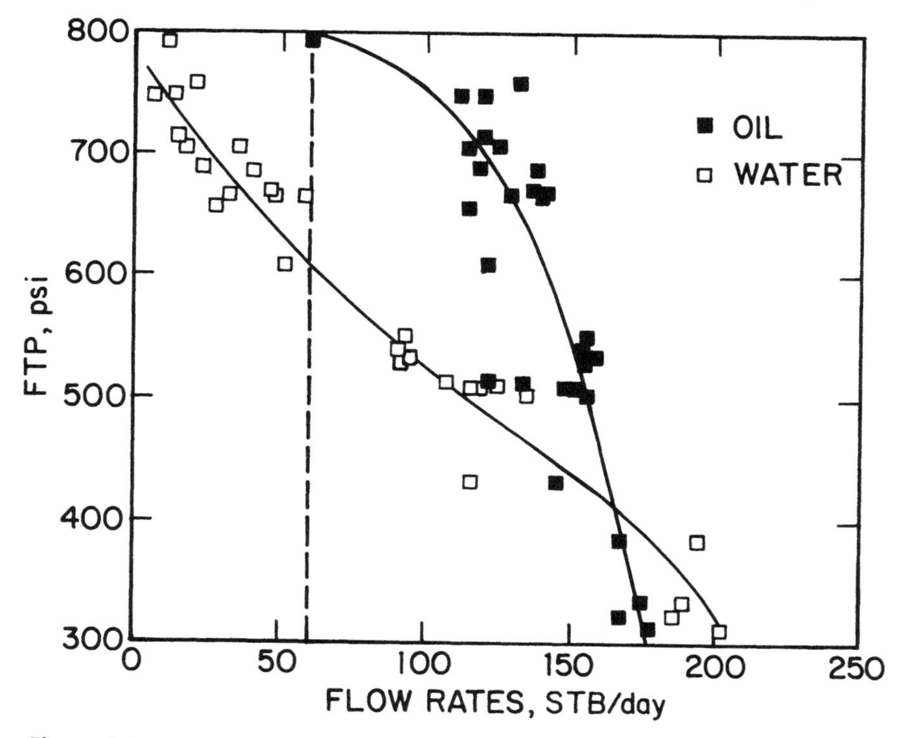

Figure 8–2 A Typical Production Testing Plot Showing Flowing Tubing Pressure (FTP) versus Flow Rates.

EXAMPLE 8–2

A production test on a vertical well in a bottomwater drive reservoir shows a critical production rate of about 170 STB/day. The reservoir is a fractured limestone. Estimate the oil column thickness. The past well buildup tests on surrounding wells have shown permeabilities from 50 to 75 md. Assume that Chaperon's critical rate correlation is valid for the reservoir.[4]

Given

$$\rho_w - \rho_o = 0.26 \text{ gm/cc} \qquad A = 40 \text{ acres}$$
$$\mu_o = 0.7 \text{ cp} \qquad \phi = 4\%$$
$$B_o = 1.1 \text{ RB/STB}$$

Solution

As shown in Equation 8–1–7, Chaperon's vertical well critical rate correlation in the U.S. oil field units is written as:

$$q_o = 4.888 \times 10^{-4} \frac{k_h\, h^2}{\mu_o} \times \frac{\Delta\rho\, q_c^*}{B_o} \qquad (8\text{–}1\text{–}7)$$

If the critical rate, $q_o = 170$ STB/day and if we assume $q_c^* \approx 1$ and substitute these values into Equation 8–1–7, the resulting equation is:

$$170 = 4.888 \times 10^{-4} \frac{k_h h^2 \times 0.26 \times 1}{0.7 \times 1.1}$$
$$k_h h^2 = 1.03 \times 10^6 \text{ md-ft}^2$$

if $k_h = 50$ md, then oil column $h = 143.5$ ft
if $k_h = 75$ md, then oil column $h = 117.2$ ft

Thus, the oil column height is between 117 to 144 ft, with an average of 131 ft. It is important to note that in solving the above example $q_c^* \approx 1$ was assumed. Table 8–1 shows that q_c^* is dependent upon the reservoir height, h and the ratio of the vertical to horizontal permeability, $\sqrt{k_v/k_h}$. The above calculated oil column height is approximate and one really needs to solve the problem by trial and error procedure until the h value used in q_c^* calculation agrees with the value obtained from Equation 8–1–7.

LIMITATIONS OF PRODUCTION TESTING

Production testing as described above works very well in many reservoirs. However, there are some reservoirs, especially some standstone reservoirs, where choking the well down results in a comparable reduction of both oil and water production. In these reservoirs, once the water cones in, the coning is irreversible. In some cases, even shutting a well down for a few days would not reduce coning when the well is put back on production again. This is probably because of the severe capillary forces that sustain high water saturation around the wellbore. In these cases production testing cannot be used to determine the critical rate.

In oil wells with a gas cap, production testing can be very difficult. For example, from production testing it may be very difficult to distinguish between the solution gas and gas from the gas cap, unless, of course, both gases have different chemical characteristics.

The production testing procedure may also be difficult in reservoirs with drawdown on the order of 5 to 10 psi. (This normally occurs in high-permeability reservoirs.) In these high-permeability reservoirs, pressure-drop changes during production testing may be too small to measure reliably.

Another important problem in production testing is an accurate estimation of the bottomhole pressure, especially when production stream has more than one phase. In many instances, only surface-well flowing pressure is measured, and the well's bottomhole pressure is calculated. For the same surface rates and the same surface pressure, different multiphase flow correlations provide different bottomhole pressure values. This makes precise estimation of bottomhole pressure difficult. One can enhance the accuracy of downhole pressure estimates by using the following two options: (1) The first option is to measure production string pressure drop by installing bottomhole and surface pressure gauges, preferably in several wells, but at least in one well. Based upon the pressure drop measurements, an appropriate two-phase correlation can be selected. It is important to note that the two phase pressure drop strongly depends upon tubing size. Thus the measured correlation can be used for a certain tubing size. (2) The second option is to use a bottomhole pressure readout device. Recently, several pressure gauges have become available which can be installed downhole permanently during well completion. Such pressure gauges are highly useful for production testing.

DECLINE CURVE ANALYSIS

Many engineers are familiar with the decline curve analysis, which was developed by Arps[6] in the 1940s and was later refined by Fetkovich.[7] The decline curves tell us that the value of the decline index b changes from 0 to 1. The decline curve equation, which is applicable during the depletion stage is given by:

$$q(t) = q_i/(1 + b\, D_i\, t)^{1/b} \tag{8–1}$$

where

q_i = oil rate at the beginning of depletion, STB/day
D_i = initial decline rate, 1/day
t = time, days
b = decline exponent, dimensionless

It has been noted that for a bottom water drive reservoir, $b = 0.5$.[6,7] In a bottom water drive reservoir, if the reservoir is operated at, or below, the critical rate, then one can have a uniform sweep of the reservoir as water moves upward, resulting in high ultimate recovery. Nevertheless, to achieve this, the wells have to be operated at, or below, critical rates at all times. This means that as more and more oil is produced, oil column height reduces, and hence, the well will be choked back and its oil production rate reduced to prevent water coning. Based on this scenario, a finite 40-acre reservoir with a bottom water drive was chosen. Various vertical well critical rate correlations listed in Table 8–1 were used to calculate critical rates.[1–5] For example, using Schols' correlation the critical rate was calculated at the first day.[5] The well was operated at that rate for a week. Then using material balance, a new oil column height was calculated at the end of one week, and accordingly a new critical rate was calculated. Once again, the new rate was held constant for a week, and at the end of the second week, a new oil column height and a new critical rate were calculated. This calculation was done by writing a simple computer program to march in time. The results of rate versus time using Chaperon's and Schols' correlations are shown in Figures 8–3a and 8–3b. It is interesting to note that numerical values of critical rates calculated by each correlation are different, but they both show a decline index of $b = 0.5$.

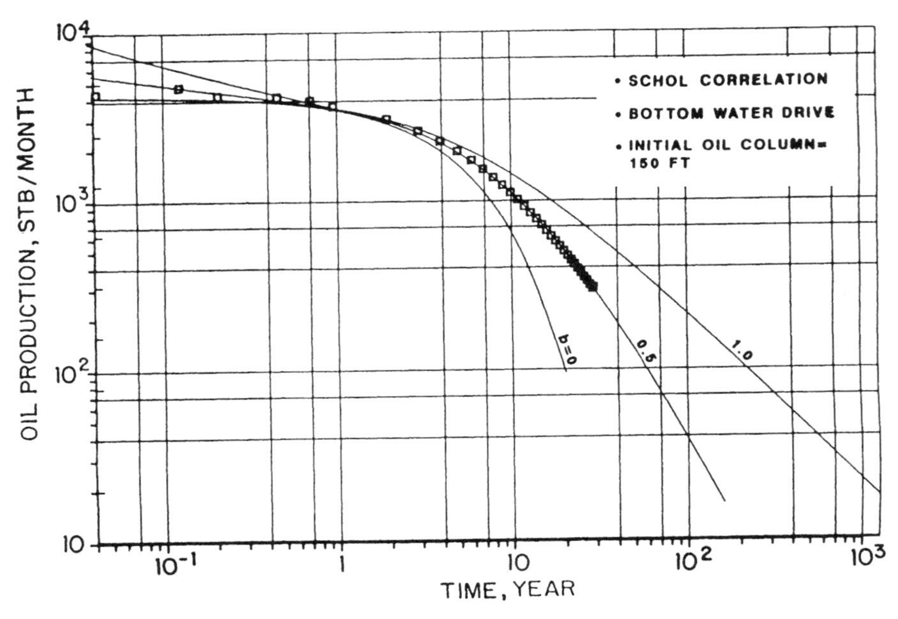

Figure 8–3a Critical Rate Variation for a Vertical Well, Schol's Correlation.

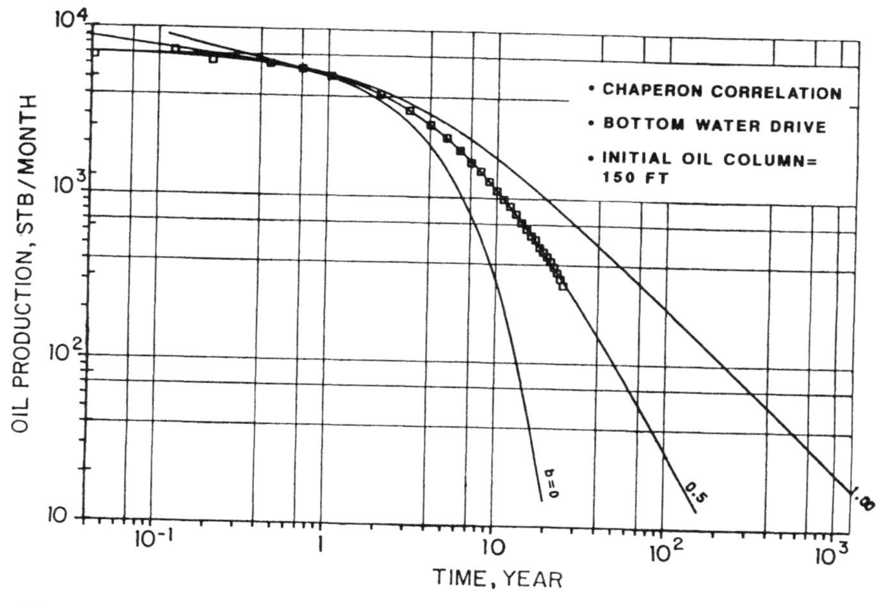

Figure 8–3b Critical Rate Variation for a Vertical Well, Chaperon's Correlation.

As noted earlier, field experiences indicate that water drive reservoirs exhibit a decline exponent of $b = 0.5$[6,7] The above numerical calculations offer justifications to the field observations of References 6 and 7. Thus, to attain maximum sweep, bottom water drive reservoirs will have to be operated at or below the critical rates. In practice, whenever production rates are above the critical rates, say by using artificial lift, the decline exponent (b value) will be less than 0.5. In other words, although we may get an initially high oil rate using a submersible pump or gas lift, the oil rate may drop rapidly. The above observations are quite useful for predicting future water cuts and estimating reserves. One can estimate maximum possible producible reserves for a given well using critical rate calculations and compare them with estimated producible reserves by the decline of oil production rate obtained by using an artificial lift. The reserves calculated using critical rates represent an upper limit of the producible reserves. (Note that total reserves by critical rate calculations are the same as those obtained by the material balance calculation using residual oil saturation.)

It is important to note that none of the critical rate correlations given in Table 8–1 are always applicable in all types of reservoirs. Field experience indicates that a certain correlation works better than others in a given field.

WATER BREAKTHROUGH IN VERTICAL WELLS

So far, critical production rates have been discussed. Many times the critical rates are too low, and for economic reasons, a well is frequently produced at rates above critical rate. This results in production of water, as well as oil. Sobocinski and Cornelius[8] and Bournazel and Jeanson[9] have reported (laboratory) experimental results on water breakthrough time; i.e., they measured the time it takes for a water cone to break through in a vertical well at a given total production rate. (Reference 8 has also reported a numerical solution to verify the laboratory results.) In their experiments, they found that the lower the vertical dimensionless cone height (defined in Table 8–2), the faster the water breakthrough. Their procedures and results are summarized in Table 8–2.

TABLE 8–2 VERTICAL WELL BREAKTHROUGH TIME CORRELATIONS

Sobocinski and Cornelius Method[8]

Dimensionless cone height is given by

$$z = \frac{0.00307\,(\rho_w - \rho_o)\,k_h\,h\,(h - h_p)}{\mu_o\,q_o\,B_o} \tag{8–2–1}$$

where

k_h = horizontal permeability, md
μ_o = oil viscosity, cp
B_o = oil formation volume factor, RB/STB
q_o = oil production rate, STB/day
ρ_w = water density, gm/cc
ρ_o = oil density, gm/cc
h_p = perforated interval, ft

and dimensionless breakthrough time, $(t_D)_{SC}^{BT}$ is given by

$$(t_D)_{SC}^{BT} = \frac{0.00137\,(\rho_w - \rho_o)\,k_v\,(1 + M^{\alpha})t_{BT}}{\mu_o\,\phi\,h} \tag{8–2–2}$$

where

t_{BT} = breakthrough time, days
ϕ = porosity, fraction

TABLE 8–2 *(Continued)*

k_v = vertical permeability, md
M = water-oil mobility ratio
= $[\mu_o(k_w)_{or}/\mu_w(k_o)_{wc}]$ where $(k_w)_{or}$ is the effective permeability to water at residual oil saturation and $(k_o)_{wc}$ is the effective permeability to oil at connate water saturation
α = 0.5 for $M < 1$; 0.6 for $1 < M < 10$.
h = oil column thickness, ft

The Sobocinski and Cornelius relationship between dimensionless cone height and dimensionless breakthrough time can be correlated as

$$(t_D)_{SC}^{BT} = \frac{z}{4}\left[\frac{16 + 7z - 3z^2}{7 - 2z}\right] \tag{8–2–3}$$

The procedure of calculating breakthrough time using the Sobocinski and Cornelius method is as follows:

1. Using Equation 8–2–1, calculate dimensionless cone height z.
2. *Calculate dimensionless breakthrough time,* $(t_D)_{SC}^{BT}$, using Equation 8–2–3.
3. Using the following equation, calculate t_{BT} (time of breakthrough in days):

$$t_{BT} = \frac{\mu_o\, \phi\, h\, (t_D)_{SC}^{BT}}{0.00137\, (\rho_w - \rho_o) k_v\, (1 + M^\alpha)} \tag{8–2–4}$$

Bournazel and Jeanson Method[9]

The Bournazel and Jeanson's laboratory results of dimensionless cone height z and dimensionless breakthrough time $(t_D)_{BJ}^{BT}$ is correlated as

$$(t_D)_{BJ}^{BT} = \frac{z}{3 - 0.7z} \tag{8–2–5}$$

The procedure of calculating breakthrough time using the Bournazel and Jeanson method is as follows:

1. Using Equation 8–2–1, calculate dimensionless cone height z.
2. *Using Equation 8–2–5, calculate dimensionless breakthrough time,* $(t_D)_{BJ}^{BT}$.
3. Finally, the breakthrough time, t_{BT} (in days), is given by:

$$t_{BT} = \frac{\mu_o\, \phi\, h\, (t_D)_{BJ}^{BT}}{0.00137\, (\rho_w - \rho_o)\, k_v\, (1 + M^\alpha)} \tag{8–2–6}$$

A close observation of equations of breakthrough times, Equations 8–2–3 and 8–2–5, provides us the guidelines to determine when water breakthrough is imminent. The equations indicate that the time corresponding to water breakthrough will be infinite if the denominators have a value equal to zero; i.e., there will be no water breakthrough. Thus Sobocinski & Cornelius' correlation,[8] Equation 8–2–3, indicates that if $z = 3.5$ or greater, there will be no water breakthrough, i.e.,

$$3.5 \leq z = \frac{0.00307\,(\rho_w - \rho_o)\,k_h\,h\,(h - h_p)}{\mu_o\,q_o\,B_o} \tag{8–2}$$

Similarly, for Bournazel and Jeanson's correlation,[9] water breakthrough will not occur if

$$4.29 \leq z = \frac{0.00307\,(\rho_w - \rho_o)\,k_h\,h\,(h - h_p)}{\mu_o\,q_o\,B_o} \tag{8–3}$$

The above two criteria, although slightly different, indicate that if dimensionless cone height z is greater than 4.29, then one should have no coning problems. This is an important observation from the reservoir operational standpoint.

EXAMPLE 8–3

Calculate the water breakthrough time using the Sobocinski and Cornelius method and the Bournazel and Jeanson method for a vertical well produced at $q_o = 200$ STB/day. The following reservoir parameters are given.

$q_o = 200$ STB/day, $h = 84$ ft
$h_p = 42$ ft, $\rho_w = 1.10$ gm/cc
$\rho_o = 0.84$ gm/cc, $B_o = 1.1$ RB/STB
$k_h = 20$ md, $k_v = 10$ md
$M = 5$, $\phi = 5\%$
$\mu_o = 1.6$ cp.

Solution

1. **Sobocinski and Cornelius Method**

Breakthrough time is calculated using the procedure given in Table 8–2. For the reservoir parameters listed above and using Equations 8–2–1 and 8–2–3, dimensionless cone height z and dimensionless breakthrough time $(t_D)_{SC}^{BT}$ are calculated, respectively, as shown below.

$$z = \frac{0.00307 \times (1.10 - 0.84) \times 20 \times 84 \times (84 - 42)}{1.6 \times 1.1 \times 200}$$

$$= 0.16$$

$$(t_D)_{SC}^{BT} = \frac{0.16}{4}\left[\frac{16 + 7 \times 0.16 - 3 \times 0.16^2}{7 - 2 \times 0.16}\right]$$

$$= 0.1$$

For $1 < M < 10$, $\alpha = 0.6$. Substituting dimensionless cone height z and dimensionless breakthrough time $(t_D)_{SC}^{BT}$ into Equation 8–2–4 gives

$$t_{BT} = \frac{\mu\phi h (t_D)_{SC}^{BT}}{0.00137 \times (\rho_w - \rho_o)\, k_v\, (1 + M^{\alpha})} \tag{8–2–4}$$

$$= \frac{1.6 \times 0.05 \times 84 \times 0.1}{0.00137 \times (1.10 - 0.84) \times 10 \times (1 + 5^{0.6})}$$

$$= 52.0 \text{ days}$$

2. **Bournazel and Jeanson Method**

Water breakthrough time can also be calculated using the Bournazel and Jeanson method (using Equation 8–2–5 instead of Equation 8–2–3 to calculate dimensionless breakthrough time). Using the same value of $z = 0.16$ as calculated earlier and substituting it in Equation 8–2–5 from Table 8–2, dimensionless breakthrough time is

$$(t_D)_{BJ}^{BT} = \frac{0.16}{3 - 0.7 \times 0.16} = 0.0554$$

$$t_{BT} = \frac{\mu_o \phi h\, (t_D)_{BJ}^{BT}}{0.00137\, (\rho_w - \rho_o)\, k_v\, (1 + M^{\alpha})} \tag{8–2–6}$$

$$= \frac{1.6 \times 0.05 \times 84 \times 0.0554}{0.00137 \times (1.1 - 0.84) \times 10 \times (1 + 5^{0.6})}$$

$$= 28.8 \text{ days} \approx 29 \text{ days}$$

Thus, the two methods give different breakthrough times. The average of these two answers is about 41 days.

EXAMPLE 8–4

Calculate the water breakthrough time for 500 and 1000 STB/day production rates using Sobocinski and Cornelius' correlation and Bournazel and Jeanson's correlation for the reservoir described in Example 8–2. Additionally, calculate the critical rate (or the maximum rate without water coning) using the above correlations. The well is perforated in the top 20 ft of the reservoir and the mobility ratio is 1.

Solution

As given in Example 8–2, density difference between oil and water is 0.26 gm/cc, oil viscosity is 0.7 cp, B_o = 1.1 RB/STB, and well spacing is 40 acres. Additionally, average oil column height from Example 8–2 is 131 ft. Thus, we have

$$\Delta\rho = 0.26 \text{ gm/cc}, \qquad h_p = 20 \text{ ft}$$
$$k_v = k_h = 50 \text{ md}, \qquad \mu_o = 0.7 \text{ cp}$$
$$B_o = 1.1 \text{ RB/STB}, \qquad h = 131 \text{ ft}$$
$$\phi = 4\%, \qquad M = 1$$

Dimensionless cone height z from Equation 8–2–1 is

$$z = \frac{0.00307\ \Delta\rho\ k_h\ h\ (h - h_p)}{\mu_o\ q_o\ B_o} \tag{8–2–1}$$

$$= \frac{0.00307 \times 0.26 \times 50 \times 131 \times (131 - 20)}{0.7 \times q_o \times 1.1}$$

$$= 753.68/q_o$$

Dimensionless breakthrough time for the Sobocinski and Cornelius method is calculated from Equation 8–2–3 while that for the Bournazel and Jeanson method is calculated using Equation 8–2–5. Substituting z values calculated above for 500 and 1000 STB/day into Equations 8–2–3 and 8–2–5 gives the following values:

DIMENSIONLESS BREAKTHROUGH TIME, $(t_D)^{BT}$

	500 STB/day	1000 STB/day
Sobocinski & Cornelius method	1.87	0.67
Bournazel & Jeanson method	0.78	0.30

Breakthrough Time in Days

Breakthrough time in days for the Sobocinski & Cornelius method can be calculated using Equation 8–2–4 given in Table 8–2 as

$$t_{BT} = \frac{\mu_o \phi h (t_D)_{SC}^{BT}}{0.00137\,(\rho_w - \rho_o) k_v (1 + M^{\alpha})} \qquad (8\text{–}2\text{–}4)$$

$$= \frac{0.7 \times 0.04 \times 131 \times (t_D)_{SC}^{BT}}{0.00137 \times 0.26 \times 50 \times (1 + 1)}$$

$$= 103\,(t_D)_{SC}^{BT}$$

Similarly, breakthrough time by the Bournazel and Jeanson method can be calculated by using Equation 8–2–6. The results are summarized below.

Water Breakthrough Time in Days

	500 STB/day	1000 STB/day
Sobocinski & Cornelius	192	69
Bournazel & Jeanson	80.3	31

In general, the Bournazel & Jeanson method gives conservative answers as compared to those calculated from the Sobocinski & Cornelius method.

Critical Rate Calculation

1. **Sobocinski and Cornelius Method**

As indicated by Equation 8–2, the breakthrough time is infinite if $z \geq 3.5$, and critical rate corresponds to $z = 3.5$. Hence,

$$z = 3.5 = 753.68/q_o$$

$$q_o = 215 \text{ STB/day}$$

2. **Bournazel & Jeanson Method**

The critical rate corresponds to $z = 4.3$. Hence,

$$z = 4.3 = 753.68/q_o$$

$$q_o = 175 \text{ STB/day}$$

It is important to note that the critical rate calculated by the Bournazel and Jeanson method of 175 STB/day is very close to field observation of 170 STB/day noted in Example 8–2.

PRACTICAL USE OF BREAKTHROUGH TIME CALCULATIONS

In practice, one can use breakthrough time to calculate effective vertical permeability. This is especially important in the development of a new field or in a field where horizontal drilling is contemplated.

In practice, many well logs show either a bottom water or a gas cap. As shown in Figure 8–4, a well log normally exhibits an oil zone and a few shale zones between oil and water zones. In a new well, sometimes it is difficult to estimate communication between the oil and water zones. In other words, it is difficult to estimate vertical permeability. Vertical per-

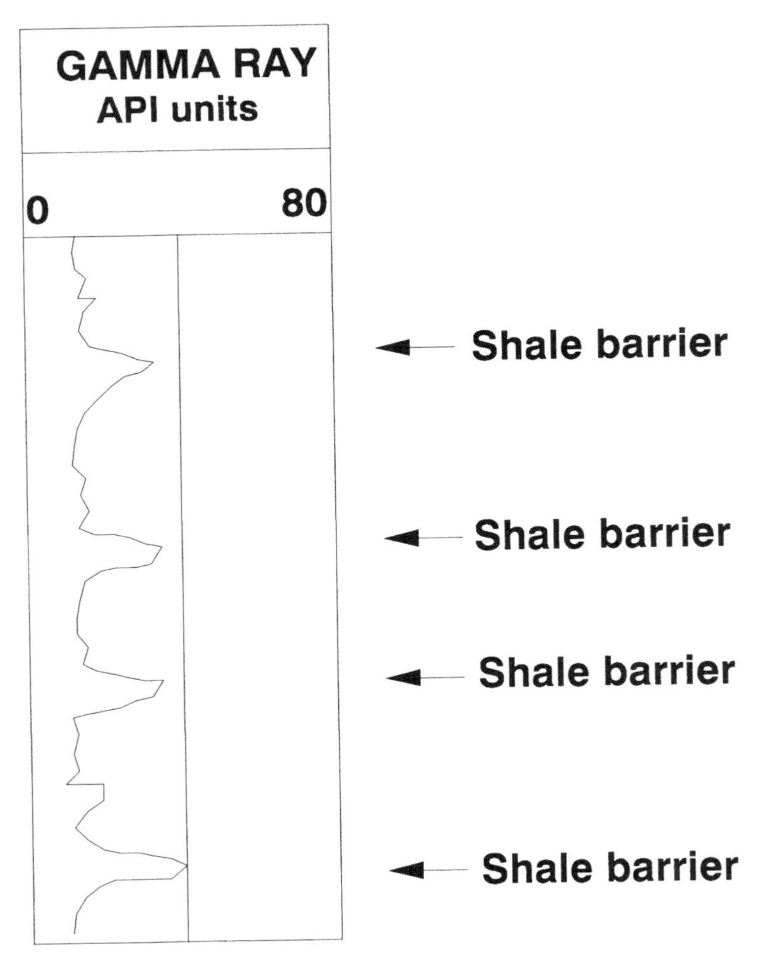

Figure 8–4 A Well Log Exhibiting an Oil Zone and a Few Shale Zones.

meability measured from the core data can provide guidelines, but actual vertical permeability may be different than that estimated from the core data. In such cases, an operator can produce an appraisal well at a certain rate for a certain length of time. If water breakthrough does not occur, then the oil production rate can be increased further and it can be once again maintained for a certain fixed time. Once again if the water breakthrough does not occur, then one can increase the rate again. The production rate can be increased until the point when breakthrough occurs. Based on the breakthrough time information, one can estimate vertical permeability of the reservoir.

The above noted approach should be integrated, at least in an appraisal well. However, in a marginal field, such an approach can be risky, especially if the reservoir rock shows irreversible coning characteristics. An operator will have to use engineering judgment while undertaking the production testing procedure.

EXAMPLE 8–5

A 150-ft-thick reservoir is underlain by a bottom water. Core testing shows average reservoir permeability of 200 md. The top 50 ft of the appraisal well is perforated. The log shows several shale streaks in the oil zone and between water and oil zones. To investigate water coning, the appraisal well was put on production testing at 5000 STB/day for about six months. The well showed no significant water breakthrough at the end of 180 days. However, on the 180th day the well started showing a trace of water production. Due to this trace of water, the test was extended for two more months. The trace of water started increasing and by the 193rd day, the well was producing 25 bbl/day of water. Assuming the 193rd day as a water breakthrough day, estimate effective vertical permeability. Given

$\phi = 12\%$, $\mu = 0.7$ cp
$B_o = 1.2$ RB/STB, $k_h = 200$ md
Mobility ratio, $M = 2$, $\rho_w - \rho_o = 0.24$ gm/cc

Solution

It is assumed that Sobocinski and Cornelius relation represents water breakthrough behavior of the reservoir. Using Equation 8–2–1, dimensionless cone height z is estimated as

$$z = \frac{0.00307 \times 0.24 \times 200 \times 150 \times (150 - 50)}{0.7 \times 5000 \times 1.2}$$
$$= 0.53$$

Using Equation 8–2–3, dimensionless water breakthrough time is estimated as

$$(t_D)_{SC}^{BT} = \frac{0.53}{4}\left[\frac{16 + 7 \times 0.53 - 3 \times 0.53^2}{7 - 2 \times 0.53}\right]$$
$$= 0.42$$

The dimensionless breakthrough time is substituted into Equation 8–2–4 to calculate effective vertical permeability.

$$t_{BT} = 193 = \frac{0.7 \times 0.12 \times 150 \times 0.42}{0.00137 \times 0.24 \times k_v \times (1 + 2^{0.6})}$$

Hence,

$$k_v = 33.1 \text{ md}$$

VERTICAL WELL POST-WATER BREAKTHROUGH BEHAVIOR

Presently, no simple analytical solution exists to calculate post-water breakthrough behavior of a vertical well. Normally, such a problem is solved using a numerical simulator. Several numerical studies have been reported in the literature. The majority of these studies are for homogeneous reservoirs with different vertical and horizontal permeabilities. (Theoretically speaking, there should be no water breakthrough problem if vertical permeability is zero.) Observation of these numerical results indicates that for a given homogeneous reservoir there is a unique relationship between water cut and percentage recovery of oil in place as shown in Figure 8–5. Regardless of the production rate, ultimate oil recovery is the same. As shown in Figure 8–6, high production rates result in high water cut at early times. According to Figure 8–5, which shows a unique relationship between water cut and recovery factors, high production rates facilitate high percentage oil recovery in a shorter time period. Because of this, under most coning situations, many operators use the most practical, high-volume artificial lift method to accelerate the oil production.

Kuo & DesBrisay[10] and Kuo[11] have correlated their numerical results to calculate water-cut behavior after breakthrough. Their correlations are summarized in Table 8–3. It is important to note that the correlation given in Table 8–3 *was developed using straight line relative permeability curves.*[10,11]

EXAMPLE 8–6

Calculate the water cut behavior over time and the recovery factor for a vertical well drilled in a reservoir with the following properties:

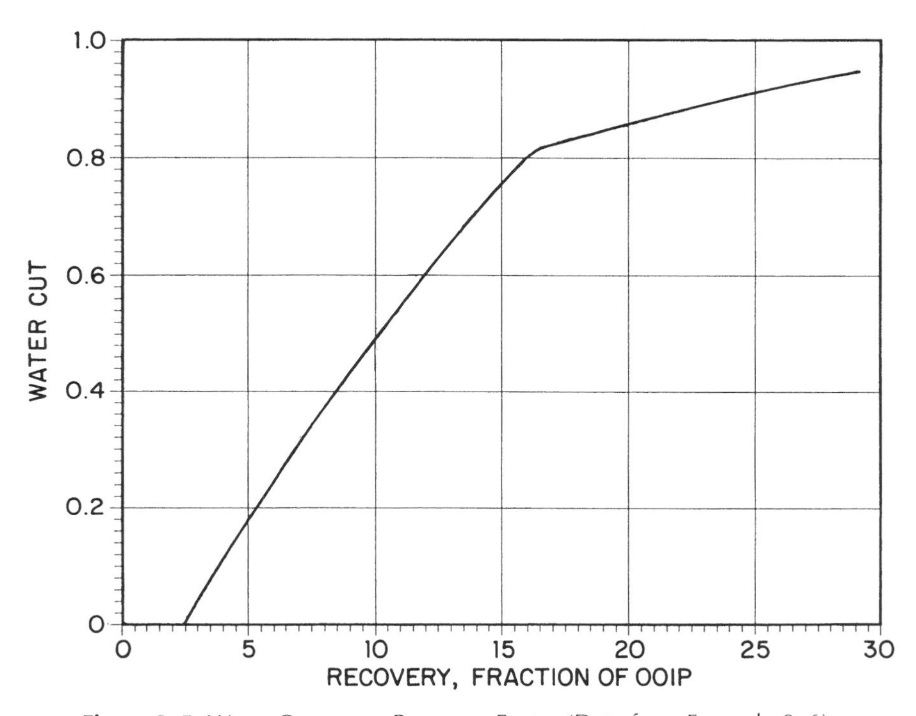

Figure 8–5 Water Cut versus Recovery Factor (Data from Example 8–6).

Initial oil column thickness,	$H_o = 84$ ft
Initial water zone thickness,	$H_w = 24$ ft
Total possible constant flow rates,	$q_t = 500$, 1000, and 5000 STB/day
Well spacing	= 80 acres
$h_p = 24$ ft,	$\rho_w = 1.095$ gm/cc
$\rho_o = 0.861$ gm/cc,	$\mu_o = 1.44$ cp
$B_o = 1.102$ RB/STB,	$k_h = 35$ md
$k_v = 3.5$ md,	$r_w = 0.29$ ft
$r_e = 1053$ ft,	$M = 3.27$
$\phi = 16.4$ %,	$S_{or} = 0.34$
$S_{wc} = 0.29$	

Solution

A calculation procedure is described here for a fixed total production rate of 500 STB/day and time elapsed from the beginning of production is 330 days.

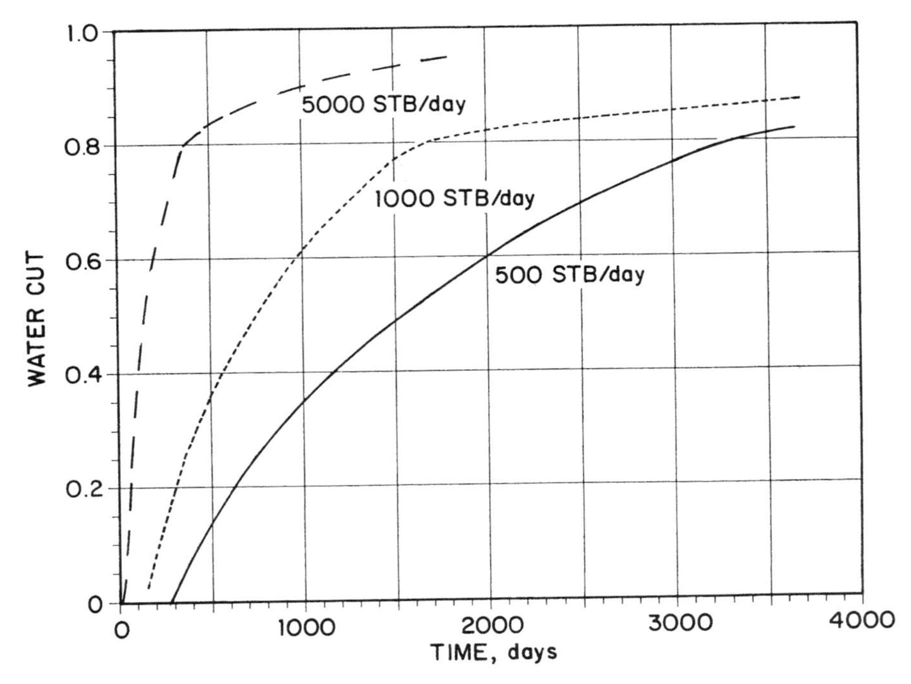

Figure 8–6 Variation in Water Cut over Time (Data from Example 8–6).

TABLE 8–3 WATER CUT BEHAVIOR OVER TIME FOR A VERTICAL WELL

The following procedure can be used to calculate rise of water cut over time for a given total fluid production rate in a vertical well. The procedure listed below is reported in References 10 and 11. The following equations are statistical correlations of the numerical simulation results.[10,11] The numerical simulation results were obtained using a straight line relative permeability curve. The calculation procedure for an oil reservoir with bottom water is listed below:

1. Calculate dimensionless breakthrough time t_{DBT}

$$t_{DBT} = t/t_{BT} \tag{8–3–1}$$

where

t is the time in days and t_{BT} is the water breakthrough time calculated either from Equation 8–2–4 or from Equation 8–2–6 as listed in Table 8–2.

TABLE 8–3 *(Continued)*

2. Calculate limiting water cut for the reservoir

$$(wc)_{\text{limit}} = \frac{Mh_w}{Mh_w + h} \quad (8\text{–}3\text{–}2)$$

where

$$h_w = H_w + H_o \times M_D \quad (8\text{–}3\text{–}3)$$
$$h = H_o (1 - M_D) \quad (8\text{–}3\text{–}4)$$
$$M_D = (N_p/N)(1 - S_{wc})/(1 - S_{or} - S_{wc}) \quad (8\text{–}3\text{–}5)$$

and

h_w = current water zone thickness, ft
h = current oil zone thickness, ft
H_w = initial water zone thickness, ft
H_o = initial oil zone thickness, ft
N_p = cumulative oil production, STB
N = initial oil in place, STB
S_{wc} = connate water saturation, fraction
S_{or} = residual oil saturation, fraction
M = mobility ratio, dimensionless

3. Calculate a new dimensionless water cut based upon dimensionless breakthrough time as below:

$$(wc)_D = 0 \quad \text{for } t_{DBT} < 0.5 \quad (8\text{–}3\text{–}6)$$
$$(wc)_D = 0.29 + 0.94 \log (t_{DBT}) \quad 0.5 \leq t_{DBT} \leq 5.7 \quad (8\text{–}3\text{–}7)$$
$$(wc)_D = 1.0 \quad t_{DBT} > 5.7 \quad (8\text{–}3\text{–}8)$$

4. Calculate actual water-cut fraction as

$$wc = (wc)_D \, (wc)_{\text{limit}} \quad (8\text{–}3\text{–}9)$$

5. Calculate water and oil rate as

$$q_w = wc \times q_T \quad (8\text{–}3\text{–}10)$$

and

$$q_o = q_T - q_w \quad (8\text{–}3\text{–}11)$$

where q_T, q_w and q_o represent total flow rate, water flow rate, and oil flow rate, respectively.

Initial oil in place

$$\begin{aligned} N &= \text{Area} \times H_o \times (1 - S_{wc}) \times \phi/(5.615\ B_o) \\ &= 80 \times 43{,}560 \times 84 \times (1 - 0.29) \times 0.164/(5.615 \times 1.102) \\ &= 5{,}508{,}432 \text{ STB} \end{aligned}$$

The water breakthrough times for three different oil rates were calculated by Sobocinski and Cornelius method (see Table 8–2) and are summarized below.[8]

RATE, q_t STB/day	z	$(t_D)_{SC}^{BT}$	t_{BT} days
500	0.160	0.102	594
1000	0.080	0.0484	282
5000	0.016	0.0092	54

As noted before, post-water breakthrough behavior is illustrated for a case when total fluid production rate is maintained at 500 STB/day. At this production rate, water breakthrough time is 594 days. The elapsed time of 330 days is shorter than the water breakthrough time of 594 days. Kuo and DesBrisay's procedure given in Table 8–3 is used to calculate water and oil rates.[10,11] A close observation of the Kuo and DesBrisay correlations (see Table 8–3 and Equation 8–3–7) indicates that even when dimensionless breakthrough time is 0.5, i.e., 297 days in this case, a small amount of water production begins.

1. Dimensionless breakthrough time

$$t_{DBT} = 330/594 = 0.56$$

2. Limiting water cut for the reservoir

$$M_D = \frac{N_p}{N} \times \frac{1 - S_{wc}}{1 - S_{or} - S_{wc}} \tag{8–3–5}$$

where

N_p = cumulative oil production, STB
N = oil in place, STB

then

$$M_D = \frac{500 \times 330}{5{,}508{,}432} \times \frac{1 - 0.29}{1 - 0.34 - 0.29}$$
$$= 0.05748$$

The present water column height is

$$\begin{aligned} h_w &= H_w + H_o \times M_D \\ &= 24 + 84 \times 0.05748 \\ &= 28.83 \text{ ft} \end{aligned} \tag{8–3–3}$$

The present oil column height is

$$\begin{aligned} h &= H_o\,[1 - M_D] \\ &= 84\,[1 - 0.05748] \\ &= 79.17 \text{ ft} \end{aligned} \tag{8–3–4}$$

The limiting water cut is

$$(wc)_{\text{limit}} = \frac{M\,h_w}{M\,h_w + h} \tag{8–3–2}$$

$$= \frac{3.27 \times 28.83}{3.27 \times 28.83 + 79.17} = 0.54$$

3. New dimensionless water cut.
   For $0.5 \leq t_{DBT} \leq 5.7$, Equation 8–3–7 should be used.

$$\begin{aligned} (wc)_D &= 0.29 + 0.94 \log (t_{DBT}) \\ &= 0.29 + 0.94 \log (0.56) \\ &= 0.053 \end{aligned} \tag{8–3–7}$$

4. The present water cut is

$$\begin{aligned} wc &= (wc)_D \times (wc)_{\text{limit}} \\ &= 0.053 \times 0.544 \\ &= 0.029 \end{aligned} \tag{8–3–9}$$

5. The present water rate is

$$\begin{aligned} q_w &= wc \times q_T \\ &= 0.029 \times 500 \\ &= 14.5 \text{ STB/day} \approx 15 \text{ STB/day} \end{aligned} \tag{8–3–10}$$

6. The present oil rate is

$$q_o = 500 - 15 = 485 \text{ STB/day}$$

Tables 8–4 through 8–6 summarize the calculations for water cut versus time for total flow rates of 500, 1000, and 5000 STB/day, respectively. The results were obtained by writing a computer program for equations listed in Table 8–3 and as per procedure outlined above. The results are plotted in Figures 8–5 and 8–6. Figure 8–5 shows water cut versus percentage oil recovery and Figure 8–6 depicts water cut as a function of producing time for different production rates. It is important to note that oil recovery (or cumulative production) is fixed at a given water-cut value regardless of total production rate.

TABLE 8–4 RESULTS OF EXAMPLE 8–6, VERTICAL WELL (Total Production Rate = 500 STB/day; Original Oil in Place = 5.5 MMSTB)

Time		Oil Rate, STB/day	Water Rate, STB/day	Water Cut, fraction	Cumul. Oil, MSTB	Oil Recovery %
days	years					
5	0.014	500	0	0.00	2.5	0.05
10	0.027	500	0	0.00	5.0	0.09
30	0.08	500	0	0.00	15.0	0.27
90	0.25	500	0	0.00	45.0	0.82
120	0.33	500	0	0.00	60.0	1.09
150	0.4	500	0	0.00	75.0	1.36
180	0.5	500	0	0.00	90.0	1.63
365	1.0	472	28	0.06	181.4	3.29
545	1.5	419	81	0.16	261.3	4.74
730	2.0	377	123	0.25	334.7	6.08
1095	3.0	312	188	0.38	459.6	8.34
1460	4.0	261	239	0.48	563.5	10.23
1825	5.0	218	282	0.56	650.6	11.81
2190	6.0	182	318	0.64	723.5	13.13
2555	7.0	151	349	0.70	784.2	14.24
2920	8.0	124	376	0.75	834.3	15.15
3285	9.0	100	400	0.80	875.1	15.89
3655	10.0	92	408	0.82	909.5	16.52

TABLE 8–5 RESULTS OF EXAMPLE 8–6, VERTICAL WELL (Total Production Rate = 1000 STB/day

Time		Oil Rate, STB/day	Water Rate, STB/day	Water Cut, fraction	Cumul. Oil, MSTB	Oil Recovery %
days	years					
10	0.027	1000	0	0.00	10.0	0.18
30	0.08	1000	0	0.00	30.0	0.54
90	0.25	1000	0	0.00	90.0	1.63
150	0.4	980	20	0.02	149.8	2.72
180	0.5	935	65	0.06	178.4	3.24
365	1.0	740	260	0.26	331.3	6.02
545	1.5	611	389	0.39	452.0	8.21
730	2.0	507	493	0.49	554.9	10.07
915	2.5	422	578	0.58	640.3	11.62
1095	3.0	351	649	0.65	709.6	12.88
1460	4.0	235	765	0.76	815.4	14.80
1645	4.5	196	804	0.80	854.5	15.51
1825	5.0	188	812	0.81	889.1	16.14
2190	6.0	173	827	0.83	954.9	17.34
2555	7.0	160	840	0.84	1015.6	18.44
2920	8.0	148	852	0.85	1071.7	19.46
3285	9.0	137	863	0.86	1123.7	20.40
3655	10.0	127	873	0.87	1172.0	21.29

CHARACTERISTICS OF WATER CUT VERSUS RECOVERY FACTOR PLOTS

The results in the previous section are for a "homogeneous" reservoir, which exhibits a unique water cut against recovery factor relationship. The curve represented in Figure 8–5 indicates the same ultimate recovery regardless of the flow rate. However, in a fractured reservoir, water cut and the recovery factor relationship may show a significantly different characteristic. For example, in a highly fractured reservoir, once water breaks through, oil production drops and water cut increases rapidly. In a short time span, water cut rises from 0 to 100%, effectively shutting down the oil flow. In such a reservoir, the ultimate oil recovery from a well does depend upon flow rate. A maximum recovery is obtained if premature water breakthrough is prevented, i.e., by operating the well at the critical rate (or below). In such fractured reservoirs, water cut can be reduced by decreasing the

TABLE 8–6 RESULTS OF EXAMPLE 8–6, VERTICAL WELL (Total Production Rate = 5000 STB/day

Time		Oil Rate,	Water Rate,	Water Cut,	Cumul. Oil,	Oil Recovery
days	years	STB/day	STB/day	fraction	MSTB	%
5	0.014	5000	0	0.00	25.0	0.45
10	0.027	5000	0	0.00	50.0	0.91
30	0.08	4849	151	0.03	149.7	2.71
90	0.25	3326	1674	0.33	384.8	6.99
120	0.33	2842	2158	0.43	475.9	8.64
150	0.41	2440	2560	0.51	553.9	10.06
180	0.5	2095	2905	0.58	621.0	11.27
365	1.0	968	4032	0.81	869.5	15.78
545	1.5	791	4209	0.84	1026.5	18.64
730	2.0	653	4347	0.87	1159.2	21.04
915	2.5	547	4453	0.89	1269.6	23.05
1095	3.0	464	4536	0.91	1360.0	24.69
1460	4.0	339	4661	0.93	1504.7	27.32
1840	5.0	250	4750	0.95	1614.1	29.33

pressure drawdown in the reservoir. This can be accomplished by increasing the wellhead and bottomhole pressures, i.e., by choking the well down.

Figure 8–7 shows different water-cut behaviors in homogeneous and highly fractured reservoirs. Thus, a plot of water cut versus recovery factor can be used to indicate fracture intensity near the wellbore. In a given reservoir, wells located at different locations may encounter different fracturing intensity, and therefore, may exhibit different water cut against recovery factor characteristics.

It is important to note that some fractured reservoirs may exhibit water cut against recovery factor characteristics similar to a homogeneous reservoir. This is true especially in highly fractured, light oil, steeply dipping reservoirs. In these reservoirs, gravity effects are strong and fluid segregation minimizes water coning.

In general, fracture intensity has a significant influence on the water-drive behavior. As shown later, horizontal wells are very useful in highly fractured reservoirs, where once the water breaks through in a well, the oil rate drops dramatically. In these reservoirs, water cones through high-permeability vertical fractures. Horizontal wells also have applications in "homogeneous" sandstone reservoirs where horizontal wells can be used to reduce water cut. This is especially important in places where water disposal

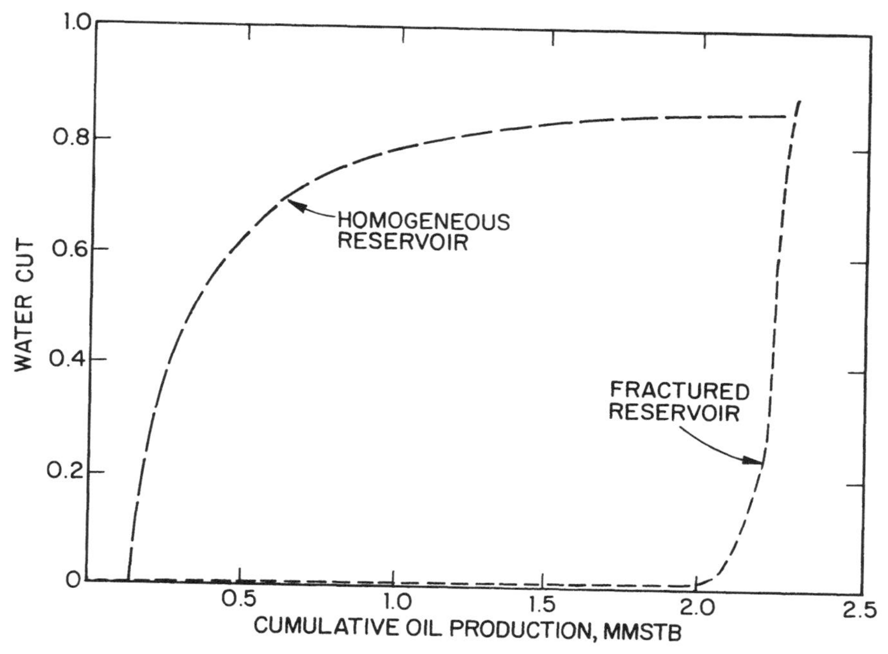

Figure 8–7 Typical Water-Cut Behaviors Observed in Homogeneous and Fractured Reservoirs.

costs are high. In addition, where pumping costs are high, reduction of water cut and enhancement of oil cut can be economically beneficial. Therefore, many horizontal wells around the world have been successful in fractured limestone as well as in sandstone reservoirs in reducing water cuts.

WATER AND GAS CONING IN HORIZONTAL WELLS

For a vertical well the majority of the pressure drawdown is consumed near the wellbore. Therefore, there is a big drawdown around the wellbore in a vertical well. In the case of horizontal wells, the pressure drop is fairly uniform throughout the reservoir, and in a small region near the wellbore, an extra pressure drop is observed. This pressure drop is, however, very small as compared to that around a vertical wellbore (see Fig. 8–1).

For horizontal wells, due to low-pressure drawdown, one expects a high oil production rate without coning. Various critical-rate correlations for horizontal wells are listed in Table 8–7.[4,12–15] In a reservoir with bottom water or top gas, rising water and downward movement of the gas cap can

TABLE 8–7 HORIZONTAL WELL CRITICAL RATE CORRELATIONS

Chaperon Method[4]

$$q_0 = 4.888 \times 10^{-4} \frac{L}{y_e} \Delta\rho \frac{(k_h h^2)}{\mu_o B_o} F \quad (8\text{–}7\text{–}1)$$

for $1 \leq \alpha'' < 70$ and $2y_e < 4L$

q_0 = critical rate, STB/day
L = horizontal well length, ft
y_e = half drainage length (perpendicular to the horizontal well), ft
$\Delta\rho$ = density difference, gm/cc
k_h = horizontal permeability, md
h = oil column thickness, ft
μ_o = oil viscosity, cp
B_o = oil formation volume factor, RB/STB
$\alpha'' = (y_e/h)\sqrt{k_v/k_h}$
F = dimensionless function tabulated below:

Function F depends upon α''.

α''	F
1	4.003
2	4.026
3	4.083
4	4.160
5	4.245
7	4.417
10	4.640
13	4.80
20	5.08
30	5.31
40	5.48
70	5.74

Note that F can be approximated as $F = 4$. This would cause a maximum error of 44% at $\alpha'' = 70$. The above tabulated results are correlated as

$$F = 3.9624955 + 0.0616438(\alpha'') - 0.000540(\alpha'')^2 \quad (8\text{–}7\text{–}2)$$

TABLE 8–7 (*Continued*)

In bottom water drive reservoirs, normally reservoir pressure is maintained constant, and therefore, pseudo-steady state or "pressure depletion state" is not reached. However, if the reservoir has a bottom water zone, but if the water zone does not provide any pressure support, reservoir pressure will decrease or deplete over time as fluids are withdrawn from the reservoir. In this pseudo-steady state, Reference 4 suggests using $y_e/2$ instead of y_e in Equation 8–7–1 to calculate critical rate.

Efros Method[12,15]

$$q_o = \frac{4.888 \times 10^{-4} k_h \Delta\rho h^2 L}{\mu_o B_o \left[2y_e + \sqrt{(2y_e)^2 + (h^2/3)}\right]} \qquad (8\text{–}7\text{–}3)$$

$2y_e$ = horizontal well spacing, ft
h = pay thickness, ft
q_o = STB/day

Note: Due to difficulty in obtaining the original reference it was not possible to check Equation 8–7–3. This author suspects that there is y_e in the denominator of Equation 8–7–3 instead of $2y_e$.

Giger and Karcher et al. Method[13,14,15]

$$q_o = 4.888 \times 10^{-4} \left[\frac{k_h}{\mu_o B_o}\right] \left[\frac{\Delta\rho h^2}{2y_e}\right] \left[1 - (1/6)\left(\frac{h}{2y_e}\right)^2\right] L \qquad (8\text{–}7\text{–}4)$$

q_o = critical rate, STB/day

Joshi Method (Gas Coning)[16]

$$q_{o,v} = \frac{1.535 \times 10^{-3} (\rho_o - \rho_g) k_h [h^2 - (h - I_v)^2]}{B_o \mu_o \ln(r_e/r_w)} \qquad (8\text{–}7\text{–}5)$$

where

$q_{o,v}$ = vertical well critical rate, STB/day
ρ_o = oil density, gm/cc
ρ_g = gas density, gm/cc
h = oil column thickness, ft
I_v = distance between the gas/oil interface and perforated top of a vertical well, ft
k_h = permeability, md

The critical rate of horizontal well, $q_{o,h}$, can be calculated via the following equation.

TABLE 8–7 *(Continued)*

$$\frac{q_{o,h}}{q_{o,v}} = \frac{[h^2 - (h - l_h)^2] \ln (r_e/r_w)}{[h^2 - (h - l_v)^2] \ln (r_e/r'_w)} \quad (8\text{–}7\text{–}6)$$

where

l_h = distance between the horizontal well and the gas/oil interface

r'_w = effective wellbore radius, ft; see Equation 3–25.

be controlled to obtain the best possible sweep of the reservoir. As shown in Figure 8–1, this is also called water cresting. With a proper operating procedure, the bottom water drive can behave similar to a waterflood from below, resulting in a very high recovery. (Some water drive reservoirs can give recoveries of the order of 60 to 65% of oil in place!)

EXAMPLE 8–7

A 1640-ft-long horizontal well is drilled in the lowest zone of an oil reservoir. The reservoir has a gas cap. Calculate critical oil-production rates for horizontal and vertical wells using the coning equations of Chaperon,[4] Efros,[12] Karcher et al.[15] and Joshi.[16] Also calculate the critical rate if the reservoir vertical permeability is one-tenth of the horizontal permeability using the Chaperon method. The wells are placed at 160-acre well spacing. The vertical well is perforated in the bottom 8 ft to minimize gas coning.

$k_h = k_v = 70$ md, $h = 80$ ft
$2x_e = 2640$ ft, $B_o = 1.1$ RB/STB
$\mu_o = 0.42$ cp, $r_w = 0.328$ ft
$\rho_o - \rho_g = 0.48$ gm/cc, $h_p = 8$ ft for vertical wells.

Solution

For 160-acre well spacing, for a square drainage area, $2x_e = 2y_e = 2640$ ft and for a circular drainage area, $r_e = 1489$ ft.

The solution assumes that the horizontal well is drilled about 8 ft from the reservoir bottom or 72 ft below the gas-oil contact. Horizontal-well critical-rate correlations are listed in Table 8–7. It is important to note that only Chaperon's correlation accounts for the difference in vertical and horizontal permeability.

Chaperon Method

1. **Horizontal Wells**

$$q_o = 4.888 \times 10^{-4} \frac{L}{y_e} \Delta\rho \frac{(k_h \times h^2)}{\mu_o B_o} F \quad (8\text{–}7\text{–}1)$$

where

$$F = 3.9624955 + 0.0616438\,(\alpha'') - 0.000540\,(\alpha'')^2 \qquad (8\text{–}7\text{–}2)$$

Case 1: Isotropic Reservoir ($k_v = k_h$)
Since the horizontal well is drilled 72 ft from the gas-oil contact, effective h for a horizontal well is 72 ft. First, α'' is calculated as

$$\alpha'' = \frac{y_e}{h}\sqrt{k_v/k_h} = \frac{1320}{72} \times 1 = 18.3$$

Substituting $\alpha'' = 18.3$ into Equation 8–7–2 gives

$$F = 3.9624955 + 0.0616438 \times 18.3 - 0.00054 \times 18.3^2$$
$$= 4.9$$

Substituting F into Equation 8–7–1 gives

$$q_o = 4.888 \times 10^{-4} \times \frac{1640}{1320} \times 0.48 \times \frac{70 \times 72^2}{0.42 \times 1.1} \times 4.9$$
$$= 1122 \text{ STB/day}$$

Case 2: Anisotropic Reservoir ($k_v/k_h = 0.1$)

$$\alpha'' = \frac{y_e}{72}\sqrt{k_v/k_h} = \frac{1320}{72}\sqrt{0.1} = 5.8$$

Substituting this value of α'' into Equation 8–7–2 yields

$$F = 3.9624955 + 0.0616438 \times 5.8 - 0.00054 \times (5.8)^2$$
$$= 4.3$$

Using this value of F in Equation 8–7–1, critical rate is calculated as

$$q_o = 4.888 \times 10^{-4} \times \frac{1640}{1320} \times 0.48 \times \frac{70 \times 72^2}{0.42 \times 1.1} \times 4.3$$
$$= 985 \text{ STB/day}$$

2. **Vertical Wells**
Case 1: Isotropic Reservoir
Similar to horizontal well calculation, the oil column height between the gas-oil contact and the top of the perforation is used as a pay zone height,

h. This h is 80 − 8 = 72 ft. Chaperon's vertical well critical rate correlation is:

$$q_o = \frac{4.888 \times 10^{-4}}{B_o} \times \frac{k_h h^2}{\mu_o} \times (\Delta\rho) q_c^* \qquad (8\text{–}1\text{–}7)$$

where q_c^* is evaluated as

$$q_c^* = 0.7311 + \frac{1.9434}{\alpha''} \qquad (8\text{–}1\text{–}8)$$

For a vertical well,

$$\alpha'' = \frac{r_e}{h}\sqrt{k_v/k_h} = \frac{1489}{72} \times 1$$
$$= 20.7$$
$$q_c^* = 0.7311 + \frac{1.9434}{20.7}$$
$$= 0.82$$

Using Equation 8–1–7, vertical well critical rate is calculated as

$$q_o = \frac{4.888 \times 10^{-4}}{1.1} \times \frac{70 \times 72^2}{0.42} \times 0.48 \times 0.82$$
$$= 151 \text{ STB/day}$$

Case 2: Anisotropic Reservoir

$$\alpha'' = \left(\frac{1489}{72}\right) \times \sqrt{0.1} = 6.54$$
$$q_c^* = 0.7311 + (1.9434/6.54)$$
$$= 1.03$$

Substituting q_c^* into Equation 8–1–7,

$$q_o = \frac{4.888 \times 10^{-4}}{1.1} \times \frac{70 \times 72^2}{0.42} \times 0.48 \times 1.03$$
$$= 189.8 \approx 190 \text{ STB/day}$$

Note that the critical rate for horizontal well increases as the vertical permeability increases, whereas the opposite is true for vertical wells. This is due to the fact that for horizontal wells, as vertical permeability decreases, the well productivity also decreases, requiring high pressure drawdown to maintain a given fluid production rate. The high drawdown reduces the critical oil-production rate for a horizontal well.

Efros Method

As shown in Table 8–7, Equation 8–7–3 is used to calculate critical oil rate

$$q_o = \frac{4.888 \times 10^{-4}\, k_h\, \Delta\rho\, h^2\, L}{\mu_o B_o\,(2y_e)\left[1 + \sqrt{1 + (h^2/3)(1/(2y_e)^2)}\right]}$$

$$= \frac{4.888 \times 10^{-4} \times 70 \times 0.48 \times (72)^2 \times 1640}{0.42 \times 1.1 \times (2640)\left[1 + \sqrt{1 + (72^2/3)(1/2640^2)}\right]}$$

$$= 57 \text{ STB/day.}$$

Giger and Karcher et al. Method

As shown in Table 8–7, Equation 8–7–4 can be used to calculate critical rate.

$$q_o = 4.888 \times 10^{-4} \times \frac{k_h\, \Delta\rho\, h^2\, L}{\mu_o B_o\,(2y_e)}\left[1 - \frac{1}{6} \times \frac{h^2}{(2y_e)^2}\right] \quad (8\text{–}7\text{–}4)$$

$$= 4.888 \times 10^{-4} \times \frac{70 \times 0.48 \times (72)^2 \times 1640}{0.42 \times 1.1 \times 2640} \times \left[1 - \frac{1}{6} \times \frac{72^2}{2640^2}\right]$$

$$= 113.9 \approx 114 \text{ STB/day}$$

Joshi Method

1. **Vertical Well Critical Rate**

$$q_{o,v} = \frac{1.535 \times 10^{-3}\, \Delta\rho\, k_h \times [h^2 - (h - I_v)^2]}{B_o \mu_o \ln(r_e/r_w)} \quad (8\text{–}7\text{–}5)$$

$$= \frac{1.535 \times 10^{-3} \times 0.48 \times 70 \times [80^2 - (80 - 72)^2]}{1.1 \times 0.42 \times \ln(1489/0.328)}$$

$$= 84 \text{ STB/day}$$

2. **Horizontal Well Critical Rate**

$$q_{o,h} = q_{o,v} \times \frac{\ln(r_e/r_w)_v}{\ln(r_e/r'_w)_h} \times \frac{h^2 - (h - I_h)^2}{h^2 - (h - I_v)^2} \quad (8\text{–}7\text{–}6)$$

The effective wellbore radius r'_w of a horizontal well can be calculated using Equation 3–25.

$$a = (L/2)\left[0.5 + \sqrt{0.25 + (2r_{eh}/L)^4}\right]^{0.5} \tag{3–11}$$

$$= (1640/2)\left[0.5 + \sqrt{0.25 + \left(\frac{2 \times 1489}{1680}\right)^4}\right]^{0.5}$$

$$= 1573 \text{ ft}$$

$$r_{eh}\,[L/(2a)] = 1489 \times 1640/(2 \times 1573) = 776$$

$$L/(2a) = 1640/(2 \times 1573) = 0.521$$

$$r'_w = \frac{r_{eh}\,[L/(2a)]}{\left[1 + \sqrt{1 - [L/(2a)]^2}\right]\,[h/(2r_w)]^{h/L}} \tag{3–25}$$

$$= \frac{776}{\left[1 + \sqrt{1 - 0.521^2}\right]\,[80/(2 \times 0.328)]^{80/1640}}$$

$$= 331 \text{ ft}$$

Since the horizontal well is drilled in the lowest zone, the distance of the horizontal well and the gas/oil contact is $I_h = 72$ ft, which is the same as $I_v = 72$ ft. Using Equation 8–7–6 from Table 8–7,

$$q_{o,h} = 84 \times \frac{\ln(1489/0.328)}{\ln(1489/331)} \times \frac{80^2 - (80 - 72)^2}{80^2 - (80 - 72)^2}$$

$$= 35 \times 5.6 \times 1$$

$$= 470 \text{ STB/day}$$

The critical oil production rates calculated by the four methods are summarized in the following table.

	$q_{o,h}$ (STB/day)	$q_{o,v}$ (STB/day)	$q_{o,h}/q_{o,v}$
Chaperon			
Isotropic	1122	151	7.18
Anisotropic ($k_v/k_h = 0.1$)	985	190	5.01
Efros	57	—	—
Giger and Karcher et al.	114	—	—
Joshi	470	84	5.6

It is seen that results from calculations for the critical rate by these methods are different, probably due to the different assumptions each of them involves. Nevertheless, Chaperon's correlation and Joshi's correlation show a significant critical oil-production rate improvement in the case of horizontal wells as compared with vertical wells, thus exhibiting reduction in water-coning tendencies. The advantage of horizontal wells over vertical wells in preventing coning becomes less significant when vertical permeability decreases.

As discussed earlier for vertical wells, one can similarly operate horizontal wells at critical rates (see discussion in the Decline Curve Analysis section). This results in a progressive drop of oil flow rate as the oil column height decreases over time. A decline curve analysis of horizontal wells operated at critical rates is shown in Figure 8–8. It is interesting to note that similar to vertical wells, it also shows a decline index of $b = 0.5$. Thus, in a water drive reservoir, either horizontal or vertical wells give a decline index of $b = 0.5$ when operated at the critical rates. It is important to note that critical rates of horizontal wells are significantly higher than those for vertical wells, resulting in an accelerated recovery of oil in place without water coning. Thus, horizontal wells not only allow one to produce wells at higher rates, but also allow one to obtain maximum recovery of oil in a short time

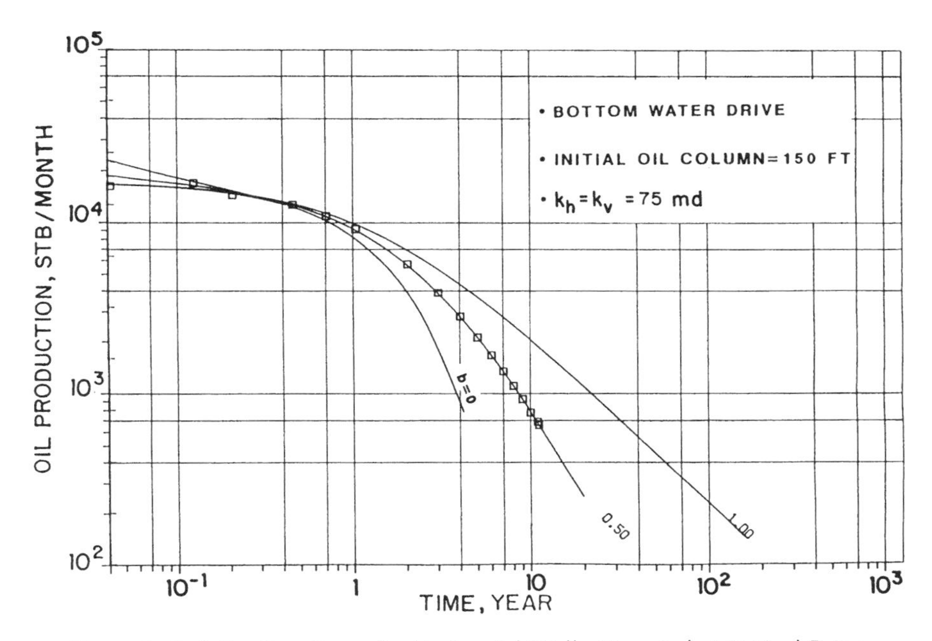

Figure 8–8 A Decline Curve for Horizontal Wells Operated at Critical Rates.

span (see Fig. 8–9). Figure 8–9 was developed using reservoir data noted in the figure. The variation of horizontal well critical rate over time is obtained by the following procedure:

1. Vertical well critical-rate correlation

$$q_{o,v} = 4.888 \times 10^{-4} k_h h^2 \Delta\rho \, q_c^* / (\mu_o B_o) \quad (8\text{–}4)$$

2. Horizontal well critical-rate correlation can be obtained using Equations 8–1–7 and 8–7–1

$$q_{o,h} = q_{o,v} F L / (y_e q_c^*) \quad (8\text{–}5)$$

where F can be calculated using Equation 8–7–2 or it can be approximated with $F \approx 4$ and $q_c^* \approx 1$.

3. The material balance equation is used to calculate the change in oil column height if a well is produced at a constant rate, q_o, for a time period of Δt. For the calculation procedure, Δt can be chosen as one week or one month, depending upon the problem. The new oil column height to the end of the time period is:

$$h = H_o - \frac{q_o \Delta t (5.615 B_o)}{\phi [1 - S_{wc} - S_{or}] A \times 43560} \quad (8\text{–}6)$$

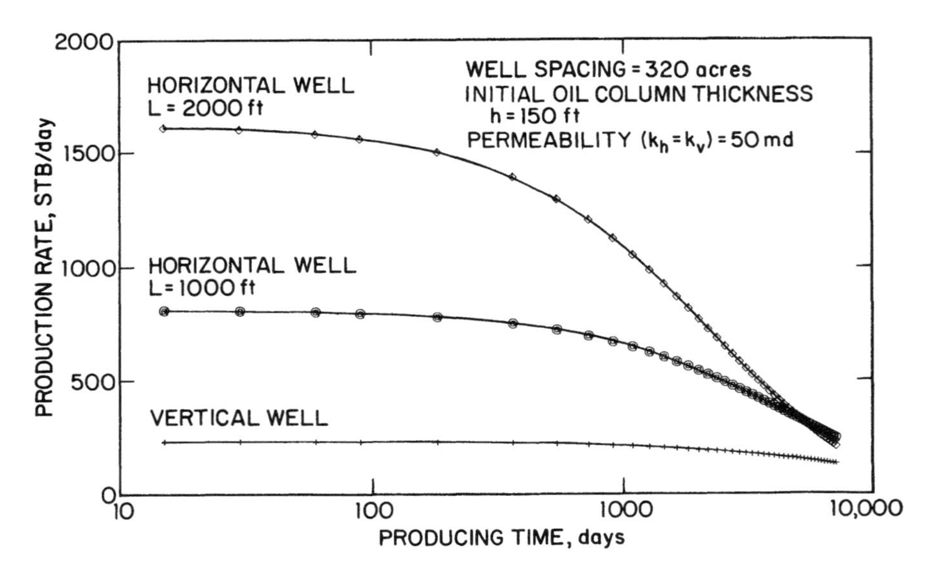

Figure 8–9 A Comparison of Vertical and Horizontal Well Critical Rates using Chaperon's Correlations.

where

H_o = initial oil column height, ft
q_o = oil rate, STB/day
Δt = production time, days
B_o = formation volume factor, RB/STB
S_{wc} = connate water saturation, dimensionless
S_{or} = residual oil saturation, dimensionless
A = drainage area, acres
h = new oil column height at the end of time period Δt, ft

It is important to note that the above calculation implicitly assumes that the bottom water drive is strong and reservoir pressure over time is constant. In other words, the problem assumes the filling of oil voidage by water.

Figure 8–9 clearly shows a distinct advantage of using horizontal wells. It is important to note that all the vertical and horizontal well rates are critical rates, and therefore, show a decline index of $b = 0.5$ on a log-log decline plot. The figure also show that the advantage of horizontal wells is to increase initial production rates without water coning. The ultimate recovery of oil in place obtained using either horizontal or vertical wells is the same. A horizontal well accelerates the recovery and improves the project economics significantly. In highly fractured reservoirs, where water coning is a serious problem, horizontal wells are highly advantageous. As noted earlier, producing a well at critical rates allows one to obtain maximum possible ultimate recovery. A horizontal well provides an option not only to enhance initial oil-production rates, but also to obtain maximum possible ultimate reserves in a shorter time span than a vertical well.

HORIZONTAL WELL BREAKTHROUGH TIME IN A BOTTOM WATER DRIVE RESERVOIR

For a bottom water drive reservoir, Ozkan and Raghavan[17] have reported a theoretical correlation to calculate water breakthrough time for a horizontal well. In their calculations, they assumed that the bottom water drive reservoir can be represented as a constant pressure boundary; i.e., pressure at the oil and water interface is constant. Their equation for the water breakthrough time is

$$t_{BT} = [f_d\, h^3\, E_s/(5.615\, q_o B_o)] \times (k_h/k_v) \qquad (8\text{–}7)$$

where

$$f_d = \phi\,(1 - S_{wc} - S_{oir}) \tag{8–8}$$

q_o = flow rate, STB/day
k_h = horizontal permeability, md
k_v = vertical permeability, md
E_s = sweep efficiency, dimensionless
f_d = microscopic displacement efficiency, dimensionless
h = oil column thickness, ft
B_o = formation volume factor, RB/STB
S_{wc} = connate water saturation, fraction
S_{oir} = residual oil saturation, fraction

The plots of sweep efficiency function, E_s, for vertical and horizontal wells are shown in Figures 8–10, 8–11a and 8–11b, respectively. Here in these

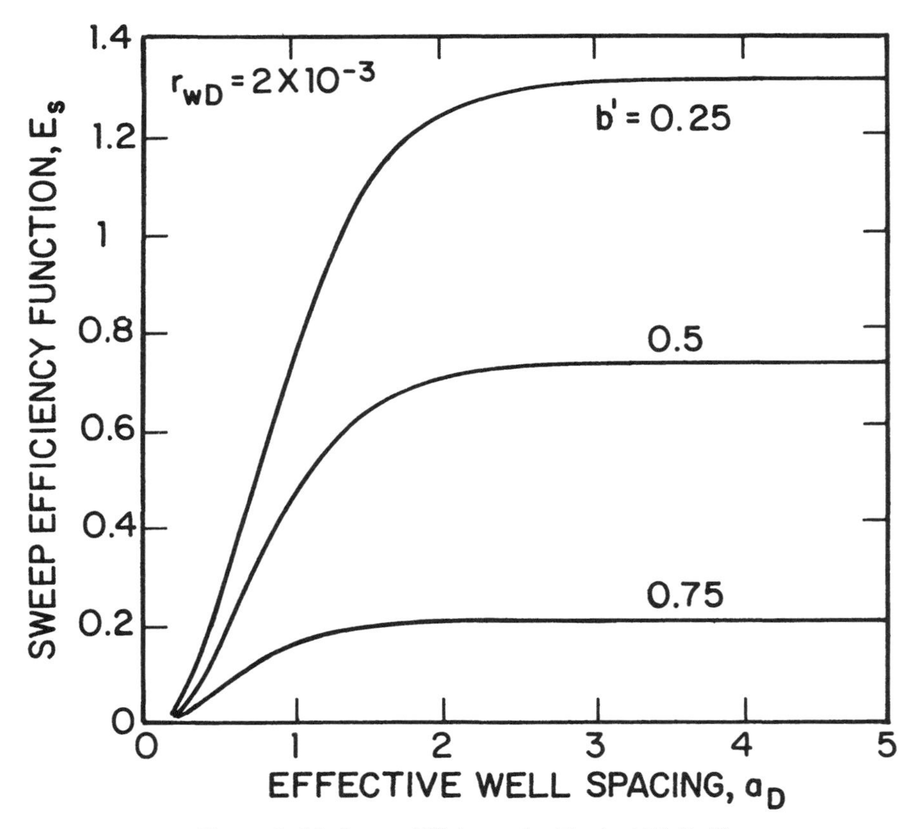

Figure 8–10 Sweep Efficiency for Vertical Wells.[17]

three figures, the effective well spacing a_D, dimensionless well length L_D, penetration ratio b', dimensionless vertical distance z_{wD}, and dimensionless wellbore radius r_{wD} are defined as

$$a_D = (2x_e/h)\sqrt{k_v/k_h} \tag{8–9}$$

$$L_D = [L/(2h)]\sqrt{k_v/k_h} \tag{8–10}$$

$$b' = h_p/h \tag{8–11}$$

$$z_{wD} = z_w/h \tag{8–12}$$

$$r_{wD} = r_w/h \tag{8–13}$$

where

x_e = half well spacing, ft
L = well length, ft
h_p = perforated interval, ft

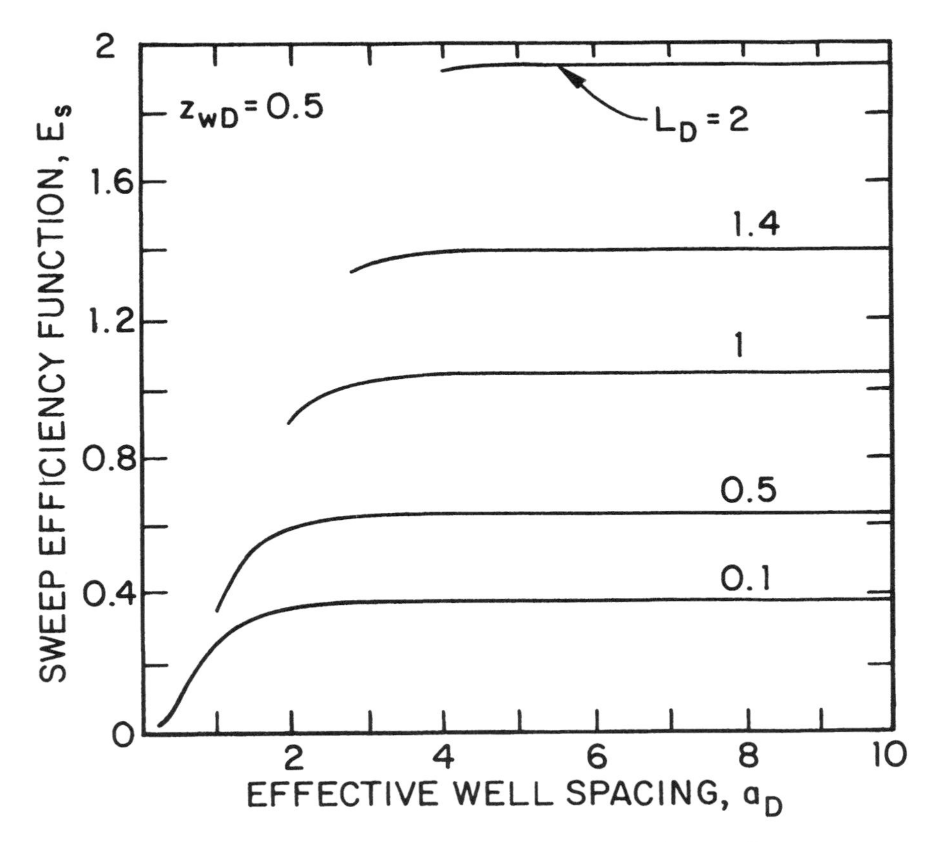

Figure 8–11a Sweep Efficiency for Horizontal Wells.[17]

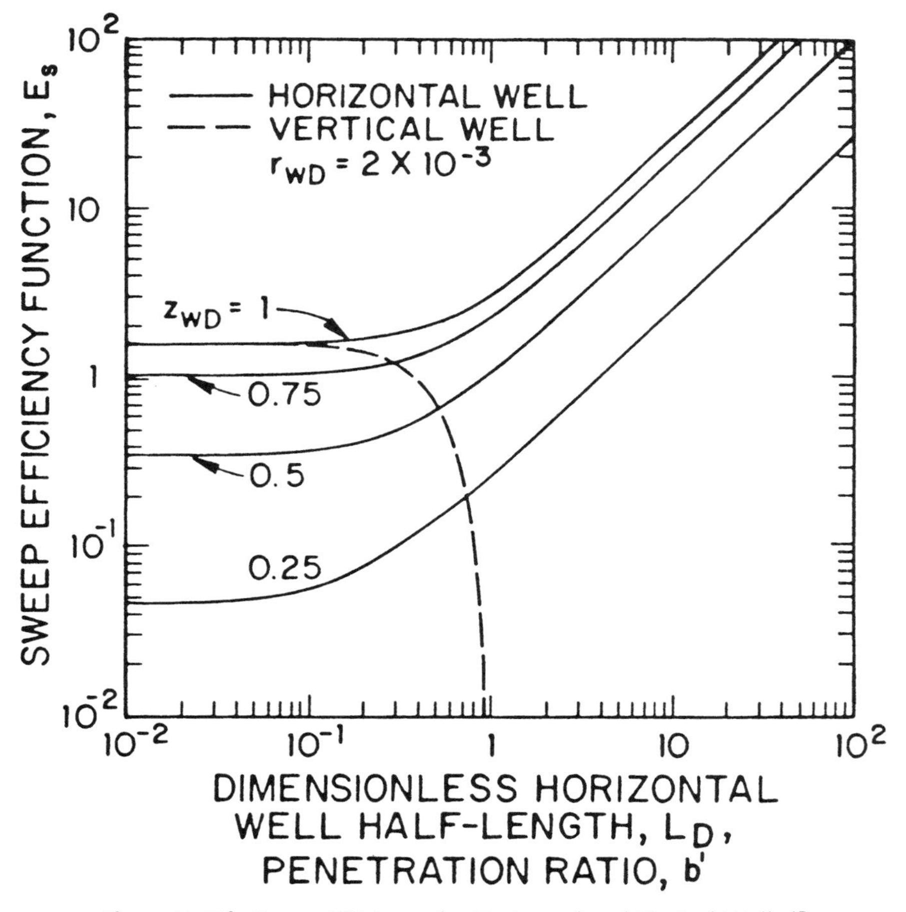

Figure 8–11b Sweep Efficiency for Horizontal and Vertical Wells.[17]

z_w = vertical distance of horizontal well from the oil-water contact at time $t = 0$, ft

r_w = wellbore radius, ft.

For horizontal wells, Figure 8–11a indicates that sweep efficiency function E_s increases with increasing well length (increasing L_D) for a fixed well spacing. This shows that increasing well length for a fixed well spacing would result in delaying water breakthrough in horizontal wells. Thus, longer wells would produce more water-free oil than shorter wells for a given well spacing or in a given acreage. Figure 8–11a also shows that if the well length is fixed, by increasing well spacing, one can delay water breakthrough. However, beyond a certain value, an increase in well spacing would not increase the water breakthrough time as long as well length is fixed. This is an important observation which allows us to optimize well spacing for horizontal wells in bottom water drive reservoirs. (Note that Figure 8–10 is for $r_{wD} = 2 \times 10^{-3}$ and Figure 8–11a is for $z_{wD} = 0.5$.) Figure 8–11b depicts sweep efficiency, E_s, for horizontal wells of different lengths located at different elevations from the oil-water contact. In Figure 8–11b, $z_{wD} = 1$ shows a horizontal well located at the top of the oil column, which obviously shows a high value of sweep efficiency, E_s. (Note that $z_{wD} = z_w/h$.)

EXAMPLE 8–8

A 2000-ft-long horizontal well is drilled in a bottom water drive reservoir. The well spacing is 160 acres. The reservoir and well characteristics are as follows:

$$
\begin{array}{llll}
h = 160 \text{ ft} & & \rho_w - \rho_o & = 0.25 \text{ gm/cc} \\
q_h = 5000 \text{ STB/day} & & q_v & = 2000 \text{ STB/day} \\
k_h = 200 \text{ md} & & k_v & = 20 \text{ md} \\
\phi = 20\% & & S_{wc} & = 0.27 \\
S_{or} = 0.25 & & z_{wD} & = 0.5 \\
B_o = 1.1 \text{ RB/STB} & & \mu_o & = 1.3 \text{ cp}
\end{array}
$$

Calculate the breakthrough times for: (1) a 2000-ft-long horizontal well, (2) a vertical well, with the top 40 ft of pay zone perforated. Vertical well spacing is 80 acres.

Solution

Ozkan and Raghavan's method is used to calculate water breakthrough.[17] Note that the horizontal well is drilled at the pay zone mid-height ($z_{wD} = 0.5$).

1. **Horizontal Well:** $L = 2000$ ft:

Dimensionless Horizontal Well Length from Equation 8–10 is

$$
\begin{aligned}
L_D &= [L/(2h)]\sqrt{k_v/k_h} = [2000/(2 \times 160)]\sqrt{0.1} \\
&= 1.98 \approx 2
\end{aligned}
$$

Dimensionless Well Spacing from Equation 8–9 is

$$a_D = (2x_e/h)\sqrt{k_v/k_h} = (2640/160)\sqrt{0.1} = 5.2$$

From Figure 8–11a, Sweep Efficiency Function $E_s = 1.95$ and f_d is calculated from Equation 8–8 as

$$f_d = \phi\,(1 - S_{wc} - S_{or}) = 0.2 \times (1 - 0.27 - 0.25) = 0.096$$

Substituting value of E_s and f_d in Equation 8–7 gives

$$\begin{aligned} t_{BT} &= [f_d h^3 E_s/(5.615\, q_o B_o)]\,(k_h/k_v) \qquad (8\text{–}7) \\ &= \frac{0.096 \times 160^3 \times 1.95}{5000 \times 1.1 \times 5.615} \times \frac{200}{20} = 248.3 \text{ days} \\ &= 0.68 \text{ years} \end{aligned}$$

Cumulative oil production before water breakthrough is calculated as $Q_{cum} = 248 \times 5000 = 1.24$ MMSTB.

2. **Vertical Well**

The drainage area for a vertical well is 80 acres.

$$\begin{aligned} 2x_e &= \sqrt{80 \times 43560} = 1867 \text{ ft} \\ a_D &= (2x_e/h)\sqrt{k_v/k_h} = (1867/160)\sqrt{0.1} \\ &= 3.69 \end{aligned}$$

For penetration $b' = 40/160 = 0.25$, $h = 160$ ft, and $a_D = 3.69$, Sweep Efficiency Function $E_s = 1.3$ (See Fig. 8–10). Thus,

$$\begin{aligned} t_{BT} &= \frac{0.096 \times 160^3 \times 1.3}{2000 \times 1.1 \times 5.615} \times \frac{200}{20} \\ &= 413.8 \text{ days} = 414 \text{ days} = 1.13 \text{ years} \end{aligned}$$

Cumulative oil production in 414 days before the water breakthrough is

$$Q_{cum} = 414 \times 2000 = 0.826 \text{ MMSTB.}$$

Thus, by drilling a 2000-ft-long horizontal well, before water breakthrough, 1.24 MMSTB can be recovered in 248 days while only 0.828 MMSTB can be recovered from a vertical well in 414 days.

EXAMPLE 8–9

What would be the breakthrough time for the horizontal well if drilling were stopped after drilling 1000 ft due to drilling difficulties? The well location and reservoir properties are the same as those given in Example 8–8.

Solution

For a 1000-ft-long horizontal well, dimensionless well length L_D is given by Equation 8–10 as

$$L_D = [L/(2h)]\sqrt{k_v/k_h} = 1000/(2 \times 160)\sqrt{0.1} = 0.99 \approx 1.$$

Dimensionless well spacing is the same as that in Example 8–8, i.e., $a_D = 5.2$. From Figure 8–11a, sweep efficiency function, $E_s = 1.00$. As in Example 8–8, $f_d = 0.096$. Substituting E_s and f_d into Equation 8–7 gives

$$t_{BT} = \frac{0.096 \times 160^3 \times 1.00}{5000 \times 1.1 \times 5.615} \times \frac{200}{20} = 127.3 \text{ days} = 127 \text{ days}$$
$$\approx 0.35 \text{ years}$$

Cumulative oil recovery in 0.35 years is 127 days × 5000 STB/day = 0.640 MMSTB. Thus, by reducing well length from 2000 to 1000 ft, water breakthrough time decreases from 0.68 years to 0.35 years. Additionally, cumulative oil recovery prior to water breakthrough decreases from 1.24 MMSTB to 0.64 MMSTB. If the horizontal well is drilled at the reservoir top instead of pay zone mid-height, then from Figure 8–11b, sweep efficiency, $E_s = 3$. This would increase water breakthrough time from 0.35 to 1.05 years.

BREAKTHROUGH TIME FOR A HORIZONTAL WELL IN A RESERVOIR WITH GAS CAP OR BOTTOM WATER

Papatzacos et al.[18,19] have also reported results on breakthrough time in a horizontal well placed in a reservoir with either bottom water or top gas. Papatzacos et al.[19] solved this problem using a semi-analytic method with the assumption that the infinitely long horizontal well is located either at the top or at the bottom of the oil zone to minimize water or gas coning. The solution was obtained by two methods. In the first method it was assumed that either the top gas or bottom water can be represented as a constant pressure boundary. With this assumption, the final correlation for the breakthrough time is

$$t_{DBT} = 1/(6q_D) \qquad (8\text{–}14)$$

where

$$q_D = \frac{325.86\, \mu_o\, q_o\, B_o}{L\sqrt{k_v k_h}\, h\, (\rho_o - \rho_g)} \qquad (8\text{–}15)$$

and t_{DBT} and t_{BT} are related by the following relationship

$$t_{DBT} = \frac{k_v(\rho_o - \rho_g)}{364.72h\, \phi\mu_o} t_{BT} \qquad (8\text{–}16)$$

In Equations 8–15 and 8–16, $\rho_o - \rho_g$ is in gm/cc, and t_{BT} is breakthrough time in days. The rest are in U.S. oil field units. For calculation of water breakthrough time, one should use density difference between oil and water, $\rho_w - \rho_o$, instead of $\rho_o - \rho_g$ in Equations 8–15 and 8–16. The above constant pressure boundary solution is similar to the one reported by Ozkan and Raghaven.[17] Papatzacos et al.[19] have also obtained semi-analytical solution by considering gravity equilibrium in the cone instead of assuming a constant-pressure boundary. The dimensionless breakthrough time for $q_D > 0.4$ is given by

$$t_{DBT} = 1 - (3q_D - 1)\ln\left[\frac{3q_D}{3q_D - 1}\right] \tag{8–17}$$

For $q_D \geq 1$, results of Equations 8–14 and 8–17 are identical. Thus, for $q_D \geq 1$, breakthrough times calculated by assuming constant pressure at the oil-water contact or by gravity equilibrium in cone are identical. Hence, bottom-water drive can be simulated as a constant-pressure boundary for producing at high rates, i.e., for $q_D \geq 1$. Papatzacos et al.[19] have also compared their semi-analytic solution with the results of the numerical solution. A comparison of their numerical and analytical breakthrough times is shown

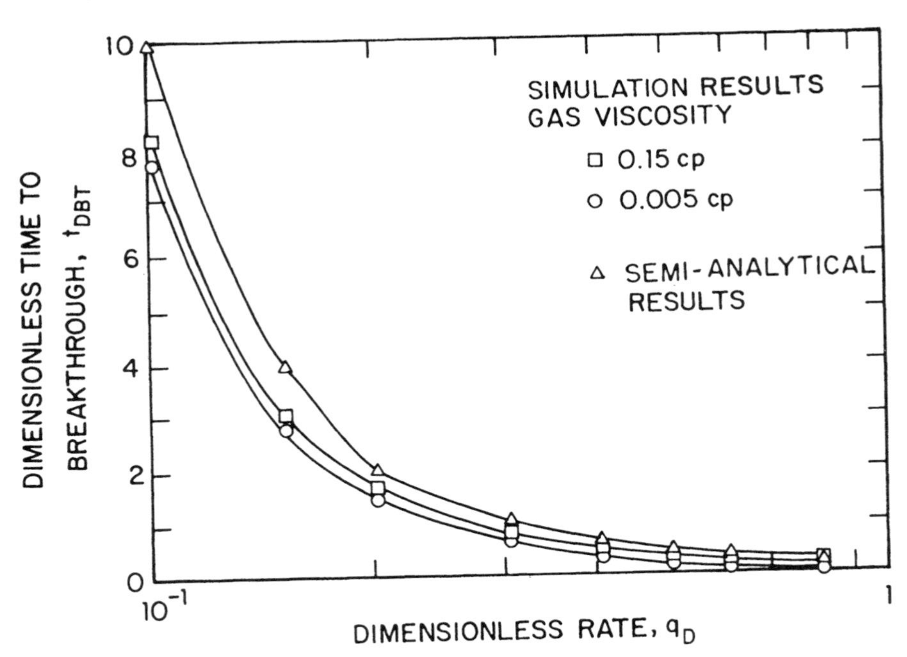

Figure 8–12 Comparison of Numerical and Analytical Breakthrough Times for Single-Cone Gas Case.[19]

in Figure 8–12 and estimated percentage error between the two solutions is depicted in Figure 8–13. The numerical breakthrough time is normally smaller than the analytic breakthrough time. The percentage error between the two solutions is smaller at higher gas viscosities than at lower gas viscosities. Thus, the analytical solution can be used with reasonable accuracy for all gas viscosities with $q_D \leq 0.3$. For gas viscosities above 0.15 cp, it can be used up to $q_D \leq 0.6$.

EXAMPLE 8–10

For the reservoir properties described in Example 8–8, calculate the water breakthrough time for a 1000-ft-long horizontal well, using Papatzacos et al.[19] single-cone model.

Solution

$$q_D = \frac{325.86\ \mu_o q_o B_o}{L\sqrt{k_v k_h}\, h\,(\rho_w - \rho_o)}$$

$$= \frac{325.86 \times 1.3 \times 5000 \times 1.1}{1000 \times \sqrt{20 \times 200} \times 160 \times 0.25}$$

$$= 0.921 \approx 0.92$$

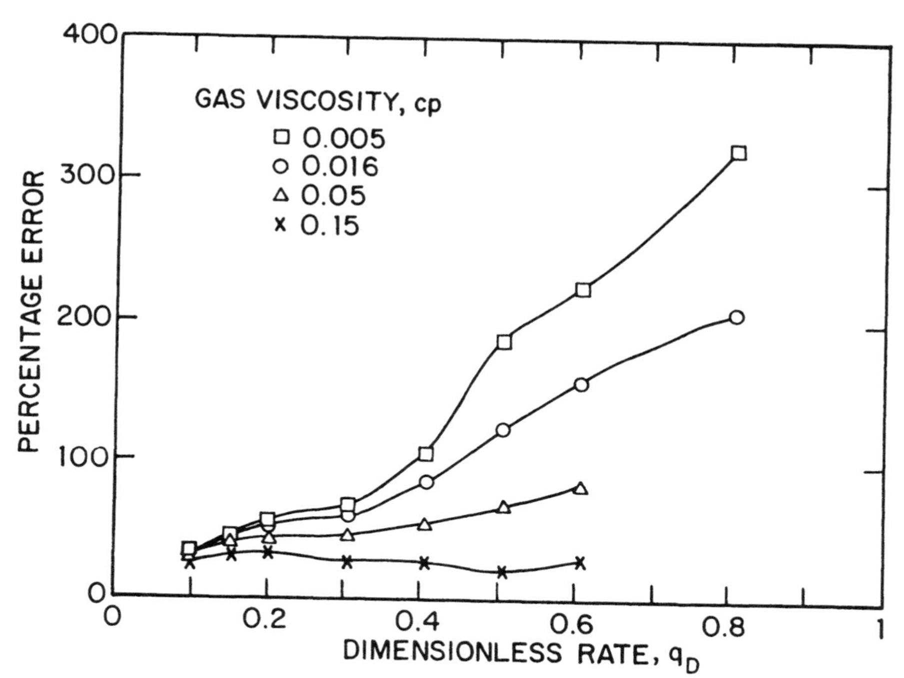

Figure 8–13 Percentage Error Between Numerical and Analytical Solutions for Single-Cone Gas Case.[19]

Since $q_D > 0.4$, Equation 8–17 can be used to calculate t_{DBT}

$$t_{DBT} = 1 - (3q_D - 1)\ln\left[\frac{3q_D}{3q_D - 1}\right]$$

$$= 1 - (3 \times 0.92 - 1)\ln\left[\frac{3 \times 0.92}{3 \times 0.92 - 1}\right]$$

$$= 0.208$$

$$t_{BT} = \frac{364.72\, h\, \phi\, \mu_o}{k_v\,(\rho_w - \rho_o)}\, t_{DBT}$$

$$= \frac{364.72 \times 160 \times 0.2 \times 1.3}{20 \times 0.25} \times 0.208$$

$$= 631.2 \text{ days} \approx 631 \text{ days} = 1.73 \text{ years}$$

This breakthrough time is more than 1.05 years calculated in Example 8–9 by the Ozkan and Raghavan method.[17] However, it should be emphasized that the equations in the Ozkan and Raghaven method do not include the effect of μ_o and $(\rho_w - \rho_o)$, since it is based on the assumption that the bottom water can be represented as a constant-pressure boundary. Also the constant-pressure boundary condition should be used for $q_D > 1$.

CONE BREAKTHROUGH TIME FOR HORIZONTAL WELLS IN RESERVOIRS WITH BOTH GAS CAP AND BOTTOM WATER

Papatzacos et al.[19] also presented a solution to calculate breakthrough time for an infinitely long horizontal well in a reservoir with both top gas and bottom water. Their procedure can also be used to determine optimum well placement. The optimum well placement in a vertical plane is the well elevation, at which both oil and gas break through at the same time.

The dimensionless breakthrough time t_{DBT} and optimum well placement, β'_{opt}, can be found using Figures 8–14 and 8–15. The values t_{DBT} and β'_{opt} can also be calculated using the following equations

$$\beta'_{opt} = C_o + C_1U + C_2U^2 + C_3U^3 \qquad (8\text{–}18)$$

$$\ln(t_{DBT}) = C_o + C_1U + C_2U^2 + C_3U^3 \qquad (8\text{–}19)$$

where

$$U = \ln(q_D) \qquad (8\text{–}20)$$

TABLE 8–8 COEFFICIENTS FOR OPTIMUM WELL PLACEMENT*

ψ	c_o	c_1	c_2	c_3
0.2	0.507	−0.0126	0.01055	−0.002483
0.4	0.504	−0.0159	0.01015	−0.000096
0.6	0.503	−0.0095	0.00624	−0.000424
0.8	0.502	−0.0048	0.00292	−0.000148
1.0	0.500	−0.0001	0.00004	0.000009
1.2	0.497	0.0042	−0.00260	0.000384
1.4	0.495	0.0116	−0.00557	−0.000405
1.6	0.493	0.0178	−0.00811	−0.000921
1.8	0.490	0.0231	−0.01020	−0.001242
2.0	0.488	0.0277	−0.01189	−0.001467

* Coefficients for Equation 8–18.

and q_D is defined in Equation 8–15. Tables 8–8 and 8–9 include a list of the coefficients used in Equations 8–18 and 8–19, respectively. As seen from Figures 8–14 and 8–15, β'_{opt} and t_{DBT} are dependent on the variable ψ, which represents a ratio of density contrasts between water, oil, and gas.

$$\psi = \frac{\rho_w - \rho_o}{\rho_o - \rho_g} \tag{8–21}$$

TABLE 8–9 COEFFICIENTS FOR BREAKTHROUGH TIME,* t_{DBT}

ψ	c_o	c_1	c_2	c_3
0.2	−2.9494	−0.94654	−0.0028369	−0.029879
0.4	−2.9473	−0.93007	0.016244	−0.049687
0.6	−2.9484	−0.9805	0.050875	−0.046258
0.8	−2.9447	−1.0332	0.075238	−0.038897
1.0	−2.9351	−1.0678	0.088277	−0.034931
1.2	−2.9218	−1.0718	0.091371	−0.040743
1.4	−2.9162	−1.0716	0.093986	−0.042933
1.6	−2.9017	−1.0731	0.094943	−0.048212
1.8	−2.8917	−1.0856	0.096654	−0.046621
2.0	−2.8826	−1.1103	0.10094	−0.040963

* Coefficients for Equation 8–19

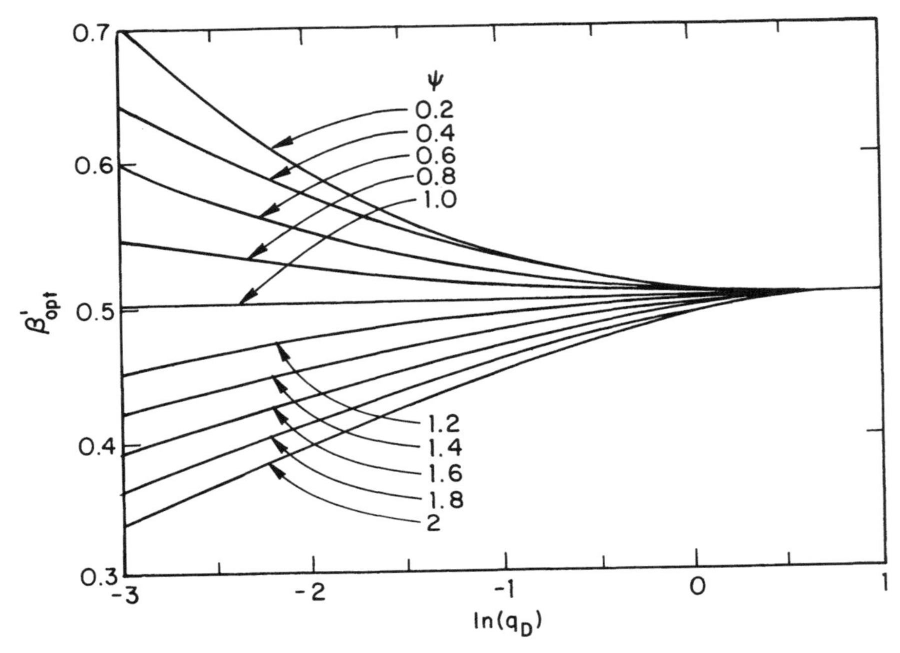

Figure 8–14 Optimum Well Placement as a Function of Dimensionless Rate (Two-Cone Case).[19]

and

$$\beta'_{opt} = c'/h \tag{8-22}$$

where, as noted in Figure 8–16,

c' = distance from well to water-oil contact, ft
d' = distance from well to gas-oil contact, ft
h = oil column thickness, $(c' + d')$, ft.

From Figure 8–14, the optimum well placement moves closer to the water-oil contact as ψ increases. Figure 8–14 also tells us that the optimum well placement, β'_{opt}, is in the center of the oil zone for all ψ at $q_D > 1$. As seen from Figure 8–15, dimensionless breakthrough times are also equal for all values of ψ at $q_D > 1$.

EXAMPLE 8–11

A 1500-ft-long horizontal well is drilled in an anisotropic oil reservoir with developing gas and water cones. The well and reservoir parameters are given below.

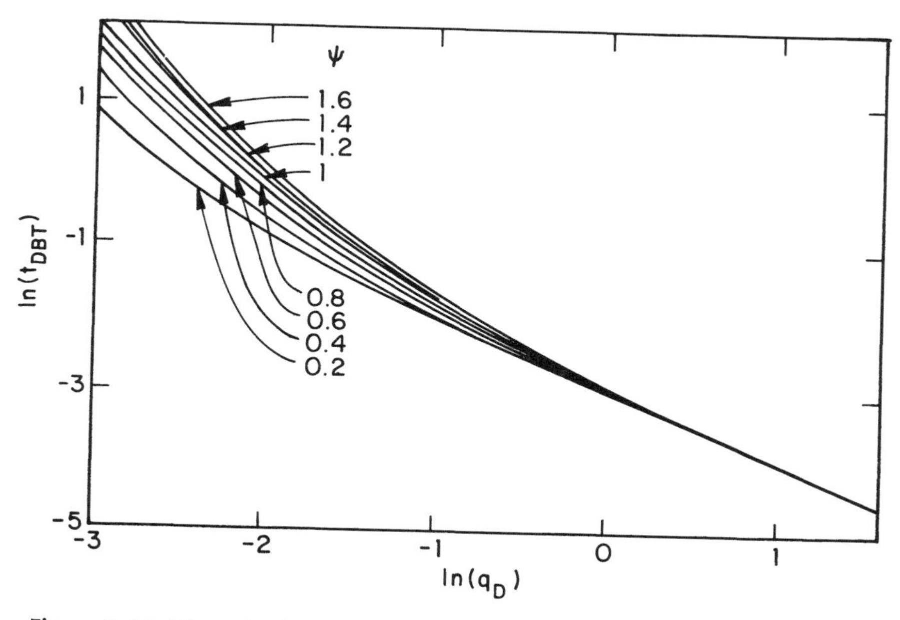

Figure 8–15 Dimensionless Time for Simultaneous Breakthrough for Water and Gas Coning (Two-Cone Case).[19]

$q_o = 8000$ STB/day	$k_v = 1.8$ md
$k_h = 5580$ md	$\phi = 31\%$
$h = 92$ ft	$\rho_o = 0.79$ gm/cc
$\rho_w = 1$ gm/cc	$\rho_g = 0.13$ gm/cc
$\mu_o = 1.6$ cp	$B_o = 1.178$ RB/STB

Determine the optimum well placement and calculate the corresponding breakthrough time. (Note the well and reservoir parameters are from Reference 19.)

Solution

First, dimensionless rate is calculated using Equation 8–15.

$$q_D = \frac{313.51\ \mu_o q_o B_o}{L\sqrt{k_v k_h}\, h(\rho_o - \rho_g)}$$

$$= \frac{325.86 \times 1.6 \times 8000 \times 1.178}{1500 \times \sqrt{1.8 \times 5580} \times 92 \times (0.79 - 0.13)}$$

$$= 0.54$$

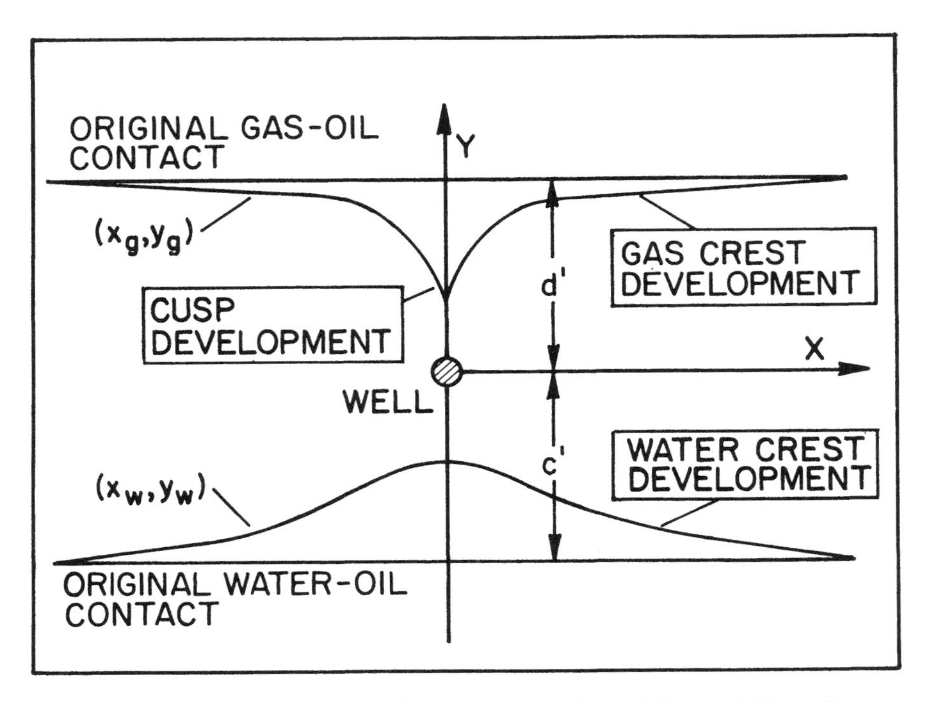

Figure 8–16 A Schematic Vertical Cross-Section View of Gas and Water Cone Development in a Horizontal Well.[19]

ψ is calculated using Equation 8–21.

$$\psi = \frac{1 - 0.79}{0.79 - 0.13} = 0.32$$

From Figure 8–14, for $q_D = 0.54$ and $\psi = 0.32$

$$\beta'_{opt} = 0.55 = c'/h$$

Hence,

$$c' = 0.55 \times h = 0.55 \times 92 = 50.6 \text{ ft} \approx 51 \text{ ft}$$

Thus, the well location in a vertical plane for simultaneous gas and oil breakthrough is 51 ft. above the original oil-water contact.

From Figure 8–15, for $q_D = 0.54$ and $\psi = 0.32$

$$t_{DBT} = 0.1$$

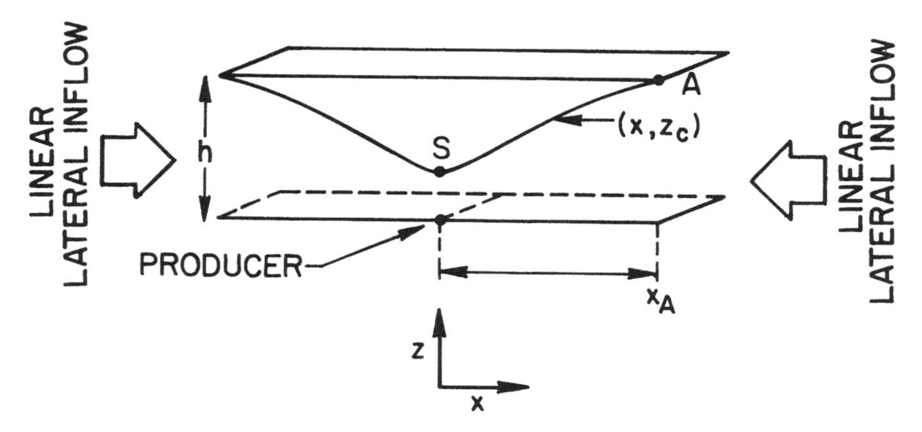

Figure 8–17 A Geometric Configuration of Well Placement in an Edge-Water Drive Reservoir.[20]

and

$$t_{BT} = \frac{364.72 \times 92 \times 0.31 \times 1.6}{1.8 \times [0.79 - 0.13]} \times 0.1$$
$$= 1401 \text{ days} \approx 3.84 \text{ years}$$

This tells us that for a well located 51 ft from the water-oil contact, the simultaneous breakthrough of water and gas will occur in approximately 3.8 years.

CRITICAL RATE FOR HORIZONTAL WELLS IN EDGE-WATER DRIVE RESERVOIR

Recently, Dikken[20] presented correlations for the critical rate for gas and water cresting for horizontal wells in lateral edge-water drive reservoirs. These correlations apply for a horizontal well either at the top or bottom of a reservoir. Figure 8–17 shows the geometric configuration of well placement in an edge-water drive reservoir. The horizontal well is placed at the bottom of the pay zone. It is assumed that the bottom water is not present and the well is only subjected to the side water drive. Dikken[20] suggested that the dimensionless critical rate can be correlated as

$$q_c^{**} = c_1 \left(\frac{x_e}{h \sqrt{k_h/k_v}} \right)^{c_2} \tag{8–23}$$

where

$$c_1 = 1.4426 \pm 0.023 \quad (8\text{–}24)$$
$$c_2 = -0.9439 \pm 0.0013 \quad (8\text{–}25)$$

and

h = oil payzone thickness, ft
k_h = horizontal permeability, md
k_v = vertical permeability, md
q_c^{**} = dimensionless critical rate per unit length, STB/day/ft
x_e = distance between horizontal well and constant pressure boundary, ft

The critical rate q_o is defined in a manner similar to one given in Reference 4.

$$q_o = 4.888 \times 10^{-4} \frac{\Delta\rho\, h \sqrt{k_h k_v}\, L}{\mu_o} q_c^{**} \quad (8\text{–}26)$$

where

h = pay zone thickness, ft
μ_o = oil viscosity, cp
$\Delta\rho$ = density difference, $\rho_w - \rho_o$ or, $\rho_o - \rho_g$, gm/cc

The correlation of the critical height z_c (in ft) is given by Equation 8–27. The critical height represents the elevation difference between the apex of the gas/water crest from the well elevation.

$$\frac{z_c}{h} = c_3 \left(\frac{x_e}{h\sqrt{k_h/k_v}} \right)^{c_4} \quad (8\text{–}27)$$

where

$$c_3 = 0.4812 \pm 0.022 \quad (8\text{–}28)$$
$$c_4 = -0.9534 \pm 0.0013 \quad (8\text{–}29)$$

The correlations of Equations 8–23 and Equation 8–27 are illustrated in Figure 8–18. Note that these correlations do not consider the effects of the increased pressure gradient due to the reduced flow area caused by the

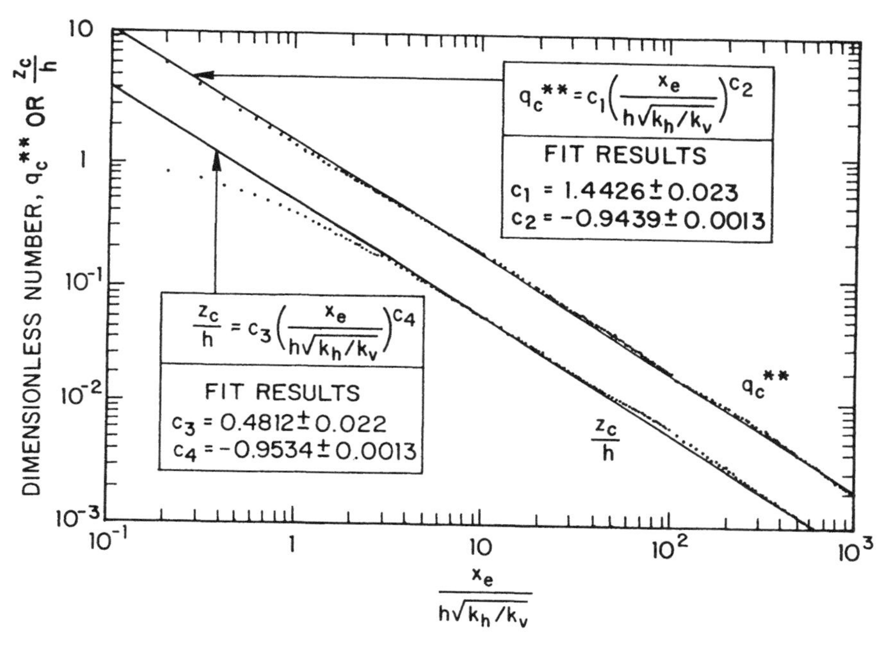

Figure 8–18 Correlation for Edge-Water Drive Critical Rate and Critical Height.[20]

presence of the crest. Figure 8–19 shows the dimensionless critical rate after incorporating this correction. The critical rates were again fitted by linear regression and the new constants in the equation for dimensionless critical rate, Equation 8–23, are

$$c_1 = 0.9437 \pm 0.01 \tag{8–30}$$

$$c_2 = -0.9896 \pm 0.0043 \tag{8–31}$$

EXAMPLE 8–12

A 1500-ft-long horizontal well is drilled on a 160-acre spacing in an edge-water drive reservoir with no active bottom-water drive. The following reservoir parameters are given:

$$h = 80 \text{ ft}, \quad k_h = 5 \text{ md}, \quad k_v = 2.5 \text{ md}$$
$$\rho_w - \rho_o = 0.4 \text{ gm/cc} \quad \text{and} \quad \mu_o = 0.7 \text{ cp}.$$

Calculate the critical oil rate: (1) neglecting the correction of the increased pressure gradient due to reduced-flow area caused by the presence of the crest, and (2) including pressure-gradient correction.

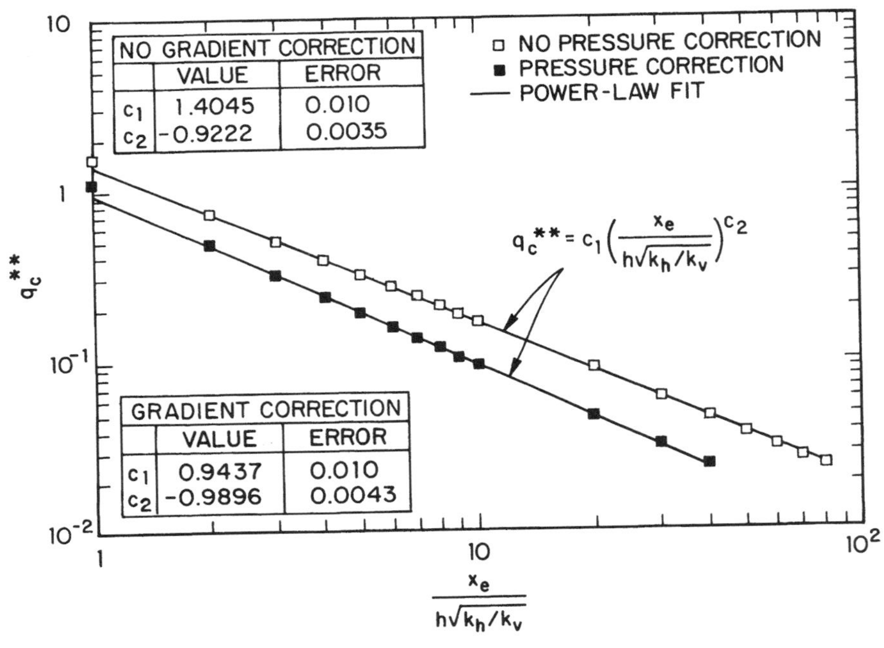

Figure 8–19 Correlation for Edge-Water Drive Critical Rate after Correction.[20]

Solution

For a 160-acre spacing assuming a square drainage area,

$$2x_e = 2y_e = 2640 \text{ ft, hence, } x_e = 1320 \text{ ft.}$$

1. **Neglecting pressure gradient correction:**

Dimensionless critical rate can be calculated using Equation 8–23 with $c_1 = 1.4426$, and $c_2 = -0.9439$.

$$q_c^{**} = c_1 \left(\frac{x_e}{h\sqrt{k_h/k_v}} \right)^{c_2} \tag{8–23}$$

$$= 1.4426 \left(\frac{1320}{80\sqrt{2}} \right)^{-0.9439} = 1.4426\,(11.67)^{-0.9439}$$

$$= 0.14192$$

Critical rate can then be calculated from Equation 8–26 as

$$q_o = 4.888 \times 10^{-4} \frac{\Delta\rho\, h \sqrt{k_h k_v}\, L}{\mu_o} q_c^{**} \tag{8–26}$$

$$= 4.888 \times 10^{-4} \frac{0.4 \times (80) \times \sqrt{5 \times 2.5} \times 1500}{0.7}$$

$$\times 0.14192$$

$$= 118.5 \times 0.14192$$

$$= 16.8 \text{ STB/day}$$

2. **Including the pressure gradient correction:**
In this case, the constants in Equation 8–23 are given by Equations 8–30 and 8–31 as $c_1 = 0.9437$ and $c_2 = -0.9896$. Substituting these constants in Equation 8–23 one can recalculate q_c^{**}

$$q_c^{**} = 0.9437\,(11.67)^{-0.9896} = 0.083.$$

Substituting this value of q_c^{**} in Equation 8–26 yields

$$q_o = 118.5 \times 0.083$$
$$= 9.83 \text{ STB/day}$$

It should be noted that correcting for the presence of the crest reduces the critical rate by more than 40%.

PRACTICAL CONSIDERATIONS

For a successful horizontal-well program, special attention should be given to the well completion. In a reservoir with a gas cap, it is important to ensure effective isolation of the wellbore entry portion from the gas cap. This isolation can be achieved either by cementing or using packers. Some horizontal wells have failed due to premature entry of gas. Special attention should also be given to the well shape. The desirable well shape has minimum bends and turns in the vertical plane. In turns and bends, liquid and gases have a tendency to accumulate, especially in low flow rate wells. The fluid accumulation in the wellbore reduces the producing ability of horizontal wells. The best shape for coning applications is a slightly tilting, upward or downward well. This will allow natural fluid segregation in the horizontal-well portion and facilitate its production.

In coning applications, let us say, in a water coning application, water may break through in a small portion of a long horizontal wellbore. In this case, a well plan should include the appropriate response for such contingencies. A brief review of completion options is given in Chapter 1. Completion details are included in Volume II.

FIELD HISTORIES

WATER CONING, OFFSHORE ITALY

Offshore Italy, in Rospo Mare field, Elf-Aquitaine has drilled several horizontal wells.[21–24] The reservoir rock is a highly fractured limestone at a depth of 4523 ft and is located offshore where the water depth is 250 ft. The gross reservoir thickness is 230 ft (70 m). The oil gravity is 12° API.

In this heavy-oil reservoir, vertical, slant, and horizontal wells were drilled initially. The 2000-ft-long horizontal well was completed using a slotted liner. On the platform, originally no water treating facilities were available. Therefore, the wells were operated at or near critical rates. The production profiles of vertical, slant, and horizontal wells are shown in Figure 8–20.[22] The figure shows that the initial horizontal-well rate is significantly higher than those for vertical and slant wells. It is also interesting to note that, after an acid treatment, the vertical well's production increased temporarily, then its rate dropped off rapidly, probably because acid may have improved the communication between the water zone and the well, resulting in high water and low oil rates. Because they had no water treating facilities,

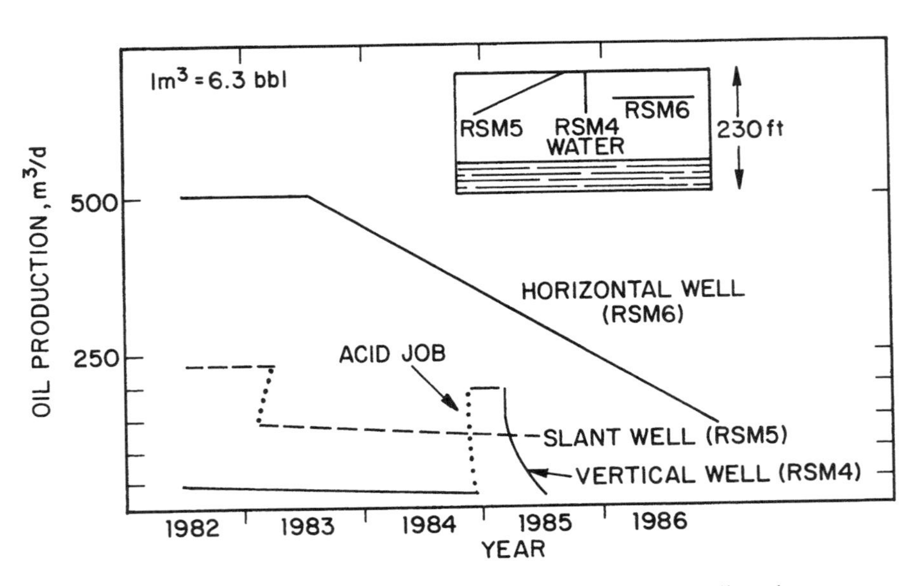

Figure 8–20 A Comparison of Horizontal, Slant and Vertical Well Performances, Rospo Mare Field, Offshore Italy.[22]

the vertical-well rate had to be reduced to minimize water production. Similarly, acid treatment did not improve the performance of the slant well.

As expected, over a long time, horizontal-well rates are approaching those of vertical-well rates. However, the horizontal well's cumulative production is significantly higher than that for slant and vertical wells. This demonstrates a successful horizontal-well application. Based on this success, Elf has drilled several more horizontal wells to develop the fractured-limestone reservoir. Without horizontal wells, it is doubtful whether field development using vertical wells would have been economical. A long-term production history for these three wells is shown in Figure 8–21.[24]

WATER CONING, HELDER FIELD, OFFSHORE NETHERLANDS[25]

In a slightly faulted anticlinal structure, in the Dutch continental shelf, UNOCAL Netherlands drilled several horizontal wells, all sidetracked from existing wells, virtually redeveloping the Helder oil field.[25] The structure is 4600 ft deep and has 72 MMSTB of 22°API gravity oil in place. The field is underlain by water over its entire 1140 acres area, and has a maximum oil column thickness of 131 ft. Due to the high viscosity oil, the flat field structure and high vertical permeability of the Vlieland sand, initial development of the field showed early water breakthrough and rapidly increasing watercuts.

To reduce water production and to obtain high well productivity, a horizontal sidetrack from an existing well, Helder A4, was drilled. A plot of the early-life cumulative oil production against cumulative gross production

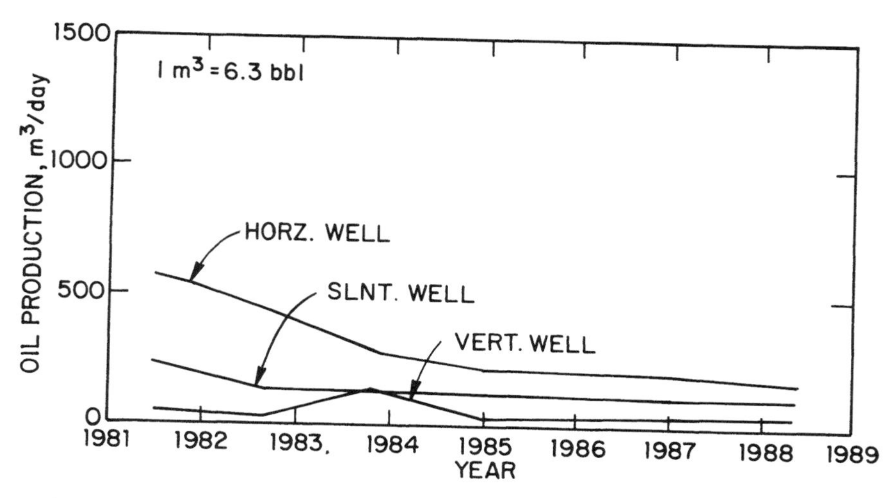

Figure 8–21 A Comparison of Horizontal, Slant and Vertical Well Production Rates, Rospo Mare Field, Offshore Italy.[24]

for both the horizontal sidetrack and the original well, is shown in Figure 8–22. This plot illustrates the improved early life oil recovery of the horizontal well over a conventional well.

Seven successful horizontal sidetracks were drilled and the expected increase in the ultimate recovery from the Helder field is 7% of oil in place and 17% in additional recovery as a direct consequence of the seven wells.[25] Horizontal wells have thus shown to provide increased oil production, reduced operating costs, and increased reserves in the redevelopment of Helder field. A summary of the Helder field data is given in Table 8–10.

GAS CONING, EMPIRE ABO UNIT, NEW MEXICO

In a fractured carbonate reservoir, ARCO drilled several 200 to 300 ft long drainholes.[26,27] The 6200-ft-deep reservoir has a porosity of 8.6% and oil gravity is 44° API. The reservoir permeability is 25 md and oil column is about 90 ft. The drainholes were drilled to reduce gas coning. The performance of one of the drainholes is shown in Figures 8–23 and 8–24. It is important to note that the drainhole is located closer to the gas-oil contact than the surrounding vertical wells. In spite of this, the gas-oil ratio (GOR) of the drainhole is smaller than that for vertical wells. Additionally, in six

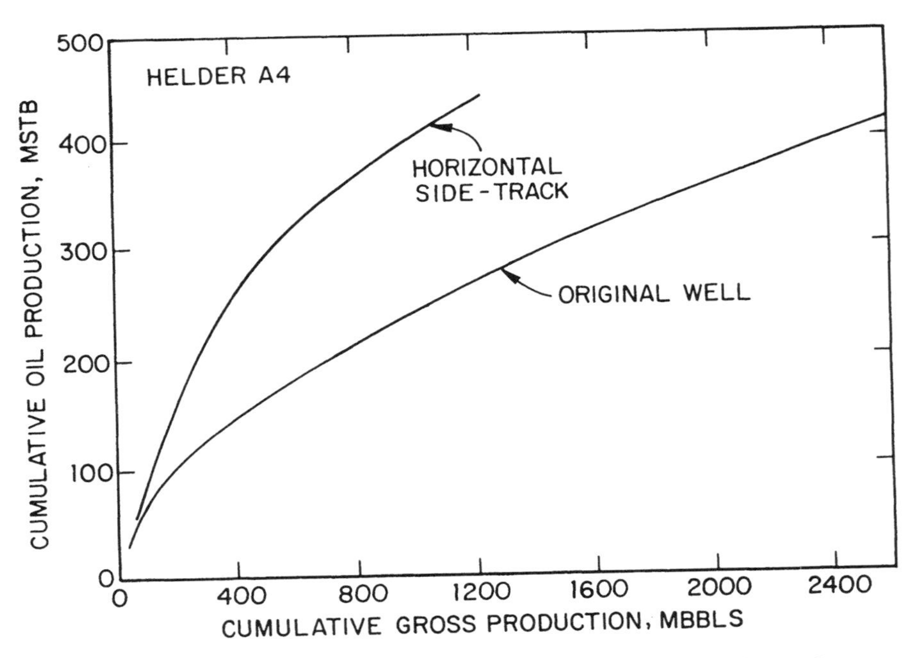

Figure 8–22 Early-Time Cumulative Oil Production versus Cumulative Gross Production, Helder Field, Offshore Netherlands.[25]

TABLE 8–10 SUMMARY OF HELDER FIELD DATA†
(Vlieland Sand)

WELLS	LENGTH, ft	h,OIL (ft)	J_h, STB/(day-psi)	J_h/L*
A-2	479	73	29	0.11
A-3	869	75	226	0.26
A-4	440	71	37	0.08
A-5	1348	53	127	0.09
A-7	1093	65	116	0.11
A-8	804	46	73	0.09

* STB/(day-psi-ft)
† 1988 data

Area = 1140 acres = 461 hectors
k = 1 to 6 darcies
μ = 30 cp, API = 22°
Bottom water drive
12 vertical wells, 6500 BOPD with 108,000 BFPD
· wc = 85 to 97%
· Submersible pumps 83 hp to 250 hp
Horizontal wells are completed with prepacked gravel pack liners
Hole cleaning: 0.1% citric acid, 1% surfactant solution
· Core experiments restores 85% of original permeability

HELDER FIELD

Well	Oil Production BOPD	Gross production BFPD	WC
Horizontal	4500	25,500	82.5%
Vertical/ original	2700	64,000	95.8%
		(35,500)	

years the drainhole has produced about 1.6 times more oil than the surrounding wells, indicating the success of the drainhole in reducing the gas coning problem.

WATER AND GAS CONING, OFFSHORE AUSTRALIA

Horizontal wells have also been drilled in offshore Western Australia, in North Herald, South Pepper and Chervil Fields.[28,29] These are sandstone reservoirs. The majority of these reservoirs have very strong active bottom water which provides the pressure support, while in some cases the gas cap

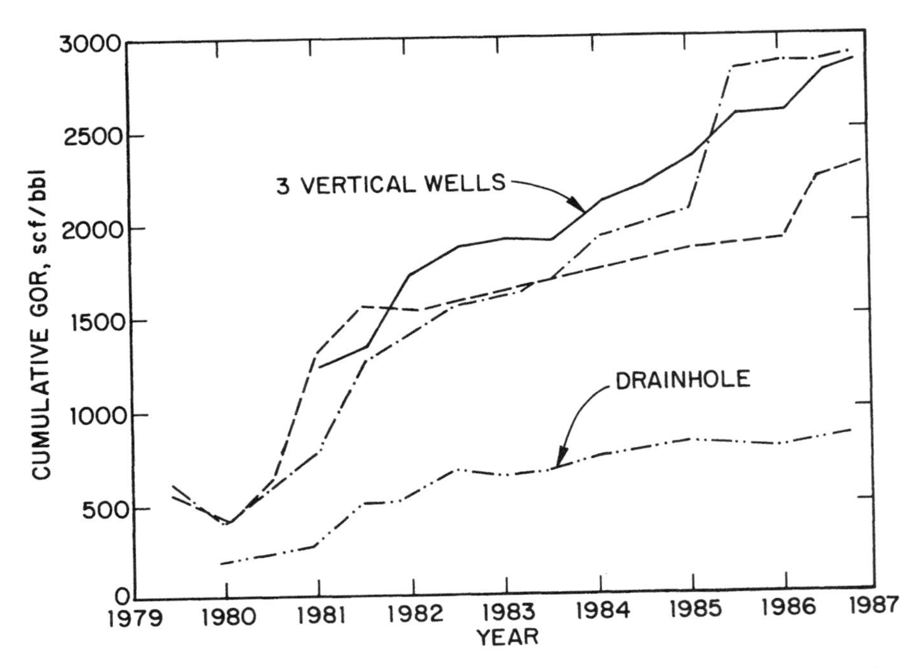

Figure 8–23 Cumulative GOR Comparison of a Drainhole and nearby Vertical Wells, Empire Abo Reef, New Mexico.[26,27]

exists. In the North Herald field, the gas cap is suspected but it has not been positively identified. Permeabilities of these reservoirs are on the order of three to ten thousand millidarcies, indicating highly permeable reservoirs. Typical reservoir data for the North Herald, South Pepper and Chervil Fields are summarized in Table 8–11 and typical productivities from the initial tests of horizontal and vertical wells are listed in Table 8–12. All of the wells are open-hole wells, and data in Table 8–12 clearly indicate that horizontal wells have been highly successful in enhancing the well productivities and reducing gas-oil ratio.

WATER AND GAS CONING, PRUDHOE BAY, ALASKA[30–32]

With horizontal well completion technology developing rapidly, and operators encountering completion problems in open-hole and slotted liner wells, more horizontal wells are being cemented and perforated or fitted with external casing packers (ECP) for zonal isolation.

British Petroleum in Alaska has developed variations of fully cemented horizontal wells and slotted liners, to reduce completion cost and enhance production by isolating zones with various water- and gas-coning problems.[30,31] Not only will the type of completion have an effect on water or

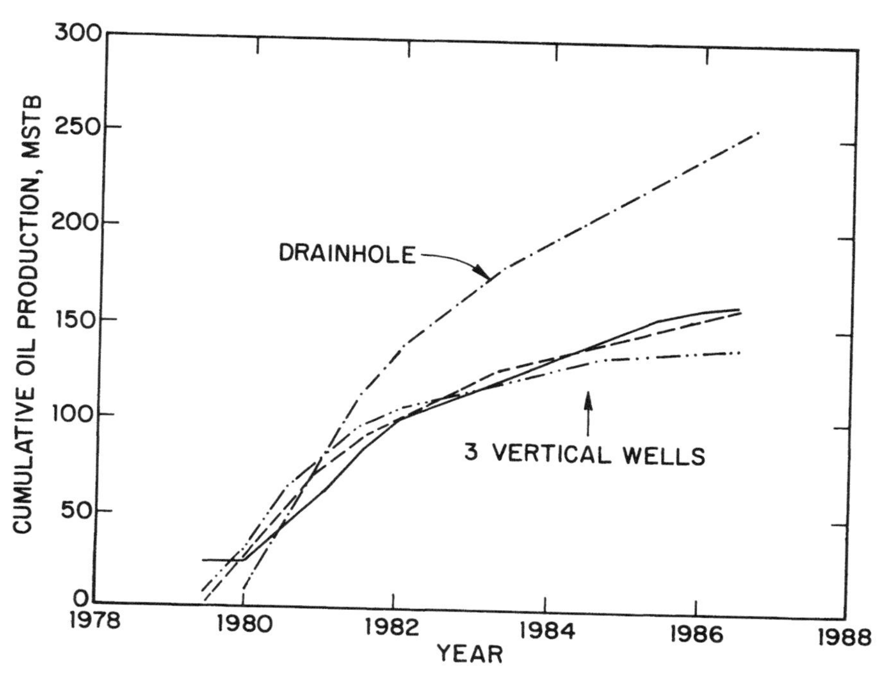

Figure 8–24 A Comparison of Oil Production from a Drainhole and nearby Vertical Wells, Empire Abo Reef, New Mexico.[26,27]

TABLE 8–11 RESERVOIR PROPERTIES OF NORTH HERALD, SOUTH PEPPER AND CHERVIL FIELDS[28,29]

	NORTH HERALD	SOUTH PEPPER	CHERVIL
Reservoir		Barrow Group	
Drive Mechanism		Natural Aquifer-Complete Pressure Support	
Datum Depth (mSS)	1169	1202	1028
Pressure (psia)	1763	1793	1548
Temp (Deg C)	83	82	81
Gas Cap	Suspected but not seen	YES	YES
Max Oil Column (m)	25	9	7.5
Porosity (%)	27	20	27
Permeability (md)	3–10,000 (Ave. 1000)	3–4,000 (Ave. 1000)	3–10,000 (Ave. 1000)
Oil API Gravity	46	44.5	46
Oil Viscosity (cp)	0.5	0.45	0.49
Gas Oil Ratio (scf/stb)	390	450	365

TABLE 8–12 INITIAL TESTS OF HORIZONTAL AND VERTICAL WELLS, NORTH HERALD, SOUTH PEPPER AND CHERVIL FIELDS

WELL	TYPE	STATUS	JSTB/(day-psi)
North Herald 1	V	P&A	11
North Herald 2	V	P&A	24
North Herald 3	H (300m)	Producer	100
South Pepper 1	V	P&A	10
South Pepper 4	V	Producer	3
South Pepper 5	V	Producer	30
South Pepper 6	H (314m)	Producer	20
South Pepper 7	H (330m)	Producer	2
South Pepper 7 Ext	H (176m)	Producer	60
South Pepper 8	H (386m)	Producer	80
Chervil 1	V	P&A	6
Chervil 4	H (233m)	Producer	100
Chervil 5	H (605m)	Producer	150–250

1m = 3.28 ft H = Horizontal Well, and
V = Vertical Well P&A = Plugged and abandoned.

gas coning, but also the placement of the well in the reservoir with respect to the gas-oil contact and water-oil contact will have a significant impact. Case histories for six wells drilled in Alaska with various completion techniques and their production performances are summarized in References 30 and 31. (Average properties of the sandstone reservoir are $k_h \approx 115$ md, $\phi = 22\%$, oil column ≈ 200 ft, and k_v/k_h can be very small).

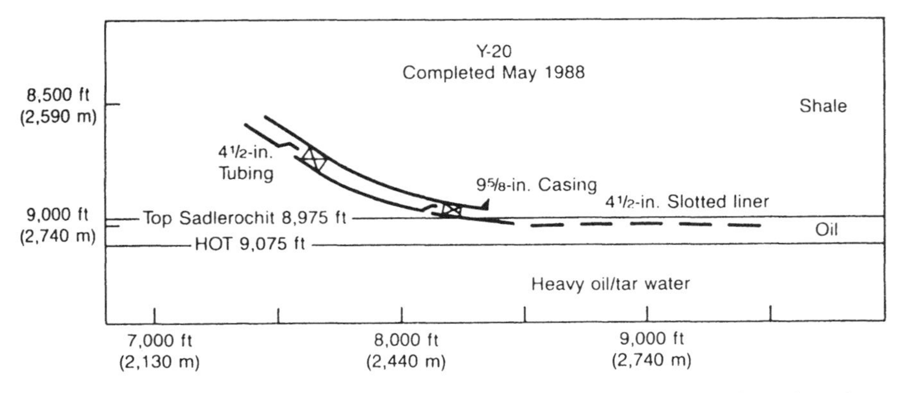

Figure 8–25 A Schematic of Peripheral Well Y–20 Completed with a Slotted Liner.[30]

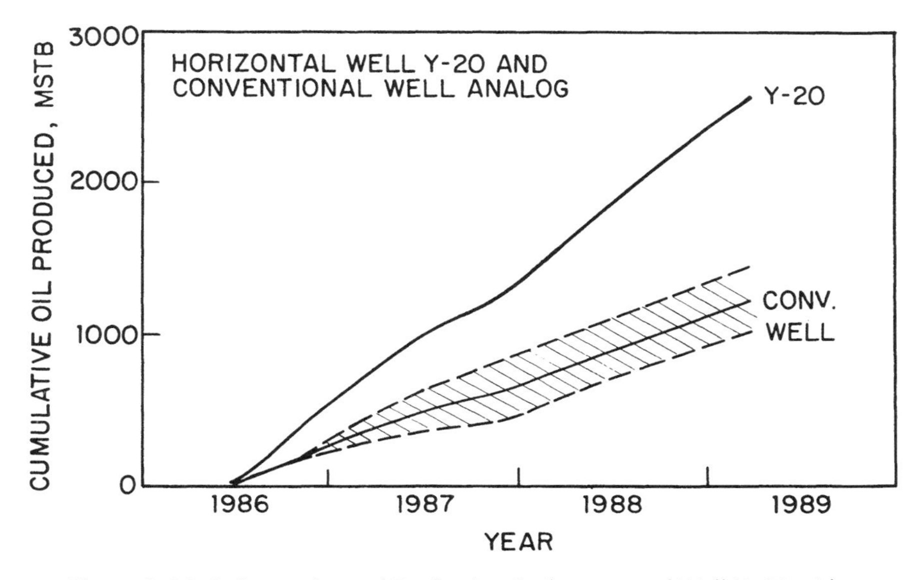

Figure 8–26 A Comparison of Production Performance of Well Y–20 with a Conventional Well.[30]

Well Y-20 was completed using a slotted liner (see Figure 8–25). The advantage of just using a slotted liner is that it is the least expensive option. The cumulative oil produced is shown in Figure 8–26. The well was placed at the top of the oil zone to reduce water coning, since a gas cap was absent. Until early 1989, water cut is minimal and no workovers have been required.

Another well design and completion technique used is to completely cement and perforate the liner in an inverted high angle well.[30] Well E-28, Figure 8–27, was drilled inverted to cross any shale barriers or restrictions in vertical permeability. Also, if gas coning occurs, it will occur at the end

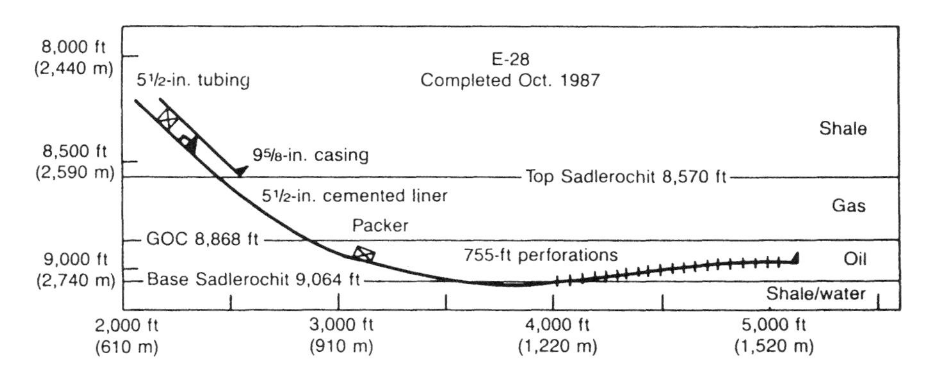

Figure 8–27 A Schematic of Well E–28, Prudhoe Bay, Alaska.[30]

of the interval, which can be plugged off without affecting oil production. Figure 8–28 shows the cumulative oil production of E-28 as compared to conventional wells.

In some horizontal wells around the world, once the gas breaks through, it is difficult to stop it, even with exotic completion schemes. The Well B-30 in Prudhoe Bay, Alaska was completed using a slotted liner. In addition to a slotted liner, the well was completed using a cemented gas isolation liner to minimize gas coning by isolating the gas cap.[30] Figure 8–29 shows the production history of Well B-30. The figure shows an excellent initial well performance followed by a slow addition in cumulative oil production after gas breakthrough.

SIMULATION STUDY OF GAS AND WATER CONING, TROLL FIELD, OFFSHORE NORWAY[33]

The Troll Field is located in the North Sea. It contains 5.7 billion STB of oil over an area of 270 square miles. The thickness of the oil column varies from 0 to 92 feet. This 9 to 10 Darcy reservoir (about 30% porosity) has gas cap and bottom water, in some cases separated by low permeability streaks. Conventional vertical wells cone gas and/or water within two to

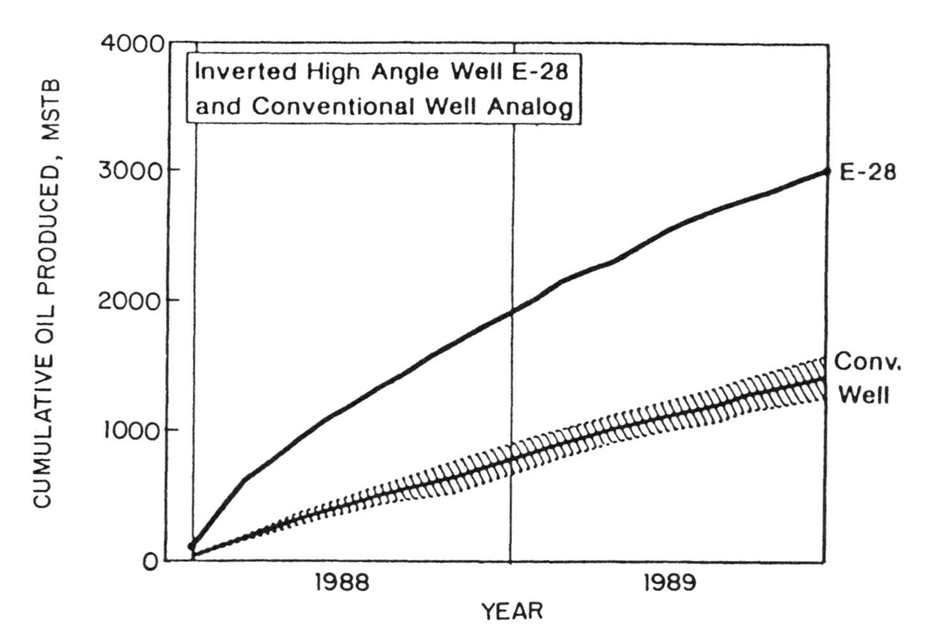

Figure 8–28 A Comparison of Production Performance of Well E–28 with a Conventional Well.[30]

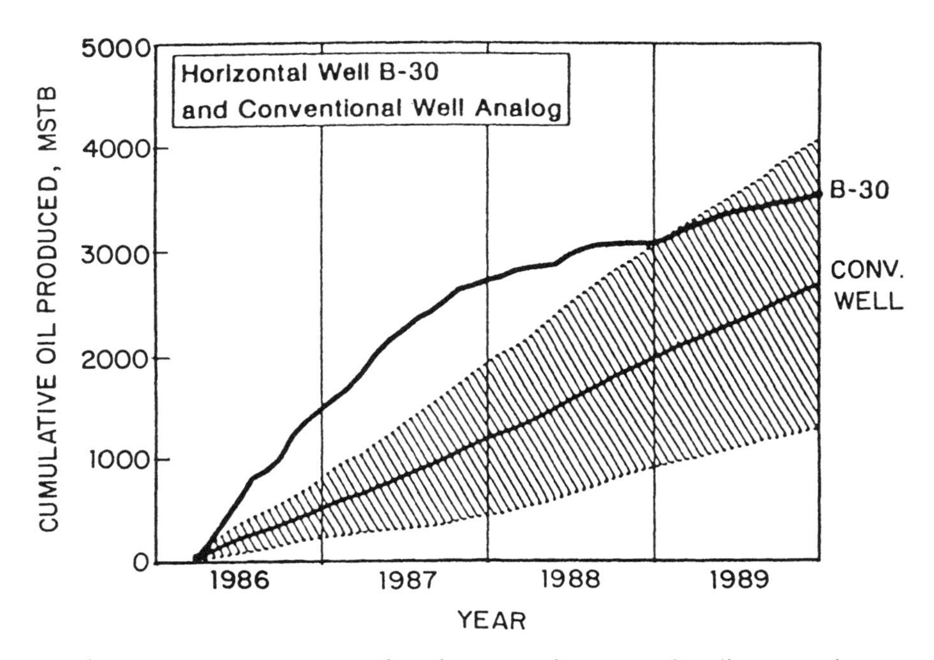

Figure 8–29 A Comparison of Production Performance of Well B–30 with a Conventional Well.[30]

three days of production while producing below the critical rates is not economical.[33] Thus the drilling of horizontal wells was proposed as a possible solution to this problem.

Simulation studies were conducted to investigate the feasibility and performances of horizontal wells. The results were compared with the performances of various patterns of conventional vertical wells. The effects of horizontal well length, the well's position relative to the fluids contacts, the change in production rate and the oil saturation below the oil-water contact were all investigated. The most significant findings were that a 1500-ft-long horizontal well would produce the same amount of oil as two vertical wells in a typical sector pattern when the sector is produced at a constant rate, and a 2000-ft-long horizontal well would perform better than three vertical wells. Recently, a successful long-term test of a 1640 ft long horizontal well drilled into a 72 ft thick oil zone in Troll gas field was reported.[34] The peak flow from a single horizontal well is more than 25,000 STB/day.

NISKU B POOL, ALBERTA, CANADA[35]

The field histories discussed in this chapter demonstrate the success of horizontal wells in reducing gas and water coning problems. However,

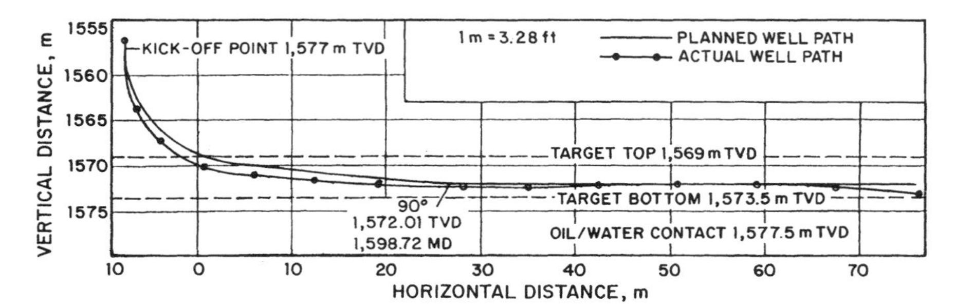

Figure 8–30 Planned and Actual Horizontal Well Paths, Nisku B Pool, Alberta, Canada.[35]

there are some cases where horizontal wells have not been effective in reducing water and gas coning. The lack of success is due to operational problems that one encounters while drilling in a very thin pay zone, where precise targeting may be difficult. Additionally, in gas coning, gas breakthrough in the entry portion of the wellbore is a common problem.

In Alberta, Canada, an approximately 212-ft (64.6 m) horizontal well was drilled in about a 15-ft (4.5 m)-thick oil zone in the Nisku B pool reservoir.[35] Figure 8–30 shows the well plan and Figure 8–31 shows the production history. The fractured vuggy dolomite reservoir exhibits about 7.5% porosity and 100 to 1000 md permeability.

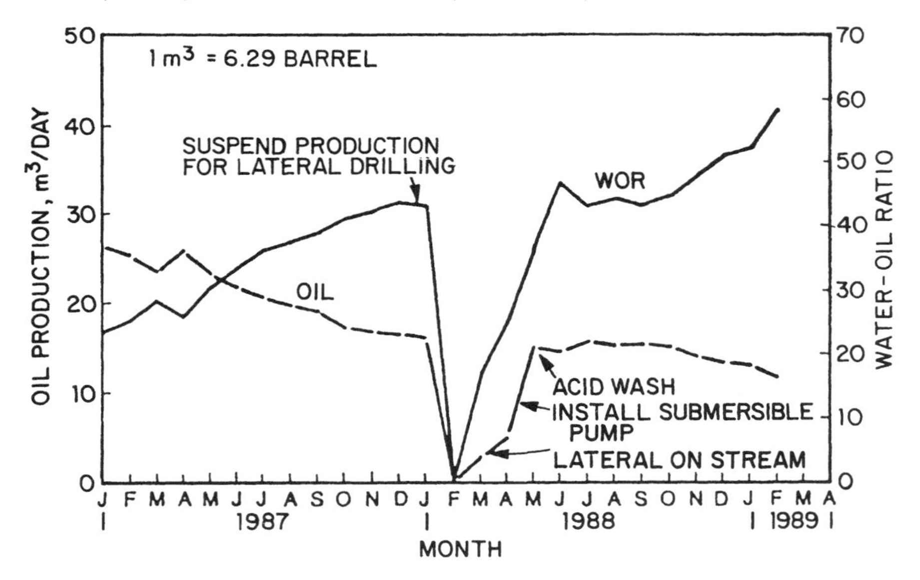

Figure 8–31 Production History of Horizontal Well, Nisku B Pool, Alberta, Canada.[35]

The short-radius well was drilled using calcium carbonate-based drilling fluids. The initial production testing, after installing a submersible pump, showed low fluid levels in the well, indicating a damaged wellbore. The well was completed as an open hole, which was treated with 182 bbls of 28% HCl acid. After the treatment, oil production increased; however, the water-oil ratio also increased significantly. The initial results are depicted in Figure 8–31. The open-hole well was drilled using a short-radius technique, and therefore, it couldn't be logged to determine points or zones of water entry into the wellbore. This resulted in an uneconomic well.

OTHER FIELD HISTORIES AND SUMMARY

Several horizontal wells and drainholes have been drilled to reduce gas and water coning, such as Nuggat-sandstone in Wyoming, offshore China, offshore Indonesia in the Bima field, and the Troll field in the North Sea. Detailed information on these projects is discussed in Volume II. The available histories suggest that horizontal wells have been successful in reducing water coning in limestone as well as sandstone reservoirs. *For gas coning applications, available histories indicate comparatively fewer successes and several failures.* Another important issue is the enhancement of recovery factors using horizontal wells. The available field data indicates an incremental recovery factor over vertical wells from 0 to 5%.[25,30] Until more long-term field histories are available, it is difficult to precisely estimate enhancement in the recovery factors.

REFERENCES

1. Craft, B. C. and Hawkins, M. F.: *Applied Petroleum Reservoir Engineering,* Englewood Cliffs, New Jersey, Prentice-Hall, 1959.
2. Meyer, H. L. and Gardner, A. O.: "Mechanics of Two Immiscible Fluids in Porous Media," *Journal of Applied Physics,* Vol. 25, No. 11, pp. 1400ff.
3. Pirson, S. J.: *Oil Reservoir Engineering,* Robert E. Krieger Publishing Company, Huntington, New York, 1977.
4. Chaperon, I.: "Theoretical Study of Coning Toward Horizontal and Vertical Wells in Anisotropic Formations: Subcritical and Critical Rates," paper SPE 15377, presented at the SPE Annual Technical Conference and Exhibition, New Orleans, Louisiana, Oct. 5–8, 1986.
5. Schols, R. S.: "An Empirical Formula for the Critical Oil Production Rate," *Erdoel Erdgas, Z.,* Vol. 88, No. 1, pp. 6–11, January 1972.

5a. Hoyland, L. A., Papatzacos, P. and Skjaeveland, S. M.: "Critical Rate for Water Coning: Correlation and Analytical Solution," *SPE Reservoir Engineering,* pp. 495–502, November 1989.

6. Arps, J. J.: "Analysis of Decline Curves," *Trans.,* AIME, Vol. 160, pp. 228–247, 1945.
7. Fetkovich, M. J.: "Decline Curve Analysis Using Type Curves," *Journal of Petroleum Technology,* pp. 1065–1077, June 1980.
8. Sobocinski, D. P. and Cornelius, A. J.: "A Correlation for Predicting Water Coning Time," *Journal of Petroleum Technology,* pp. 594–600, May 1965.

9. Bournazel, C. and Jeanson, B.: "Fast Water-Coning Evaluation Method," paper SPE 3628, presented at the 46th SPE Annual Meeting, New Orleans, Louisiana, Oct. 3–6, 1971.
10. Kuo, M. C. T. and DesBrisay, C. L.: "A Simplified Method for Water Coning Predictions," paper SPE 12067, presented at the 58th Annual Technical Conference and Exhibition, San Francisco, California, Oct. 5–8, 1983.
11. Kuo, M. C. T.: "Correlations Rapidly Analyze Water Coning," *Oil and Gas Journal*, pp. 77–80, Oct. 2, 1989.
12. Efros, D. A.: "Study of Multiphase Flows in Porous Media," (in Russian) *Gastoptexizdat*, Leningrad, 1963.
13. Giger, F.: "Evaluation theorique de l'effet d'arete d'eau sur la production par puits horizontaux," *Revue de l'Institut Francais du Petrole*, Vol. 38, No. 3, May–June 1983 (in French).
14. Giger, F. M.: "Analytic 2-D Models of Water Cresting Before Breakthrough for Horizontal Wells," SPE Reservoir Engineering, pp. 409–416, November, 1989.
15. Karcher, B. J., Giger, F. M., and Combe, J.: "Some Practical Formulas to Predict Horizontal Well Behavior," paper SPE 15430, presented at the SPE 61st Annual Technical Conference and Exhibition, New Orleans, Louisiana, Oct. 5–8, 1986.
16. Joshi, S. D.: "Augmentation of Well Productivity Using Slant and Horizontal Wells," *Journal of Petroleum Technology*, pp. 729–739, June 1988.
17. Ozkan, E. and Raghavan, R.: "Performance of Horizontal Wells Subject to Bottom Water Drive," paper SPE 18545, presented at the SPE Eastern Regional Meeting, Charleston, West Virginia, Nov. 2–4, 1988.
18. Papatzacos, P., Gustafson, S. A. and Skjaevelan, S. M.: "Critical Time for Cone Breakthrough in Horizontal Wells," presented at Seminar on Recovery from Thin Oil Zones, Norwegian Petroleum Directorate, Stavanger, Norway, April 21–22, 1988.
19. Papatzacos, P., Herring, T. U., Martinsen, R. and Skjaeveland, S. M.: "Cone Breakthrough Time for Horizontal Wells," paper SPE 19822, presented at the 64th SPE Annual Technical Conference and Exhibition, San Antonio, Texas, Oct. 8–11, 1989.
20. Dikken, B. J.: "Pressure Drop in Horizontal Wells and Its Effect on Their Production Performance," paper SPE 19824, presented at the 64th SPE Annual Conference and Exhibition, San Antonio, Texas, Oct. 8–11, 1989.
21. Giger, F. M. and Jourdan, A. P.: "The Four Horizontal Wells Producing Oil in Western Europe," proceedings of the AOSTRA's Fifth Annual Advances in Petroleum Recovery and Upgrading Technology Conference, Calgary, Alberta, Canada, June 14–15, 1984.
22. Reiss, L. H.: "Horizontal Wells' Production After Five Years," *Journal of Petroleum Technology*, pp. 1411–1416, November 1987.
23. Jourdan, A.: "Drilling of Horizontal Wells," *Petrole et Techniques*, No. 294, December 1982.
24. Reiss, L. H.: "Producing the Rospo Mare Oil Field By Horizontal Wells," presented at Seminar on Recovery from Thin Oil Zones, Stavanger, Norway, April 21–22, 1988.
25. Murphy, P. J.: "Performance of Horizontal Wells in the Helder Field," *Journal of Petroleum Technology*, pp. 792–800, June 1990.
26. Detmering, T. J.: "Update on Drainhole Drilling—Empire Abo," proceedings of 31st Annual Southwestern Petroleum Short Course, Lubbock, Texas, April 1984.
27. Stramp, R. L.: "The Use of Horizontal Drainholes in Empire Abo Unit," paper SPE 9221, presented at the Annual Meeting, Dallas, Texas, Sept. 21–24, 1980.

28. "Horizontal Wells Were a Success," *Oil & Gas Australia,* pp. 3–9, February 1988.
29. Anderson, Barry, personal communication, April 1990.
30. Stagg, T. O. and Reiley, R. H.: "Horizontal Well Completions in Alaska," *World Oil,* pp. 37–44, March 1990.
31. Broman, W. H., Stagg, T. O. and Rosenzweig, J. J.: "Horizontal Well Performance Evaluation at Prudhoe Bay," paper CIM/SPE 90–124, presented at CIM/SPE International Technical Meeting, Calgary, Canada, June 10–13, 1990.
32. Sherrard, D. W., Brice, B. W. and MacDonald, D. G.: "Application of Horizontal Wells at Prudhoe Bay," *Journal of Petroleum Technology,* pp. 1417–1425, November 1987.
33. Kossack, C. A., Kleppe, L. and Assen, T.: "Oil Production From the Troll Field: A Comparison of Horizontal and Vertical Wells," paper SPE 16869 presented at the SPE 62nd Annual Technical Conference and Exhibition, Dallas, Texas, Sept. 27–30, 1987.
34. "Norsk Hydro Taps Oil Pay in Troll Gas Field," *Oil & Gas Journal,* pp. 34–35, May 14, 1990.
35. Malone, M. F. and Hippman A.: "Short Radius Horizontal Well Fails to Improve Production," *Oil & Gas Journal,* pp. 41–45, Oct. 16, 1989.

CHAPTER 9

Horizontal Wells in Gas Reservoirs

INTRODUCTION

Horizontal wells are suitable for gas reservoirs. They are applicable in low-permeability reservoirs as well as in high-permeability gas reservoirs.

In the low-permeability reservoirs, it is difficult to drain large volumes using vertical wells. For example: reservoirs with permeabilities of less than 0.01 md may need vertical wells with less than 40-acre spacing to drain them effectively in a reasonable time frame. Fracturing vertical wells does help drainage, but creating long fractures in a tight reservoir is difficult. Horizontal wells provide an alternative to achieve long penetration lengths in the formations.

In high-permeability gas reservoirs, wellbore turbulence limits the deliverability of a vertical well. To reduce turbulence near the wellbore, the only alternative is to reduce the gas velocity around the wellbore. This can be partly achieved by fracturing a vertical well. However, fracturing is not very effective in a high-permeability reservoir, because proppants themselves have a limited flow capacity, which may be comparable to that of the reservoir rock. The most effective way to reduce gas velocity around the wellbore is to reduce the amount of gas production per unit well length. This can be accomplished by using horizontal wells. The long wells may produce less gas per unit well length than a vertical well, but total horizontal-well production can be higher than for a vertical well because of the long length. Thus, horizontal wells provide an excellent method to minimize near-wellbore turbulence, and at the same time, enhance total gas production from a well.

GAS RESERVE ESTIMATION

Before drilling a well, either vertical or horizontal, it is important to estimate gas in place and recoverable gas. This section includes a brief description of reserve estimation. More details can be found in other references.[1,2]

One acre-ft of reservoir rock is capable of holding volume V_g at reservoir pressure and temperature conditions. The gas volume V_g is

$$V_g = 43560\, \phi\, [1 - S_w] \tag{9–1}$$

where

V_g = gas in place in ft^3/(acre-ft)
ϕ = porosity, fraction
S_w = water saturation, fraction

The gas volume V_g exists at reservoir pressure p and temperature T. Using the gas law, the gas volume can be converted to any standard condition which is utilized for gas measurement, as

$$G = 43.56\, \phi(1 - S_w)\, p\, T_{st}\, z_{st}/(z\, p_{st}\, T) \tag{9–2}$$

where

G = gas volume, Mcf/(acre-ft)
p = reservoir pressure, psia
p_{st} = standard condition pressure, psia
T_{st} = standard condition temperature, °R
z_{st} = compressibility factor at standard conditions, dimensionless

z = gas compressibility at reservoir condition, dimensionless
T = reservoir temperature, °R(°R = °F + 460°).

In the United States, 14.65 psia and 60°F, i.e., 520°R, are used as standard conditions in gas measurement. Substituting these standard condition values in Equation 9–2

$$G = 1546.2\,\phi\,(1 - S_w)\,p/(zT) \tag{9–3}$$

where

G = gas volume at standard conditions, Mcf/(acre-ft).

It is important to note that Equation 9–3 includes an assumption that the gas compressibility at standard conditions is one, i.e., $z_{st} = 1$. The total gas in place is calculated as

$$G_i = 1546.2\,\phi\,(1 - S_w)\,p_i A\,h/(z_i T) \tag{9–4}$$

where

G_i = initial gas in place at standard conditions, Mcf
T = reservoir temperature, °R
p_i = initial reservoir pressure, psia
z_i = gas compressibility at initial reservoir condition, dimensionless
A = area, acres
h = reservoir height, ft.

A procedure for calculating the gas compressibility factor is shown in Appendix A. A detailed discussion is given in Appendix B.

If a vertical well drains area A, then, a 1000-ft-long horizontal well could drain area $2A$, twice the area drained by a vertical well. A 2000-ft-long horizontal well could drain $3A$, three times the area drained by a vertical well. (Please note that these are rules of thumb. Additionally, these thumb rules indicate minimum well spacing desired for horizontal wells. Actual well spacing can be even larger than that given by the rules of the thumb.) Thus a horizontal well can be targeted to drain a larger reservoir volume than a vertical well. In other words, horizontal well spacing should be larger than that used in vertical well development.

EXAMPLE 9–1

Calculate initial gas in place for a vertical well spaced at 40 acres. Estimate initial gas in place for 1000 ft and 2000 ft long horizontal wells. The reservoir properties are given as

initial reservoir pressure, p_i = 2000 psia
water saturation, S_w = 0.46
reservoir height, h = 50 ft
gas compressibility, z_i = 0.91
reservoir temperature, T = 105°F = 565°R
porosity, ϕ = 0.12

Solution

For a vertical well, gas in place in 40 acres is calculated using Equation 9–4.

$$\begin{aligned} G_i &= 1546.2\,\phi\,(1 - S_w)\,p_i\,A\,h/(z_i T) \\ &= 1546.2 \times 0.12 \times (1 - 0.46) \times 2000 \times 40 \times 50/(0.91 \times 565) \\ &= 779{,}490.5 \text{ Mscf} \\ &= 779 \text{ MMscf} \end{aligned}$$

Using the rules of thumb for a horizontal well drainage area, we can assume that a 1000-ft-long well drains twice the vertical well giving a drainage area of 80 acres. Similarly, a 2000-ft-long well can be assumed to drain three times the vertical well spacing, i.e., 120 acres.

Hence, gas in place for a 1000-ft-long well is

$$\begin{aligned} [G_i]_{1000\text{ ft}} &= \text{Gas in place for a 40-acre vertical well} \times 2 \\ &= 779 \times 2 \\ &= 1558 \text{ MMscf.} \end{aligned}$$

Similarly, gas in place for a 2000-ft-long well

$$\begin{aligned} [G_i]_{2000\text{ ft}} &= \text{Gas in place for a 40-acre vertical well} \times 3 \\ &= 779 \times 3 \\ &= 2339 \text{ MMscf.} \end{aligned}$$

ESTIMATION OF GAS IN PLACE USING p/z METHOD

Material balance equations can be utilized to predict gas in place. This is normally referred to the p/z method. The method is based on the reservoir material balance, which shows

$$G_p = G_i - G_r \tag{9–5}$$

G_p = produced gas
G_i = initial gas in place
G_r = remaining gas in place

Gas law is

$$p_i V_p/(z_i T_i) = p_{st}\, G_i/T_{st} \tag{9–6}$$

V_p = reservoir pore volume
p_{st} = reference pressure (normally 14.65 psia)
T_{st} = reference temperature (normally 60°F, i.e., 520°R)

In Equation 9–6, the implicit assumption is that $z_{st} = 1$. Substituting gas law into Equation 9–5 to reflect various G values as shown in Equation 9–6 we have

$$G_p = \frac{T_{st}}{p_{st}}\frac{p_i}{z_i}\frac{V_p}{T} - \frac{T_{st}}{p_{st}}\frac{pV_p}{zT} \tag{9–7}$$

where T is the reservoir temperature. We can rearrange Equation 9–7 as

$$\log\left[(p_i/z_i) - (p/z)\right] = \log G_p + \log\left[\frac{p_{st} \times T}{T_{st} \times V_p}\right]. \tag{9–8a}$$

or

$$\frac{p}{z} = \left[1 - \frac{G_p}{G_i}\right]\frac{p_i}{z_i} \tag{9–8b}$$

Equation 9–8b gives a straight line when plotted on Cartesian coordinates. Thus if one plots p/z on the y axis and produced gas on the x axis, one can estimate initial gas in place by extrapolating the curve as shown in Figure 9–1. Similarly, most wells have abandonment pressure p_a. A similar plot can be used to estimate produced gas at the abandonment pressure (see Figure 9–2). Thus, producible gas up to abandonment in Mcf is

$$G_p = G_i - G_a = 1546.2\,\phi\,\frac{(1 - S_w)Ah}{T}\left[\frac{p_i}{z_i} - \frac{p_a}{z_a}\right]. \tag{9–9}$$

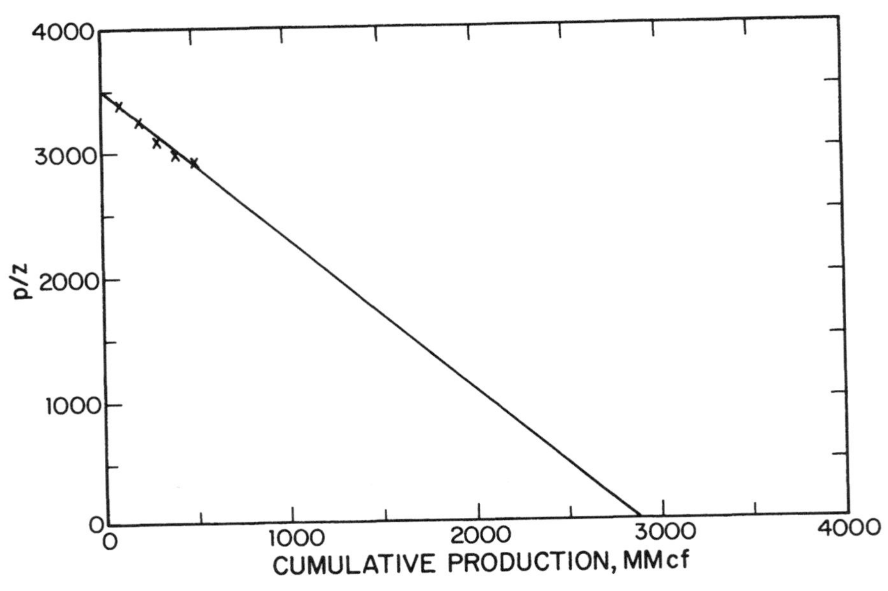

Figure 9–1 A Typical *p/z* versus Cumulative Production Plot.[1]

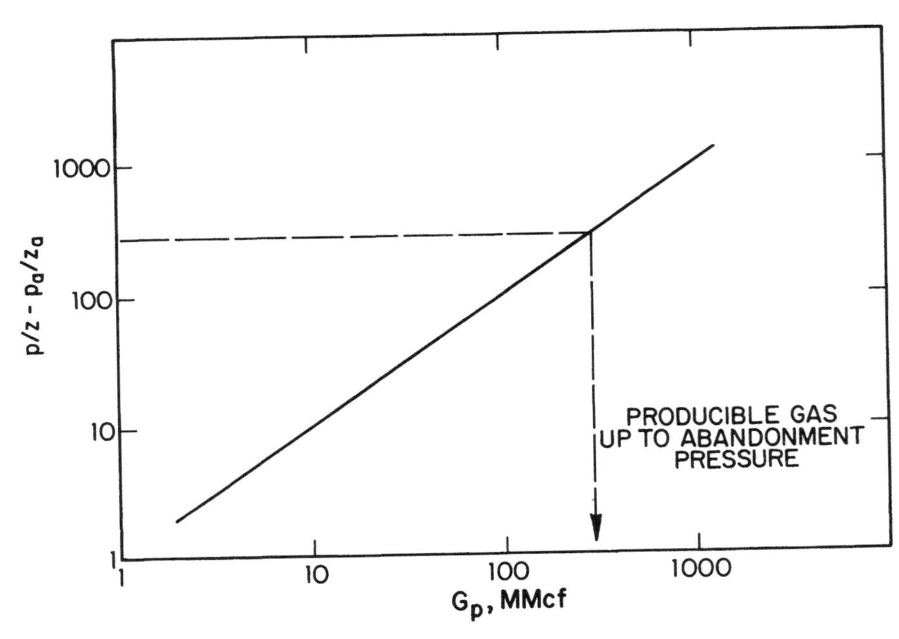

Figure 9–2 *p/z* Plot Showing Producible Reserve up to Abandonment Pressure.[2]

The total producible gas up to the reservoir abandonment pressure can also be obtained from Equation 9–4 by substituting for initial and abandonment pressures.

EXAMPLE 9–2

For the following reservoir data, calculate the gas in place and producible gas up to abandonment pressure.

$T = 120°F$	$p_i = 4000$ psia
$\gamma_g = 0.70$	$\phi = 0.12$
$h = 50$ ft	$A = 40$ acres
$S_w = 0.36$	$p_a = 800$ psia

Solution

Gas in place G_i can be calculated using Equation 9–4.

$$G_i = 1546.2\,\phi\,(1 - S_w)\,p_i\,A\,h/(z_i T)$$

$$G_i = \frac{1546.2(0.12)(1 - 0.36)\,4000\,(40)\,(50)}{z_i(120 + 460)}$$

$$G_i = 1{,}637{,}906/z_i \text{ Mscf} = 1.638/z_i \text{ Bscf}$$

where

z_i can be calculated from the Standing and Katz correlation using Equations A–24 through A–27 and Figure A–8 in Appendix A.

$$z_i = 0.86$$
$$G_i = 1.9 \text{ Bscf.}$$

The producible gas up to abandonment pressure (G_p) can be calculated using Equation 9–9.

$$G_p = G_i - G_a = 1546.2\,\phi\,\frac{(1 - S_w)Ah}{T}\left[\frac{p_i}{z_i} - \frac{p_a}{z_a}\right], \text{ Mscf}$$

z_a is calculated in the same procedure as z_i, but at p_a instead of p_i.

$$z_a = 0.89$$
$$G_p = 1.54 \text{ Bscf.}$$

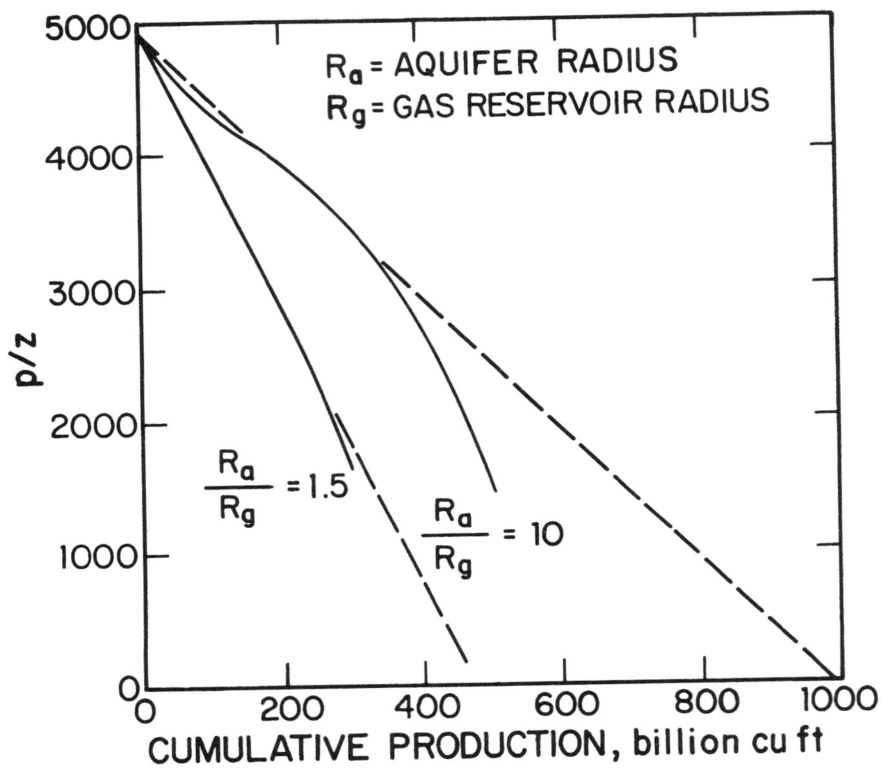

Figure 9–3 *p/z* Plot Showing Possible Over-estimation of Original Gas in Place Depending Upon Aquifer Size[3].

LIMITATION OF *p/z* METHOD

It is important to note that *p/z* against cumulative production plots can be used to estimate initial gas in place for a closed reservoir. If the reservoir has water intrusion via aquifer, then changes in reservoir pressure over gas production will be affected by aquifer size. Large aquifers will cause a large water intrusion resulting in a minimum pressure loss from the original reservoir pressure, resulting in gross overestimation of the original gas in place (see Fig. 9–3).[1,3,4] Thus, *p/z* curves should be used carefully and should be used only for the closed reservoirs.

GAS FLOW THROUGH POROUS MEDIA

This section includes steady-state and pseudo-steady state equations for gas flow through a reservoir. For each subsection, mathematical equations for vertical wells are described first, followed by mathematical equations for horizontal wells. In the gas wells, normally two different methods are used to describe the relationship between pressure and flow rate.

Method I: In this method, the gas flow rate is proportional to the pressure square terms. This method is generally employed when reservoir pressures are less than 2500 psia.
Method II: In this method, the gas pseudo-pressure $m(p)$ is defined.[5] The gas flow rate is directly proportional to the pseudo-pressures. The pseudo-pressure is defined as

$$m(p) = 2 \int_{p_{ref}}^{p} \frac{p}{\mu Z} dp \tag{9–10}$$

in Equation 9–10, p_{ref} is a reference pressure. At the reference pressure, pseudo-pressure is assigned a datum value of zero. Although choice of reference pressure can be arbitrary, many times 0.0 psia is chosen as the reference pressure. Figure 9–4 depicts a comparison of p^2 and $m(p)$. The

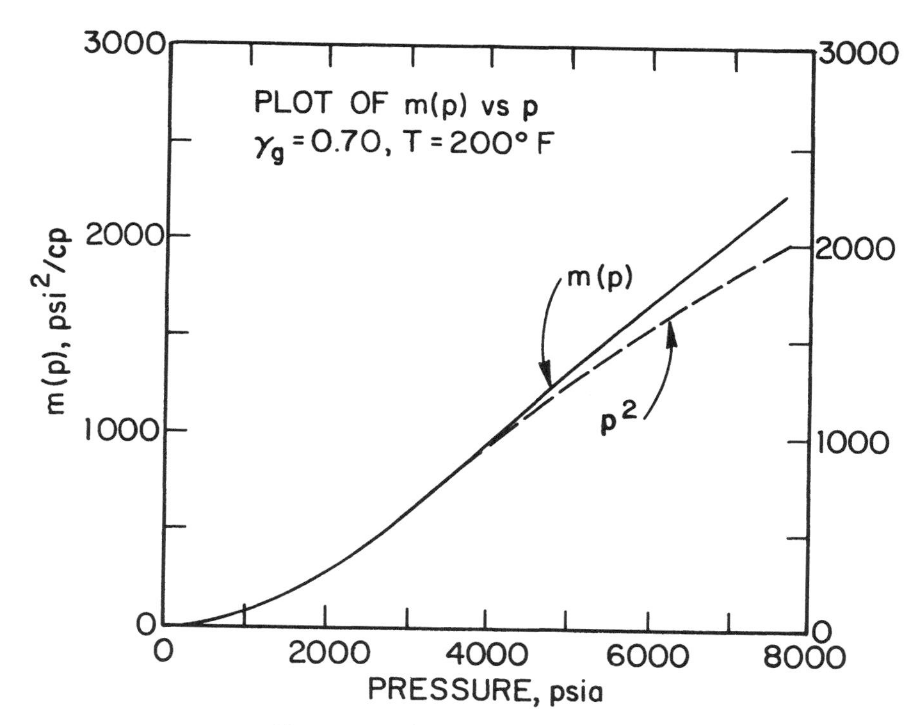

Figure 9–4 Plots of p^2 and $m(p)^2$.

figure shows that p^2 and $m(p)$ have identical values up to 2500 psia. Above 2500 psia, p^2 and $m(p)$ exhibit different values. Thus, below 2500 psia, either p^2 or $m(p)$ can be used. Above 2500 psia, $m(p)$ should be used.

As noted earlier in the text, steady state assumes a constant pressure at the drainage boundary. Now, if the wellbore pressure is fixed, this would result in a constant pressure drop from the reservoir to the wellbore. In practice this pressure drop may change over time. In the case of pseudo-steady state, a well's drainage area is fixed and there is no flow across the drainage boundaries. Therefore, as time progresses, reservoir pressure decreases as more and more fluid is withdrawn from the reservoir. In the following subsections, steady and pseudo-steady state equations for vertical and horizontal gas wells are described.

STEADY-STATE GAS FLOW

A steady-state equation for gas flow through porous media is written as

$$q = \frac{0.0007027\, kh\,(p_e^2 - p_{wf}^2)}{\mu z T \ln (r_e/r_w')} \quad (9\text{–}11)$$

or

$$q = \frac{0.000305\, kh\,(p_e^2 - p_{wf}^2)}{\mu z T \log (r_e/r_w')} \quad (9\text{–}12)$$

also

$$q = \frac{0.0007027\, kh\,(m(p_e) - m(p_{wf}))}{T \ln (r_e/r_w')} \quad (9\text{–}13)$$

or

$$q = \frac{0.000305\, kh\,(m(p_e) - m(p_{wf}))}{T \log (r_e/r_w')} \quad (9\text{–}14)$$

where

q = gas flow rate, Mscfd at 14.65 psia and 60°F
k = permeability, md
h = thickness, ft
p_e = pressure at external radius, r_e, psia
p_{wf} = wellbore flowing pressure, psia

μ = average viscosity, cp
z = average compressibility factor, dimensionless
T = reservoir temperature, °R, (°R = °F + 460)
r_e = drainage radius, ft
r'_w = effective wellbore radius, ft

Although Equations 9–11 through 9–14 are for vertical wells, they can be used to calculate steady-state gas production rate from a horizontal well. This is accomplished by substituting effective wellbore radius r'_w, of a horizontal well in the above equations. The Equations in Chapter 3, namely Equations 3–25 through 3–27, can be used to calculate effective wellbore radius r'_w for a horizontal well.

EXAMPLE 9–3

Calculate steady-state gas flow rate from a well with a 100-ft-long infinite-conductivity fracture. The 50-ft-thick reservoir has a pressure of 1800 psia. The wellbore diameter is 8½ in.

Given

reservoir permeability, k_h = 0.1 md
vertical permeability, k_v = 0.1 md
gas viscosity, μ = 0.02 cp
gas compressibility, z = 0.9
reservoir temperature, T = 105°F
well flowing pressure, p_{wf} = 350 psia
drainage area, A = 40 acres
drainage radius, r_e = 745 ft

Solution

$$q = \frac{0.0007027\, k_h h\, [(p_e^2) - (p_{wf}^2)]}{\mu z T \ln (r_e/r'_w)} \tag{9–11}$$

For an infinite-conductivity fracture of total length 100 ft, half-length x_f is 50 ft. The effective wellbore radius r'_w of an infinite-conductivity fracture is

$$r'_w = x_f/2 = 50/2 = 25 \text{ ft}$$

substituting r'_w in Equation 9–11

$$q = \frac{0.0007027 \times 0.1 \times 50\,(1800^2 - 350^2)}{0.02 \times 0.9 \times (105 + 460)\ \ln (r_e/25)}$$

$$\begin{aligned} q &= 1077/\ln (r_e/25) \\ &= 1077/\ln(745/25) \\ &= 317.3 \text{ Mscfd} \\ &\approx 317 \text{ Mscfd} \end{aligned}$$

EXAMPLE 9–4

If we drill a 500-ft-long horizontal well, what is the expected gas flow rate for a steady-state condition? (Assume reservoir properties and drainage area are the same as those noted in Example 9–3.)

Solution

The steady-state gas flow can be calculated using Equation 9–11.

$$q = \frac{0.0007027\ kh\ (p_e^2 - p_{wf}^2)}{\mu z T \ln (r_e/r'_w)} \tag{9–11}$$

Using the same reservoir properties as Example 9–3, we have

$$q = 1077/\ln(r_e/r'_w)$$

For a horizontal well, r'_w is calculated from Equation 3–25

$$r'_w = \frac{r_{eh}\ (L/2)}{a[1 + \sqrt{1 - (L/2a)^2}]\ [h/(2r_w)]^{h/L}} \tag{3–25}$$

where

$$a = 0.5L\ [0.5 + [0.25 + (2\ r_{eh}/L)^4]^{0.5}]^{0.5} \tag{3–11}$$

The problem is solved using two different methods to calculate horizontal well's drainage radius r_{eh}.

Method I:

Assume

$$\begin{aligned} r_{eh} &= r_{ev} \\ a &= 0.5(500)\ [0.5 + [0.25 + (1490/500)^4]^{0.5}]^{0.5} \\ a &= 766 \text{ ft} \end{aligned}$$

Substituting the value of a into Equation 3–25 we have

$$r'_w = \frac{745\ (500/2)}{766\ [1 + \sqrt{1 - (500/1532)^2}]\ [50/0.708]^{50/500}}$$

$$r'_w = 81.66 \text{ ft}$$

Solving for flow rate in Equation 9–11

$$q = 1077/\ln\ (745/81.66)$$

$$q = 487 \text{ Mscfd}$$

In the above calculation we have assumed an equal drainage area for horizontal and vertical wells. Another way to solve the problem is to estimate drainage area of the horizontal well as shown in Chapter 2.

Method II: Assume elliptical drainage area

$$\text{drainage area} = \pi r_{eh}^2 = \pi a'b'$$

where a' is half the major axis and b' is half the minor axis.

For a 500-ft-long wellbore with a 40-acre vertical well drainage area, we have

$$a' = L/2 + r_{ev} = (500/2) + 745 = 995 \text{ ft}$$

$$b' = r_{ev} = 745 \text{ ft}$$

Substituting the values of a' and b' in Equation 9–14b, r_{eh} is calculated as

$$r_{eh} = \sqrt{995 \times 745} = 861 \text{ ft}$$

$$a = 879.3 \text{ ft (from Eq. 3–11)}$$

Then using $r_{eh} = 861$ ft and $a = 879.8$ ft in Equation 3–25, the effective wellbore radius of a 500-ft long horizontal well is

$$r'_w = 81.68 \text{ ft}$$

and for the horizontal well, $q = 1077/\ln\ (r_{eh}/r'_w)$

$$q = 1077/\ln\ (861/81.68)$$

$$q = 457.3 = 457 \text{ Mscfd}$$

The flow rates calculated by Methods I and II are 487 and 457 Mscfd, respectively. This shows that for the given drainage area, the calculation method has a small effect on the production rate under steady-state conditions. Generally, the difference of the rates calculated using the two methods is small, as long as the well length L is small (especially when the well length is less than the radius of a vertical well's circular drainage area). When the well length exceeds the drainage radius of a vertical well, one should proportionately increase the drainage volume of a horizontal well. This is important for long horizontal wells. As noted earlier, if a vertical well drains A acres, a 1000-ft and a 2000-ft-long well would drain $2A$ and $3A$ acres, respectively (see Example 2–8). This is the basis for Example 9–5.

EXAMPLE 9–5

Calculate the steady-state production rate for 1000- and 2000-ft-long horizontal wells if they drain 80 and 120 acres, respectively. (All the reservoir properties are the same as those noted in Example 9–3.)

Solution

For 80 acres, $r_{eh} = 1053$ ft

Therefore for a 1000-ft-long well from Equation 3–11

$$a = 1114 \text{ ft}$$

Substituting this into Equation 3–25

$$r'_w = \frac{1053\ (1000/2)}{1114\ [1 + \sqrt{1 - (1000/2228)^2}]\ [50/0.708]^{50/1000}}$$

$$r'_w = 201.7 \text{ ft}$$

For a 1000-ft-long horizontal well with 80-acre well spacing, one can write

$$a = 1077/\ln\ (r_{eh}/r'_w)$$
$$q = 1077/\ln\ (1053/201.7)$$
$$= 651.7 \simeq 652 \text{ Mscfd}$$

For 120 acres, $r_{eh} = 1290$ ft.

For a 2000-ft-long well, with 120-acre well spacing, using the procedure as shown above

$$a = 1495.8 \text{ ft}$$
$$r'_w = 444.6 \text{ ft}$$
$$q = 1011 \text{ Mscfd}$$

Thus, comparing the results of Examples 9–3 and 9–5, one can see that for a steady-state, a fractured vertical well with 40-acre spacing can produce at a rate of 317 Mscfd. A 1000-ft-long horizontal well at 80-acre spacing can produce 652 Mscfd. A 2000-ft-long horizontal well at 120-acre well spacing can produce 1011 Mscfd.

PSEUDO-STEADY STATE GAS FLOW

The pseudo-steady state equation for vertical wells, fractured vertical wells and horizontal gas wells are

$$q = \frac{0.0007027\, kh\, (\bar{p}^2 - p_{wf}^2)}{T\tilde{z}\tilde{\mu}[\ln(r_e/r_w) - 0.75 + s + s_m + s_{CA} - c' + Dq]} \tag{9–15}$$

$$q = \frac{0.0007027\, kh\, [m(\bar{p}) - m(p_{wf})]}{T[\ln(r_e/r_w) - 0.75 + s + s_m + s_{CA} - c' + Dq]} \tag{9–16}$$

$$D = \frac{2.222 \times 10^{-15} \times \gamma_g\, k_a h\, \beta'}{\mu_{pwf}\, r_w\, h_p^2} \tag{9–17}$$

$$\beta' = 2.73 \times 10^{10}\, k_a^{-1.1045} \tag{9–18a}$$

or

$$\beta' = 2.33 \times 10^{10}\, k_a^{-1.201} \tag{9–18b}$$

s = equivalent negative skin factor due to either well stimulation or due to horizontal well
s_m = mechanical skin damage, dimensionless
s_{CA} = shape related skin factor, dimensionless
c' = shape factor conversion constant, dimensionless
k = permeability, md
h = reservoir height, ft
$\bar{p}$ = average reservoir pressure, psia
p_{wf} = well flowing pressure, psia
$m(p)$ = pseudo pressure, psia^2/cp
q = gas rate, Mscfd
T = reservoir temperature, °R
$\tilde{u}$ = gas viscosity evaluated at some average pressure between $\bar{p}$ and pwf
μ_{pwf} = gas viscosity at well flowing conditions, cp
$\tilde{z}$ = gas compressibility factor evaluated at some average pressure between $\bar{p}$ and pwf
β' = high velocity flow coefficient, 1/ft
γ_g = gas gravity, dimensionless
r_w = wellbore radius, ft

h_p = perforated interval, ft
k_a = permeability in the near wellbore region, md

Equation 9–18a for β' is given in Reference 6 while Equation 9–18b is given in Reference 2. Depending upon β' definition used, a somewhat different answer will be obtained.

The above noted pseudo-steady state equations, namely Equations 9–15 and 9–16, are based upon circular drainage area as a reference area. We can write similar equations on the basis of square drainage area as

$$q = \frac{0.0007027\, kh\, (\bar{p}^2 - p_{wf}^2)}{T\bar{z}\bar{\mu}\, [\ln(r_e/r_w) - 0.738 + s + s_m + s_{CA} - c' + Dq]} \quad (9\text{–}19)$$

and

$$q = \frac{0.0007027\, kh\, (m(\bar{p}) - m(p_{wf}))}{T\, [\ln(r_e/r_w) - 0.738 + s + s_m + s_{CA} - c' + Dq]} \quad (9\text{–}20)$$

Equations 9–19 and 9–20 are similar to Equations 9–15 and 9–16, except for shape-related skin factor values (s_{CA}). In Equations 9–15 and 9–16 the s_{CA} is based upon circular drainage area. In Equations 9–19 and 9–20, the s_{CA} is based upon square drainage area. In the above equations, definition s_{CA} and c' depends upon the type of the well as listed below:

Vertical Well: $c' = 0$, s_{CA} from Table 7–1

Fractured Vertical Well: $c' = 1.386$, and $s_{CA} = s_{CA,f}$
$s_{CA,f} = \ln\sqrt{30.88/C_f}$
C_f is obtained from Tables 7–2 through 7–4

Horizontal Wells: $c' = 1.386$ and $s_{CA} = s_{CA,h}$
$s_{CA,h}$ is obtained from Figures 7–5 through 7–7 or Table 7–5.

Another important term in Equations 9–15, 9–16, 9–19, and 9–20, is the turbulence term, Dq. This is also called turbulence skin, or rate dependent skin factor.[2,6] This term accounts for additional pressure drop in the near wellbore region due to high gas velocity. In high-permeability reservoirs, near wellbore gas velocities are high, resulting in a large pressure drop. The near wellbore turbulence can be reduced by reducing gas velocities around the wellbore. A horizontal well provides a large surface area for gas production, reduces gas velocities near wellbore, and minimizes near wellbore turbulent pressure drop. A detailed discussion on the turbulence is included in the section titled horizontal well applications in high-permeability reservoirs.

In contrast to high-permeability reservoirs, in low-permeability (tight) reservoirs, gas velocities near wellbore are small and near wellbore turbulence is rarely a problem. However, in tight reservoirs, transition flow time is very long and it may take years before the well begins pseudo-steady state.

The pseudo-steady state equations, namely Equations 9–15, 9–16, 9–19, and 9–20, can also be used to calculate horizontal-well rates. This can be accomplished by either substituting effective wellbore radius r'_w of a horizontal well in these equations or using equivalent negative skin factor for the horizontal wells. Additionally an appropriate shape-related skin factor s_{CA} should be used. The discussion on horizontal well's shape-related skin factor is included in Chapter 7.

HORIZONTAL WELL APPLICATION

In this section, first we will examine horizontal-well application in tight reservoirs followed by examination of its application in high-permeability reservoirs.

TIGHT GAS RESERVOIRS

To deplete a given gas reservoir, it is important to space wells so as to drain the reservoir effectively. As noted earlier in Chapter 2, in tight gas reservoirs, the time to start pseudo-steady state can be very large. In such cases, vertical gas wells can be drilled at close spacing to efficiently drain the reservoir. This would require a large number of vertical wells. One alternative is to stimulate vertical wells by fracturing, not only to enhance their productivity, but also to increase drainage along the length of the fracture. However, in many instances, it is difficult to create long fractures in tight reservoirs, especially if the reservoir is overlain or underlain by a weak cap or base rock. During the fracture jobs, when pumping pressure exceeds the formation parting pressure for an effective proppant placement, the high pumping pressure may open up a large portion of the weak cap or base rock, resulting in an excessive fracture height and fracture growth in unproductive zones. The excessive fracture height results in a less-than-desired fracture extension in the reservoir. In such reservoirs, horizontal wells provide an alternative to obtain long fracture extension, since horizontal wells represent a long fracture with a height equal to the wellbore diameter.

Other types of tight reservoirs where fracturing is difficult are those underlain or overlain by highly fractured rock. In these cases, during fracturing, if the fracture establishes communication with the highly fractured cap or base zone, there could be excessive fluid loss during the fracturing operation, resulting in a screen out. In these cases, horizontal wells provide an alternative for creating long extensions in the reservoir.

Similarly, vertical wells located near faults can also screen out due to excessive fluid loss, giving short fracture penetrations. A fracture job that establishes communication with a natural open fault may, in fact, result in a high-productivity well. This is because an open fault itself represents a high-conductivity fracture.

As noted in Example 2–7, if a vertical well drilled at 20-acre spacing takes 0.91 years to reach pseudo-steady state, then a 1000-ft and 2000-ft-long horizontal well would also take 0.91 years to start pseudo-steady state in 44 and 68 acres, respectively. As shown in Figure 9–5, a long horizontal well can drain larger volumes than a single vertical well in the same time interval. This has an impact on field economics, because one can develop the field using a fewer number of horizontal wells than vertical wells.

As noted in Chapter 2, horizontal wells can be used effectively to

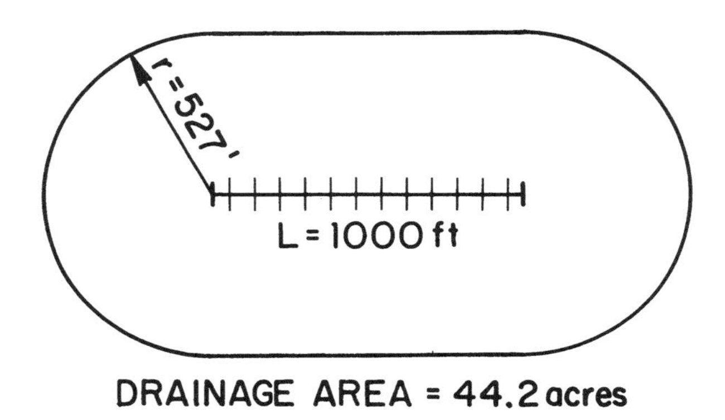

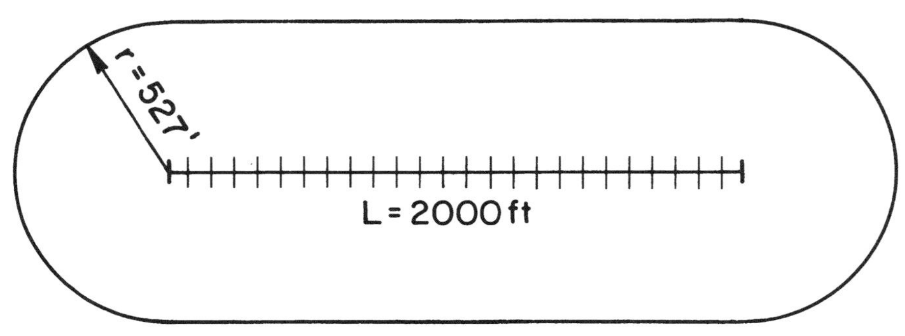

Figure 9–5 Horizontal Well Drainage Areas Assuming that the Vertical Well Spacing is 20 Acres, i.e., r_{ev} = 527 ft.

enhance drainage in an anisotropic reservoir, i.e., in reservoirs where permeability k_x in the x direction is different than permeability k_y in the y direction. (Note that x and y are in the horizontal plane.) Such situations are quite common in naturally fractured reservoirs where one can drill a long horizontal well along the low permeability direction to enhance drainage volume by intersecting natural fractures.

In the United States, efforts to develop tight gas reservoirs, such as Devonian shale reservoirs in the Eastern United States using horizontal wells are based on the principles noted above. In the tight reservoirs, sometimes even the horizontal wells are stimulated, so as to create fractures perpendicular to the horizontal well, resulting in a higher drainage volume.

Recently, the Department of Energy (DOE) conducted a field test of a fractured horizontal well in Devonian shale, a naturally fractured low-permeability reservoir (on the order of 50 microdarcies). The field test consisted of drilling a 2000-ft-long horizontal wellbore into the Devonian shale formation and isolating eight different sections of the wellbore with external casing packers (Fig. 9–6). Fractures were then initiated using a variety of fracture fluids (see Table 9–1). The stimulation of various zones along the well length resulted in production improvements ranging from 4- to 25-fold.[7,8] The results are summarized in Table 9–1.

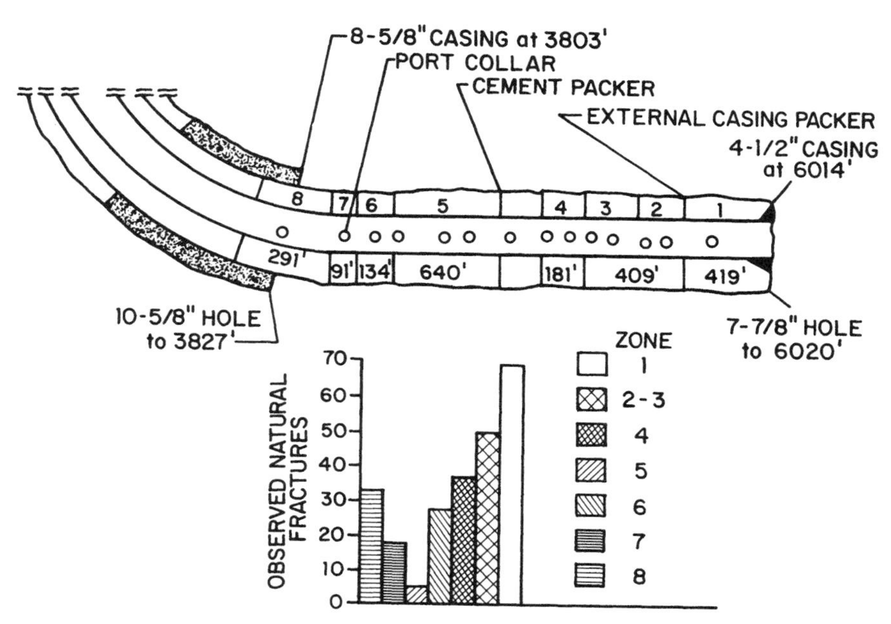

Figure 9–6 A Fractured Horizontal Well in the Devonian Shale Formation.[7]

TABLE 9–1 SUMMARY OF STIMULATION TEST SERIES FOR A WELL IN THE DEVONIAN SHALE FORMATION[7,8]

Test No.	Zone	Fluid	Rate	Volume	Frac Diagnostics
1	6	N_2 (Gas)	5 BPM	37 MCF	None
2	6	N_2 (Gas)	15 BPM	212 MCF	None
3	6	N_2-Foam	5 BPM	100 BBLS	Iodine 131
4	6	N_2-Foam	12 BPM	300 BBLS	Scandium 46
5	1	N_2 (Gas)	8–16 BPM	3745 BBLS	Tilt Meters
6	1	CO_2 (Liq)	12 BPM	200 BBLS	Iodine 131
7	1	CO_2 (Liq)	20 BPM	400 BBLS	Scandium 46
8	1	N_2-Foam	10 BPM	166 BBLS	Antimony 124
9	1	N_2-Foam	10 BPM	595 BBLS	Iridium 192
10	2–3, 4	N_2-Foam	40 BPM	905 BBLS	None
11	2–3, 4	N_2-Foam	30 BPM	2142 BBLS	Scandium 46

SUMMARY OF RESULTS OF STIMULATION TESTS TO INJECT INTO OLD FRACTURES OR CREATE NEW ONES

Test Number	Zone	Natural Fractures Detected	Fractures Pumped Into	Production Improvement
1	6	6	6 (Assumed not measured)	4.1
2	6	6	6 (Assumed not measured)	4.1
3	6	6	14	4.1
4	6	6	14	4.1
5	1	69	12 (Based on Test 6 results)	5.0
6	1	69	27 (Over 4 zones: 1, 2, 3, 4)	25.0
7	1	69	67 (Over 4 zones: 1, 2, 3, 4)	25.0
8	1	69	17 (Over 3 zones: 1, 2, 3)	15.5
9	1	69	69 (Over 4 zones: 1, 2, 3, 4)	15.5
10	2–3, 4	72	Not determined	(N.D)
11	2–3, 4	72	54 (Over 3 zones: 2, 3, 4)	(N.D)

HIGH-PERMEABILITY RESERVOIRS

As noted earlier, horizontal wells are also useful in high permeability gas wells, especially in those wells where near-wellbore turbulence is very high. The near wellbore turbulence is inversely proportional to the well's perforated interval. By drilling a horizontal well, one can increase productive length and therefore, decrease the near-wellbore turbulence and enhance well productivity.

Turbulent Flow

Darcy's law of flow through porous media is valid only for laminar flow through the reservoir

$$dp/dr = a\,v \qquad (9\text{–}21)$$

where

v is velocity, a is a constant, and dp/dr is pressure gradient. Forchheimer modified Darcy's law to account for turbulence effects as

$$dp/dr = av + bv^2 \qquad (9\text{–}22)$$

Equation 9–22 can be rewritten as either

$$\bar{p}^2 - p_{wf}^2 = aq + bq^2 \qquad (9\text{–}23)$$

or

$$m(\bar{p}) - m(p_{wf}) = a'q + b'q^2 \qquad (9\text{–}24)$$

where

$$a = \frac{T\bar{z}\bar{\mu}}{0.0007027\,kh}\left[\ln(r_e/r_w) - 0.75 + s + s_{CA} - c'\right] \qquad (9\text{–}25)$$

$$b = DT\bar{z}\bar{\mu}/(0.0007027\,kh) \qquad (9\text{–}26)$$

$$a' = [T/(0.0007027\,kh)]\,[\ln(r_e/r_w) - 0.75 + s + s_{CA} - c'] \qquad (9\text{–}27)$$

$$b' = [DT/(0.0007027\,kh)] \qquad (9\text{–}28)$$

D = turbulence factor defined in Equation 9–17, 1/Mscfd.

In the above equations, definitions s_{CA} and c' depends upon the type of the well as listed below:

Vertical Well: $c' = 0$, s_{CA} from Table 7–1

Fractured Vertical Well: $c' = 1.386$, and $s_{CA} = s_{CA,f}$
$s_{CA,f} = \ln\sqrt{30.88/C_f}$
C_f is obtained from Tables 7–2 through 7–4

Horizontal Wells: $c' = 1.386$ and $s_{CA} = s_{CA,h}$
$s_{CA,h}$ is obtained from Figures 7–5 through 7–7 or Table 7–5.

Equations 9–23 and 9–24 tell us that the influence of turbulence is to increase the pressure drop, or pressure drawdown required to produce the given gas production rate. Thus, the presence of turbulence reduces net production from a well. The turbulence effect can be minimized by reducing fluid velocity near the wellbore. (The highest fluid velocity occurs near the wellbore, where flow converges. Hence, if there is no turbulence near the wellbore, there will be no turbulence in the reservoir.) The fluid velocity near the wellbore can be minimized by increasing perforated producing length h_p (see Equation 9–17). A horizontal well provides a means to significantly enhance perforated interval and reduce near wellbore turbulence.

Turbulent Identification

A multirate test can be used to confirm the existence of high velocity effects in a well. Flow tests are conducted with different surface pressures. At each surface pressure a stabilized gas flow rate q is recorded. Based on surface pressure and flow rate, downhole well flowing pressure p_{wf} is estimated. The data is correlated as

$$q = c\,(\bar{p}^2 - p_{wf}^2)^n \tag{9–29}$$

where c is a constant and n is a dimensionless constant ($\frac{1}{2} \le n \le 1$). *Equation 9–29 is rewritten as*

$$\ln q = \ln c + \ln (\bar{p}^2 - p_{wf}^2)^n \tag{9–30}$$

If slope $n = 1$ then there is no turbulence. However, if $n < 1$, turbulence does exist. The lower the value of n, the higher is the turbulence effect, and when $n = \frac{1}{2}$ turbulence is dominant. It is important to use pseudo pressure ($m(p)$ values) when pressures are above 2500 psia. As shown in Figure 9–7, a log-log plot of rate against $\bar{p}^2 - p_{wf}$ can be used to detect turbulence.

EXAMPLE 9–6

Calculate the influence of turbulence in a vertical well drilled in a 40-ft-thick sandstone reservoir with permeability of 10 md. The initial reservoir pressure is 2000 psia and well spacing is 640 acres. The well could be operated with a minimum bottomhole pressure of 300 psia. The bottomhole temperature is 120°F, gas viscosity is 0.02 cp, gas compressibility factor is 0.9, $\gamma_g = 0.7$, $r_w = 0.25$ ft, and perforated length $h_p = 40$ ft. (Use the p^2 equation to calculate the flow rates.)

Solution

Assuming the well is centrally located in the drainage area, shape-related skin factor $s_{CA} = 0$. Additionally, mechanical skin damage $s_m = 0$, and since the well is not stimulated, $s = 0$.

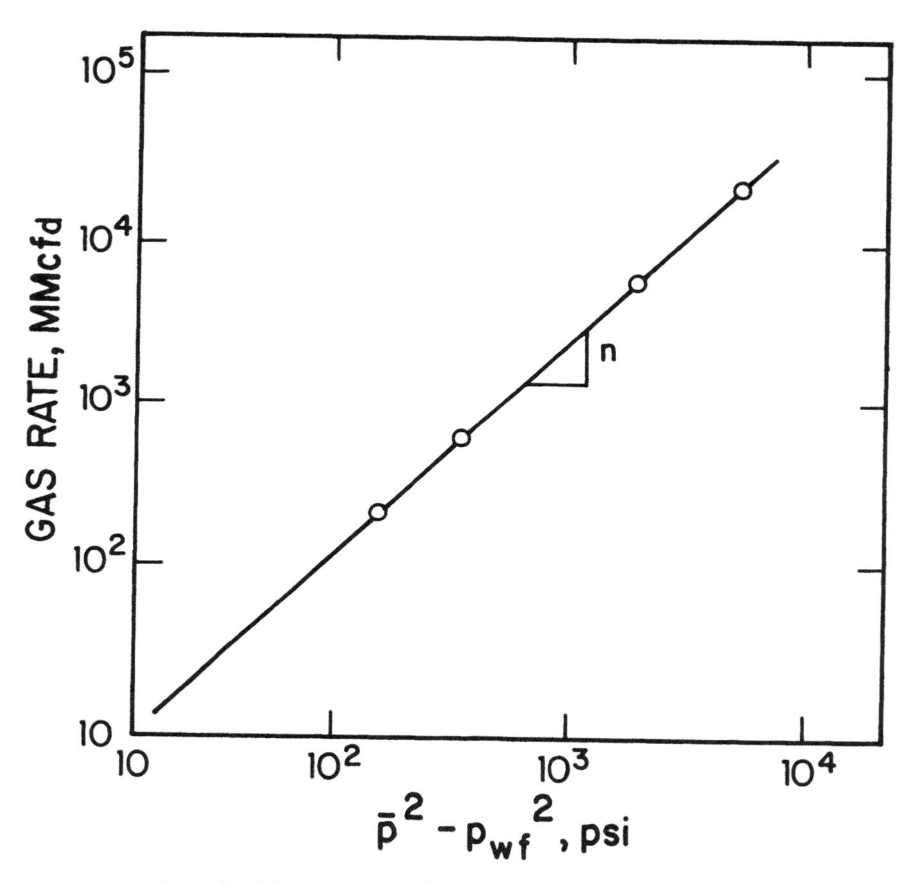

Figure 9–7 Log-Log Plot of Gas Flow Rate against $\bar{p}^2 - p_{wf}^2$.

Substituting these parameters in Equation 9–25 we have

$$a = \frac{T\tilde{z}\tilde{\mu}}{0.0007027\,kh}\,[\ln(r_e/r_w) - 0.75 + s + s_m + s_{CA}]$$

$$= \frac{[(460 + 120) \times 0.9 \times 0.02]}{0.0007027\,(10 \times 40)} \times$$

$$[\ln(2978/0.25) - 0.75 + 0 + 0 + 0]$$

$$= 321$$

value of b can be calculated using Equation 9–26 as

$$b = DT\tilde{z}\tilde{\mu}/(0.0007027\,kh) \qquad (9\text{–}26)$$

To calculate b, first estimate D and β' as shown below using Equations 9–17 and 9–18a.

$$D = \frac{2.222 \times 10^{-15} \times \gamma_g k_a h \beta'}{\mu_{pwf} r_w h_p^2} \qquad (9\text{–}17)$$

and

$$\beta' = 2.73 \times 10^{10} k_a^{-1.1045}, \ 1/\text{ft} \qquad (9\text{–}18a)$$

Assume near-wellbore permeability, k_a, is the same as the reservoir permeability, $k = 10$ md. Using Equation 9–18, β' is

$$\beta' = 2.146 \times 10^9, \ 1/\text{ft}$$

and substituting β' value in Equation 9–17,

$$D = \frac{2.222 \times 10^{-15} \times 0.7 \times 10 \times 40}{0.02 \times 0.25 \times 40 \times 40} \times 2.146 \times 10^9$$
$$= 1.669 \times 10^{-4}, \ 1/\text{Mcfd}$$

Substituting the value of D into Equation 9–26, b is calculated as

$$b = \frac{1.669 \times 10^{-4} (580) \times 0.9 \times 0.02}{0.0007027 \times 10 \times 40}$$
$$= 6.199 \times 10^{-3}, \ 1/(\text{Mscfd})^2$$

Substituting values of a and b into Equation 9–23,

$$\bar{p}^2 - p_{wf}^2 = 321q + 6.199 \times 10^{-3} q^2$$

The above quadratic equation is rearranged as

$$6.199 \times 10^{-3} q^2 + 321q - (\bar{p}^2 - p_{wf}^2) = 0$$

Solving the quadratic equation, the value of q is calculated as

$$q = \frac{-321 + \sqrt{103041 + 4 \times 6.199 \times 10^{-3} (\bar{p}^2 - p_{wf}^2)}}{2 \times 6.199 \times 10^{-3}}$$

TABLE 9–2 EFFECT OF TURBULENCE ON VERTICAL WELL PRODUCTIVITY (EXAMPLE 9–6)

p_{wf} (psi)	$(\bar{p}^2 - p_{wf}^2)$	No Turbulence $D = 0$ q, MMscfd	With Turbulence q, MMscfd
1700	111×10^4	3.48	3.25
1500	175×10^4	5.49	4.97
1000	300×10^4	9.40	8.08
500	375×10^4	11.75	9.82

Calculated values of q, both with and without turbulence for various values of p_{wf} are summarized in Table 9–2. This table indicates a significant effect of turbulence on well productivity. As shown in the following example a horizontal well can be used to minimize turbulence effects.

EXAMPLE 9–7

An offshore reservoir concession was recently signed by an oil company. The lease concession lasts only for a period of five years. The gas is to be delivered to a pipeline which is operated at 300 psia. To meet this line pressure requirement, it is important to maintain a wellhead pressure of 500 psia. Before testing, the test well, which is vertical, was cemented, perforated, and cleaned using acid. The perforated interval in a vertical well was 60 ft. The reservoir has bottom water zone separated by a 10-ft-thick layer of shale (k_v/k_h = ?). It appears that the reservoir is not in communication with the bottom water.

An engineer suggested drilling a 2000-ft-long horizontal well not only to reduce near-wellbore turbulence but also to ensure against water coning. Develop inflow performance curves for vertical and horizontal wells. Assume that reservoir is not in communication with the bottom water zone. The following reservoir and gas properties are given. Tubing pressure drop data are given in Table 9–3.

RESERVOIR PROPERTIES (EXAMPLE 9–7)

Reservoir	= Sandstone
Depth	= 7870 ft
Reservoir pressure	= 3400 psia
Reservoir thickness	= 60 ft
Average reservoir permeability	= 6 md
Vertical permeability (assumed)	= 0.6 md
Estimated well spacing	= 640 acres
Average porosity	= 14%
Water saturation	= 30%

GAS PROPERTIES (EXAMPLE 9–7)

Reservoir Temperature = 185° F
Gas Gravity (AIR = 1.0) = 0.605
Pseudo-Critical Temperature = 354.4° R
Pseudo-Critical Pressure = 664.5 psia

Pressure psia	Viscosity cp	z Factor dim	Gas Comp psia^{-1}	$m(p)$ psia^2/cp	p/z Psia
100.0	0.01269	0.9921	0.1008E−01	0.000000E+00	100.8
500.0	0.01313	0.9629	0.2071E−02	0.188818E+08	519.3
1000.0	0.01401	0.9327	0.1056E−02	0.769886E+08	1072.2
1500.0	0.01486	0.9115	0.7019E−03	0.171120E+09	1645.7
2000.0	0.01585	0.9009	0.5113E−03	0.296924E+09	2220.1
2500.0	0.01711	0.9011	0.3880E−03	0.448470E+09	2774.4
3000.0	0.01847	0.9111	0.3018E−03	0.619257E+09	3292.6
3400.0	0.01959	0.9250	0.2505E−03	0.765793E+09	3675.5

Solution

The reservoir has a permeability of 6 md, and hence a well drilled at 640 acre well spacing will begin pseudo-steady state in about 25 days. Therefore, the initial transient flow portion is ignored in the following calculations. The inflow performance curve is based upon a pseudo-steady state solution, i.e., Equation 9–16. Since the vertical well is centrally located in the drainage plane, $s_{CA} = 0$. For turbulence calculation, Equation 9–17 and 9–18b are used. For horizontal well turbulence the perforated length h_p is simply replaced by well length L in Equation 9–17.

1. **Calculation of Inflow Performance Curve**

The generalized pseudo-steady equation for a gas well, Equation 9–16 is rewritten as

$$q = \frac{kh\,[m(p_i) - m(p_{wf})]}{1422\,T\,[\ln(r_e/r_w) - 0.75 + s + s_m + s_{CA} + Dq - c']} \tag{9–29}$$

q = gas flow rate, Mscfd
k = formation permeability, md
h = pay zone thickness, ft.
$m(p_i)$ = real gas potential at initial pressure p_i, psia^2/cp
$m(p_{wf})$ = real gas potential at bottomhole pressure, p_{wf}, psia^2/cp
T = temperature, °R
r_e = drainage radius, ft
r_w = wellbore radius, ft
D = turbulence coefficient, 1/Mcfd; = 0 if wellbore turbulence is ignored

TABLE 9–3 TUBING PRESSURE DROP DATA FOR EXAMPLE 9–7

Beggs & Brill—5½″ Tubing—Only Gas Flow (p_{surf} = 500 psi)

	FLOW RATE		PRESSURE		
No.	MMscf/D	Water (STB/D)	Inlet (psia)	Surface (psia)	Drop (psi)
1	10	.00	623.58	500.00	123.58
2	20	.00	693.39	500.00	193.39
3	30	.00	803.97	500.00	303.97
4	40	.00	936.07	500.00	436.07
5	50	.00	1082.95	500.00	582.95
6	60	.00	1239.66	500.00	739.66
7	70	.00	1397.67	500.00	897.67
8	80	.00	1561.74	500.00	1061.74
9	90	.00	1727.90	500.00	1227.90
10	100	.00	1893.51	500.00	1393.51
11	110	.00	2061.99	500.00	1561.99
12	120	.00	2231.39	500.00	1731.39

Beggs & Brill Correlation—5½″ Tubing—Gas along with 20 BW/MMscf

	FLOW RATE		PRESSURE		
No.	MMscf/D	Water (STB/D)	Inlet (psia)	Surface (psia)	Drop (psi)
3	30	600.00	1079.26	500.00	579.26
4	40	800.00	1295.13	500.00	795.13
5	50	1000.00	1525.06	500.00	1025.06
6	60	1200.00	1761.77	500.00	1261.77
7	70	1400.00	2001.83	500.00	1501.83
8	80	1600.00	2243.46	500.00	1743.46
9	90	1800.00	2486.25	500.00	1986.25
10	100	2000.00	2730.03	500.00	2230.03
11	110	2200.00	2974.81	500.00	2474.81
12	120	2400.00	3220.93	500.00	2720.93

s = equivalent horizontal well or stimulated vertical well skin factor

s_m = mechanical skin damage

s_{CA} = shape related skin factor

c' = shape factor conversion constant = 1.386

For a well drainage of 640 acres

$$\text{drainage area} = \pi r_e^2 = 640 \text{ acres} \times 43{,}560 \text{ ft}^2/\text{acre}$$
$$r_e = 2979 \text{ ft}$$

To develop a non-turbulence IPR curve for a vertical well, the turbulent term in Equation 9–29 is ignored, i.e., $D = 0$. Then Equation 9–29 can be solved explicitly as

$$q = \frac{6 \times 60\,[m(3400) - m(p_{wf})]}{1422\,(185 + 460)\,[\ln(2979/0.25) - 0.75]} \tag{9–30}$$
$$= 0.000045\,[7.658 \times 10^8 - m(p_{wf})] \tag{9–31}$$

Equation 9–31 can be used for different well flowing pressures to calculate gas flow rates. The results are shown in Table 9–4 and Figure 9–8. For turbulent flow, Equation 9–30 is rewritten as

$$q = \frac{6 \times 60\,[m\,(3400) - m(p_{wf})]}{1422 \times (185 + 460)\,[\ln(2979/0.25) - 0.75 + Dq]} \tag{9–32}$$
$$q = \frac{0.0003925\,[7.658 \times 10^8 - m(p_{wf})]}{8.636 + Dq} = \frac{C}{B + Dq}$$

This is a quadratic equation

$$Dq^2 + Bq - C = 0 \tag{9–33}$$

TABLE 9–4 IPR CALCULATIONS FOR VERTICAL WELL (Example 9–7)

p_{wf} psi	$m(p_{wf})$ psi²/cp	$\Delta m(p)$*	μ_{pwf} cp	D 1/Mcfd	q MMscfd No Turb	q MMscfd Turb
500	0.19×10^8	7.469×10^8	0.0131	1.112×10^{-4}	34.4	25.5
1000	0.77×10^8	6.888×10^8	0.0140	1.041×10^{-4}	31.7	24.2
1500	1.71×10^8	5.948×10^8	0.0149	0.978×10^{-4}	27.4	21.7
2000	2.97×10^8	4.688×10^8	0.0159	0.916×10^{-4}	21.6	17.9
2500	4.49×10^8	3.173×10^8	0.0171	0.852×10^{-4}	14.6	12.8
3000	6.19×10^8	1.468×10^8	0.0185	0.787×10^{-4}	6.8	6.3

* $\Delta m(p) = m(\bar{p}) - m(p_{wf})$

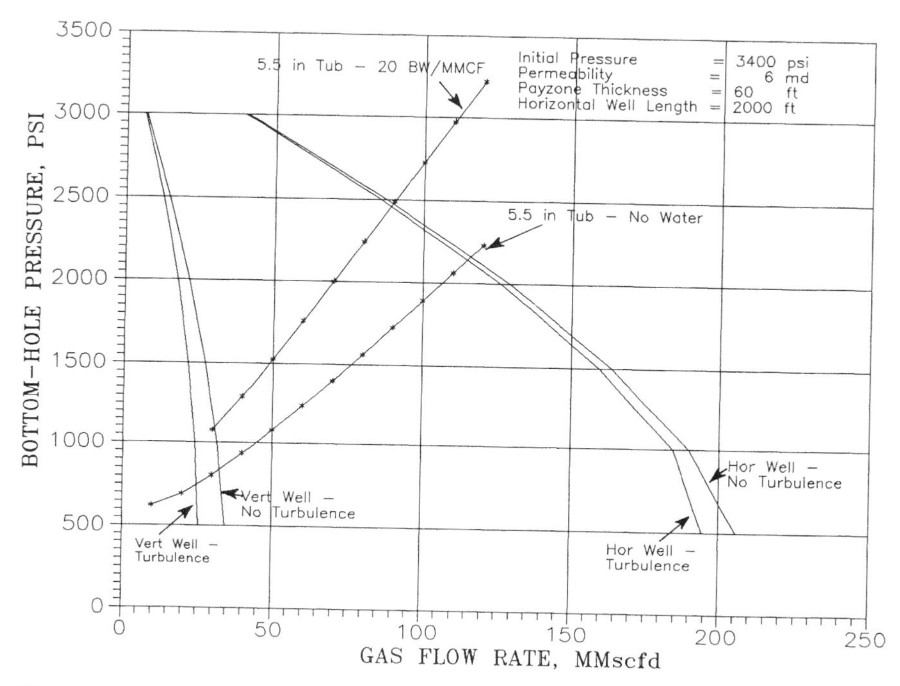

Figure 9–8 Effect of Turbulence on Inflow Performance of Vertical and Horizontal Wells.

and

$$q = \frac{-B + \sqrt{(B)^2 + 4\,DC}}{2D} \tag{9–34}$$

$$q = \frac{-8.636 + \sqrt{(8.636)^2 + 4 \times D \times 0.0003925\,[7.658 \times 10^8 - m(p_{wf})]}}{2D} \tag{9–35}$$

To solve Equation 9–35 we need to calculate turbulence factor D using Equations 9–17 and 9–18b and assuming $k = k_a$,

$$\beta' = \frac{2.33 \times 10^{10}}{6^{1.201}} = 2.709 \times 10^9$$

$$D = \frac{2.222 \times 10^{-15}\,(2.709 \times 10^9) \times 0.605 \times 6 \times 60}{\mu_{pwf} \times 0.25 \times 60 \times 60}$$

$$= 1.457 \times 10^{-6}/\mu_{pwf}$$

The flow rate results are summarized in Table 9–4 and Figure 9–8. Table 9–4 shows an effect of turbulence on well flow rates.

For horizontal calculations, Equation 9–29 is written as

$$q = \frac{6 \times 60\,[m(3400) = m(p_{wf})]}{1422 \times (185 + 460)[\ln(2979/0.25) - 0.75 + s + s_m + s_{CA} + Dq - c']} \quad (9\text{–}36)$$

For a 2000-ft-long horizontal well, skin factor s and shape-related skin factor s_{CA} need to be calculated.

2. **Calculation of Skin Factor *s* for Horizontal Well**

$$r'_w = L/4 = 2000/4 = 500 \text{ ft} \quad (9\text{–}37)$$

$$r_w = 0.25 \text{ ft}$$

$$s = -\ln(r'_w/r_w) = -\ln(500/0.25) \quad (9\text{–}38)$$

$$= -7.6$$

3. **Calculation of s_{CA}**

Assuming a square drainage area with each side being $2x_e$ for 640 acres, we have

$$2x_e = \sqrt{640 \times 43560} = 5280 \text{ ft} \quad (9\text{–}39)$$

$$L/(2x_e) = 2000/5280 = 0.38 \quad (9\text{–}40)$$

$$k_v/k_h = 0.6/6 = 0.1$$

$$L_D = (L/(2h))\sqrt{k_v/k_h} = (2000/(2 \times 60))\sqrt{0.1} = 5.3 \quad (9\text{–}41)$$

TABLE 9–5 IPR CALCULATIONS FOR HORIZONTAL WELL (Example 9–7)

					q MMscfd	
p_{wf} psi	$m(p_{wf})$ psi²/cp	$\Delta m(p)$*	μ_{pwf} cp	D 1/Mcfd	No Turb	Turb
500	0.19×10^8	7.469×10^8	0.0131	1.001×10^{-7}	205.6	199.4
1000	0.77×10^8	6.888×10^8	0.0140	0.936×10^{-7}	189.6	184.3
1500	1.71×10^8	5.948×10^8	0.0149	0.880×10^{-7}	163.7	159.5
2000	2.97×10^8	4.688×10^8	0.0159	0.825×10^{-7}	129.1	126.0
2500	4.49×10^8	3.173×10^8	0.0171	0.767×10^{-7}	87.3	85.5
3000	6.19×10^8	1.468×10^8	0.0185	0.709×10^{-7}	40.4	39.7

* $\Delta m(p) = m(\bar{p}) - m(p_{wf})$

From Figure 7–5, corresponding to $L_D = 5.3$ and $L/2x_e = 0.38$,

$$s_{CA} = 1.8 \tag{9–42}$$

Rewriting Equation 9–36 to calculate gas flow rate

$$q = \frac{0.0003925\,[m(3400) - m(p_{wf})]}{(8.636 + s_m + s + s_{CA} + Dq - c')} \tag{9–43}$$

Substituting for s and s_{CA} in Equation 9–43 from Equations 9–38 and 9–42, respectively

$$q = \frac{0.0003925\,[m(3400) - m(p_{wf})]}{(8.636 + 0 - 7.60 + 1.8 + Dq - 1.386)} \tag{9–44}$$

Equation 9–44 assumes that mechanical skin $s_m = 0$.

For a non-turbulence case, D = 0 and Equation 9–44 is solved explicitly for various values of p_{wf}. The final results are summarized in Table 9–5 and Figure 9–8.

For a turbulence case, D is calculated by substituting 2000 ft as perforated length instead of 60 ft (for vertical well) as used in Equation 9–17.

$$D = \frac{2.222 \times 10^{-15} \times (2.709 \times 10^{9}) \times 0.605 \times 6 \times 60}{\mu_{pwf} \times 0.25 \times 2000 \times 2000}$$

$$= 1.311 \times 10^{-9}/\mu_{pwf} \tag{9–45}$$

For turbulence calculation, the value of D from Equation 9–45 is substituted into Equation 9–44.

$$q = \frac{0.0003925\,[m(3400) - m(p_{wf})]}{1.45 + Dq} = \frac{C}{B + Dq} \tag{9–46}$$

Equation 9–46 is a quadratic equation, which can be solved as

$$q = \frac{-B + \sqrt{(B)^2 + 4DC}}{2D} \tag{9–47}$$

and

$$q = \frac{-1.45 + \sqrt{1.45^2 + 4 \times D \times 0.0003925\,[m(\bar{p}) - m(p_{wf})]}}{2D} \tag{9–48}$$

by substituting appropriate D value, Equation 9–48 is solved for various well flowing pressure (p_{wf}) values.

In Figure 9–8 inflow performance relationships (IPR curves) for vertical and horizontal wells are shown. The plots are for two cases: (1) without turbulence and (2) with turbulence. The plot also includes tubing data given in Table 9–3. Figure 9–8 shows a drop in the vertical well's productivity due to turbulence. The figure also shows that turbulence has no effect on horizontal wells, i.e., horizontal wells minimize turbulence-related pressure drops. Thus, in high-permeability gas reservoirs, horizontal wells provide a method to minimize wellbore turbulence.

Recently Celier et al.[9] presented the following formula to calculate ratio of pressure drop due to turbulence, i.e., nondarcy flow, in horizontal and vertical wells:

$$\frac{(\Delta p)_{h,t}}{(\Delta p)_{v,t}} = \frac{2\beta^2}{(1 + \beta)}\left[\frac{h}{L}\right]^2 \tag{9–49}$$

where $\beta = \sqrt{k_h/k_v}$, h is reservoir height and L is well length. It is important to note that the above equation assumes that h ft of vertical well is perforated in a h-ft-thick reservoir, similarly L ft of horizontal well is open to flow.

EXAMPLE 9–8

Calculate reduction in turbulence related pressure drop in a 60-ft-thick, 6-md reservoir by drilling a 2000-ft-long horizontal well. What would be the pressure drop ratio if $k_v/k_h = 0.01, 0.1$, and 1.0?

Solution

A ratio of horizontal and vertical well pressure drops due to nondarcy flow is given in Equation 9–49.

$$\frac{(\Delta p)_{h,t}}{(\Delta p)_{v,t}} = \frac{2\beta^2}{1 + \beta}\left[\frac{h}{L}\right]^2$$

if $k_v/k_h = 1, \beta = \sqrt{k_h/k_v} = 1$
if $k_v/k_h = 0.1, \beta = \sqrt{10} = 3.16$
if $k_v/k_h = 0.01, \beta = \sqrt{100} = 10$

Assuming the vertical well is fully penetrating, we have

$$\frac{(\Delta p)_{h,t}}{(\Delta p)_{v,t}} = \frac{2\beta^2}{1 + \beta}\left[\frac{60}{2000}\right]^2$$

$$= \frac{0.0018\,\beta^2}{(1 + \beta)}$$

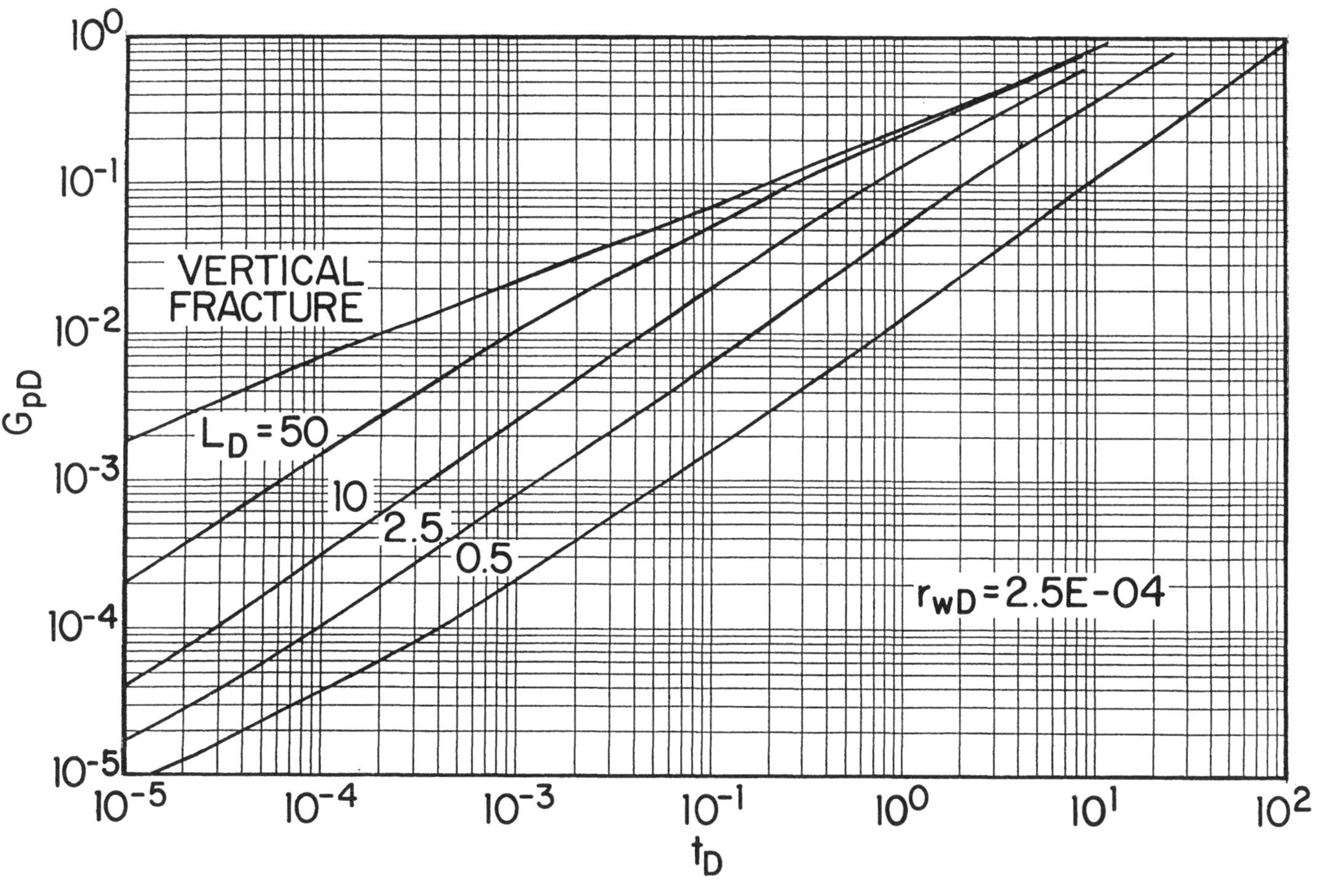

Figure 9–9 A Production Type Curve for Horizontal Gas Wells in an Infinite Reservoir.[10]

Thus, $(\Delta p)_{h,t}/(\Delta p)_{v,t}$ values due to non-turbulence for different k_v/k_h ratios are

1. 0.0009 if $k_v/k_h = 1$
2. 0.0043 if $k_v/k_h = 0.1$
3. 0.016 if $k_v/k_h = 0.01$

As expected, reduction in turbulence-related pressure drop by drilling a horizontal well is highest in a high vertical permeability reservoir, and is lowest in a low vertical permeability reservoir.

PRODUCTION TYPE CURVES

Recently Duda[10] and Aminian and Ameri[11] reported production type curves for horizontal gas wells. Type curves for two dimensionless wellbore radius, namely, $r_{wd} = 2.5 \times 10^{-4}$ and 5.0×10^{-4}, where $r_{wd} = \sqrt{4r_w^2/L^2}$. The type curves shown in Figures 9–9 through 9–15 are for production in 320- and 640-acre reservoirs for a square drainage area with no flow across the drainage boundary. Additionally, production type curves are also reported for an infinite reservoir. In the type curves various dimensionless terms are defined as

$$L_D = [L/(2h)]\sqrt{k_v/k_h} \quad (9\text{–}50)$$

$$G_{pD} = 9.009 G_p T/[h\phi\mu c_t L^2\, \Delta m(p)] \quad (9\text{–}51)$$

$$t_D = 0.001055\, kt/(\phi\mu c_t L^2) \quad (9\text{–}52)$$

where k is permeability in md, G_p is cumulative gas in Mcf and time t is in hours. Additionally, well length L is in ft, viscosity μ is in cp, reservoir thickness is in ft, and reservoir temperature, T, is in °R. These production-type curves can either be used for production feasibility or for estimating reservoir properties from the production histories. It is important to note that these type curves are developed for a constant bottomhole flowing pressure.

Figure 9–15 shows a production-type curve in an anisotropic reservoir, where a horizontal well is drilled along the x direction. The figure clearly indicates the benefits of drilling a horizontal well along the low-permeability direction. The figure also notes loss in production by drilling a horizontal well along the high-permeability direction.

The production-type curve discussion so far was either for infinite reservoirs or for square drainage areas. Production-type curves also have been developed for rectangular drainage areas.[11] As shown in Figure 9–16, a rectangular drainage area initially shows higher cumulative production

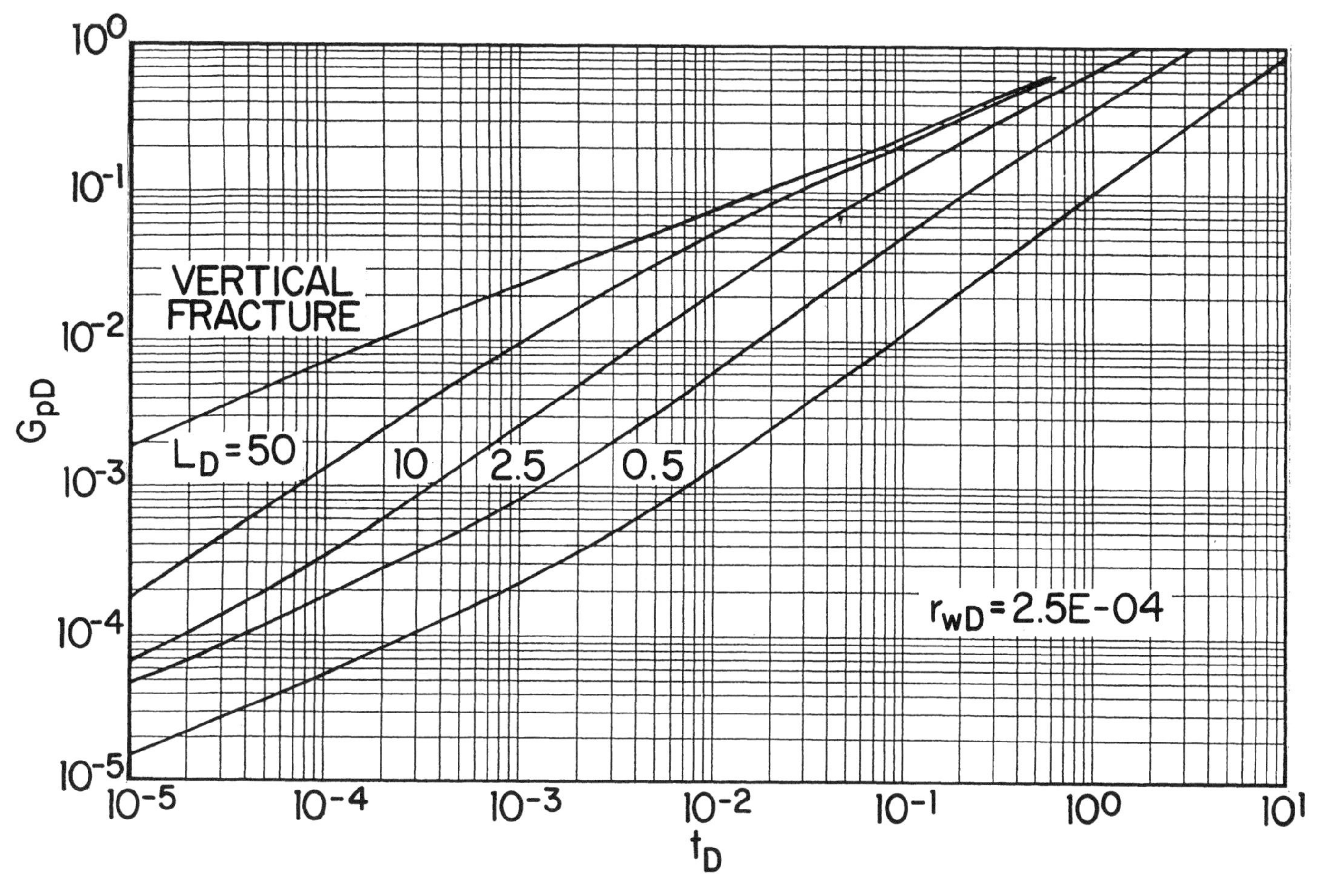

Figure 9–10 A Production Type Curve for Horizontal Gas Wells in a 640-Acre Drainage Area.[10]

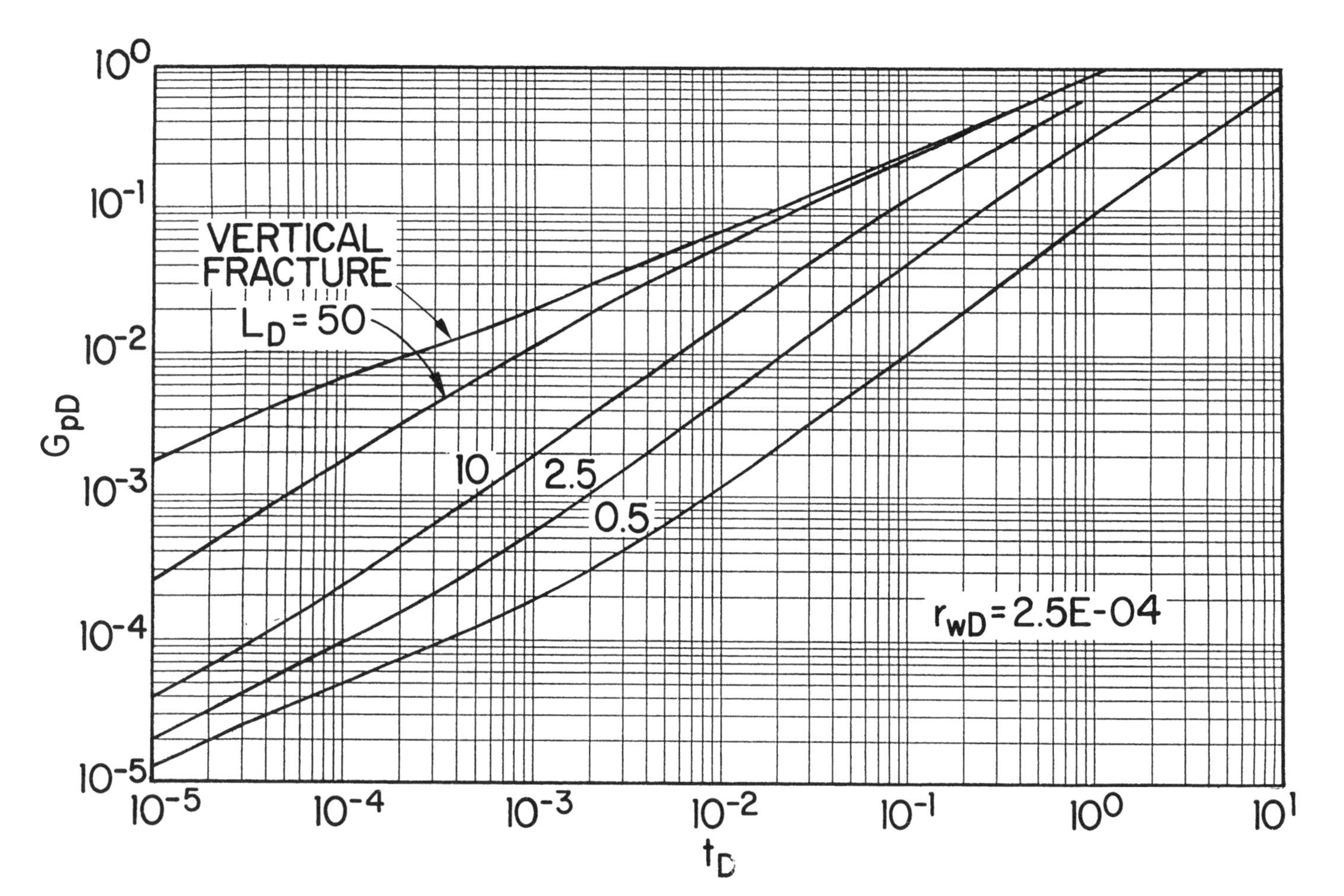

Figure 9–11 A Production Type Curve for Horizontal Gas Wells in a 320-Acre Drainage Area.[10]

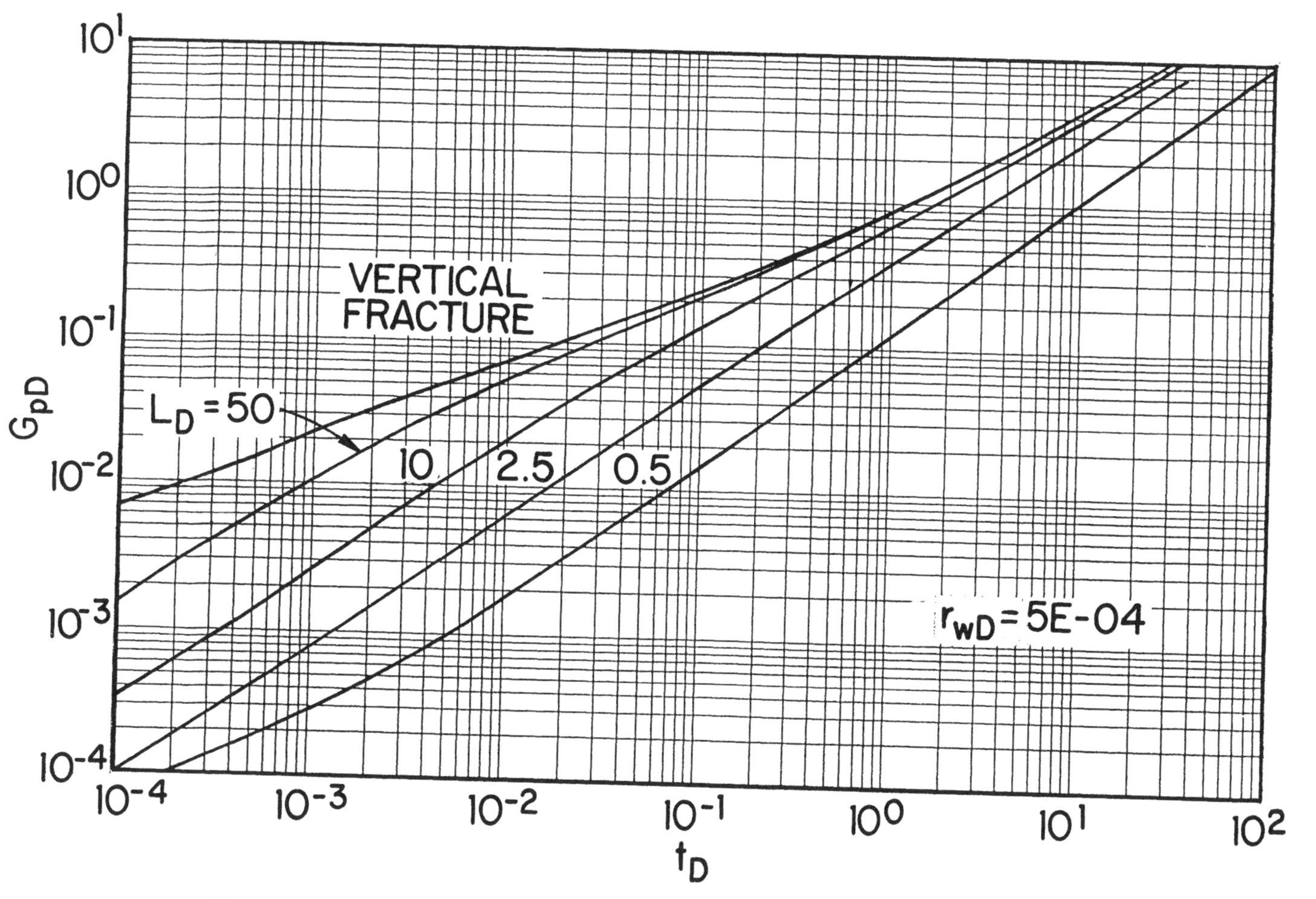

Figure 9–12 A Production Type Curve for Horizontal Gas Wells in an Infinite Reservoir.[10]

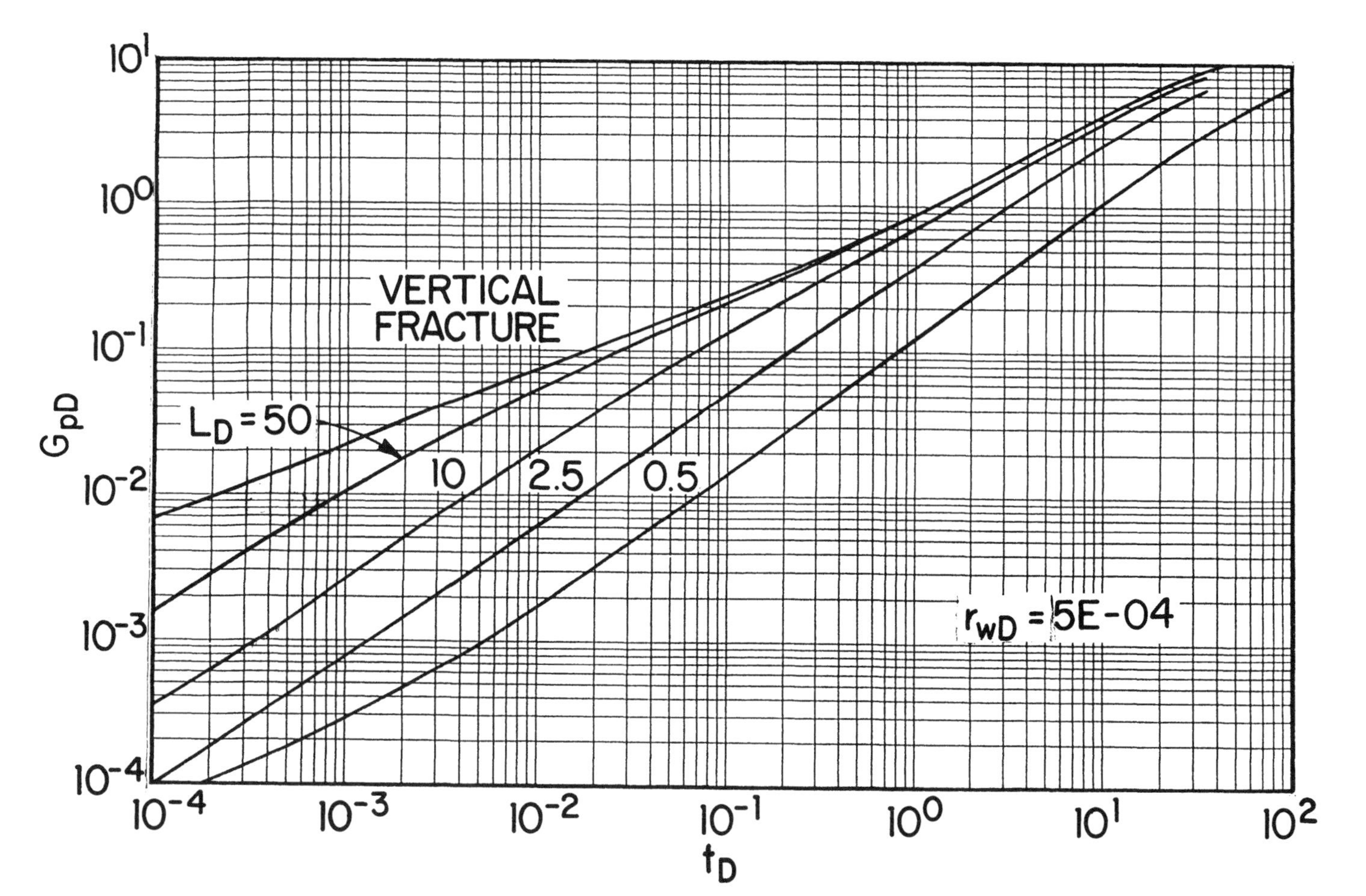

Figure 9–13 A Production Type Curve for Horizontal Gas Wells in a 640-Acre Drainage Area.[10]

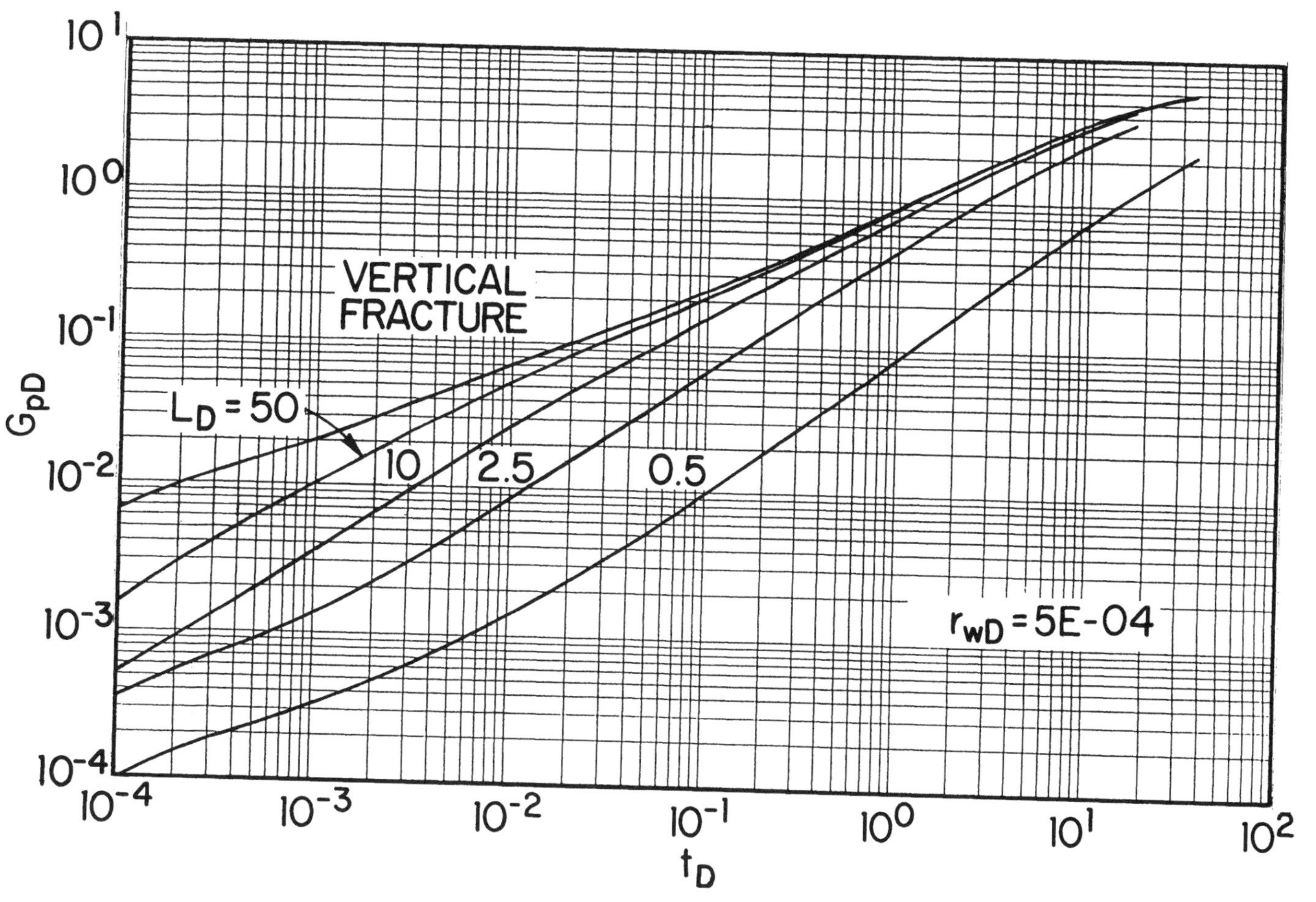

Figure 9–14 A Production Type Curve for Horizontal Gas Wells in a 320-Acre Drainage Area.[10]

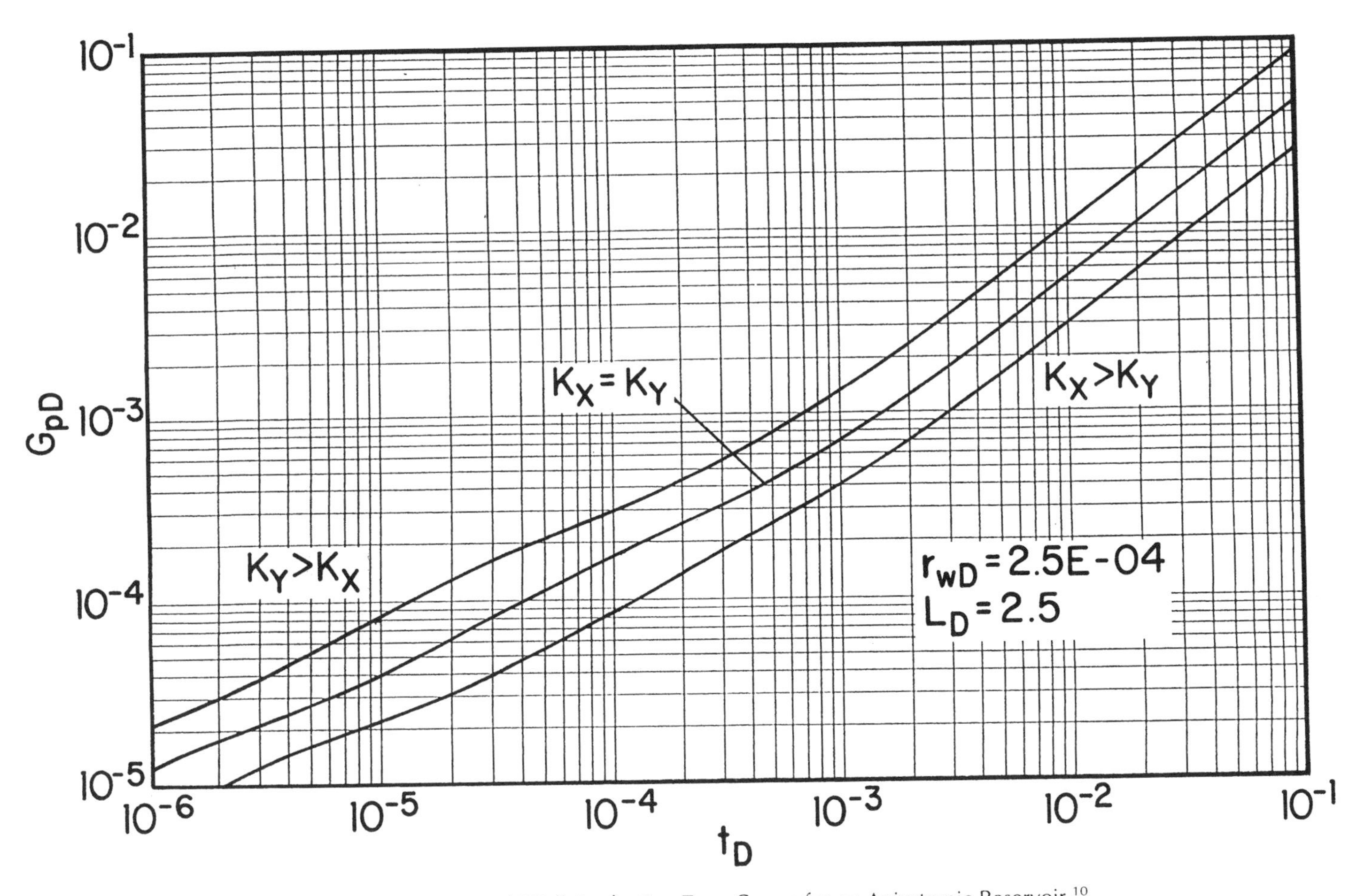

Figure 9–15 Horizontal Well Production Type Curves for an Anisotropic Reservoir.[10]

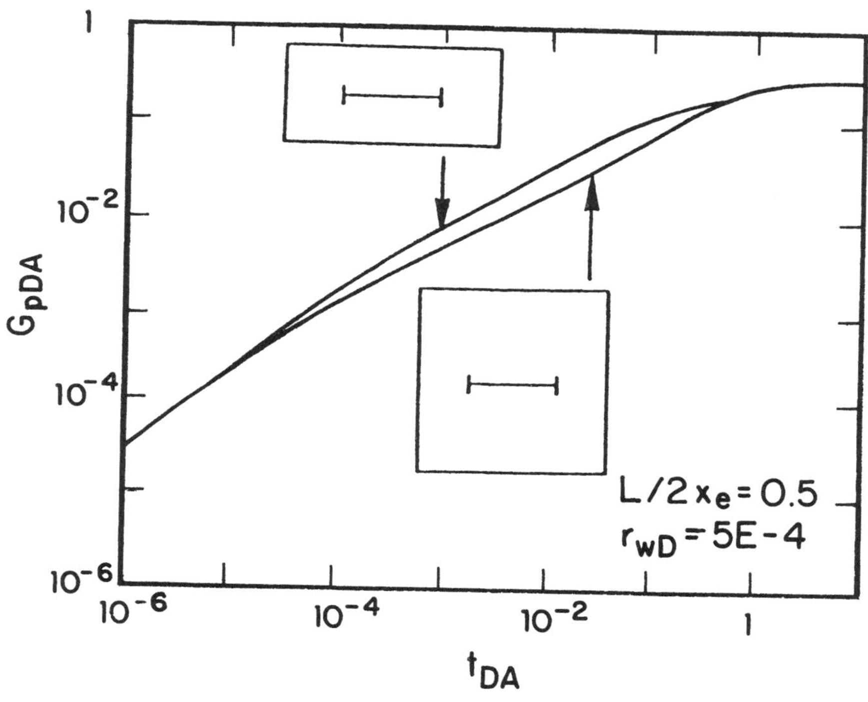

Figure 9–16 A Production Type Curve for a Horizontal Well in a Rectangular Reservoir.[11]

than a square drainage area. In Figure 9–16, the dimensionless time, t_{DA}, is defined as

$$t_{DA} = \left[\frac{0.001055\, k_h}{\phi \mu c_t A}\right] t \tag{9–53}$$

Dimensionless cumulative production based on area is defined as

$$G_{pDA} = \left[\frac{36\, T}{h \phi \mu c_t A\, m(\bar{p})}\right] G_p \tag{9–54}$$

k_h = horizontal permeability, md
ϕ = porosity, fraction
μ = viscosity, cp
$\bar{p}$ = reservoir pressure, psi
c_t = total compressibility, 1/psi

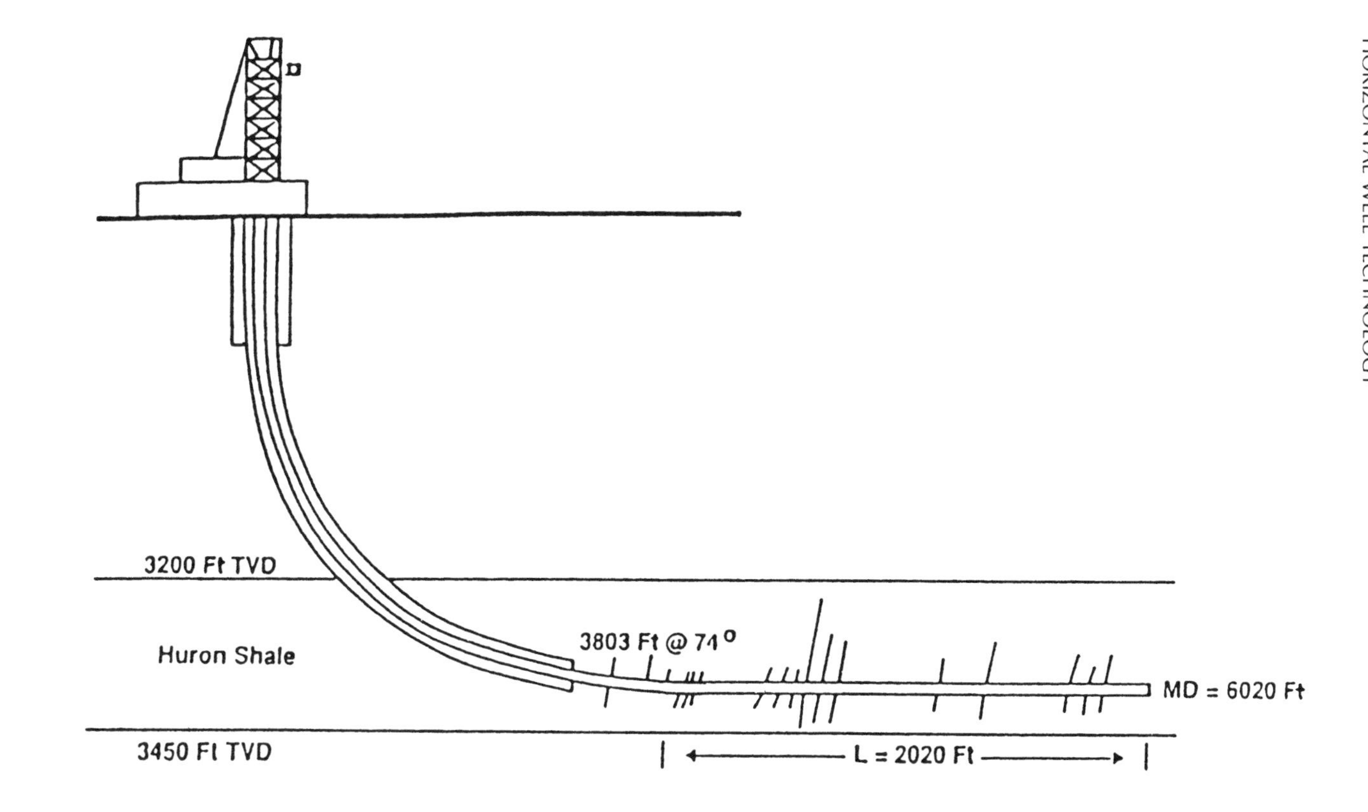

Figure 9–17 A Schematic View of Horizontal Well Ret #1, Huron Shale.[11]

A = Drainage area, ft^2
$m(\bar{p})$ = pseudo-pressure, psi^2/cp
t = time, hours
G_p = cumulative production, Mscf
T = temperature, degrees Rankin

CASE HISTORIES

This section includes discussion of two case histories. One case is for a low-permeability reservoir, while the other is for a high-permeability reservoir. These two cases demonstrate horizontal well applications in low-, as well as high-permeability reservoirs.

HURON SHALE, WEST VIRGINIA, U.S.A.

In 1985, U.S. Department of Energy (USDOE) sponsored a drilling of a horizontal well in the Devonian shale formation in southwest West Virginia. The 2020-ft-long well, Ret #1, was drilled in the lower Huron Shale formation. The shale formation exhibits very low permeability, about 0.03 md, (about 30 microdarcies) and also exhibits a low porosity, typically 2 to 3%. The reservoir is naturally fractured. Figure 9–17 shows the well schematic and Figure 9–18 shows a match between the cumulative production and

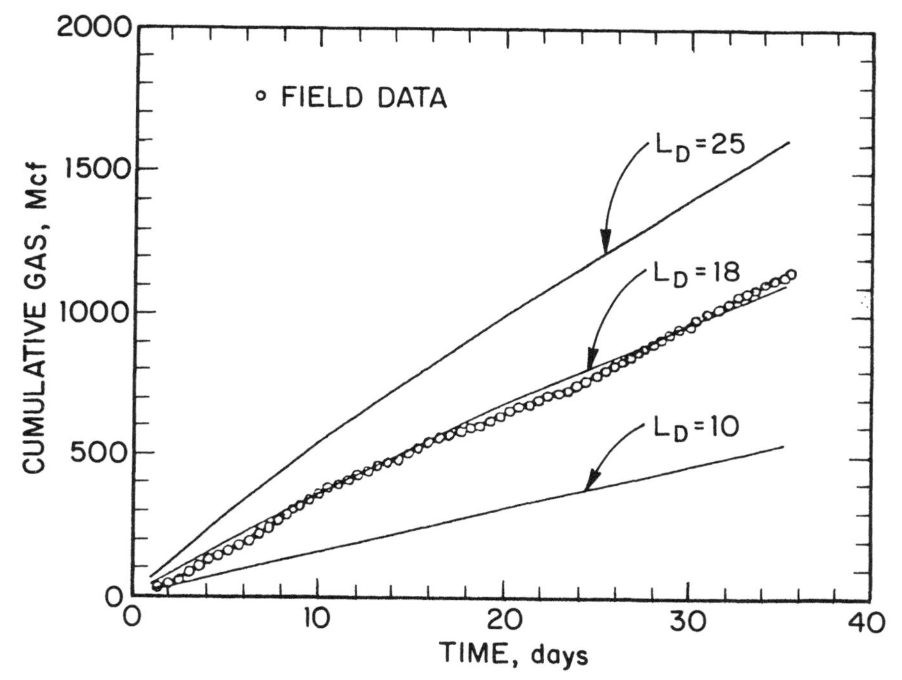

Figure 9–18 A Type Curve Match of Production Data from Well Ret #1.[11]

the well production for about 35 days. A fairly good match is obtained for $L_D = 18$, where dimensionless length L_D is defined as

$$L_D = [L/(2h)] \sqrt{k_v/k_h}$$

The match was obtained assuming uniform horizontal permeability (no areal anisotropy) and vertical permeability equal to the horizontal permeability.[12] In a fractured reservoir, vertical permeability is probably equal to the horizontal permeability. Nevertheless, in practice, one has to estimate vertical permeability and net reservoir thickness. As noted in Reference 11, it is difficult to precisely estimate these parameters. Additionally, as noted in Reference 11, the mathematical model used to generate cumulative production-type curves is based upon a single porosity reservoir assumption, while the Huron Shale Formation is naturally fractured. In spite of these discrepancies, a fairly good match with the production data is seen in Figure 9–18. The other reservoir parameters are listed in Table 9–6.

It is interesting to note that a 2020-ft-long horizontal well produced about 1300 Mscf in about 35 days, giving an average daily production rate of about 37 Mscfd. At this rate, a horizontal well is uneconomical and requires stimulation to enhance production. This well was later stimulated to enhance production.[12] The well stimulation results are shown in Figure 9–6 and Table 9–1.

ZUIDWAL FIELD, THE NETHERLANDS

Zuidwal Field is in The Netherlands, in the middle of Waddenzee, an inland sea. Thus, although the field is "onshore," it is developed using a platform.[9] The reservoir is 6037 ft deep and consists of layers of clay and sand. The overall thickness is about 328 ft with average porosity of about 10 to 15% and reservoir permeability of 1 to 10 md. As noted earlier, a permeability on the order of 5 md constitutes high permeability for a gas well. In these reservoirs, near-well turbulence restricts productivity of vertical wells.

TABLE 9–6 BASIC WELL AND RESERVOIR DATA FOR RET #1[11]

Well Length	2020 ft
Well Radius	0.328 ft
Gross Pay Thickness	250 ft
Porosity	2%
Permeability	0.03 md
Reservoir Pressure	200 psia
Reservoir Temperature	553°R
Flowing Well Pressure	45 psia
Gas Gravity (air = 1.0)	0.72

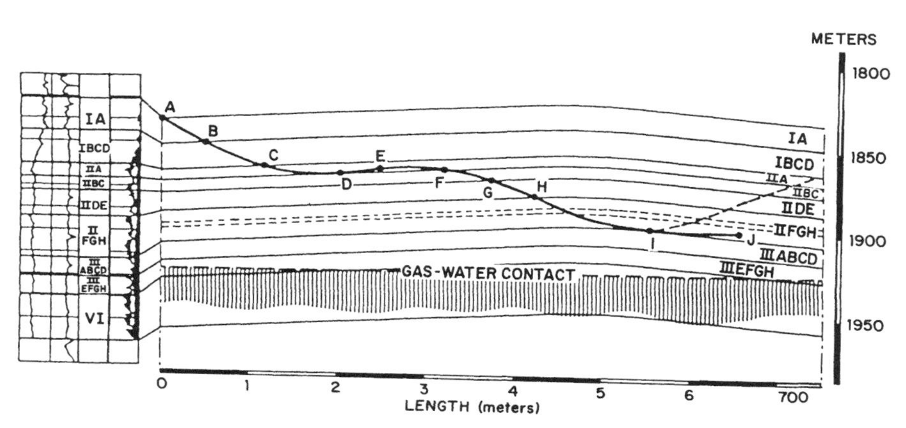

Figure 9–19 A Proposed Horizontal Well Profile in Zuidwal Field.[9]

The main productive zones are 20-ft-thick zones. These zones, IIA and IIF, are located at different elevations. Therefore, higher angle wells (angle about 85°) were drilled to contact these zones. The drilling, completion, and reservoir details for three horizontal wells are given in Table 9–7.

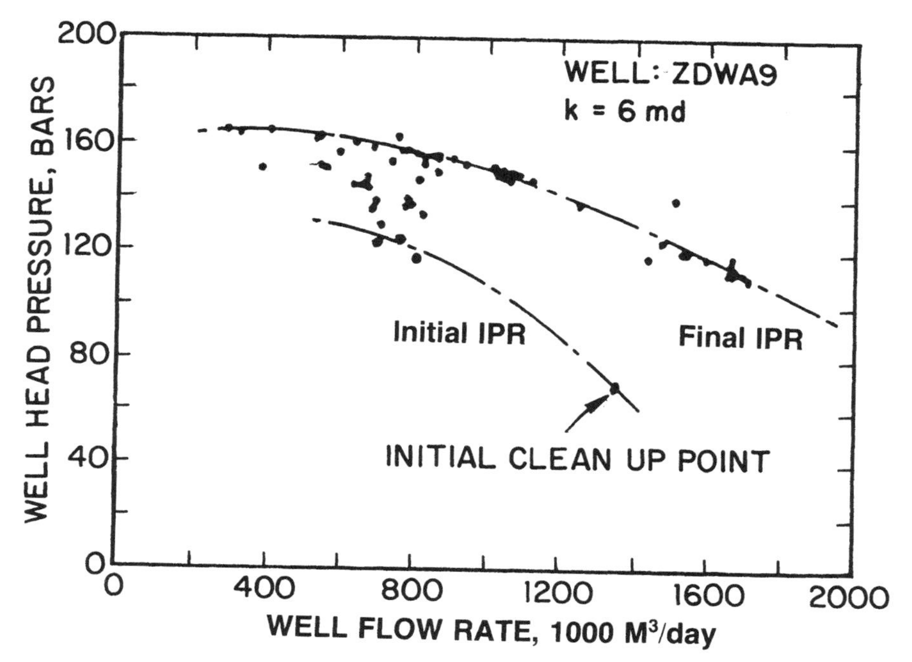

Figure 9–20 A Horizontal Well Production Profile Showing Mud Clean-Up Effects, Zuidwal Field, Netherlands.[9]

TABLE 9–7 COMPLETION, LOGGING AND RESERVOIR DATA FOR ZUIDWAL FIELD HORIZONTAL WELLS[9]

		ZDWA6	ZDWA8	ZDWA9	ZDWA4 (DEVIATED)
DRILLING	Completion date	4/24/1988	1/23/1989	1/17/1989	2/17/1988
	Total drilling and compl. days	66	69	62	59
	Vertical depth (meters)	1896	1900	1900	2108
	Departure from vertical (meters)	1507	1942	1022	1725
	Maximum inclination (degrees)	91	96.2	93.4	52
	Maximum buildup rate (deg/10m)	2.1	3.7	3.5	1.3
	Top drive	Yes	Yes	Yes	Yes
	Type of mud in reservoir	Polymer	Polymer	Polymer	Polymer
COMPLETION	open hole log running tool	TLC Simphor	TLC Simphor	TLC Simphor	Wire line
	open hole logs completed	1:LDT,CNL,NGT DIL,SLS,GPIT	1:LDT,CNL,NGT DIL,SLS,GPIT 2:FMS,GR	1:DIL,SLS,GR,GPIT 2:LDT,CNL,NGT,DIL SLS,GPIT	1:LDT,CNL,NGT,DIL SLS. 2:EVA.
	Completion type	Slotted liner	Slotted liner	Slotted liner	TCP Perforations
	Liner diameter	7 in.	7 in.	$4\frac{1}{2}$ in.	Casing 7 in.
	Tubing diameter	$5\frac{1}{2}$ in.	$4\frac{1}{2}$ in.	$4\frac{1}{2}$ in.	$4\frac{1}{2}$ in.
	Production logging running tool	Coiled tubing	Coiled tubing	Coiled tubing	Wire line
	Production logging completed	PLT:22 stations	PLT:19 stations + 1 run	PLT:16 stations + 1 run	PLT:6 runs TDP-P

R	Total reservoir thickness (meters)	43	62	56	57.9
E	thickness IIA + IIF (meters)	10.2	12.0	9.80	11.5
S E	drilled in the reservoir (meters)	665	605	546.5	70.5
R V	drilled in IIA + IIF (meters)	375	235	190.5	14.0
O	Ratio length IIA + IIF/total length	.564	.39	.349	.190
I R	AOFP (Std M3/d)	4814100	5319000	4158300	1376800

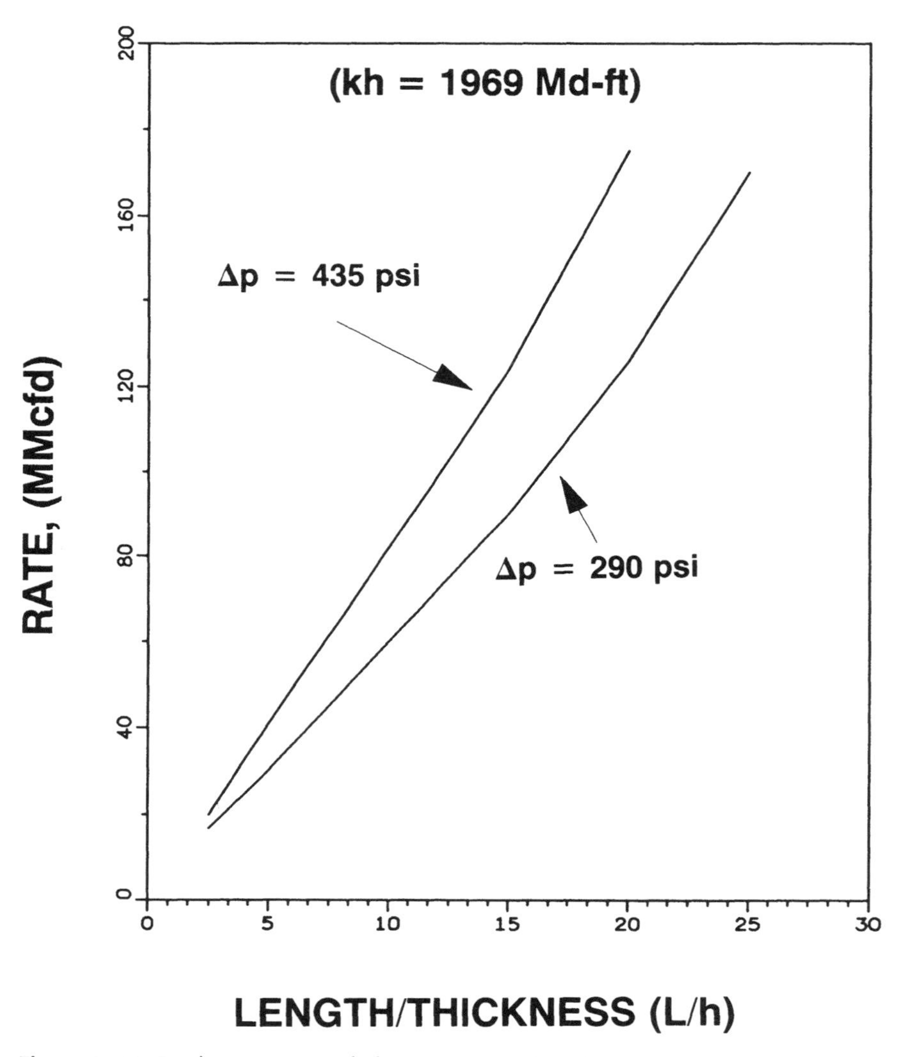

Figure 9–21 Production Data of Three Horizontal Wells Showing the Relationship between Well Length and Production Rate.[9]

A typical well profile is shown in Figure 9–19. It is interesting to note that the wells were drilled using polymer mud and were completed using slotted liner. As noted in Chapter 2, in such a high-permeability reservoir, the expected drilling damage is relatively small as compared to one in a tight zone. Even then, wells seem to exhibit a limited damage, despite polymer mud. The wells initially show progressive increase in well productivity over time, indicating well clean up (see Fig. 9–20). The horizontal wells here were found to reduce near-wellbore turbulence and enhance well productivity. Length versus rate for three horizontal wells is shown in Figure 9–21.

SUMMARY

Chapter 9 includes a discussion of horizontal wells in gas reservoirs. The discussion and case histories presented indicate the advantages of horizontal wells in low-, as well as high-permeability reservoirs. In low-permeability reservoirs, horizontal wells enhance the drainage area in a given time period. In high-permeability reservoirs, horizontal wells reduce near-wellbore turbulence and enhance well deliverability. Horizontal wells have high potential in gas reservoirs.

REFERENCES

1. Smith, R. V.: *Practical Natural Gas Engineering,* Tulsa, Oklahoma, PennWell Publishing Co., 1983.
2. Brown, K. E.: *The Technology of Artificial Methods,* Tulsa, Oklahoma, PennWell Publishing Co., 1984.
3. Burns, J. R. and Fetkovich, M. J., and Meitzen, V. C.: "The Effect of Water Influx on *p/z*—Cumulative Gas Production Curves," *Journal of Petroleum Technology,* p. 287, March 1965.
4. Agarwal, R. G. and Al-Hussainy, R. and Ramey Jr., H. J.: "The Importance of Water Influx in Gas Reservoirs," *Trans.* AIME, v. 234, p. 1336, 1965.
5. Al-Hussainy, R., Ramey Jr., H. J., and Crawford, P. B.: "The Flow of Real Gases Through Porous Media," *Journal of Petroleum Technology,* pp. 624–636; *Trans.,* AIME, v. 237, 1966.
6. Golan, M. and Whitson, C. H.: *Well Performance,* International Human Resources Development Corporation, Boston, 1986.
7. Yost II, A. B. and Overbey Jr., W. K. and Wilkens, D. A. and Locke, C. D.: "Hydraulic Fracturing of a Horizontal Well in a Naturally Fractured Reservoir: Gas Study for Multiple Fracture Design," paper SPE 17759, presented at the SPE Gas Technology Symposium, Dallas, Texas, June 13–15, 1988.
8. Overbey Jr., W. K., Yost II, A. B., and Wilkins, D. A.: "Inducing Multiple Hydraulic Fractures From a Horizontal Wellbore," paper SPE 18249, presented at the SPE 63rd Technical Conference and Exhibit Annual Meeting, Houston, Texas, Oct. 2–5, 1988.
9. Celier, G. C. M. R., Jouault, P., de Montigny, O. A. M. C.: "Zuidwal: A Gas Field Development With Horizontal Wells," paper SPE 19826, presented at the SPE 64th Annual Technical Conference and Exhibition of the Society of Petroleum Engineers, San Antonio, Texas, Oct. 8–11, 1989.
10. Duda, J. R.: "Type Curves For Predicting Production Performance From Horizontal Wells in Low Permeability Gas Reservoirs," paper SPE 18993, Richardson, Texas.
11. Aminian, K. and Ameri, S.: "Predicting Horizontal Well Production Performance Using Type Curves," paper SPE 19342 presented at the SPE Eastern Regional Meeting, Morgantown, West Virginia, Oct. 24–27, 1989.
12. Yost, Albert B.: "Horizontal Gas Well Promises More Devonian Production," *American Oil & Gas Reporter,* July 1988.

CHAPTER 10

Pressure Drop Through a Horizontal Well

INTRODUCTION

From the reservoir engineering standpoint, a horizontal wellbore is considered as an infinite-conductivity fracture, i.e., the pressure drop along the well length is very small and is negligible. Thus, the horizontal well represents a long wellbore where well pressure throughout the wellbore is constant. In practice, some pressure drop from the tip of the horizontal wellbore to the producing end is essential to maintain fluid flow within the wellbore. Consequently, the producing end of the horizontal well will be at a lower pressure than the other tip (see Fig. 10–1). Nevertheless, from an engineering standpoint, the question to be addressed is, What is the expected pressure drop along the well length? What is the magnitude of wellbore pressure drop, Δp, as compared to the pressure drop from the reservoir to the wellbore, $\bar{p}$-p_{wf}. If the horizontal wellbore pressure drop is very small as compared to the pressure drawdown from the reservoir to the wellbore,

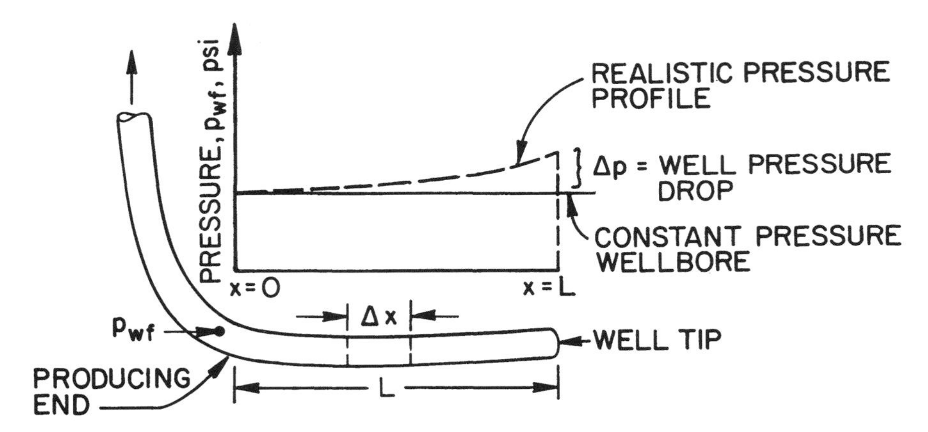

Figure 10–1 A Schematic Diagram of Pressure Loss along the Well Length.

then for all practical purposes, a horizontal well can be considered as an infinite-conductivity wellbore, i.e., a wellbore at a constant pressure. In contrast, if the pressure drop through the wellbore is significant as compared to the reservoir drawdown, then reservoir drawdown along the well length would change, and therefore, production along the well length would also change. To calculate the changing production rate along the well length, one will have to simultaneously solve pressure drop along the pipe equations with reservoir flow equations as shown in Figure 10–2.

In general, pressure drop along the well length is very small and can be ignored. However, under certain circumstances, such as those involving high flow rates of light oil (greater than 10,000 RB/day) or flow of highly viscous crudes, it is possible to have a large pressure drop in the wellbore.

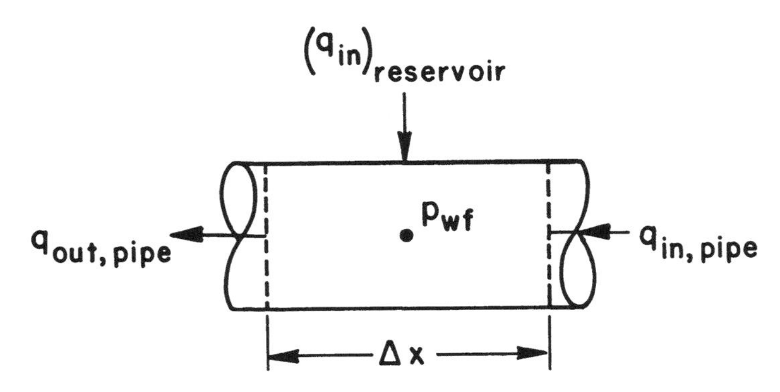

Figure 10–2 A Schematic Diagram of Pressure Loss and Flow Relationships between Reservoir and Pipe Flow.

In these situations, an optimum well length can be calculated by considering pipe pressure drops and reservoir production rates for the changing drawdown along the well length.

In practice it is important to estimate pressure drop along the well length. This pressure drop not only has impact on the production behavior of the well, but also influences well completion and well profile design. In the following sections this is discussed in detail.

INFLUENCE OF HIGH PRESSURE DROPS

A high pressure drop along the horizontal well length is possible in the case of high viscosity fluids such as heavy oils and tar sands. High pressure drop is also possible for light oils if flow rates are in excess of a few thousand barrels per day, i.e., flow rates on the order of 10,000 to 30,000 RB/day. Such high rates are possible only in high-permeability reservoirs where permeability is on the order of 1000 md or more. In these reservoirs, pressure drawdown from the reservoir to the wellbore can be very small, and can be comparable to pressure drop through the horizontal wellbore. In such cases, beyond a certain length, drilling a longer well would not yield any additional production.[1]

In reservoirs with gas and water coning problems, an excessive pressure drop through the wellbore may enhance the tendency of gas and water to cone rapidly at a point of minimum pressure in the horizontal well, i.e., at the producing end of the wellbore.

As noted earlier, a large pressure drop through the horizontal section would occur mainly in high-permeability reservoirs. In such reservoirs, flow rates are not restricted by well productivity, but rather by flow-string pressure drop limitations. In such high-productivity reservoirs, the reason for drilling a horizontal well should be critically reviewed. In high-permeability reservoirs, water and gas coning problems are minimal. In these reservoirs, a horizontal well can be drilled to reduce coning and enhance oil cuts, but the gain in performance by drilling horizontal wells instead of vertical wells will not be as significant as in a low-permeability reservoir (see Chapter 8).

REMEDIES TO MINIMIZE HIGH WELLBORE PRESSURE DROPS

Several different steps can be taken to minimize pressure drop through the wellbore.

1. High pressure drop occurrence is mainly due to turbulent flow in the wellbore. To minimize the wellbore pressure drop, it is desirable to have laminar flow through the wellbore, or at least to have the minimum

possible flow velocities through the wellbore. One way is to consider drilling the largest possible size hole. For example, in the case of a medium-radius well, one can drill as small as a 4½-in. hole to as large as 9⅞-in. hole. In the case of a long-radius well, one can even drill a 12¼-in. hole. After choosing the largest possible hole size, one can also choose the largest possible liner sizes that can be safely inserted in a hole without getting stuck (see Tables 10–1 and 10–2).[2,3] For a given production rate, by increasing the well diameter twofold, the pressure drop can be reduced by at least thirty-two-fold. This is because, at least for single phase flow, the pressure drop is inversely proportional to the fifth power of the diameter ($\Delta p \propto 1/d^5$).

TABLE 10–1 CASING DIMENSIONS[3]

O.D. (in)	Weight (lbm/ft)	I.D. (in)
4	5.65	3.607
4	9.50	3.500
4	11.60	3.428
4½	6.75	4.216
4½	9.50	4.090
4½	10.50	4.052
4½	11.00	4.026
4½	11.60	4.000
4½	12.60	3.958
4½	13.50	3.920
4½	15.10	3.826
4½	16.60	3.754
4½	18.80	3.640
4¾	9.50	4.364
4¾	16.00	4.082
4¾	18.00	4.000
5	8.00	4.696
5	11.50	4.560
5	13.00	4.494
5	15.00	4.408
5	18.00	4.276
5	20.30	4.184
5	20.80	4.156
5	21.00	4.154
5	23.20	4.044
5	24.20	4.000

TABLE 10–1 (Continued)

$5\frac{1}{4}$	8.50	4.944
$5\frac{1}{4}$	10.00	4.886
$5\frac{1}{4}$	13.00	4.768
$5\frac{1}{4}$	16.00	4.648
$5\frac{1}{2}$	9.00	5.192
$5\frac{1}{2}$	13.00	5.044
$5\frac{1}{2}$	14.00	5.012
$5\frac{1}{2}$	15.00	4.974
$5\frac{1}{2}$	15.50	4.950
$5\frac{1}{2}$	17.00	4.892
$5\frac{1}{2}$	20.00	4.778
$5\frac{1}{2}$	23.00	4.670
$5\frac{1}{2}$	25.00	4.580
$5\frac{1}{2}$	26.00	4.548
$5\frac{3}{4}$	14.00	5.290
$5\frac{3}{4}$	17.00	5.190
$5\frac{3}{4}$	19.50	5.090
$5\frac{3}{4}$	20.00	5.090
$5\frac{3}{4}$	22.50	4.990
$5\frac{3}{4}$	25.20	4.890
6	10.50	5.672
6	12.00	5.620
6	15.00	5.524
6	16.00	5.500
6	17.00	5.450
6	18.00	5.424
6	20.00	5.352
6	23.00	5.000
6	26.00	5.140
$6\frac{5}{8}$	12.00	6.287
$6\frac{5}{8}$	13.00	6.255
$6\frac{5}{8}$	17.00	6.135
$6\frac{5}{8}$	20.00	6.049
$6\frac{5}{8}$	22.00	5.989
$6\frac{5}{8}$	24.00	5.921
$6\frac{5}{8}$	26.00	5.855
$6\frac{5}{8}$	28.00	5.791
$6\frac{5}{8}$	29.00	5.761
$6\frac{5}{8}$	32.00	5.675
$6\frac{5}{8}$	34.00	5.595
7	13.00	6.520

TABLE 10–1 (Continued)

7	17.00	6.538
7	20.00	6.456
7	22.00	6.398
7	23.00	6.366
7	24.00	6.336
7	26.00	6.276
7	28.00	6.214
7	29.00	6.184
7	30.00	6.154
7	32.00	6.094
7	33.70	6.048
7	35.00	6.004
7	38.00	5.920
7	40.00	5.836
7⅝	14.75	7.263
7⅝	20.00	7.125
7⅝	24.00	7.025
7⅝	26.40	6.969
7⅝	29.70	6.875
7⅝	33.70	6.765
7⅝	39.00	6.625
7⅝	45.00	6.445
7⅝	45.30	6.435
8	16.00	7.628
8	20.00	7.528
8	26.00	7.386
8⅛	28.00	7.485
8⅛	32.00	7.385
8⅛	35.50	7.285
8⅛	39.50	7.185
8⅛	42.00	7.125
8⅝	20.00	8.191
8⅝	24.00	8.097
8⅝	28.00	8.017
8⅝	32.00	7.921
8⅝	36.00	7.825
8⅝	38.00	7.775
8⅝	40.00	7.725
8⅝	43.00	7.651
8⅝	44.00	7.625
8⅝	48.00	7.537

TABLE 10–1 (Continued)

$8\frac{5}{8}$	49.00	7.511
9	34.00	8.290
9	38.00	8.196
9	40.00	8.150
9	45.00	8.032
9	50.20	7.910
9	55.00	7.812
$9\frac{5}{8}$	29.30	9.063
$9\frac{5}{8}$	32.30	9.001
$9\frac{5}{8}$	36.00	8.921
$9\frac{5}{8}$	40.00	8.835
$9\frac{5}{8}$	43.60	3.775
$9\frac{5}{8}$	47.00	8.681
$9\frac{5}{8}$	53.50	8.535
$9\frac{5}{8}$	58.40	8.435
$9\frac{5}{8}$	61.10	8.375
$9\frac{5}{8}$	71.80	8.125
10	33.00	9.384
10	41.50	9.200
10	45.50	9.120
10	50.50	9.016
10	55.50	8.908
10	61.20	8.690
$10\frac{3}{4}$	32.75	10.192
$10\frac{3}{4}$	35.75	10.136
$10\frac{3}{4}$	40.50	10.050
$10\frac{3}{4}$	45.50	9.950
$10\frac{3}{4}$	51.00	9.850
$10\frac{3}{4}$	54.00	9.784
$10\frac{3}{4}$	55.00	9.760
$10\frac{3}{4}$	60.70	9.660
$10\frac{3}{4}$	65.70	9.560
$10\frac{3}{4}$	71.10	9.450
11	26.75	10.552
$11\frac{3}{4}$	38.00	11.150
$11\frac{3}{4}$	42.00	11.084
$11\frac{3}{4}$	47.00	11.000
$11\frac{3}{4}$	54.00	10.880
$11\frac{3}{4}$	60.00	10.772
$11\frac{3}{4}$	65.00	10.682
12	31.50	11.514

TABLE 10–1 (Continued)

12	40.00	11.384
$12\frac{3}{4}$	43.00	12.130
$12\frac{3}{4}$	53.00	11.970
13	36.50	12.482
13	40.00	12.438
13	45.00	12.360
13	50.00	12.282
13	54.00	12.220
$13\frac{3}{8}$	48.00	12.715
$13\frac{3}{8}$	54.50	12.615
$13\frac{3}{8}$	61.00	12.515

TABLE 10–2 FLOW STRING WEIGHTS AND SIZES[2]

OD (in)	Nominal Size (in)	Weight (lbm/ft)	ID (in)
1.660	$1\frac{1}{4}$	2, 3 or 2.4	1.380
1.900	$1\frac{1}{2}$	2, 3 or 2.748	1.610
2.375	2	4.00	2.041
2.375	2	2.5 or 4.7	1.995
2.875	$2\frac{1}{2}$	5.897	2.469
2.875	$2\frac{1}{2}$	6.25 or 6.5	2.441
3.500	3	7.694	3.068
3.500	3	8.50	3.018
3.500	3	9.30	2.992
3.500	3	10.2	2.922
4.000	$3\frac{1}{2}$	9.26 or 9.50	3.548
4.000	$3\frac{1}{2}$	11.00	3.476
4.500	4	10.98	4.026
4.500	4	11.75	3.990
4.500	4	12.75	3.958
4.750	$4\frac{1}{2}$	16.00	4.082
4.750	$4\frac{1}{2}$	16.50	4.070
5.000	$4\frac{3}{4}$	12.85	4.500
5.000	$4\frac{3}{4}$	13.00	4.494
5.000	$4\frac{3}{4}$	15.00	4.408
5.000	$4\frac{3}{4}$	18.00	4.276
5.000	$4\frac{3}{4}$	21.00	4.154

TABLE 10–2 (Continued)

5.250	—	16.00	4.648
5.500	$5\frac{3}{16}$	17.00	4.892
5.500	$5\frac{3}{16}$	20.00	4.778
5.750	—	14.00	5.290
5.750	—	17.00	5.190
5.750	—	19.50	5.090
5.750	—	22.50	4.990
6.000	$5\frac{5}{8}$	20.00	5.350
6.625	$6\frac{1}{4}$	20.00	6.049
6.625	$6\frac{1}{4}$	24.00	5.921
6.625	$6\frac{1}{4}$	26.00	5.855
6.625	$6\frac{1}{4}$	28.00	5.791
6.625	$6\frac{1}{4}$	29.00	5.761
7.000	$6\frac{5}{8}$	20.00	6.456
7.000	$6\frac{5}{8}$	22.00	6.398
7.000	$6\frac{5}{8}$	24.00	6.336
7.000	$6\frac{5}{8}$	26.00	6.276
7.000	$6\frac{5}{8}$	28.00	6.214
7.000	$6\frac{5}{8}$	30.00	6.154
7.625	—	34.00	6.765
8.000	$7\frac{5}{8}$	26.00	7.386
8.125	—	28.00	7.485
8.125	—	32.00	7.385
8.125	—	35.50	7.285
8.125	—	39.5 or 40.00	7.185
8.125	—	42.00	7.125
8.625	$8\frac{1}{4}$	24.00	8.097
8.625	$8\frac{1}{4}$	28.00	8.017
8.625	$8\frac{1}{4}$	32.00	7.921
8.625	$8\frac{1}{4}$	32.00	7.907
8.625	$8\frac{1}{4}$	36.00	7.825
8.625	$8\frac{1}{4}$	38.00	7.775
8.625	$8\frac{1}{4}$	43.00	7.651
8.625	$8\frac{1}{4}$	44.85	7.625
9.000	$8\frac{5}{8}$	34.00	8.290
9.000	$8\frac{5}{8}$	38.00	8.196
9.000	$8\frac{5}{8}$	40.00	8.150
9.000	$8\frac{5}{8}$	45.00	8.032
9.000	$8\frac{5}{8}$	54.00	7.812
9.625	—	43.80	8.755
9.625	—	47.20	8.681

TABLE 10–2 (Continued)

9.625	—	53.60	8.535
9.625	—	57.40	8.451
9.625	$9\frac{1}{4}$	36.00	8.921
10.000	$9\frac{5}{8}$	33.00	9.384
10.000	$9\frac{5}{8}$	60.00	8.780
10.750	10	32.75	10.192
10.750	10	35.75	10.136
10.750	10	40.00	10.054
10.750	—	40.50	10.050
10.750	10	45.00	9.960
10.750	—	45.50	9.950
10.750	10	48.00	9.902
10.750	—	51.00	9.850
10.750	10	54.00	9.784

2. Pressure drop along the well length can be minimized by controlling fluid production rates along the well length. This can be accomplished by manipulating the area open for fluid entry into a wellbore. If the well is to be completed using a slotted or predrilled liner, one can vary the hole or slot sizes along the well length so as to minimize pressure drop along the well length. In the case of a cemented hole, one can change not only shot density, but also perforated interval length to minimize pressure drop along the length.
3. In a high-permeability formation, where pressure drop through a horizontal wellbore is comparable to the reservoir pressure drawdown, a gravel pack will probably be used to complete the well. In such cases, if the well is completed with a perforated liner, then fluid entry points into the wellbore, i.e., slots, should be placed as far apart as possible. This will let the gravel pack act as a "choke" for each slot and facilitate maintaining minimum pressure drops along the well length.

Thus, if wellbore pressure drop is found excessive, then in the well planning stage an appropriate completion scheme can be designed to minimize wellbore pressure drop. Hence, before finalizing horizontal-well drilling and completion plans, it may be worthwhile to calculate wellbore pressure drop.

In the following section, procedures are suggested to calculate pressure drops in horizontal wellbores.

PRESSURE DROP THROUGH A HORIZONTAL WELL

Assuming that a horizontal wellbore can be represented as a horizontal pipe, the equation for pressure drop calculation in a pipe can be written using the laws of conservation, mass, momentum, and energy as

$$\frac{dp}{dL} = \left(\frac{dp}{dL}\right)_{gravity} + \left(\frac{dp}{dL}\right)_{friction} + \left(\frac{dp}{dL}\right)_{acceleration} \quad (10\text{–}1)$$

where dp represents pressure drop and dL represents incremental length. Furthermore, assuming the gravity and acceleration terms are negligible in a horizontal section of pipe and the flow is fully developed, the equation would reduce to

$$\frac{dp}{dL} = \left(\frac{dp}{dL}\right)_{friction} = -\frac{f_m \rho v^2}{2 g_c d} \quad (10\text{–}2)$$

or

$$\Delta p = -f_m \rho v^2 L/(2g_c d) \quad (10\text{–}3)$$

where

f_m = friction factor, dimensionless
ρ = density of fluid, lbm/ft^3
v = velocity of fluid, ft/sec
g_c = gravitational constant, 32.2 lbm-ft/(sec^2-lbf)
d = diameter of pipe, ft
Δp = pressure drop, lbf/ft^2
L = well length, ft

Equation 10–3 represents a single-phase flow pressure drop calculation through a pipe. For single-phase flow of oil through a horizontal wellbore, Equation 10–3 can be rewritten in terms of U.S. field units as

$$\Delta p = (1.14644 \times 10^{-5}) f_m \rho q^2 L/d^5 \quad (10\text{–}4)$$

(Note that in Equation 10–4 Δp, the pressure drop is defined as a positive quantity along the flow direction.)

where

f_m = Moody's friction factor, dimensionless
ρ = fluid density, gm/cc
q = flow rate at reservoir conditions, RB/day
L = horizontal length, ft
d = pipe diameter, inches
Δp = pressure drop, psia

In Equation 10–4, d represents the internal pipe diameter. A list of tubing and casing sizes and their corresponding internal diameters are listed in Tables 10–1 and 10–2.

In addition to pipe or liner diameter, a second parameter of importance in Equation 10–4 is the dimensionless friction factor f_m. The dimensionless friction factors for circular pipes are shown in Figures 10–3 and 10–4.[4] The figures show that the friction factor depends upon flow regime, i.e., whether flow is laminar or turbulent. The figures also show that in turbulent flow, the friction factor strongly depends upon pipe roughness, ϵ/d. For flow through

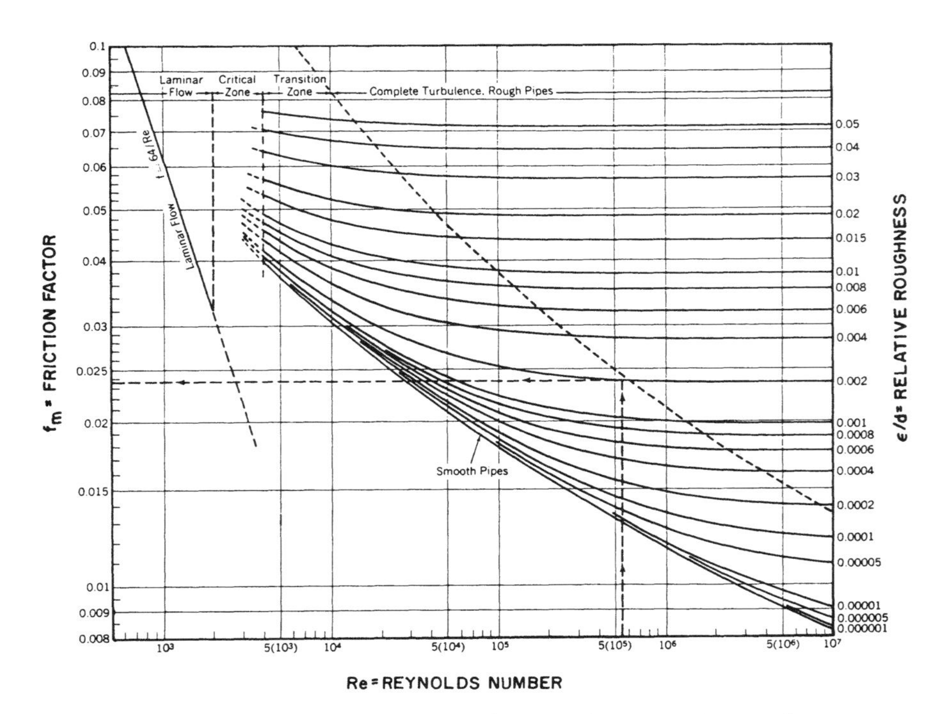

Figure 10–3 Moody's Friction Chart for Flow through Pipes.[4]

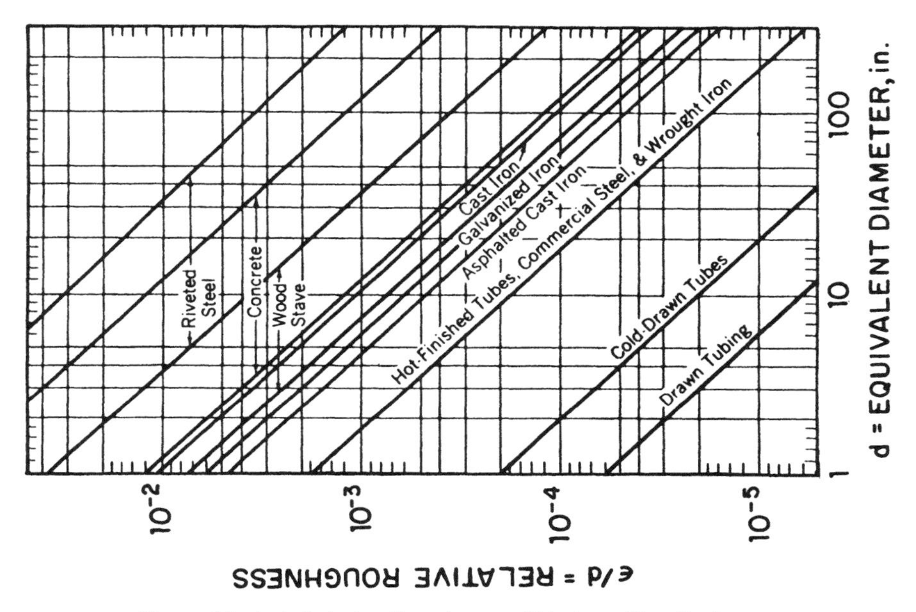

Figure 10–4 A Relative Roughness of Various Pipe Surfaces.

a circular pipe, laminar flow[5] occurs when the Reynolds number (Re) is less than 2300. Thus

$$\text{for laminar flow:} \quad Re < 2300 \tag{10–5}$$

$$\text{for turbulent flow:} \quad Re > 4000 \tag{10–6}$$

$$\text{for transition region:} \quad 2300 < Re < 4000 \tag{10–7}$$

where

$$Re = \text{Reynolds number} = \rho dv/\mu \tag{10–8}$$

The Reynolds number is a dimensionless number, which represents a ratio of inertial and viscous forces. In U.S. field units, Equation 10–8 is rewritten as

$$Re = 92.23\, \rho q/(\mu d) \tag{10–9}$$

where

ρ = fluid density, gm/cc
q = flow rate, RB/day

μ = viscosity, cp
d = internal pipe diameter, inches

One can calculate the Reynolds number using Equation 10–9 and then Figures 10–3 and 10–4 can be used to calculate friction factor f_m. The calculated friction factor can then be substituted in Equation 10–4 to calculate single-phase flow pressure drop through a horizontal wellbore. Instead of Figures 10–3 and 10–4, one can also use the following equations to calculate friction factor f_m for rough as well as smooth pipes.

for laminar flow:

$$f_m = 64/Re, \text{ for } Re < 2300 \tag{10–10}$$

for turbulent flow:

for $4000 < Re < 10^8$, $10^{-8} \leq \epsilon/d \leq 0.1$

$$f_m = \{1.14 - 2\text{Log}[(\epsilon/d) + 21.25\, Re^{-0.9}]\}^{-2} \tag{10–11}$$

For turbulent flow, there are several correlations in the literature, an extensive list of which is given in Reference 6. Equation 10–11 is suggested by Jain,[7] which shows less than 3.19% error with the classical turbulent-flow equation suggested by Prandtl, Karman and Nikurdase.[6] Additionally, Equation 10–11 allows an explicit calculation of the friction factor, facilitating speedy calculations, and hence, it is recommended here.

For the transition region, i.e., when $2300 < Re < 4000$, no explicit correlation is given here. One can use Figures 10–3 and 10–4 to calculate friction factors in the transition region.

CRITICAL RATE DEFINITION

Critical rate is defined as a rate below which flow will be laminar. Initial experiments of Reynolds[8] in 1883 showed two different values of critical rates. One was a lower value of critical rate while the other was an upper value. The actual critical value depends upon the disturbance present at the pipe inlet. Ekman[9] and later Pfenninger[10] experimentally noted the highest critical value to be $Re = 1.001 \times 10^5$, by eliminating all the disturbance at the pipe inlet. The critical value of $Re = 1.001 \times 10^5$ is recognized as the highest critical value of Reynolds number recorded to date.

In 1921, Schiller[11] performed careful experiments and introduced the concept of a lower limit for critical Reynolds number, which is defined as the value of Re at which laminar flow remains laminar no matter how large a magnitude of disturbances are present at the duct inlet. Schiller's experiments defined this lower critical value of Reynolds number to be 2320, which is rounded off to 2300. Recently, Simonek[12] theoretically obtained a value of 2295 as the lower limit of critical Reynolds number, which shows excellent agreement with Schiller's and other experimental data.

The above results clearly indicate that regardless of disturbance at the duct inlet, the flow will always remain laminar, as long as the Reynolds number is less than 2300. In a horizontal wellbore, fluid flowing horizontally is disturbed by a fluid entering into a wellbore at various points along its length. *This fluid entry in a direction perpendicular to the main wellbore flow may create flow disturbance, but as long as the Reynolds number is less than 2300, the flow will remain laminar. This is an important observation for practical field calculations.*

COMMENT ON FULLY DEVELOPED FRICTION FACTORS

It is important to note that Moody's friction factors, given in Figures 10–3 and 10–4, and Equation 10–10 are for fully developed flow; i.e., they assume that the velocity profile of the fluid is stabilized and does not change as fluid travels downstream in a pipe. In laminar flow, the pipe length x required for stabilizing velocity profile is proportional to the pipe diameter and the Reynolds number and is given as[5]

$$x = 0.0565d(Re) \qquad (10\text{–}12)$$

In a horizontal well, depending upon the completion method, fluid may enter the wellbore at various locations along the well length. For example, in a cemented and perforated well, oil may enter at a perforated section D in the wellbore. The next fluid entry into the wellbore will be at the next set of perforation, say, at section C. The distance between perforation sets D and C may not be sufficient to achieve a stabilized velocity profile (see Fig. 10–5). The unstabilized section is referred to as *entry flow* or *developing flow* in the classical fluid mechanics literature.

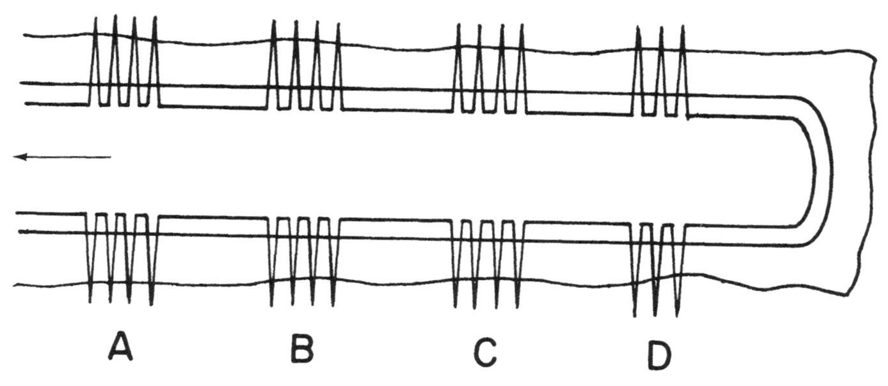

Figure 10–5 A Schematic of a Perforation Scheme in a Cemented Liner.

Laminar Developing Flow

In the flow development region, the apparent friction factor f_{app} incorporates the combined effect of wall shear and the change in momentum due to the developing velocity profile. The total pressure drop from the inlet ($x = 0$) to the point of interest is given as

$$\Delta p = (4f_{app})\, \rho\, v^2\, L/(2g_c d) \qquad (10\text{–}13)$$

in U.S. oil field units

$$\Delta p = (1.14644 \times 10^{-5})(4f_{app})\, \rho\, q^2\, L/d^5 \qquad (10\text{–}14)$$

where dimensionless apparent friction factor f_{app} is given as[5]

$$f_{app} = \Delta p^* d/(4x) \qquad (10\text{–}15)$$

and

$$\Delta p^* = 13.74(x^*)^{0.5} + \frac{1.25 + 64x^* - 13.74(x^*)^{0.5}}{1.0 + 0.00021(x^*)^{-2}} \qquad (10\text{–}16)$$

and

$$x^* = \frac{x/d}{Re} \qquad (10\text{–}17)$$

Table 10–3 shows a tabulation of Δp^* for various x^* values. Additionally, Figure 10–6 shows a variation of $f_{app}Re$ against $1/x^*$. It is important to note that for a fully developed laminar flow

$$f_m = 4f_{app} \qquad (10\text{–}18)$$

Thus, due to flow development, one can have as high as three to four times more pressure drop than that calculated using fully developed laminar friction factor values.

Turbulent Entry-Level Flow

As shown in Figure 10–7 and Table 10–4, the turbulent entry flow region[6] is about two to 10 times the pipe diameter and hence, can be neglected in total pressure drop calculations in a long horizontal well.

Entry Flow Summary

In general, the average laminar friction factors in the developing region f_m are 1.5 to 4 times more than those calculated using stabilized flow con-

TABLE 10–3 LAMINAR FLOW PRESSURE DISTRIBUTION IN THE HYDRODYNAMIC ENTRANCE REGION OF A CIRCULAR DUCT[5]

x^*	Pressure Drop Δp^*
0.00000	0.0000
0.00050	0.3220
0.00125	0.5034
0.00250	0.7204
0.00375	0.8960
0.00500	1.0506
0.00750	1.3212
0.01000	1.5610
0.01250	1.7822
0.01750	2.1900
0.02250	2.5692
0.03000	3.1064
0.04000	3.7894
0.05000	4.4520
0.06250	5.2688

$$x^* = \frac{x/d}{Re}$$

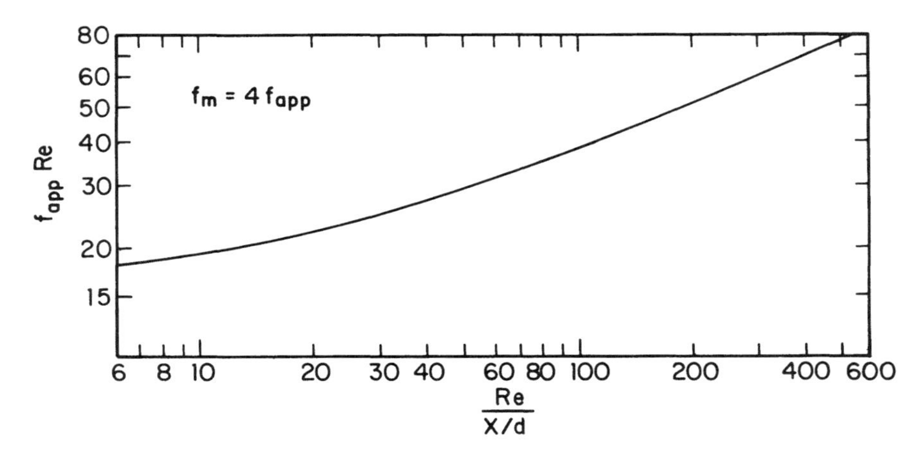

Figure 10–6 Laminar Flow Friction Factors in the Hydrodynamic Entry Length of a Circular Pipe with Uniform Inlet Velocity.[5]

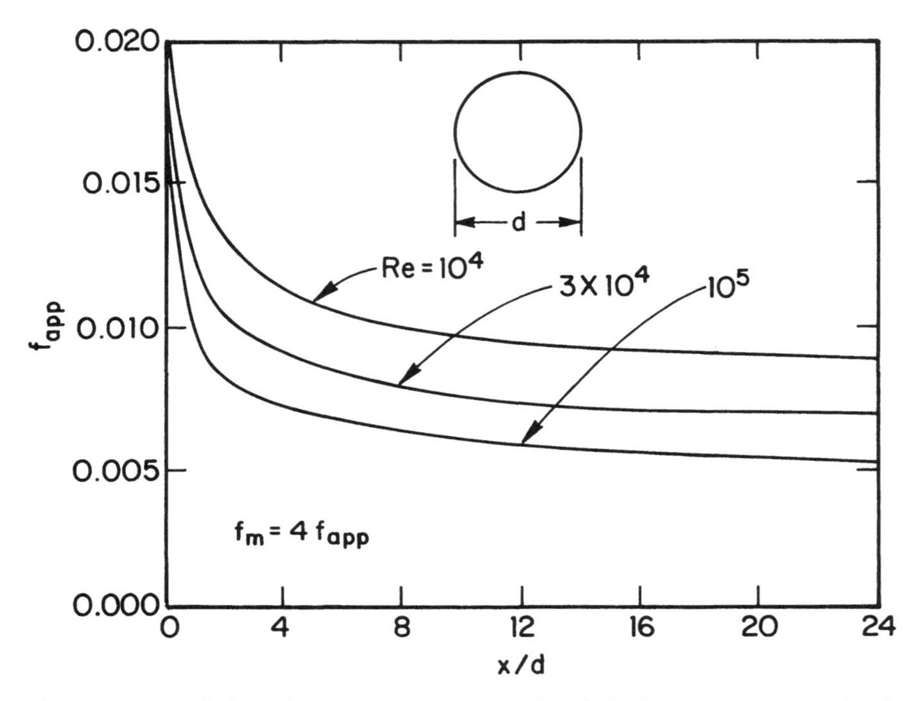

Figure 10–7 Turbulent Flow Friction Factors in the Hydrodynamic Entry Length of a Smooth Circular Duct with Uniform Inlet Velocity.[6]

ditions. In this book, all calculations are made assuming the flow is fully developed or stabilized. Thus, actual pressure drops for laminar flow may be 1.5 to 3 times larger than those calculated here.

PRESSURE DROP IN A CURVED WELLBORE SECTION

Another important parameter is frictional pressure drop through the curved section of the wellbore, i.e., the section where a well turns from the vertical to the horizontal direction. Recently Joshi and Shah[13] have reviewed the literature to compare various methods to calculate friction factors in bends and curves. Their results show that if $2R/d > 50$, where R = radius of curvature and d = diameter of pipe, then the pressure drop through a bend is almost equal to the pressure drop through a straight pipe, with pipe length equal to the distance along the curve. This tells us that for short-, medium-, and long-radius horizontal wells, the influence of curvature on the pressure drop is negligible. Thus, for pressure-drop calculation, the curved

TABLE 10–4 TURBULENT MOMENTUM TRANSFER RESULTS IN THE HYDRODYNAMIC ENTRANCE REGION OF A SMOOTH CIRCULAR DUCT

$\frac{x/d}{Re^{1/4}}$	Δp^*	$f_{app}Re^{1/4}$
0.0000	0.0000	
0.0800	0.0505	0.1578
0.1923	0.1018	0.1325
0.3213	0.1537	0.1196
0.4615	0.2058	0.1115
0.6093	0.2577	0.1057
0.7616	0.3091	0.1015
0.9153	0.3594	0.0982
1.0679	0.4082	0.0956
1.2167	0.4551	0.0935
1.3590	0.4994	0.0919

section of the wellbore can be divided into several segments of straight pipe with each segment having a different angle with the vertical axis.

DRILLED WELLBORE SIZES AND LINER SIZES

In domestic U.S. operations, it is common practice to drill 8½- to 8¾-in. horizontal wellbores.[14] This size is mainly chosen because of the limitations of pump sizes and ratings available on the rigs. Liner size is based upon three considerations:

1. *Well path:* A horizontal well is rarely horizontal and it rarely exhibits a perfectly smooth well path. Rather, the wells exhibit several bends and turns in the vertical and lateral planes. A liner diameter should be chosen so that it can be inserted through this "nonperfect" hole.
2. *Well cleaning:* Sometimes it is difficult to remove all the cuttings from a horizontal well. The cuttings left behind may accumulate on the bottom side of the hole and reduce effective hole size. A horizontal well liner size should be selected so as to not get stuck in this "reduced" size hole.
3. *Well cementing:* Normally, for vertical wells, one needs at least ¾-in. clearance between the drilled hole and a liner for proper cementing. It may be difficult to properly centralize a horizontal well. Therefore, one

may need at least 1-in. clearance between the hole and the liner to obtain proper cementing. A larger clearance in horizontal wells than vertical wells is used to mitigate the problem of improper centralizing.

In horizontal wells many operators use a liner diameter which is about 2 in. smaller than the hole size.[14] Therefore, a $6\frac{1}{2}$-in. liner would be the largest size liner that can be inserted in a $8\frac{1}{2}$-in. hole. Since a $6\frac{1}{2}$-in. liner is not commonly available, $5\frac{1}{2}$-in. liners ranging from 17 to 23 lb/ft are commonly used.

In offshore and remote area applications, where well flow rates are usually high, $9\frac{1}{2}$-in. drilled holes with 7-in. liners are commonly used.[15] These larger hole sizes are limited to applications where rig sizes are large enough and pump capacities are high enough to circulate the increased volume of drilling fluid.

SINGLE-PHASE PRESSURE DROP THROUGH A HORIZONTAL WELL

As noted earlier, in a horizontal well, fluid enters the wellbore along its length. A maximum possible pressure drop can be obtained by assuming that all flow enters the wellbore at the nonpumping end of the wellbore. On the other hand, minimum pressure drop can be calculated by assuming that the total flow enters the wellbore in the last foot of the pumping end. These two would give a range of numbers within which actual pressure drop would occur. It is easier to calculate maximum pressure drop first and compare it with the drawdown. If the maximum pressure drop is very small as compared to the reservoir drawdown, then pressure drop through the wellbore can be ignored. In contrast, if wellbore pressure drop is significant as compared to the reservoir drawdown additional calculations are necessary.

EXAMPLE 10–1

A 2000-ft-long well produces at a rate of 10,000 RB/day. Oil viscosity is 1.6 cp and density is 0.84 gm/cc. The well is completed using a $5\frac{1}{2}$-in. (20 lb/ft) liner. Calculate pressure drop in the horizontal wellbore.

Solution

Horizontal liners are made from casings. For a $5\frac{1}{2}$-in. (20 lb/ft) flow string, internal diameter d = 4.778 inches (see Table 10–1). Reynolds number Re is calculated using Equation 10–9 as

$$Re = 92.23\ \rho q/(\mu d)$$

$$= \frac{92.23 \times 0.84 \times 10000}{1.6 \times 4.778}$$

$$Re = 101{,}341$$

Since $Re > 2300$, the wellbore shows a turbulent flow. For turbulent flow, the pressure drop depends upon pipe roughness. For a cast iron pipe of $5\frac{1}{2}$ in., from Figure 10–4, roughness, $\epsilon/d = 0.002$. Now using Figure 10–3 for $Re = 101{,}341$, $\epsilon/d = 0.002$ one obtains a friction factor $f_m = 0.025$. Using Equation 10–4 to calculate pressure drop,

$$\Delta p = \frac{1.14644 \times 10^{-5} \times 0.025 \times 0.84 \times (10{,}000)^2 \times L}{(4.778)^5}$$

$$= 0.00967 \times L \text{ psia}$$

$$(\Delta p)_{max} = 0.00967 \times 2000$$

$$= 19.3 \text{ psia}$$

Thus, maximum pressure drop is 19.3 psia, assuming all fluids enter the wellbore at the nonpumping tip. In practice, actual pressure drop would be smaller than 19.3 psia. To minimize this pressure drop further, one can choose a larger liner size.

EXAMPLE 10–2

Calculate the pressure drop in a horizontal well, assuming only frictional pressure drop. The well is completed with a 7-in. (23 lb/ft) liner. The well length, fluid properties, and production rate are the same as that in Example 10–1.

Well length, L = 2000 ft
Flow rate, q = 10,000 RB/day
Density, ρ = 0.84 gm/cc
Viscosity, μ_o = 1.6 cp

Solution

The first step is to determine flow regimes (whether laminar or turbulent flow), which requires calculation of the Reynolds number. For a 23 lb/ft 7-in. pipe, the internal diameter is 6.366 inches (see Table 10–1). The Reynolds number is calculated using Equation 10–9 as

$$Re = 92.23\, \rho q/(\mu d)$$

$$Re = \frac{92.23\ (.84 \text{ gm/cc})\ 10{,}000 \text{ bbl/day}}{1.6 \text{ cp}\ (6.366 \text{ in})}$$

$$Re = 76{,}061$$

Since $Re > 2300$, the flow is turbulent. The friction factor and pressure drop can be calculated from Equations 10–11 and 10–4, respectively.

Assuming a 7-in. cast-iron pipe from Figure 10–4, the value of roughness parameter ϵ/d is 0.0016. Using this roughness and a Reynolds number of 76,061, and substituting into Equation 10–11 yields

$$f_m = \{1.14 - 2\log[(0.0016) + 21.25\,(76061)^{-0.9}]\}^{-2}$$
$$f_m = 0.0247 \approx 0.025$$

Similarly, from Fig. 10–3, for $Re = 76{,}061$ and $\epsilon/d = 0.0016$, friction factor $f_m = 0.025$. Substituting this into Equation 10–4,

$$\Delta p = \frac{(1.14644 \times 10^{-5}) \times 0.025 \times 0.84 \times (10000)^2 \times 2000}{(6.366)^5} = 4.6 \text{ psia}$$

These results indicate that even for turbulent flow through a 2000-ft-long rough pipe (cast iron) at a high rate (10,000 bbl/day), the pressure drop across the entire length of pipe is small (4.6 psia), assuming that the entire flow enters the wellbore at the well tip. In practice, actual pressure drop would be much smaller, since all the fluid would not enter the wellbore at the tip, but along the well length. Thus, for practical purposes the pressure drop through the horizontal wellbore can be neglected.

Examples 10–1 and 10–2 show that by choosing appropriate wellbore and liner sizes, one can minimize pressure drop through the liner.

EXAMPLE 10–3 HEAVY OIL FLOW

Calculate the pressure drop for a flow of 1000 RB/day of 15°API oil through a 1000-ft section of a horizontal wellbore. The gas-oil ratio is 50 and reservoir temperature is 190°F. The well is completed using a 5½-in. (20 lb/ft) liner.

Solution

First, calculate the density and viscosity of the crude to be able to determine the Reynolds number. These values can be calculated from Appendix A.

$$\gamma_o = 141.5/(131.5 + °\text{API})$$

For 15° API crude, $\gamma_o = 0.966$

$$\rho_o = 0.966 \times (1 \text{ gm/cc}) = 0.966 \text{ gm/cc}$$

$$\mu_o \text{ (from Figures A–1 and A–2)} = 35 \text{ cp}$$

Reynolds number can be calculated using Equation 10–9 as

$$Re = 92.23\ \rho q/(\mu d) \tag{10–9}$$

$$Re = 92.23\ \frac{(0.966)\ 1000}{35\ (4.778)}$$

$$Re = 532.8 < 2300$$

Therefore, the flow regime is laminar and the friction factor is $64/Re = 64/533 = 0.12$.

Pressure drop is then calculated using Equation 10–4

$$\Delta p = \frac{1.14644 \times 10^{-5} \times 0.12 \times 0.966 \times (1000)^2 \times L}{(4.778)^5}$$

$$= 0.0005 \times L \text{ psia}$$

$$(\Delta p)_{max} = 0.0005 \times 1000$$

$$= 0.5 \text{ psia}$$

As seen from the results, there is no significant pressure drop at these conditions.

The results in Examples 10–1 to 10–3 demonstrate that the pressure drop through the horizontal section can be controlled by both the flow rates and the proper selection of liner and hole sizes.

CALCULATION OF PRODUCING RATE BY COMBINING RESERVOIR AND WELLBORE PRESSURE DROPS—LAMINAR FLOW

Recently, Dikken[1] presented a solution to calculate oil production rate in a horizontal wellbore by including wellbore pressure loss in the reservoir calculations. For laminar flow

$$q_w(x) = \Delta p_R \sqrt{\frac{J'_h}{c'_w}} \times \frac{\sinh\left(\sqrt{J'_h\, c'_w}\,(L - x)\right)}{\cosh\left(\sqrt{J'_h\, c'_w}\, L\right)} \tag{10–19}$$

For small x

$$q_w(x) = J'_h\, \Delta p_R\, (L - x) \tag{10–20}$$

For x close to L

$$q_w(x) = \frac{J'_h\, \Delta p_R\, (L - x)}{\cosh\left(\sqrt{J'_h\, c'_w}\, L\right)} \tag{10–21}$$

Thus

$$(\Delta p_R)_{x=0}/(\Delta p_R)_{x=L} = \frac{1}{\cosh\left(\sqrt{J_h' c_w'}\, L\right)} \tag{10–22}$$

where

$$\Delta p_R = \bar{p} - p_{wf}(x), \text{ psia} \tag{10–23}$$

J_h' = horizontal well productivity per unit length, RB/(day-psi-ft)
c_w' = wellbore pressure drop coefficient, (psia/ft)/(RB/day)
$\bar{p}$ = reservoir pressure, psia
$p_{wf}(x)$ = well flowing pressure at point x, psia
x = distance measured along well length, ft
q_w = well rate at any point along the well length, RB/day

The wellbore pressure drop coefficient c_w' can be obtained by combining pressure drop, Reynolds number, and friction factor equations, namely Equations 10–4, 10–9, and 10–10. A combination of these equations yields

$$\Delta p = (1.14644 \times 10^{-5})\frac{64\,\mu d}{92.23\,\rho q} \times \frac{\rho q^2 L}{d^5} \tag{10–24}$$

$$(\Delta p/L) = (7.955 \times 10^{-6}\,\mu/d^4)q \tag{10–25}$$

The terms in parenthesis in Equation 10–25 represent the value of c_w'. Thus, c_w' is

$$c_w' = 7.955 \times 10^{-6}\,\mu/d^4 \tag{10–26}$$

where

μ = viscosity, cp
d = diameter, inches
q = flow rate, RB/day
c_w' = wellbore coefficient, (psia/ft)/(RB/day)

Equation 10–22 tells us that pressure drop along the well length is negligible when

$$[(\Delta p_R)_{x=0}]/[(\Delta p_R)_{x=L}] = 1 = \frac{1}{\cosh\left(\sqrt{J_h' c_w'}\, L\right)} \tag{10–27}$$

or

$$\cosh^{-1} 1 = 0 = \sqrt{J_h' c_w'}\, L \tag{10–28}$$

This corresponds to $c_w' = 0$.

The above laminar solution is calculated assuming well rate,

$$q_w = 0 \text{ at } x = L$$

and

$$\frac{d}{dx}(q_w) = J'_h(\bar{p} - p_{wf}) \text{ at } x = 0 \qquad (10\text{–}29)$$

EXAMPLE 10–4

Calculate wellbore pressure drop and the ratio of reservoir pressure drawdown at each well end experienced by a 1000-ft-long well with properties given in Example 10–3. In this heavy oil reservoir, horizontal well productivity is estimated to be 0.01 RB/(day-psi-ft). The reservoir pressure drawdown is 50 psia.

Solution

Equation 10–25 can be used to calculate wellbore pressure drop as

$$\begin{aligned} \Delta p &= (7.955 \times 10^{-6}\ \mu/d^4)qL \qquad (10\text{–}25) \\ &= (7.955 \times 10^{-6} \times 35/(4.778)^4)\ 1000 \times 1000 \\ &= 0.53 \text{ psia} \end{aligned}$$

The above-calculated wellbore pressure drop of 0.53 psia is practically the same as that calculated as 0.5 psia in Example 10–3. Thus, as expected, both methods give similar results for the wellbore pressure. This wellbore pressure drop assumes that all the fluid enters at the nonpumping end of the wellbore. The actual pressure drop will be smaller than 0.5 psia. Additionally, 0.5 psia drop is very small as compared to reservoir drawdown of 50 psia. Using Equation 10–22 we can compare change in pressure drawdown from one end of the well to the other.

$$\begin{aligned} \text{The well productivity} &= J'_h = 0.01 \text{ RB/(day-psi-ft)} \\ \text{wellbore coefficient} &= c'_w = 7.955 \times 10^{-6}\ \mu/d^4 \qquad (10\text{–}26) \\ &= 7.955 \times 10^{-6} \times 35/(4.778)^4 \\ &= 5.342 \times 10^{-7} \text{ (psia/ft)/(RB/day)} \end{aligned}$$

Using Equation 10–22, the ratio of the reservoir pressure drawdown at each end can be calculated as

$$[(\Delta p_R)_{x=0}]/[(\Delta p_R)_{x=L}] = \frac{1}{\cosh[\sqrt{J'_h\, c'_w}\, L]} \qquad (10\text{–}22)$$

$$= \frac{1}{\cosh[\sqrt{5.342 \times 10^{-7} \times 0.01} \times 1000]}$$

$$= 1/1.003 = 0.997 \approx 1$$

Therefore, there is no change in reservoir pressure drawdown from one end to the other end of a 1000-ft-long wellbore.

CALCULATION OF PRODUCTION RATE BY COMBINING RESERVOIR AND WELLBORE PRESSURE DROP—TURBULENT FLOW

For turbulent flow through a wellbore, Blasius equation for turbulent flow in a smooth pipe was modified as[1]

$$f_m = 4f = 0.3164\,(Re)^{-\alpha'} \qquad 4000 < Re < 10^5 \qquad (10\text{–}30)$$

with $\alpha' = 0$ for a rough surface and 0.25 for a smooth surface. (Note that turbulent flow Equation 10–11 has a wider range of application than Equation 10–30.) For rough surface, if $\alpha' = 0$ then from Equation 10–30, friction factor, f_m, is a constant equal to 0.3164. This value of friction factor is at least four fold higher than those observed in Moody's friction factor charts for rough pipes (see Figures 10–3 and 10–4 and Equation 10–11). For turbulent flow in a finite horizontal wellbore, the differential equation is[1]

$$\frac{d^2}{dx_d^2} q_{wd} = \frac{6 - 2\alpha'}{(1 - \alpha')^2} (q_{wd})^{2-\alpha'} \qquad (10\text{–}31)$$

with boundary conditions of

$$q_{wd} = 0 \text{ at } x_d = L'_d \qquad (10\text{–}32)$$

$$\frac{d}{dx_d} q_{wd} = 2/(\alpha - 1) \text{ at } x_d = 0 \qquad (10\text{–}33)$$

also

$$q_{wd} = q_{td} \text{ at } x_d = 0 \qquad (10\text{–}34)$$

where

$$q_{wd} = q_w\, K^{(2/(1-\alpha'))} \qquad (10\text{–}35)$$

$$x_d = \frac{1 - \alpha'}{2K} \sqrt{\frac{2J'_h c'_w}{3 - \alpha'}}\, x \tag{10–36}$$

$$K = \left[\Delta p_R \sqrt{\frac{(3 - \alpha')\, J'_h}{2\, c'_w}} \right]^m \tag{10–37}$$

and

$$m = (\alpha' - 1)/(3 - \alpha') \tag{10–38}$$

Additionally, L'_d is obtained by substituting $x = L$ in Equation 10–36 and q_{td} represents total dimensionless flow at the wellbore pumping end. It is not possible to solve Equation 10–31 analytically for a finite wellbore. Therefore, it is solved numerically. It is important to note that with a turbulent flow in a rough pipe ($\alpha' = 0$), the equation is solved until well flow rate at a certain length becomes zero due to a large friction pressure drop. A typical turbulent flow problem in a rough pipe, as given in Reference 1, is shown in Example 10–5.

EXAMPLE 10–5 EFFECT OF PRESSURE DROP IN AN EDGE WATER DRIVE RESERVOIR

Determine the effect of pressure drop in a horizontal well on the effective horizontal well length. The reservoir data and fluid properties are given as

$k_h = 1000$ md
$k_v = 1000$ md
$h = 65.6$ ft
$x_e = 820$ ft (half the size of a drainage distance)
$\mu_o = 0.5$ cp
$\rho_o = 0.7$ gm/cc
$\rho_w = 1.1$ gm/cc

Additionally, productivity indices for a horizontal well based on critical-rate calculations are given as

d **(inch)**	J'_h, **RB/(day-psi-ft)**	Δp_R **(psia), reservoir drawdown**
$4\frac{1}{2}$	0.5808	6.19
$5\frac{1}{2}$	0.5845	6.16
7	0.5895	6.10
$9\frac{5}{8}$	0.5996	6.00

Assume a turbulent flow in a rough pipe, i.e., $\alpha' = 0$

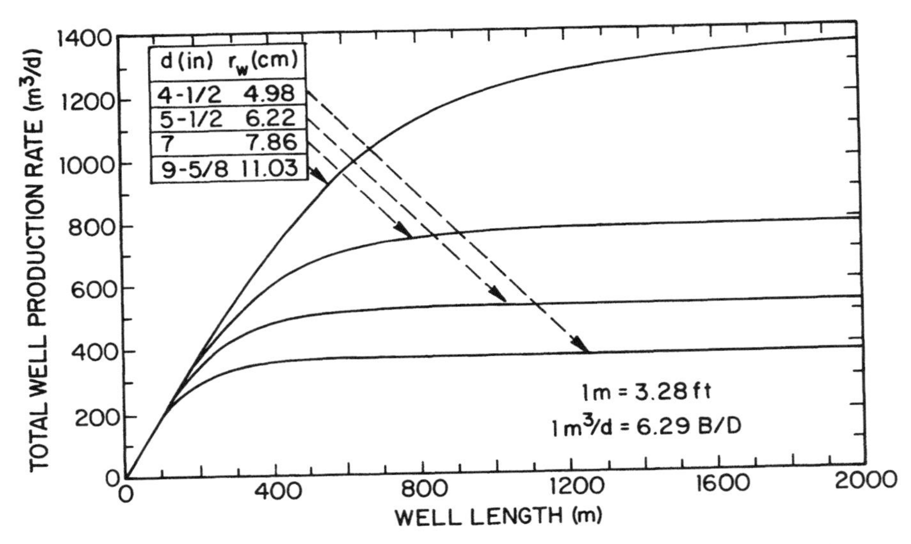

Figure 10–8 A Variation in Well Flow Rate with Well Flow Length and Well Diameter for $\alpha' = 0$, i.e., for a Maximum Roughness (Example 10–5).[1]

Solution

As seen from the above data, reservoir drawdown required to prevent water coning is very small. The problem is solved numerically using Equations 10–31 through 10–38. The plot was generated assuming $\alpha' = 0$, i.e, friction factor is independent of the flow rate.[1] The constant value of friction factor is 0.3164 (see Equation 10–30). Due to large frictional pressure drop in the well, for a given liner size, beyond certain well length, drilling a longer well does not yield any higher production rate (see Figure 10–8). Figure 10–8 depicts a plot of production versus well length for the various sizes. The figure shows that by using a liner size larger than 4½ in. one can extend the productive length of a horizontal well. Thus, a selection of hole size and liner size is important. An additional factor which has significant influence on well productivity is pipe roughness. As shown in Figure 10–9, $\alpha = 0$ represents rough pipe and $\alpha' = 0.25$ represents a smooth pipe. Even for a small liner such as the 4½-in. size, a reduction from a very rough pipe to a smooth pipe would more than triple the productive length of a horizontal well. *This tells us that estimation of pipe roughness is important. Figures 10–3 and 10–4 are recommended to estimate liner roughness, unless direct roughness-friction factor data are available from a liner manufacturer.*

It is important to note that in Example 10–5, wellbore pressure losses are important because reservoir drawdown is very small. If reservoir drawdown would have been larger, wellbore pressure losses would not have exhibited such a strong influence on the well's productive length. Addi-

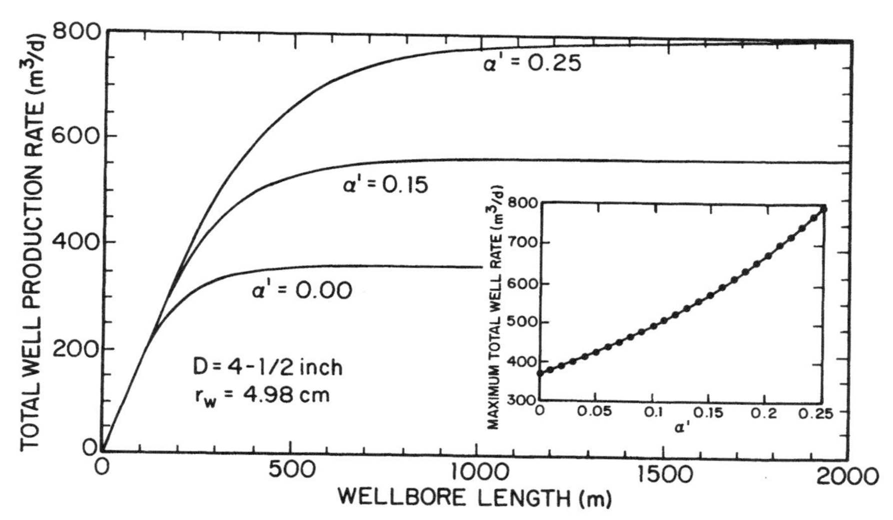

Figure 10–9 An Influence of Pipe Roughness on Productive Well Length for the Data Given in Example 10–5. Insert in the Figure Shows the Variation of Asymptotic Rates in the Main Figure with Roughness Parameter.[1]

tionally, for a turbulent flow in a rough pipe, instead of using $\alpha' = 0$ and $f_m = 0.3164$ using Figures 10–3 and 10–4 (or Equation 10–11) gives significantly lower values of friction factors than $f_m = 0.3164$. These lower friction factors values show less severe penalty on a horizontal well flow rate than that depicted for $\alpha' = 0$ in Figures 10–8 and 10–9.

PRESSURE DROP THROUGH A HORIZONTAL WELL FOR GAS FLOW

Pressure drop calculations for gas wells are more complicated than those for oil wells. This is because in gas wells, due to friction, the temperature of gas may increase as it travels through the wellbore. Additionally, gas properties such as density and viscosity are strongly dependent upon gas pressure and temperature. This may result in a changing pressure drop per foot length of a well along the entire well length. Moreover, in high-rate gas wells, one needs to ensure that gas velocities are subsonic, i.e., the Mach number is less than one.

$$\text{Mach no.} = M = v/C \tag{10–39}$$

where

v = gas velocity, ft/sec
C = sonic velocity at gas pressure and temperature, ft/sec

The Mach number equal to one represents a choke flow, i.e., a maximum flow that a given diameter pipe can carry.

As noted above, gas properties may change along the well length; therefore, to estimate pressure drop along the well length, one may have to divide a long well length into several sections to estimate the pressure drop in each section and sum them together. One of the simplest equations to estimate pressure drop due to flow of dry gas in a horizontal pipe is the Weymouth equation.

$$q_g = 15320 \left[\frac{(p_1^2 - p_2^2)\, d^{16/3}}{\gamma_g\, T\, z\, L} \right]^{0.5} \tag{10–40}$$

where

q_g = gas flow rate, scfd
p_1 = pipe inlet pressure, psia
p_2 = pipe outlet pressure, psia
L = pipe length, miles
T = average temperature, °R
z = average gas compressibility factor
γ_g = gas gravity (air gravity = 1)
d = pipe diameter, in

In addition to the Weymouth equation, several other correlations are available in the literature. For example, various multi-phase flow correlations for horizontal pipe are also applicable for a single-phase flow of either gas or oil. The example below also illustrates that pressure drop in horizontal gas wells can be reduced significantly by choosing appropriate liner size.

EXAMPLE 10–6 PREDICTION OF PRESSURE DROP FOR GAS FLOW

Calculate the pressure drop through a horizontal wellbore using the Weymouth formula and a multiphase flow correlation for the following conditions:

Pipe length, L = 2000 ft
Inside diameter, d
for possible liner sizes = 4.026, 5.00, 6.336 in.
Roughness, ϵ = 0.0005 ft
Gas gravity, γ_g = 0.605
Inlet temperature, T = 185° F
Inlet pressure, p_1 = 2500 psi
Gas flow rate, q_g = 40 MMscfd

Gas Properties

Reservoir Temperature = 185, °F
Gas Gravity (AIR = 1.0) = 0.605
Pseudo-Critical Temp = 354.4, °R
Pseudo-Critical Pres, psia = 664.5

Pressure psia	Viscosity cp	z Factor dimensionless	*p/z* *psia*
2000	0.01585	0.9009	2220.1
2200	0.01635	0.8997	2445.3
2400	0.01686	0.9002	2666.0
2500	0.01711	0.9011	2774.4
2600	0.01736	0.9024	2881.3
2800	0.01790	0.9061	3090.3
3000	0.01847	0.9111	3292.6

Solution

The pressure drop is calculated assuming that the entire flow enters the well at the 'non-pumping' end. Thus, pressure drop is calcualted for the total well flow through a 2000 ft long well. This represents the maximum possible pressure drop through the well. The actual pressure drop will be smaller than that calculated here.

Calculating pressure drop through the horizontal wellbore, Δp, using the Weymouth formula is an iterative process, as shown below.

The Weymouth formula is:

$$q_g = 15320 \left[\frac{(p_1^2 - p_2^2)\, d^{16/3}}{\gamma_g\, T\, z\, L} \right]^{0.5} \tag{10–40}$$

or

$$p_1^2 - p_2^2 = \left[\left(\frac{q_g}{15320} \right)^2 \left[\frac{\gamma_g z T L}{d^{16/3}} \right] \right]$$

Solving for the outlet pressure, p_2

$$p_2 = \left[p_1^2 - \frac{\gamma_g z L T}{d^{16/3}} \left(\frac{q_g}{15320} \right)^2 \right]^{0.5}$$

for

$$p_1 = 2500 \text{ psi}$$
$$L = 2000/5280 = 0.379 \text{ miles}$$
$$T = 185 + 460 = 645° \text{ R}$$
$$\gamma_g = 0.605$$
$$q_g = 40{,}000{,}000 \text{ scfd}$$
$$d = 4.026 \text{ in}$$
$$p_2 = [6{,}250{,}000 - z(593{,}784)]^{1/2}$$

Since z is a function of p_2 an iterative process is needed to calculate p_2.

1. Estimate Δp and calculate $p_2 = p_1 - \Delta p$.
2. Calculate average pressure $\bar{p} = (2/3)[p_1 + p_2 - (p_2 p_1/(p_1 + p_2))]$.
3. Determine z at $\bar{p}$ from Gas Properties.
4. Solve for p_2 and compare with p_2 in Step 1.

Repeat steps 2–4 until convergence within an acceptable tolerance is reached. The following table is this iterative process for calculating pressure drop through a 4.026-in. pipe.

PASS	$\Delta p_{(est)}$	$p_{2(est)}$	$\bar{p}$	z	p_2 (calculated)
1	100	2400	2450.3	0.90065	2390
2	110	2390	2445.4	0.90061	2391
3	109	2391	2445.9	0.90061	2391

Since p_2 remained the same from 2 to pass 3, convergence has been achieved resulting in a pressure drop Δp of 109 psi.

Pressure drop calculations for various diameter pipes using both the Weymouth formula and the Beggs and Brill correlation[17] are tabulated below.

Pipe I.D. in	Δp Weymouth psi	Δp Beggs & Brill psi
4.026	109	117.9
5.000	44	37.2
6.336	9.5	10.7

From the results, it can be seen that the pipe diameter plays an important role in the magnitude of the pressure drop. This also tells us that drilling large holes is important for horizontal wells that are expected to produce at high rates.

MULTIPHASE PRESSURE DROP THROUGH A HORIZONTAL WELL

In practice, a horizontal well will produce more than single phase; i.e., it can produce oil, water, and gas. To calculate multiphase pressure drop, one still can assume that a horizontal wellbore can be represented as a horizontal pipe. This allows use of several multiphase flow correlations that are available in the literature.[17] The multiphase flow equations are more complex than for single-phase flow because of slip velocities between the phases. To calculate pressure drops, one can either solve Equation 10–1 or as noted above, use pipe flow correlations.

In general, for the same flow conditions and for the same pipe length, different multiphase correlations may give different values of pressure drop. This problem is quite common even with vertical well correlations. The best approach is probably to measure pressure drop in a wellbore and compare it with various multiphase correlations to see which correlation gives the best fit of the data. However, it is difficult to insert pressure transducers at both ends of a horizontal well and calibrate the data. Unless some similar laboratory data are available, it may be advisable to use various two-phase correlations to estimate wellbore pressure drop and take the average of all results but the highest and lowest values.

EXAMPLE 10–7

A 1000-ft-long horizontal well is producing 1000 STB/day from a solution gas drive reservoir. Oil gravity is 33° API, reservoir pressure is 1500 psia, reservoir temperature is 200° F and gas gravity is 0.75. Calculate pressure drop in the horizontal wellbore. (Given: wellbore pressure is about 800 psi, pipe roughness = 0.0005 ft.)

Solution

Assumption: The total flow of 1000 STB/day of oil and the associated gas enters the wellbore at the 'non-pumping' end. Thus, pressure drop is calculated for the total flow through a 1000 ft long well.

If we assume that the producing gas-oil ratio is equal to the solution gas-oil ratio, then the solution gas-oil ratio can be calculated by various correlations given in Appendix A. Using the Standing correlation, the producing gas oil ratio is found to be 293 SCF/STB.

To predict the pressure drops, a nodal analysis scheme was used wherein pressure was assumed to be 800 psia at the end of the well and pressure profile and holdup as well as the flow pattern is calculated for every short node using a pressure loss correlation. Several correlations are available in the literature to predict the pressure drops.[17] The pressure drops obtained using two different correlations are listed below:

Linear size (in)	Pressure drop (psi)	
	Beggs and Brill	Dukler et al.
3.5	1.4	1.4
4.778	0.3	0.2

The above results represent the maximum possible pressure drop. The actual pressure drop will be smaller than the numbers noted above.*

INFLUENCE OF FLUID ENTRY PROFILE ON PRESSURE DROP

Pressure drop through a horizontal wellbore depends upon the fluid-entry profile. In this text, most of the single-phase calculations assumed that all the fluid entered at the nonpumping end of the horizontal wellbore. This, of course, would give an uppermost limit of the expected pressure drop. The other entry profile is uniform-flux profiles, which assume the same amount of fluid entry per unit length of a horizontal well. As shown in Figure 10–10, depending upon the well-boundary condition either infinite-conductivity or uniform-flux, one would have different flow profiles for the fluid entry into the wellbore. Additionally, several other fluid entry profiles are possible, depending upon the reservoir heterogeneity along the well length and the pipe frictional pressure drop. For example, two types of triangular entry profiles are shown in Figure 10–10. In Example 10–8 the effect of different fluid-entry flow profiles on wellbore pressure drop is examined. A computer model for horizontal pipe flow at the University of Tulsa was used to calculate these results.

EXAMPLE 10–8 MULTIPHASE FLOW IN A HORIZONTAL WELL

Calculate pressure drop due to multiphase flow in a horizontal well using the following three different fluid inlet profiles:

* The author would like to thank Tulsa University Fluid Flow project for allowing use of their computer programs for multiphase calculations. The authors would also like to thank G. Perez for his assistance.

a. UNIFORM WELLBORE PRESSURE (INFINITE-CONDUCTIVITY)

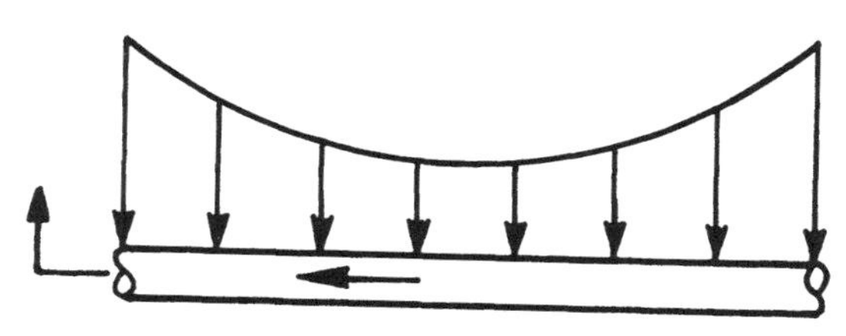

b. UNIFORM FLUX ENTRY

$$q(x) = \frac{q_{total}}{L}$$

L

x

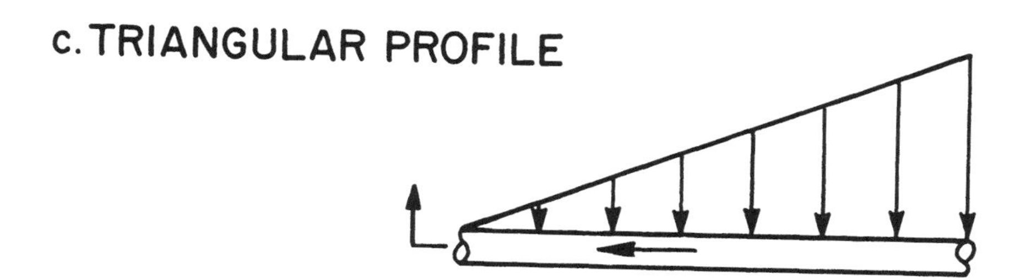

d. TRIANGULAR PROFILE

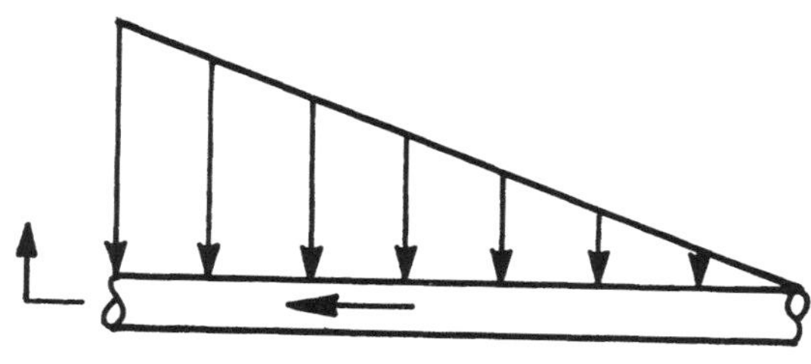

Figure 10–10 Some of the Possible Fluid Entry Profiles into a Horizontal Well.

Case 1: uniform flow distribution, (Fig. 10–10b)

$$q(x) = q_{\text{total}}/L$$

Case 2: flow distribution linearly increases with distance x (Fig. 10–10c)

$$q(x) = 2xq_{\text{total}}/L^2$$

Case 3: flow distribution linearly decreases with distance x (Fig. 10–10d)

$$q(x) = \frac{2(L - x)q_{\text{total}}}{L^2}$$

Using the following data, calculate the pressure drop due to acceleration and due to friction in the linear as well as the total pressure drops in the vertical, curved, and horizontal parts of the wellbore.

Length of vertical part, L	= 4000 ft
Length of curved part (long radius), L	= 2000 ft
Length of horizontal part, L	= 2000 ft
Length of the slotted liner, L	= 2000 ft
Tubing and the slotted liner, I.D.	= 4.5 in
Roughness of tubing and slotted liner, ϵ	= 0.00006 ft
Inclination angle gradient, ang	= 4.5°/100 ft
Oil production rate, q_o	= 4000 RB/day
Water production rate, q_w	= 2000 RB/day
Producing gas oil ratio, GOR	= 2000 scF/STB
Oil gravity, °API	= 28.5
Water specific gravity, γ_w	= 1.1
Gas specific gravity, γ_g	= 0.78
Separator pressure, p_{sep}	= 200 psia
Separator temperature, T_{sep}	= 120° F
Fluid temperature, T	= 150° F
Pressure at dead end of slotted liner, p_{wf}	= 2500 psia
Constant Term	$\alpha'' = 1$ turbulent flow
	$\alpha'' = 0.5$ laminar flow

Solution

For multiphase isothermal flow in a horizontal pipe, the energy equation for a well length dx is

$$\frac{dp}{dx} + \frac{f_{TP}\,\rho_{TP}V_m^2}{2g_c d} + \frac{\rho_{TP}V_m dV_m}{\alpha''\,g_c dx} = 0 \qquad (10\text{–}41)$$

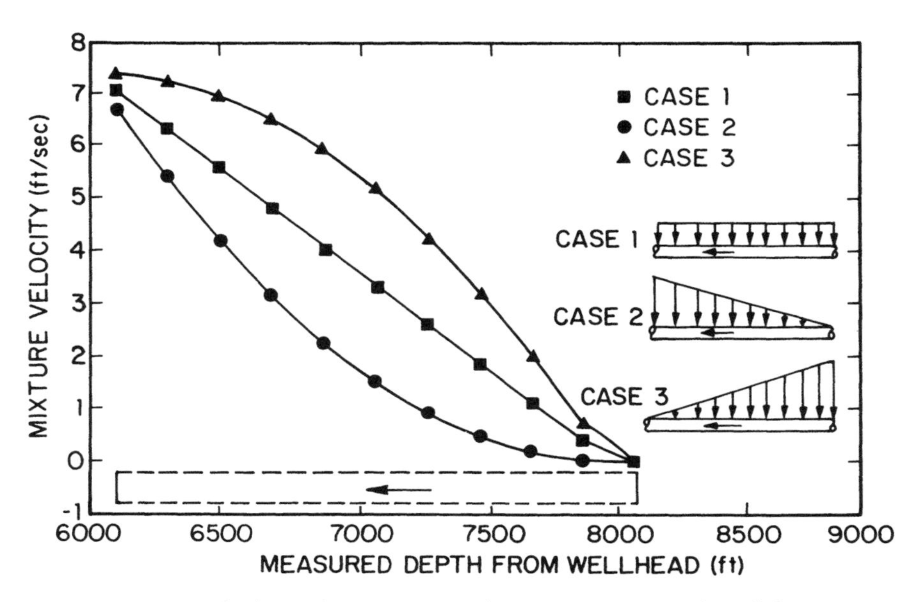

Figure 10–11 Multiphase Flow Mixture Velocity in a Horizontal Well for Data in Example 10–8.[18]

The second and the third terms of the above equation represent contributions due to friction and acceleration, respectively. The frictional loss is calculated using the Beggs & Brill correlation. The acceleration term is a function of mixture velocity and change of mixture velocity, which in turn depends upon pressure, temperature, and flow rate $q(x)$. The above differential equation was solved numerically to determine the pressure drop along the slotted liner. The results show that[18]

1. As shown in Figure 10–11, mixture velocities do not change significantly for the three cases.
2. As shown in Figure 10–12, for the first case, pressure drop due to friction is much larger than that due to acceleration.
3. As shown in Figure 10–13, depending upon the fluid entry profile total pressure drop in the well varies from a minimum of 6 psi to a maximum of 14.5 psi.
4. As shown in Figure 10–14, the wellhead pressures are almost the same even though fluid entry distributions along the well length are different for these cases.

SUMMARY OF EXAMPLE RESULTS

The results of Examples 10–1 to 10–8 indicate that even with low to moderate flow rates, pressure drops through horizontal wells are generally

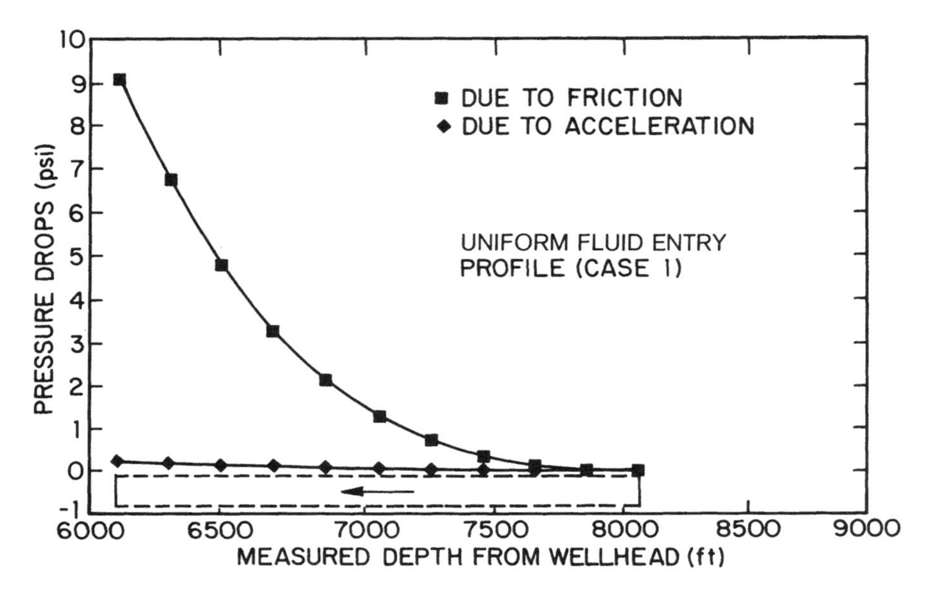

Figure 10–12 Acceleration and Friction Pressure Drop for Data Given in Example 10–8.[18]

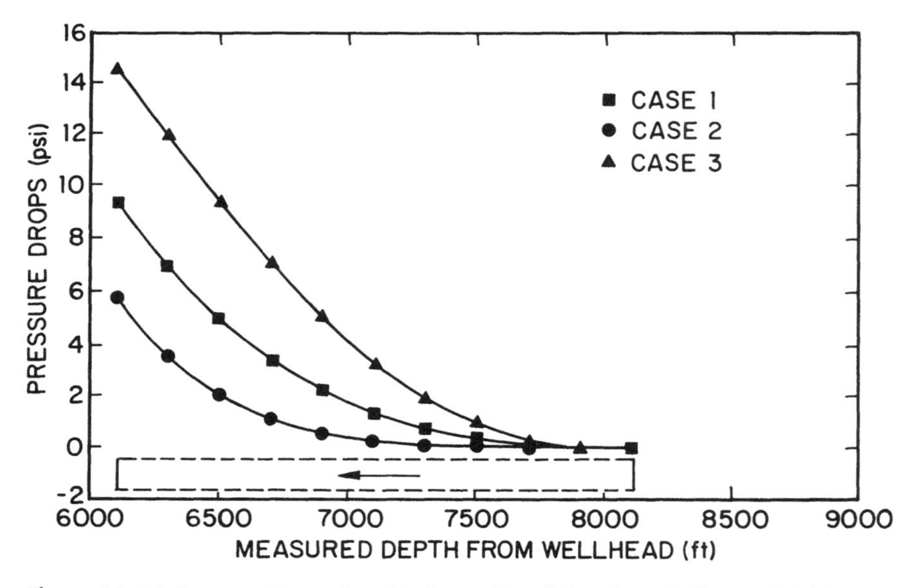

Figure 10–13 Pressure Drops in a Horizontal Well for Three Different Fluid Entry Profiles for the Data Given in Example 10–8.[18]

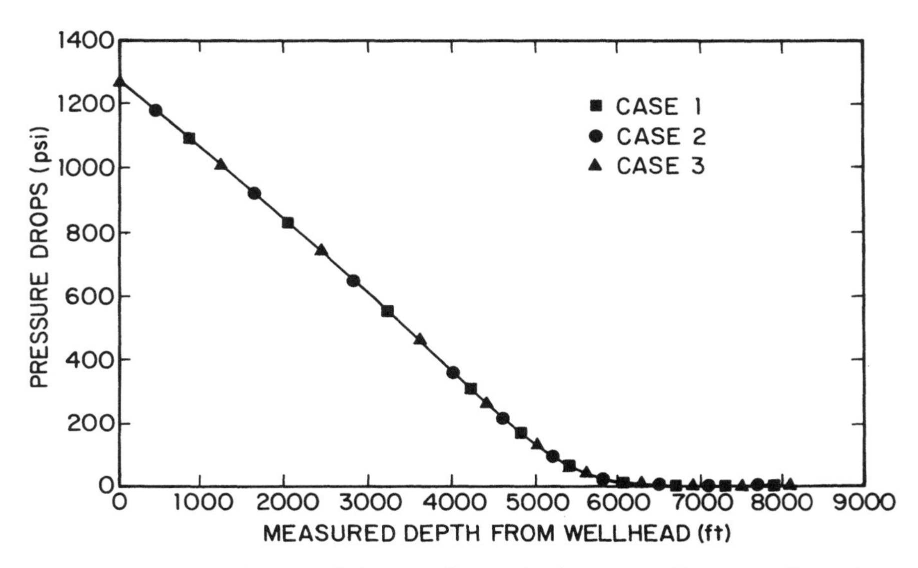

Figure 10–14 Influence of Three Different Fluid Entry Profiles on Well-Head Pressure for the Data Given in Example 10–8.[18]

on the order of 1 to 10 psia. This magnitude of wellbore pressure drop would have a significant influence on well productivity, provided reservoir drawdown is also on the same order of magnitude as the wellbore pressure drop. This may occur in high-permeability (about 1000 md and above) reservoirs. Even in these reservoirs, *the influence of wellbore pressure loss or well productivity can be minimized by choosing appropriate hole size, length and proper completion geometry*. The influence of fluid entry profile on the total wellbore pressure drop can be measurable, but it may not be large enough to change the wellhead pressures.

PRACTICAL CONSIDERATIONS

Generally, the reserves recovered from a well depend upon minimum bottomhole pressure that can be attained in a producing wellbore. In a horizontal well which is on an artificial lift, the minimum pressure attainable in the wellbore depends upon location of the pumping system. At present, many operators choose to locate the artificial lift system in the vertical section of a horizontal well. (Submersible pumps have been used in the build curve and also in the horizontal portions of horizontal wells.) Assuming that the artificial lift is located in the vertical section, as shown in Figure 10–15, depending upon the turning radius, different pressure drops would be required to lift fluids through curved sections of the wellbore.[19] For example,

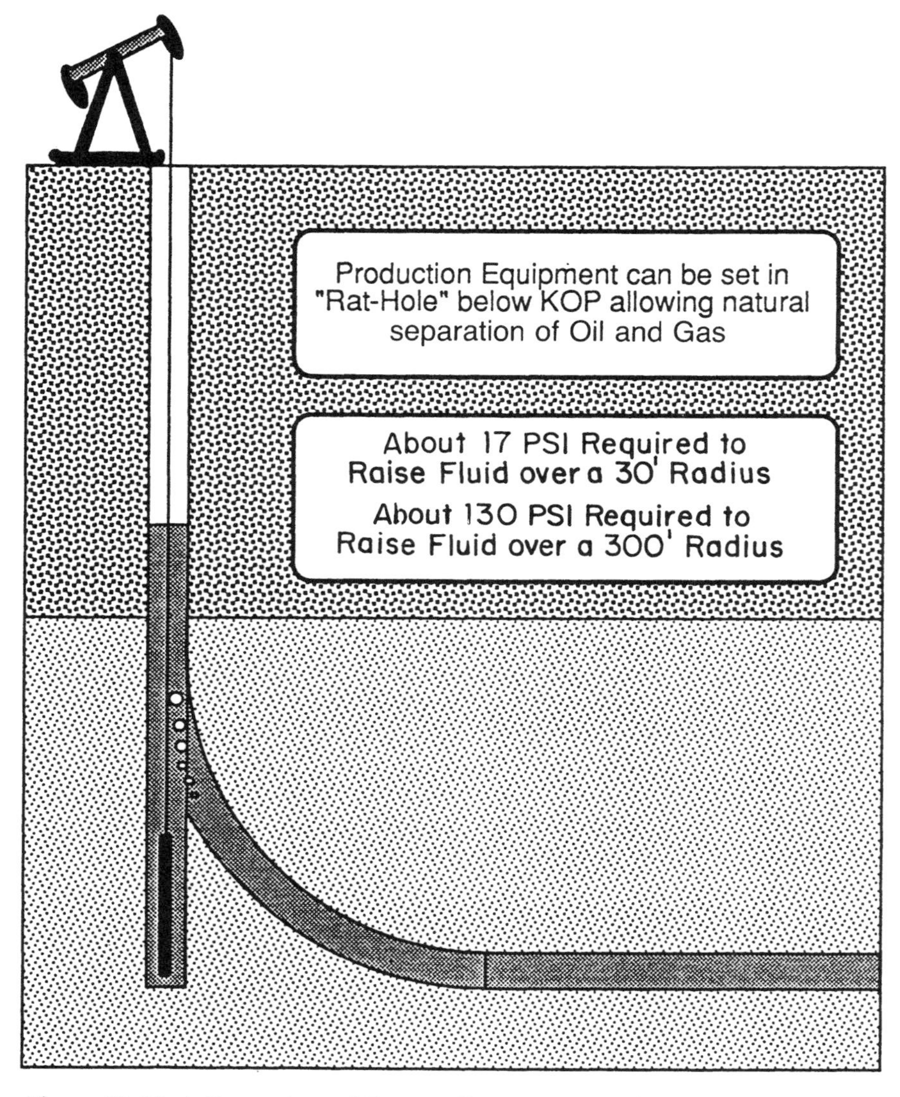

Figure 10–15 A Comparison of Pressure Drop in a Curved Section in Horizontal Wells.[19]

in a short radius system, with 40 ft turning radius, about 17 psi would be required to lift fluid through the curved section assuming that the pressure drop is close to hydrostatic head of 0.43 psi/ft. For a medium radius system with a turning radius of 300 ft, about 129 psi would be required to raise fluid through the curved section. Thus, if one locates the pumping system in the vertical section, then depending upon the turning radius, a certain amount of reservoir drawdown will be lost; and hence, producible reserves from a well may also change. This is especially important in low pressure reservoirs.

The minimum wellbore pressure drop has an impact on gas and condensate reservoirs. In tight gas wells, which are normally fractured using proppants, careful attention should be given to the amount of fluid injected. This is because the frac fluid may accumulate in the low spots of the wellbore and it may be difficult to unload it. Similarly, in condensate reservoirs, proper attention should be given in designing the drilled wellbore path so as to minimize well loading and excessive fluid accumulations in low spots.

REFERENCES

1. Dikken, B. J.: "Pressure Drop in Horizontal Wells and Its Effect on Their Production Performance," paper SPE 19824, presented at the SPE 64th Annual Technical Conference, San Antonio, Texas, 1989.
2. Bradley, Howard B.: *Petroleum Engineering Handbook,* 1st Printing, Society of Petroleum Engineers, Richardson, Texas, 1987.
3. *Field Data Handbook,* Dowell Schlumberger, Central Region Office, Tulsa, Oklahoma.
4. The American Society of Mechanical Engineers, *ASME Transactions,* Vol. 66, November 1944.
5. Kays, W. M.: *Convective Heat and Mass Transfer,* McGraw-Hill Book Company, pp. 60–63, 1966.
6. Bhatti, M. S. and Shah, R. K.: "Turbulent and Transition Flow Convective Heat Transfer in Ducts," Chapter 4, *Handbook of Single-Phase Convective Heat Transfer,* Edited by S. Kakac, R. K. Shah, and W. Aung, John Wiley and Sons, Inc., New York, 1987.
7. Jain, A. K.: "An Accurate Explicit Equation for Friction Factor," *J. Hydraulics Div. ASCE,* Vol. 102, pp. 674–677, May 1976.
8. Reynolds, O.: "On the Dynamic Theory of Incompressible Viscous Fluids and the Determination of the Criterion," *Philos. Trans. Roy. Soc.,* Vo. T186A, pp. 123–164, 1894.
9. Ekman, V. W.: "On the Change from Steady to Turbulent Motion of Liquids," *Ark. Mat. Astron. OCH Fys.,* Vol. 6, No. 12, pp. 1–16, 1910.
10. Pfenninger, W.: "Experiments with Laminar Flow in the Inlet Length of a Tube at High Reynolds Numbers With and Without Boundary Layer Suction," Technical Report, Northrop Aircraft Inc., Hawthorne, California, May 1952.
11. Schiller, L.: "Experimentelle Untersuchungen zum Turbulenzproblem," *Z. Angew. Math. Mech.,* Vol. 1, pp. 436–441, 1921.
12. Simonek, J.: "Turbulent Transport in the Transition Flow Region," *Int. J. Heat Mass Transfer,* Vol. 27, pp. 2415–2420, 1984.
13. Joshi, S. D. and Shah, R. K.: "Convective Heat Transfer in Bends and Fittings," Chapter 10, *Handbook of Single-Phase Convective Heat Transfer,* Edited by S. Kakac, R. K. Shah, and W. Aung, John Wiley and Sons, Inc., New York, 1987.
14. Schuh, F. J., personal communication, December 1989.

15. Reiley, R. H., Black, J. W., Stagg, T. O., and Walters, D. A., and Atol, F. R.: "Improving Liner Cementing in High-Angle/Horizontal Wells," *World Oil,* July 1988.
16. MacDonald, R. R.: "Drilling the Cold Lake Horizontal Well Pilot No. 2," paper SPE 14428, presented at the 60th Annual Technical Conference and Exhibition of the Society of Petroleum Engineers, Las Vegas, Nevada, Sept. 22–25, 1985.
17. Brill, J. P. and Beggs, D. H.: *Two-Phase Flow in Pipes,* 6th Edition, University of Tulsa, Tulsa, Oklahoma. December 1988.
18. Unpublished report on Horizontal Well Analysis by J. J. Xiao, University of Tulsa, Tulsa, OK, Summer of 1988 Petroleum Engineering class project report.
19. Eastman Christensen, Horizontal Drilling and Completion Symposium, Bakersfield, California, January 1989.

APPENDIX A

Petroleum Fluid Properties

Fluid properties are important for both reservoir and production system calculations.

1) For reservoir performance studies, fluid properties are needed at reservoir temperatures and pressures.
2) For wellbore hydraulics, fluid properties are needed at various temperatures and pressures.

This appendix is divided into the following sections:

1) Evaluation of PVT properties
2) Oil properties and correlations
3) Gas properties and correlations
4) Water properties and correlations

EVALUATION OF PVT PROPERTIES:

Fluid properties can be evaluated by the following two methods:

1) *Laboratory PVT Analysis*
   Disadvantages—Usually conducted at reservoir temperatures.
   —May not be available early in life of reservoir (for economic reasons).
2) *Empirical Correlations*
   —Based on laboratory PVT analyses from various fields around the world.
   —Can be applied to wide ranges of pressures, temperatures and oil properties.

OIL PROPERTIES AND CORRELATIONS

1. OIL DENSITY

Symbol: ρ_o

Units lbm/ft³, gm/cc, °API

Definition

$$\text{Oil Density} = \frac{\text{Mass of oil}}{\text{Oil volume}}$$

CORRELATION

$$\rho_o = \frac{350\ \gamma_o + 0.0764\ \gamma_g R_s}{5.615\ B_o} \qquad \text{(A–1)}$$

where:

ρ_o = oil density, lbm/ft³
γ_o = oil specific gravity, dimensionless
γ_g = gas specific gravity, dimensionless
R_s = solution or dissolved gas, SCF/STB,
B_o = oil formation volume factor, RB/STB,

If $p > p_b$, the bubble-point pressure, then

$$\rho_o = \rho_{ob} \exp\left[c_o(p - p_b)\right] \qquad \text{(A–2)}$$

where

ρ_o = oil density at p, T, gm/cc
ρ_{ob} = oil density at p_b, T, gm/cc
p = pressure, psia,
T = temperature of interest, °F
p_b = bubble-point pressure at temperature T, psia,
c_o = oil isothermal compressibility at T, psi^{-1}

c_o can be calculated using Equation A–23 which is listed in the latter section of this Appendix.

γ_o = Oil specific gravity = oil density/water density

$$\gamma_o = \frac{141.5}{131.5 + \gamma_{API}} \qquad \text{(A–3)}$$

where γ_o is oil specific gravity, and γ_{API} is oil gravity, °API. The API gravity is normally measured for the stock tank oil at 14.7 psia and 60°F.

EXAMPLE A–1

A well produces oil with gas at a producing gas-oil ratio of 600 SCF/STB. Calculate the oil density in lbm/ft^3 at reservoir conditions.

Given:

Stock tank oil gravity = 28° API.
Gas gravity = 0.76.
Oil formation volume factor, B_o = 1.25 RB/STB.
Reservoir pressure, p = 2500 psi.
Bubble-point pressure, p_b = 2600 psi.

Solution

Since $p < p_b$, density is calculated using Equations A–3 and A–1.

$$\gamma_o = \frac{141.5}{131.5 + 28} = 0.887$$

$$\rho_o = \frac{350 \times 0.887 + 0.0764 \times 0.76 \times 600}{5.615 \times 1.25}$$

$$= 49.2 \text{ lbm/ft}^3$$

2. OIL VISCOSITY

Symbol: μ or μ_o

Units poise, centipoise, Pascal-sec, lbm/(ft-sec)

Units Conversion 1 cp = 0.01 poise = 0.001 Pascal-sec
= 6.72×10^{-4} lbm/(ft-sec).

Definition

- Viscosity is a measure of oil resistance to flow.
- Kinematic viscosity $= \dfrac{\text{Absolute Viscosity}}{\text{Density}}$

Factors Affecting Oil Viscosity

1. *Composition:* μ increases with a decrease in API gravity
2. *Temperature:* μ increases with a decrease in temperature
3. *Dissolved gas:* Lightens the oil and thus decreases molecular weight and viscosity
4. *Pressure:* An increase in pressure on an undersaturated oil compresses the oil and causes the viscosity to increase.

CORRELATIONS

1. **Saturated oil**
   Saturated oil is oil in equilibrium with gas at bubble-point pressure.
   a. *Chew and Conally's Correlation*[1] (see Fig. A–2):
      i. Calculate gas-free or dead oil viscosity.
         —Dead oil viscosity depends on API gravity of stock tank oil and temperature of interest. Dead oil viscosity is calculated from Figure A–1, which is based upon Beal's Correlation.[2]
      ii. Modify the dead oil viscosity to include effect of dissolved gas using Figure A–2 (Chew and Connally's Correlation).
   b. *Beggs and Robinson's Correlation*[3]
      This correlation can be used for both dead and saturated oils.
      i. Dead Oil Viscosity

$$\mu_{od} = 10^{X} - 1.0 \qquad \text{(A–4)}$$

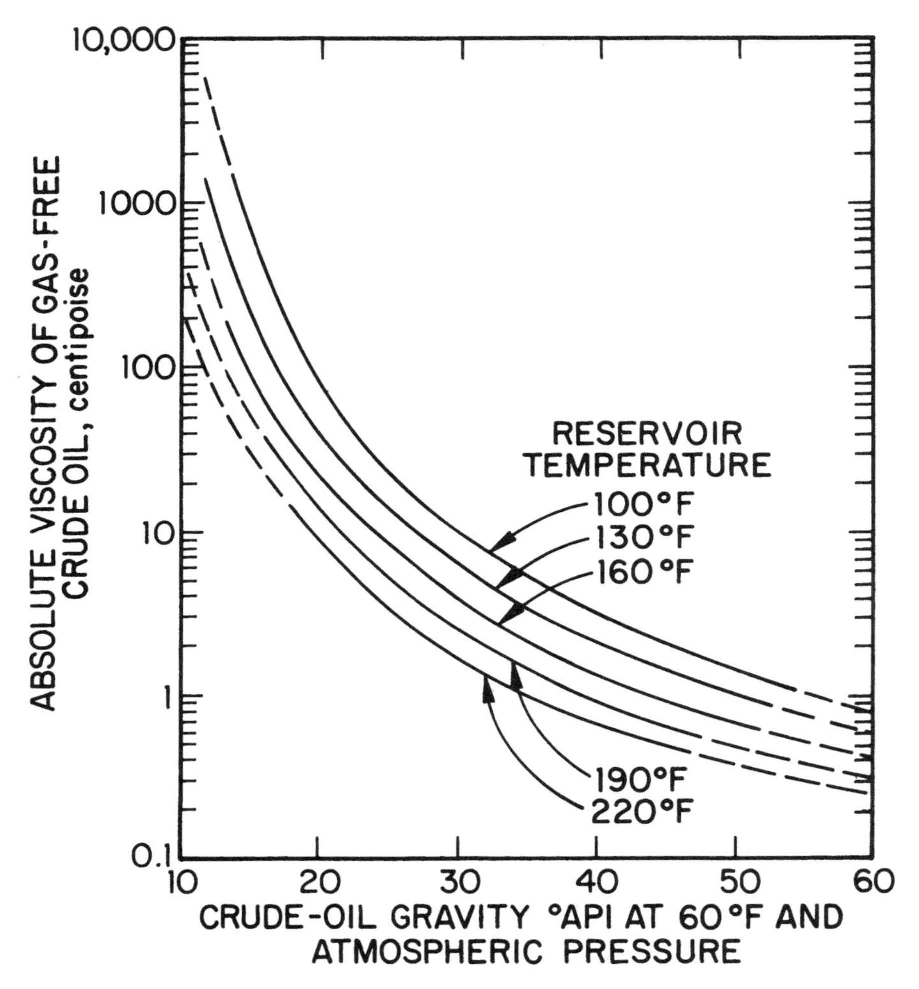

Figure A–1 Variation in Viscosity of Gas-Free Crude with Stock-Tank Crude Gravity.[2]

where

$$X = T^{-1.163} \exp(6.9824 - 0.04658\ \gamma_{API}) \tag{A–5}$$

μ_{od} = dead-oil vicosity, cp,

T = temperature of interest, °F, and

γ_{API} = stock-tank oil gravity, °API

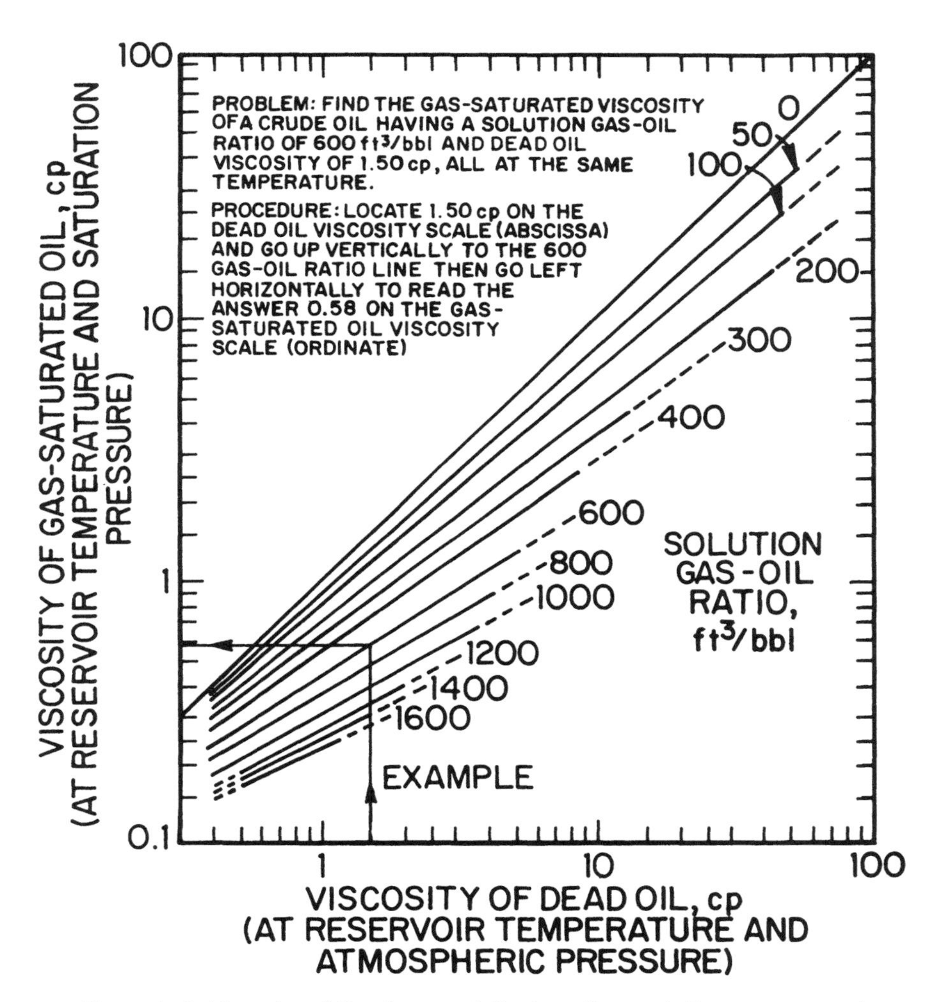

Figure A–2 Viscosity of Gas-Saturated Crude at Reservoir Temperature and Pressure; Chew and Connally's Correlation.[1]

ii. *Saturated Oil Viscosity* (μ_{os})

$$\mu_{os} = A(\mu_{od})^B \qquad \text{(A–6)}$$

where

μ_{os} = saturated-oil viscosity, cp

$$\mu_{od} = \text{dead-oil viscosity, cp}$$

$$A = 10.715\,(R_s + 100)^{-0.515} \tag{A–7}$$

$$B = 5.440\,(R_s + 150)^{-0.338} \tag{A–8}$$

where R_s is the solution GOR in SCF/STB.

2. **Undersaturated Oil System** (i.e. pressure > p_b).
   Above the bubble-point pressure only one phase exists in the reservoir, the liquid oil. This oil is called undersaturated oil and is capable of holding additional dissolved gas in solution. Increasing the pressure above p_b compresses the oil and increases the viscosity.
   a. *Beal's Correlation*
      Oil viscosity is calculated using Figure A–3, which can be used to calculate the rate of increase of viscosity above bubble-point pressure.

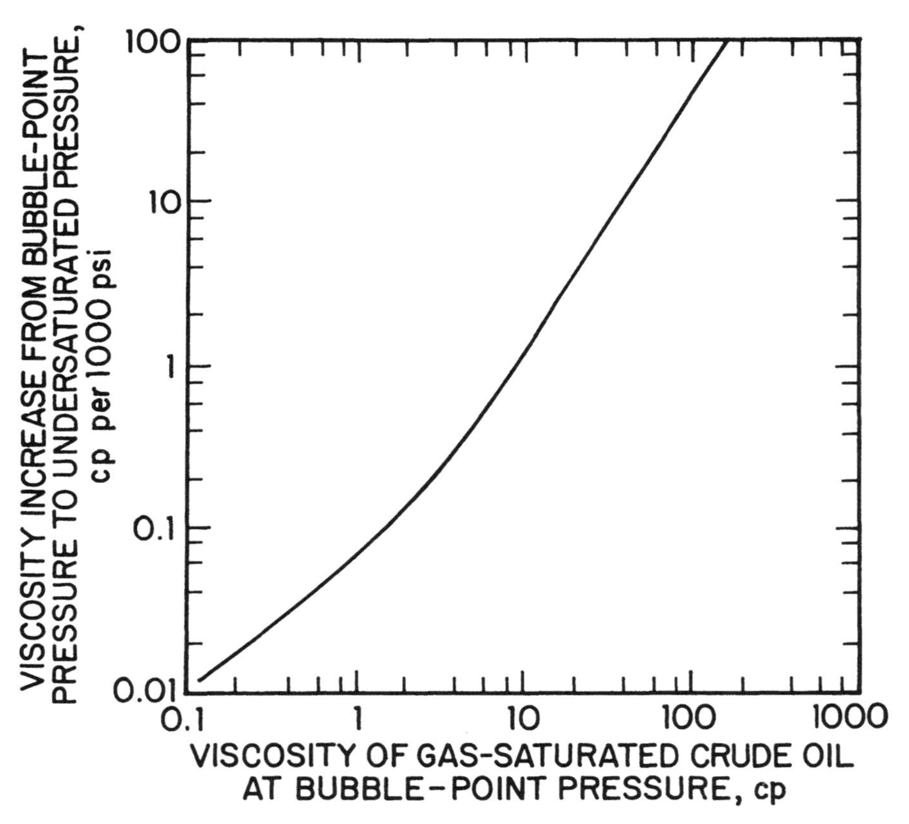

Figure A–3 Effect of Pressure on Viscosity of Gas-Saturated Crude Oils; Beal's Correlation.[2]

b. *Vasquez and Beggs' Correlation*[4]

$$\mu_o = \mu_{ob}(p/p_b)^m \qquad \text{(A–9)}$$

where

μ_o = viscosity at $p > p_b$, cp
μ_{ob} = viscosity at p_b, cp,
p = pressure of interest, psia
p_b = bubble-point pressure, psia

The exponent m is pressure dependent and is calculated from

$$m = C_1 p^{C_2} \exp(C_3 + C_4 p) \qquad \text{(A–10)}$$

where

p = pressure of interest, psia,
C_1 = 2.6,
C_2 = 1.187,
C_3 = −11.513, and
C_4 = -8.98×10^{-5}

EXAMPLE A–2

For the oil in Example A–1, calculate the oil viscosity at bubble-point pressure p_b of 2600 psia. (Given: Reservoir temperature = 200°F)

Solution

Since μ_o is needed at p_b, one can use μ_o correlations for saturated oil.

Method 1

1. Calculate dead oil viscosity, μ_{od}, using Beal's correlation. From Figure A–1, μ_{od} = 2.8 cp.
2. Calculate saturated oil viscosity, μ_{os}. From Figure A–2, μ_{os} = 0.69 cp.

Method 2

1. Calculate dead oil viscosity using Equations A–4 and A–5:

$$X = T^{-1.163} \exp(6.9824 - 0.04658\,\gamma_{API}) \qquad \text{(A–5)}$$
$$= 200^{-1.163} \exp(6.9824 - 0.04658 \times 28).$$
$$= 0.6164$$
$$\mu_{od} = 10^X - 1 = 10^{0.6164} - 1 = 3.13 \text{ cp} \qquad \text{(A–4)}$$

2. Calculate saturated oil viscosity, μ_{os}. Using Equations A–6, A–7, and A–8

$$A = 10.715\,(R_s + 100)^{-.515} \tag{A–7}$$

$$= 10.715 \times (600 + 100)^{-.515}$$

$$= 0.3670$$

$$B = 5.44\,(R_s + 150)^{-0.338} \tag{A–8}$$

$$= 5.44\,(600 + 150)^{-0.338}$$

$$= 0.5805$$

$$\mu_{os} = A\,(\mu_{os})^{B} = 0.367 \times (3.13)^{0.5805} \tag{A–6}$$

$$= 0.71 \text{ cp.}$$

Thus, Method 1, using charts, gives viscosity of 0.69 cp and Method 2, using correlations, shows viscosity of 0.71 cp.

3. BUBBLE-POINT PRESSURE

Symbol: p_b

Units: psia

Definition

- The pressure at which the first bubble of gas evolves, as the oil pressure is reduced. This is also called *saturation pressure*. At this pressure, oil is saturated with gas.

Factors Affecting Bubble-Point Pressure

1. Reservoir temperature, T_R.
2. Stock-tank oil gravity, γ_{API}
3. Dissolved gas gravity, γ_g
4. Solution gas-oil ratio (GOR) at initial reservoir pressure, R_{sb}.

CORRELATIONS

1. **Standing's Correlation** (see Fig. A–4)

$$p_b = 18 \times \left(\frac{R_{sb}}{\gamma_g}\right)^{0.83} \times 10^{y_g} \quad \text{(A–11)}$$

$$y_g = 0.00091\,(T_R) - 0.0125\,\gamma_{API} \quad \text{(A–12)}$$

$$= \text{mole fraction of gas}$$

$$R_{sb} = \text{solution GOR at } p_b\text{, SCF/STB}$$

p_b is bubble-point pressure in psia, T_R is reservoir temperature in °F and γ_{API} is API oil gravity.

2. **Lasater's Correlation**[5]
   a. Using Figure A–5, find M_o, the effective molecular weight of the stock-tank oil from API gravity.
   b. Calculate y_g, the mol fraction of the gas in the system.

$$y_g = \frac{R_{sb}/379.3}{R_{sb}/379.3 + 350\gamma_o/M_o} \quad \text{(A–13)}$$

   c. Calculate bubble-point pressure factor, $[p_b\gamma_g/T_R]$ using Figure A–6.
   d. Calculate p_b.

$$p_b = \left(\frac{p_b\gamma_g}{T_R}\right)\frac{T_R}{\gamma_g}$$

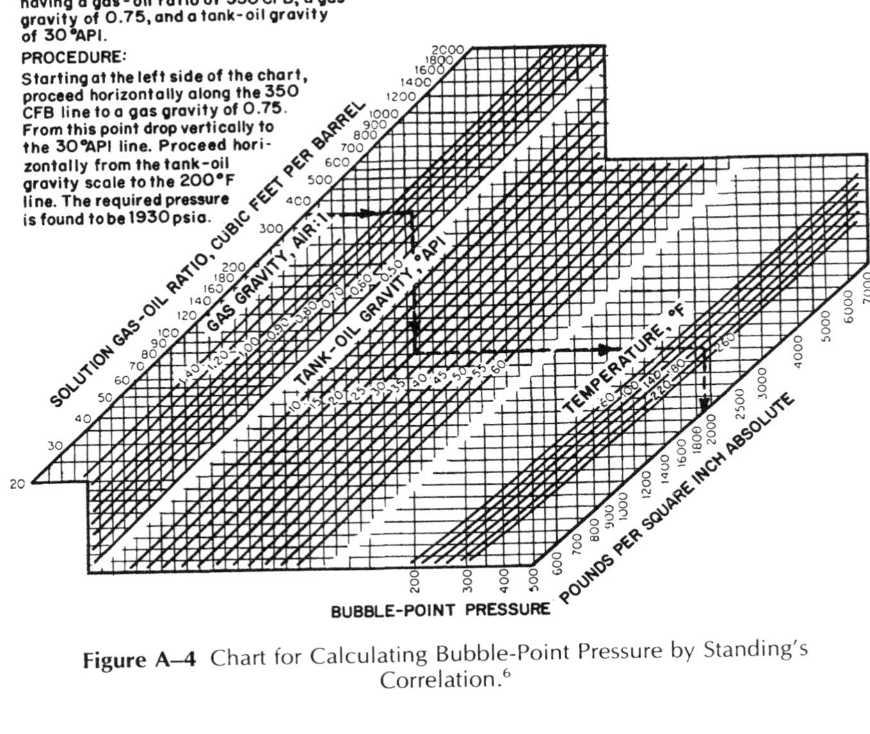

Figure A–4 Chart for Calculating Bubble-Point Pressure by Standing's Correlation.[6]

Where T_R is reservoir temperature, °R or (°F + 460)

p_b = bubble-point pressure, psia
γ_o = oil specific gravity (see Equation A–3)

3. **Vasquez and Beggs' Correlation**[4]

$$p_b = \left[\frac{R_{sb}}{C_1\gamma_g \exp[C_3\ \gamma_{API}/(T_R + 460)]}\right]^{1/C_2} \qquad \text{(A–14)}$$

where

p_b = bubble-point pressure, psia,
R_{sb} = solution GOR at p_b, SCF/STB,
γ_g = gas gravity,
γ_{API} = oil gravity, °API, and
T_R = reservoir temperature, °F.

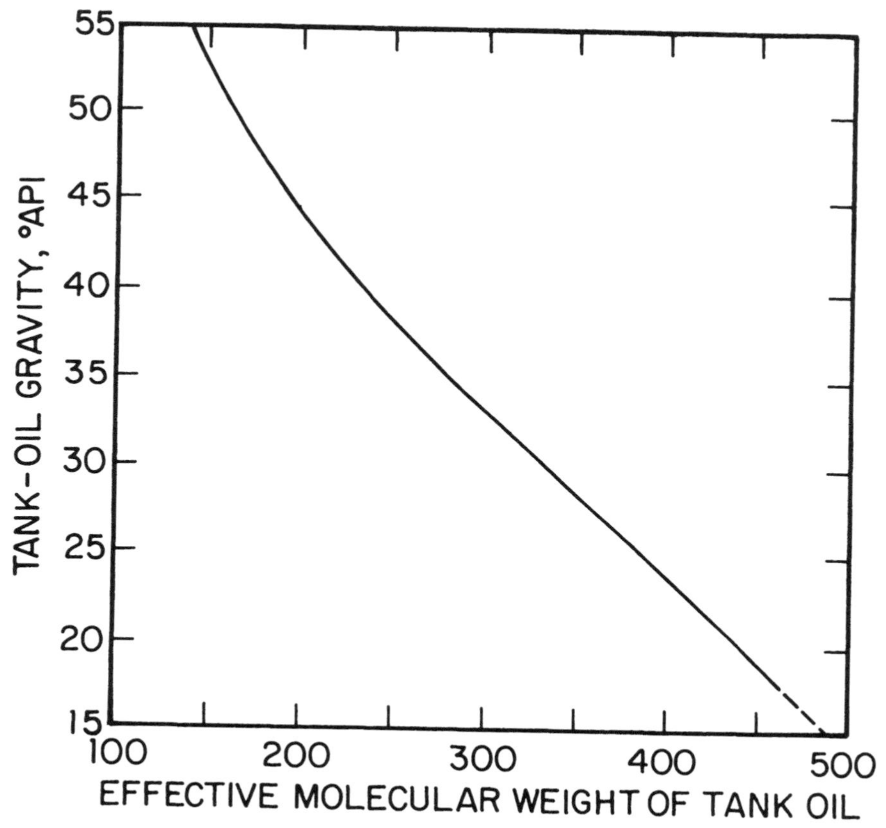

Figure A–5 Effective Molecular Weight Related to Stock-Tank Oil Gravity.[5]

Additionally, C_1 and C_2 are constants which are listed in Table A–1.

TABLE A–1 BUBBLE-POINT EQUATION CONSTANTS

	°API ≤ 30	°API > 30
C_1	0.0362	0.0178
C_2	1.0937	1.1870
C_3	25.724	23.9310

EXAMPLE A–3

For the following data, estimate bubble-point pressure, using Lasater's and Vasquez and Beggs' Correlations.

R_{sb} = 600 SCF/STB, T_R = 250°F, γ_g = 0.76, γ_{API} = 30° and γ_o = 0.876

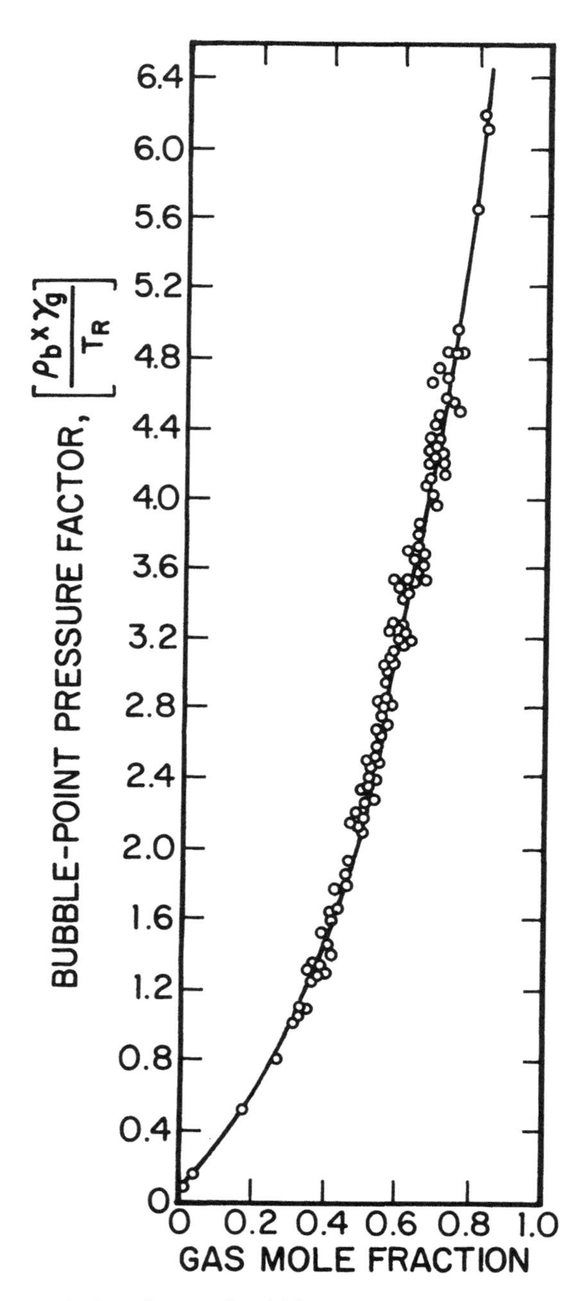

Figure A–6 Lasater's Correlation of Bubble-Point Pressure Factor with Gas-Mole Fraction.[5]

Solution

$$T_R = 250°F = 250 + 460 = 710° R.$$

1. **Lasater's Correlation**
   a. Calculate M_o, the effective mol wt. of stock-tank oil. From Figure A–5, $M_o = 330$.
   b. Calculate y_g, mol fraction of gas.

$$y_g = \frac{R_{sb}/379.3}{R_{sb}/379.3 + 350\,\gamma_o/M_o}$$

$$= \frac{600/379.3}{600/379.3 + 350 \times 0.876/330}$$

$$= 0.63$$

   c. Calculate bubble-point pressure factor, $(p_b\gamma_g/T_R)$. From Figure A–6, $(p_b\gamma_g/T_R) = 3.6$.
   d. Calculate p_b

$$p_b = \left(\frac{p_b\,\gamma_g}{T_R}\right) \times \frac{T_R}{\gamma_g} = 3.6 \times \frac{710}{0.76}$$

$$= 3363 \text{ psia}$$

2. **Vasquez and Beggs' Correlation**
   p_b is obtained from Equation A–14 using constants from Table A–1.

$$p_b = \left[\frac{600}{0.0362 \times (0.76) \exp\,[25.724\,(30)/710]}\right]^{1/1.0937}$$

$$= 3430 \text{ psia}$$

RANGE OF VALIDITY OF DIFFERENT CORRELATIONS

Correlation	Date	No of Tests	Comments
Standing	1947	105	Based only on gas-crude systems from California
Lasater	1958	158	Developed from data on black oil systems produced in Canada, western & mid-continental U.S., & South America
Vazquez & Beggs	1976	5008	Based on more than 600 PVT analyses from fields all over the world

4. SOLUTION GAS OIL RATIO

Symbol R_s

Units SCF/STB

Definition

R_s represents the amount of dissolved gas that will evolve from the oil as pressure is reduced from reservoir pressure to the atmospheric pressure.

CORRELATIONS

1. **Standing's Correlation**

$$R_s = \gamma_g \left(\frac{p}{18 \times 10^{y_g}} \right)^{1.204} \quad \text{(A–15)}$$

where

$$y_g = 0.00091(T) - 0.0125\,(\gamma_{API}) \quad \text{(A–16)}$$

R_s = solution GOR, SCF/STB
p = pressure, psia
γ_g = gas gravity
γ_{API} = oil gravity, API
T = temperature of interest, °F

2. **Lasater's Correlation**
   a. Using Figure A–5, find M_o, the effective molecular weight of the stock tank oil from the API gravity.
   b. Calculate y_g, the gas mol fraction.

For $\frac{p\gamma_g}{T} < 3.29$:

$$y_g = 0.359 \ln \left[\frac{1.473 p\gamma_g}{T} + 0.476 \right] \quad \text{(A–17)}$$

For $\frac{p\gamma_g}{T} > 3.29$:

$$y_g = \left(\frac{0.121 p\gamma_g}{T} - 0.236 \right)^{0.281} \quad \text{(A–18)}$$

where T is in °R in Equations A–17 and A–18.

$$R_s = \frac{132755\gamma_o y_g}{M_o(1 - y_g)} \quad \text{(A–19)}$$

3. **Vasquez and Beggs' Correlation**

$$R_s = C_1 \gamma_g p^{C_2} \exp\left[\frac{C_3 \gamma_{API}}{T + 460}\right] \quad \text{(A–20)}$$

where

R_s = gas in solution at p and T, SCF/STB
γ_g = gas gravity
p = pressure of interest, psia
γ_{API} = stock-tank oil gravity, °API
T = temperature of interest, °F
C_1, C_2, *and* C_3 are listed in Table A–1.

5. OIL FORMATION VOLUME FACTOR (FVF)

Symbol: B_o

Units RB/STB

Definition

- B_o accounts for the shrinkage of oil due to evolution of gas as oil is brought from the reservoir (reservoir pressure and temperature) to stock tank conditions (atmospheric pressure and temperature).
- It is the volume occupied by 1 STB oil plus its solution of gas at reservoir pressure and temperature.

CORRELATIONS

1. **Saturated Systems**
   a. *Standing Correlation:*
   Knowing R_s, T, p, γ_{API}, γ_g, use Figure A–7 to obtain B_o.
   b. *Vasquez and Beggs' Correlation*[4]

$$B_o = 1 + C_1R_s + C_2(T - 60)(\gamma_{API}/\gamma_{gc}) + C_3R_s(T - 60)(\gamma_{API}/\gamma_{gc}) \qquad \text{(A–21a)}$$

where

B_o = oil FVF at p and T, RB/STB
R_s = solution GOR at p, T, SCF/STB
T = temperature of interest, °F
p = pressure of interest, psia
γ_{API} = oil gravity, °API,
y_{gc} = gas gravity corrected (y_{air} = 1).
C_1, C_2, C_3 are constants which are listed in Table A–2.
γ_{gc} = corrected gas gravity as shown below:

$$\gamma_{gc} = \gamma_g[1.0 + 5.912 \times 10^{-5}\gamma_{API}T_s\log(p_s/14.7) \qquad \text{(A–21b)}$$

where

T_s = separator temperature, °F
p_s = separator pressure, psia

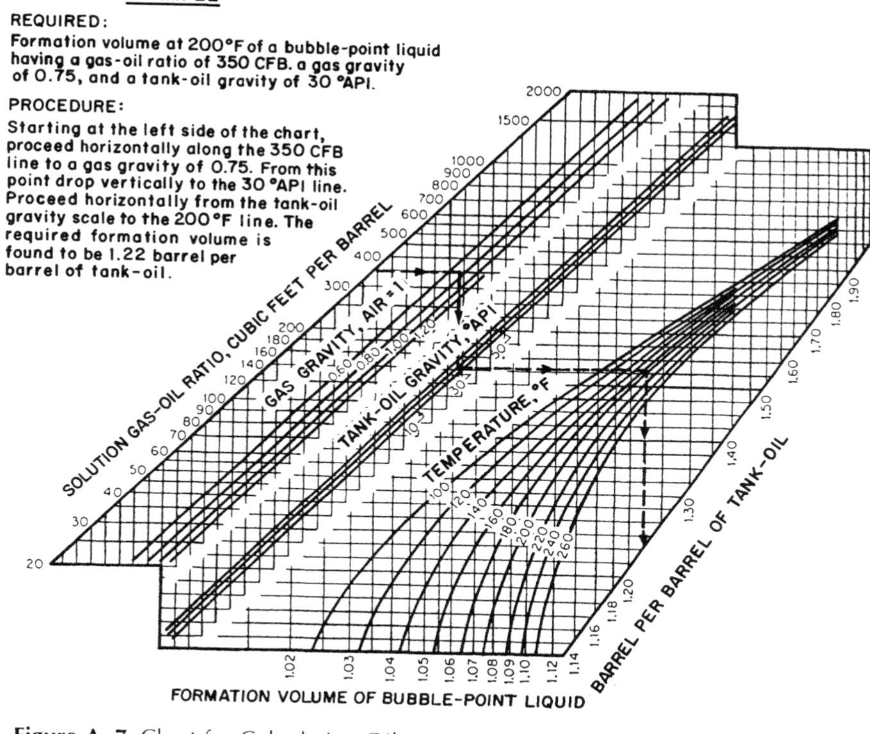

Figure A–7 Chart for Calculating Oil-Formation Volume Factor using Standing's Correlation.[6]

TABLE A–2 CONSTANTS FOR OIL FVF CORRELATION

	API ≤ 30	*API* > 30
C_1	4.677×10^{-4}	4.670×10^{-4}
C_2	1.751×10^{-5}	1.100×10^{-5}
C_3	-1.811×10^{-8}	1.337×10^{-9}

2. **Undersaturated System** (i.e. $p > p_b$)
The formation volume factor decreases with increase in pressure at $p > p_b$

$$B_o = B_{ob} \exp\left[c_o(p_b - p)\right] \tag{A–22}$$

and

B_{ob} = oil formation volume factor at the bubble-point pressure, p_b, RB/STB

c_o is in psi^{-1}, p and p_b are in psia

EXAMPLE A–4

Determine the oil formation volume factor, FVF, of oil with the following properties using Standing and Vasquez and Beggs correlations.

$$p_b = 2700 \text{ psia},\ R_{sb} = 600 \text{ SCF/STB},\ \gamma_g = 0.75$$
$$\gamma_{API} = 30° \text{ API},\ T = 250° \text{ F},\ p_s = 14.7 \text{ psia},\ T_s = 70°\text{F}.$$

Solution

1. **Standing's Correlation**
   From Figure A–7

   $B_o = 1.38$ bbl/STB

2. **Vasquez and Beggs' Correlation:**

$$\gamma_{gc} = \gamma_g = 0.75$$
$$B_o = 1 + 4.677 \times 10^{-4} \times 600 + 1.751 \times 10^{-5} \times (250 - 60) \times (30/0.75) + (-1.811 \times 10^{-8}) \times 600 \times (250 - 60) \times (30/0.75)$$
$$= 1.331 \text{ RB/STB}.$$

6. OIL COMPRESSIBILITY

Symbol c_o

Units 1/psi

Importance

- Oil compressibility is required to predict oil formation volume factors of undersaturated crude oils. Additionally, it is also needed to calculate oil density at different pressures and temperatures (see Equation A–2).

CORRELATION
Vasquez and Beggs' Correlation

$$c_o = \left[\frac{5R_{sb} + 17.2T - 1180\,\gamma_g + 12.61\,\gamma_{API} - 1433}{p \times 10^5}\right] \qquad \text{(A–23)}$$

where

T is in °F, p is in psi, and
R_{sb} is the solution gas-oil ratio at bubble-point pressure in SCF/STB.

EXAMPLE A–5

For the oil described in Example A–4, calculate the oil compressibility and hence the oil formation volume factor at 3000 psia. The following data are given:

$$p_b = 2700 \text{ psia},\ R_{sb} = 600 \text{ SCF/STB},\ \gamma_g = 0.75$$
$$\gamma_{API} = 30° \text{ API},\ T = 250° \text{ F},\ B_o = 1.331 \text{ RB/STB}$$

Solution

The oil compressibility and formation volume factor are calculated using Equations A–23 and A–22, respectively.

$$c_o = \left[\frac{5 \times 600 + 17.2 \times 250 - 1180 \times 0.75 + 12.61 \times 30 - 1433}{3000 \times 10^5}\right]$$
$$= 1.787 \times 10^{-5} \text{ psi}^{-1}$$
$$B_o = 1.331 \exp [1.787 \times 10^{-5} (2700 - 3000)]$$
$$= 1.324 \text{ RB/STB}$$

GAS PROPERTIES AND CORRELATIONS

7. GAS COMPRESSIBILITY FACTOR

Symbol z

Units Dimensionless

Definition

Gas compressibility factors account for the deviation of a real gas from the ideal gas behavior.

CORRELATIONS

In this section only one correlation is listed. For additional correlations see Appendix B.

Standing and Katz's Correlation[6]

1. Calculate pseudo-critical pressure p_{pc} (in psia) and temperature, T_{pc} (in °R) from free gas gravity, γ_g.

$$p_{pc} = 708.75 - 57.5\,\gamma_g \qquad \text{(A–24)}$$
$$T_{pc} = 169 + 314\,\gamma_g \qquad \text{(A–25)}$$

To account for the presence of gases such as CO_2, N_2 and H_2S, the following equations can be used.

$$p_{pc} = 689.60 - 30.48\,\gamma_g + 4.35\,(\%\ CO_2) - 1.68\,(\%\ N_2) + 4.44\,(\%\ H_2S) \qquad \text{(A–24a)}$$
$$T_{pc} = 152.89 + 339.29\,\gamma_g - 1.68\,(\%\ CO_2) - 2.69\,(\%\ N_2) \qquad \text{(A–25a)}$$

2. Calculate pseudo-reduced pressure p_{pr} and pseudo-reduced temperature T_{pr}.

$$p_{pr} = \frac{p}{p_{pc}} \qquad \text{(A–26)}$$
$$T_{pr} = \frac{T}{T_{pc}} \qquad \text{(A–27)}$$

3. Calculate z from Standing and Katz's z factor chart (Fig. A–8). For $p_{pr} < 1.5$, use Figure A–9,[7] for $p_{pr} < 0.007$, use Figure A–10.[7]

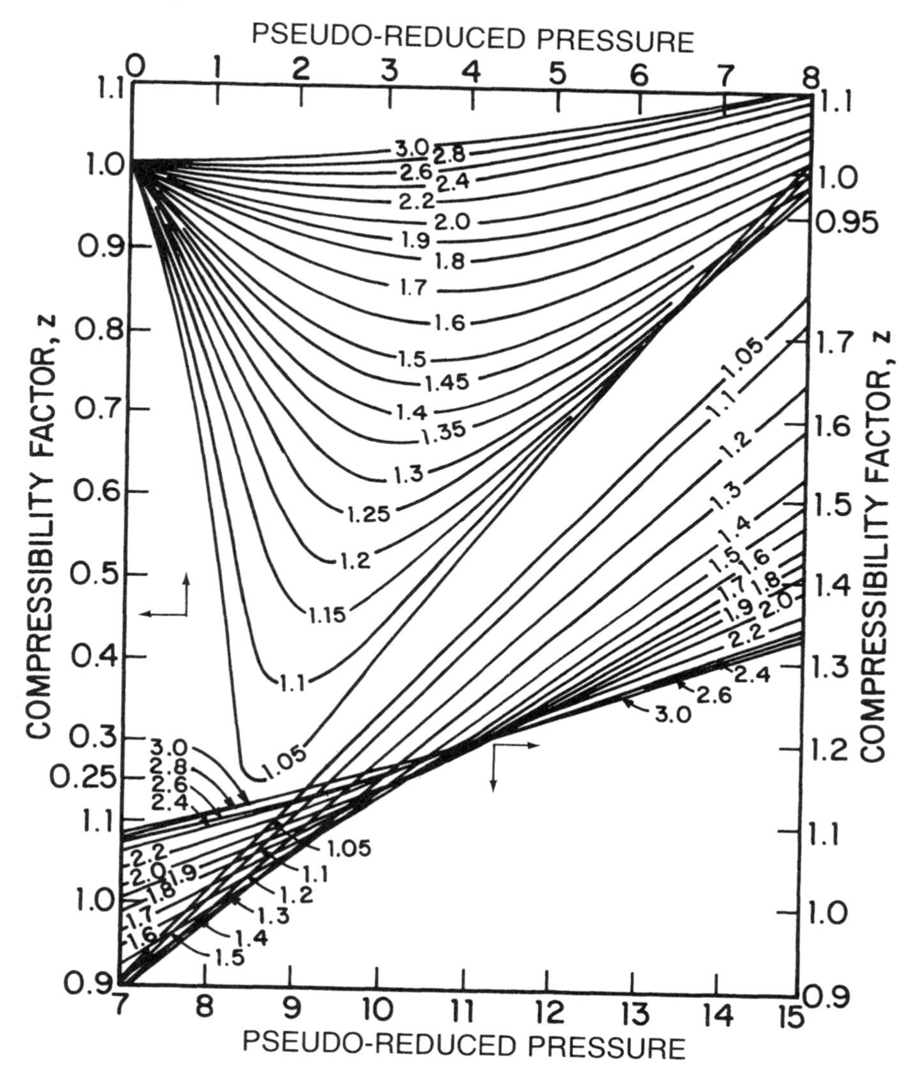

Figure A–8 Compressibility Factor for Natural Gases.[6]

EXAMPLE A–6

A gas has an average molecular weight of 17.4. Calculate the compressibility factor of gas at 1000 psia and 100°F.

Solution

$$\text{Specific gravity of gas, } \gamma_g = \frac{\text{Average Mol Wt.}}{\text{Average Mol Wt. of Air}} = \frac{17.41}{28.90} = 0.6$$

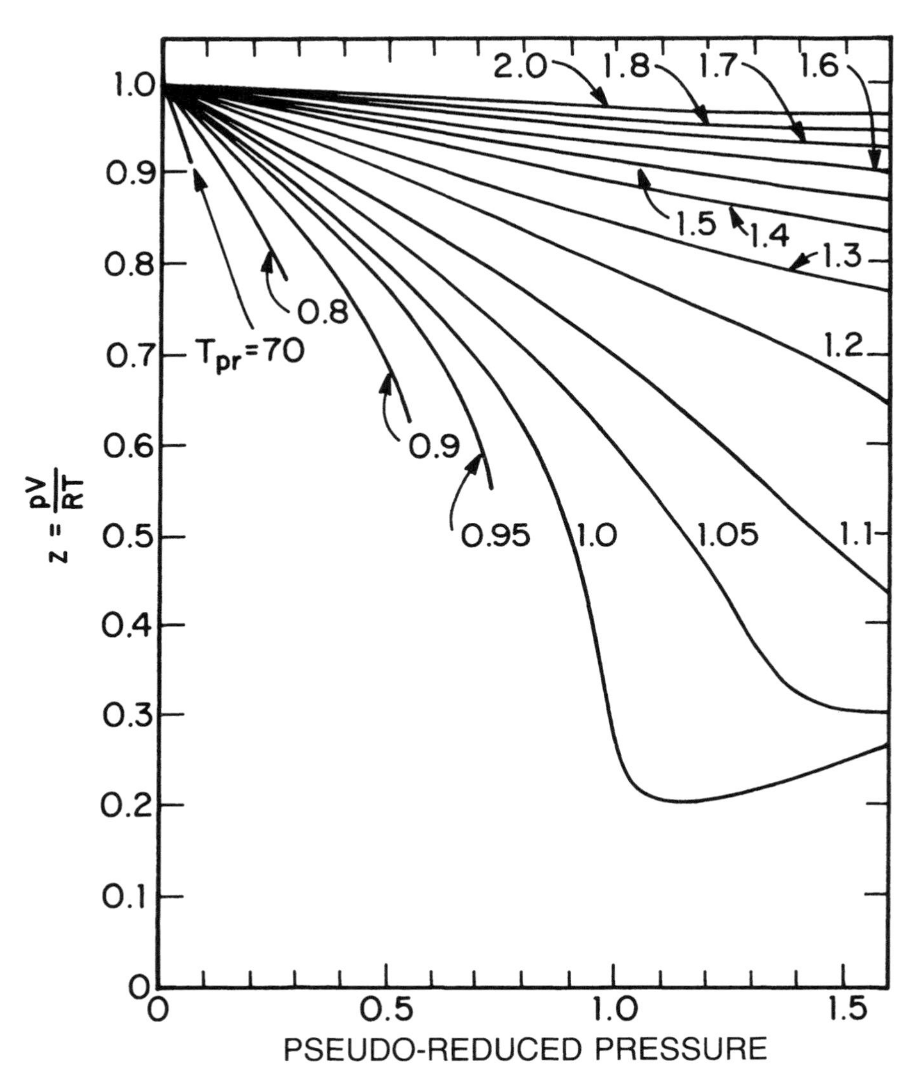

Figure A–9 Compressibility Factors for Natural Gases at Low Pseudo-reduced Pressures.[6]

The following parameters can be calculated using Equations A–24, A–25, A–27, and A–28.

$$p_{pc} = 708.75 - 57.5 \times 0.6 = 674.25 \text{ psia}$$
$$T_{pc} = 169 + 314 \times 0.6 = 357.4°\text{R}$$

$$p_{pr} = p/p_{pc} = 1000/674.25 = 1.483$$
$$T_{pr} = T/T_{pc} = (100 + 460)/357.4 = 1.567$$

From Figure A–9, $z = 0.89$

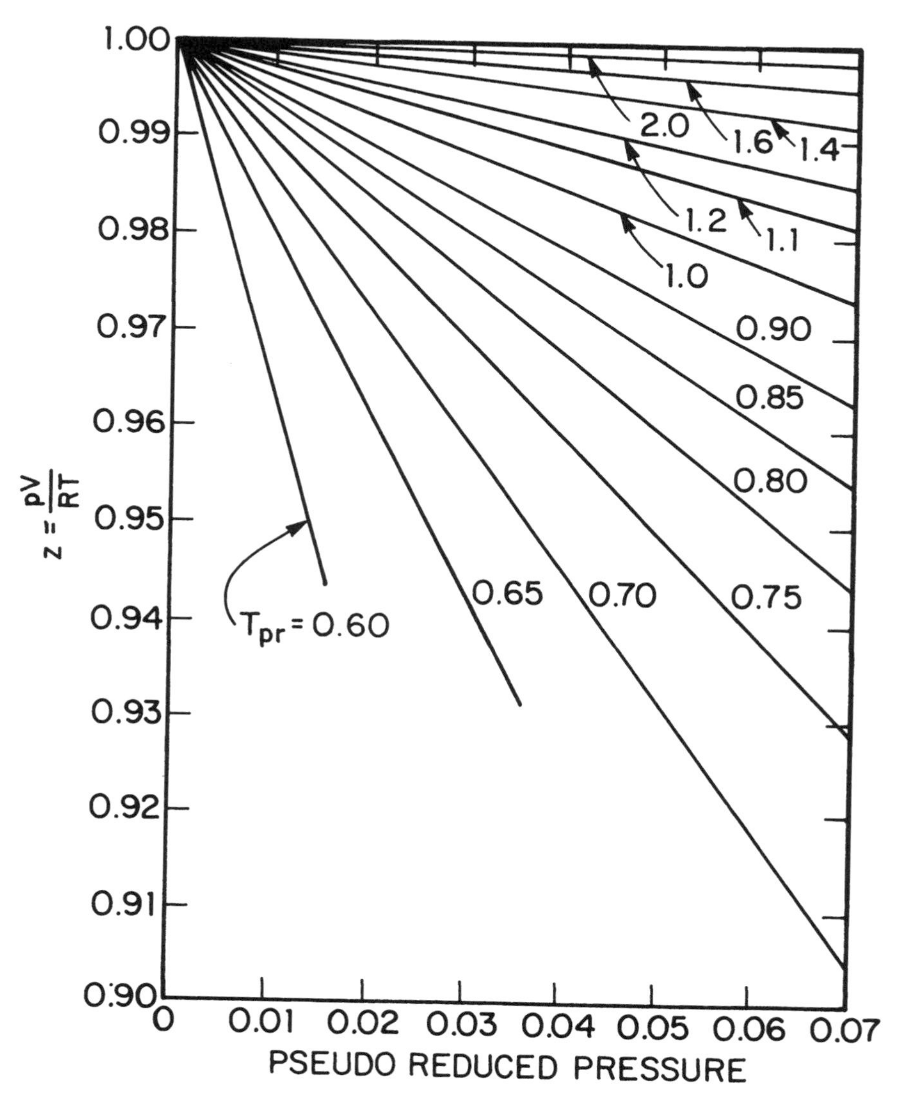

Figure A–10 Compressibility Factors for Natural Gases near Atmospheric Pressures.[6]

8. GAS FORMATION VOLUME FACTOR

Symbol B_g

Units ft^3/SCF, bbl/SCF

Definition

It is the reservoir volume in ft^3 occupied by 1 SCF of gas.

CORRELATION

$$B_g = \frac{0.0283\ zT}{p} \qquad \text{(A–28)}$$

B_g is in ft^3/SCF, T is in °R, and p is in psia.

EXAMPLE 8–7

Calculate the formation volume factor for the gas in Example 7 at 1000 psia and 100°F.

Solution

Using Equation A–28, B_g can be calculated

$$B_g = 0.0283 \times 0.89 \times \frac{(100 + 460)}{1000}$$

$$= 0.01410 \text{ ft}^3\text{/SCF}$$

9. GAS DENSITY

Symbol ρ_g

Units lbm/ft^3

CORRELATION

1.

$$\rho_g = 2.7 \frac{p \gamma_g}{z T} \tag{A–29}$$

p is in psia, ρ_g is in lbm/ft^3 and T is in °R.

2. If B_g is known,

$$\rho_g = \rho_{gsc}/B_g \tag{A–30}$$

where ρ_{gsc} represents the gas density at standard conditions of 14.7 psia and 60°F

$$\rho_{gsc} = \gamma_g \times (0.0764) \tag{A–31}$$

EXAMPLE A–8

Calculate the density of a 0.7 specific gravity gas at a pressure of 2000 psia and a temperature of 150°F.

Given: z at 2000 psia and 150°F = 0.745

Solution

$$\rho_g = \frac{2.7\, p\, \gamma_g}{zT} = \frac{2.7 \times 2000 \times 0.7}{0.745\,(150 + 460)}$$

$$= 8.318 \text{ lbm/ft}^3$$

10. GAS COMPRESSIBILITY

Symbol c_g

Units 1/psi

Definition

The isothermal compressibility of a gas is defined as the change in pressure.

$$c_g = -\frac{1}{V}\left(\frac{\partial V}{\partial p}\right)_T \qquad \text{(A–32)}$$

$$\text{For an ideal gas, } c_g = 1/p \qquad \text{(A–33)}$$

$$\text{For a real gas, } c_g = \frac{1}{p} - \frac{1}{z}\left(\frac{\partial z}{\partial p}\right)_T \qquad \text{(A–34)}$$

CORRELATION

1. If z values are known as a function of pressure, Equation A–34 can be used to estimate c_g by plotting z against p.

$$\text{Slope of plot} = \frac{\Delta z}{\Delta p} = \left(\frac{\partial z}{\partial p}\right)$$

The slope of the plot can be substituted in Equation A–34 to calculate gas compressibility.

2. **Trube's Method**[8]
   a. Calculate pseudo-reduced pressure, p_{pr} and pseudo-reduced temperature, T_{pr} from Equations A–24 to A–27.
   b. Use Figure A–11 or A–12 to estimate the pseudo-reduced compressibility c_{pr}.
   For low p_{pr}, use Figure A–11.
   For high p_{pr} ($p_{pr} > 3$), use Figure A–12.
   c. $$c_g = \frac{c_{pr}}{p_{pc}} \qquad \text{(A–35)}$$

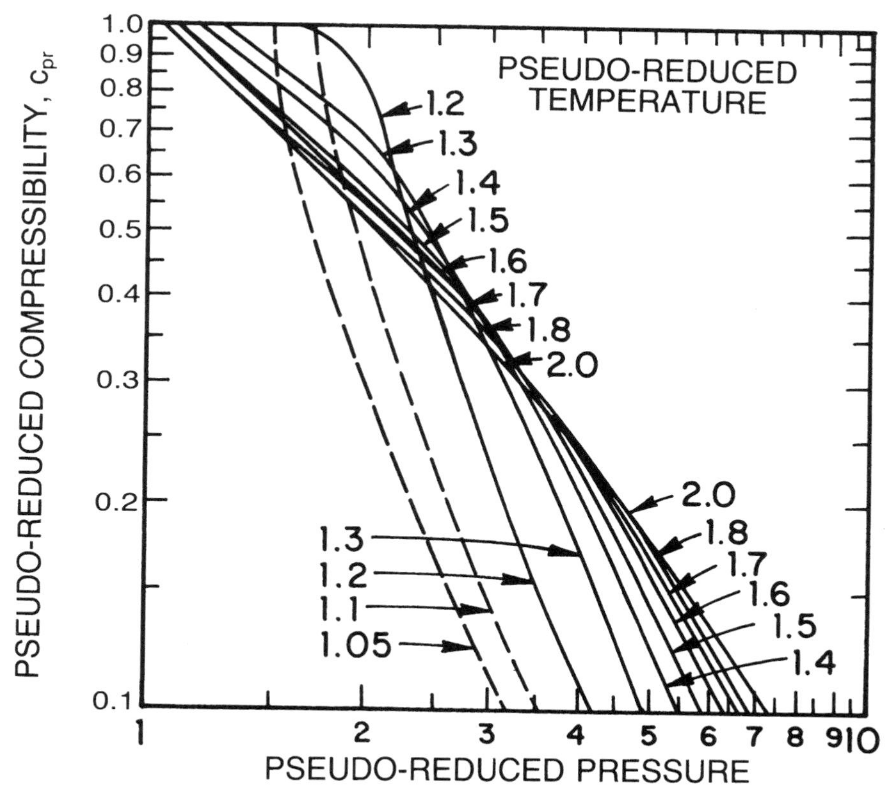

Figure A–11 Reduced Compressibility Coefficients for Low Pseudo-reduced Pressures and Fixed Pseudo-reduced Temperatures.[8]

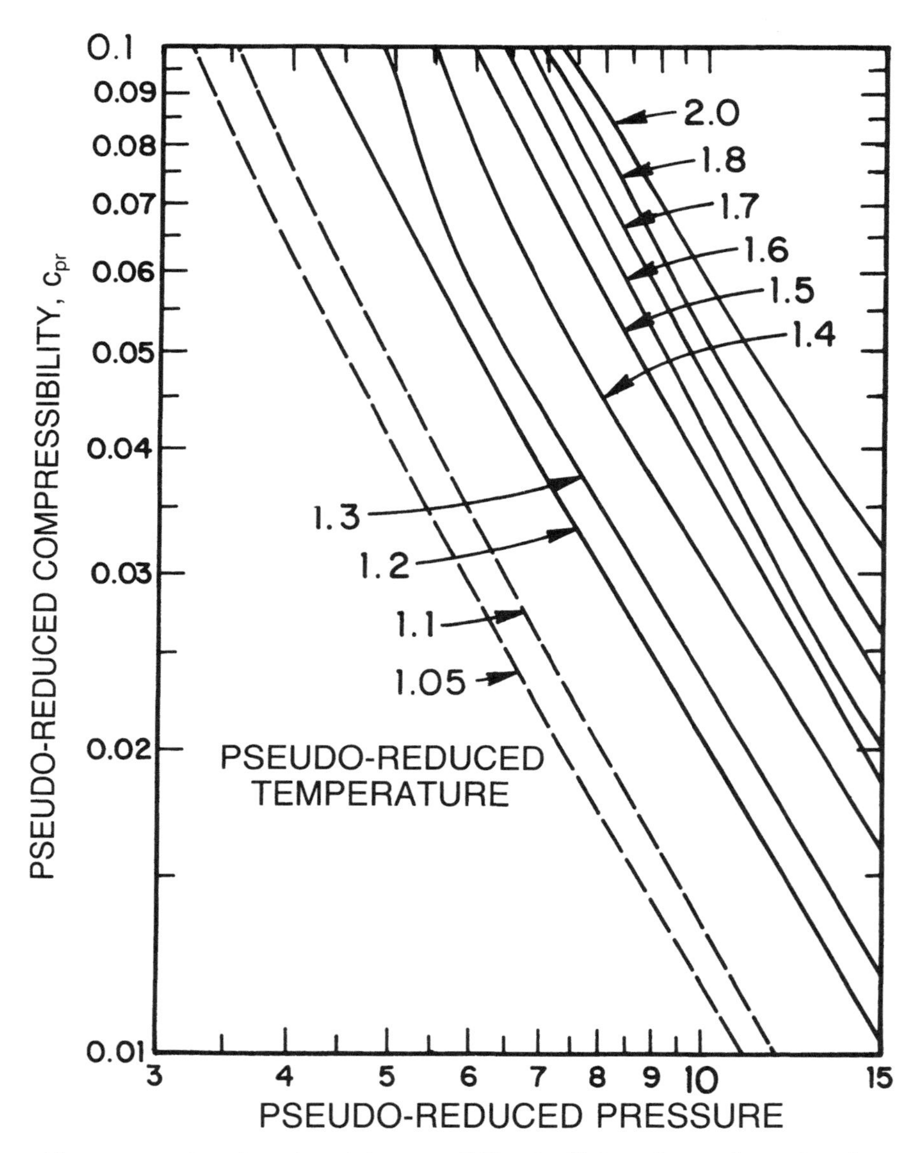

Figure A–12 Pseudo-reduced Compressibility Coefficients for Moderate Pseudo-reduced Pressures and Fixed Pseudo-reduced Temperatures.[8]

11. GAS VISCOSITY

Symbol μ_g

Units cp

Importance

To determine the resistance to flow during production and marketing of gas.

CORRELATION

Lee et al.'s Correlation[9]

$$\mu_g = K\, 10^{-4} \exp\,(X\, \rho_g^y) \qquad \text{(A–36)}$$

where

$$K = \frac{(9.4 + 0.02M)\, T^{1.5}}{209 + 19\, M + T} \qquad \text{(A–37)}$$

$$X = 3.5 + \frac{986}{T} + 0.01M \qquad \text{(A–38)}$$

$$y = 2.4 - 0.2X \qquad \text{(A–39)}$$

$$\rho_g = 0.0433\, \gamma_g \frac{p}{zT} \qquad \text{(A–40)}$$

T is in °R, μ_g is in cp, ρ_g is in gm/cc, p is in psia. M is molecular weight of the gas.

EXAMPLE A–9

Calculate the viscosity of a 0.75 specific gravity gas at 2000 psia and 150°F using Lee et al. correlation.

Given: 2000 psia and 150°F, $z = 0.745$

Solution

Equations A–36 through A–40 are used to calculate gas viscosity.

$$\text{Gas Mol. Wt.} = 0.75 \times 28.96 = 21.72$$

$$K = \frac{(9.4 + 0.02 \times 21.72)\,(610)^{1.5}}{209 + 19 \times 21.72 + 610} = 120.33$$

$$X = 3.5 + \frac{986}{610} + 0.01 \times 21.72 = 5.33$$

$$y = 2.4 - 0.2\,(5.33) = 1.33$$

$$\rho_g = 0.0433 \times 0.75 \times \frac{2000}{0.745 \times 610} = 0.143 \text{ gm/cc}$$

$$\mu_g = 120.33 \times 10^{-4} \exp\,[5.33 \times 0.143^{1.33}]$$

$$= 0.01797 \text{ cp}$$

12. TOTAL FORMATION VOLUME FACTOR FOR OIL

Symbol B_t

Units RB/STB

Definition

Total formation volume factor for oil, FVF, is the volume occupied by one stock tank barrel of oil, its remaining solution gas, and the free gas $(R_{si} - R_s)$ that has evolved from the oil.

$$B_t = B_o + B_g (R_{si} - R_s) \quad \text{(A–41)}$$

R_{si} = Initial solution GOR, SCF/STB.

R_s = Solution GOR, SCF/STB.

B_o = Oil formation volume factor, RB/STB

CORRELATION

Standing's Correlation

Knowing the values of GOR and gas and oil gravities, Figure A–13 can be used to estimate B_t at a given pressure and temperature condition.

EXAMPLE

REQUIRED:

Formation volume of the gas plus liquid phases of a 1500 CFB mixture, gas gravity = 0.80. tank-oil gravity = 40°API, at 200°F and 1000 psia.

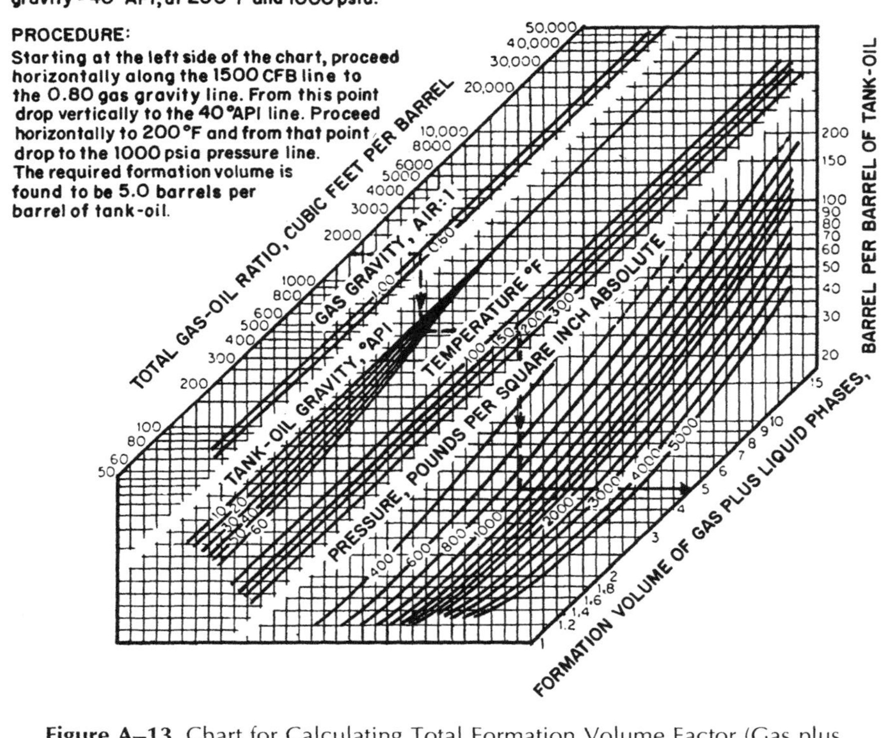

Figure A–13 Chart for Calculating Total Formation Volume Factor (Gas plus Liquid Phase) by Standing's Correlation.[6]

WATER PROPERTIES AND CORRELATIONS

13. WATER FORMATION VOLUME FACTOR

Symbol B_w

Unit RB/STB of water

CORRELATION

1. Estimation of B_W below bubble point

   Method 1

   a. Calculate B'_w, the formation volume factor of pure, dead (no dissolved gas) water at a desired pressure and temperature using Figure A–14.

   b. From Figure A–15, calculate ΔB_w, the correction for the effect of solution gas in the pure water.

   c. Calculate the parameter y, to account for the effect of salinity on the amount of gas in the solution using Fig. A–16.

   d. Calculate B_w

$$B_w = B'_w + y \, \Delta \, B_w \tag{A–42}$$

   McCain's Method

$$B_w = (1 + \Delta \, V_{wp})(1 + \Delta \, V_{wT}) \tag{A–43}$$

   ΔV_{wp} = change in volume due to pressure reduction to atmospheric pressure. (see Fig. A–17).

   ΔV_{wT} = at atmospheric pressure, change in volume due to reduction in temperature to 60°F (see Fig. A–18).

2. Estimation of B_W above bubble point

$$B_w = B_{wb} \exp \left[c_w \, (p - p_b)\right] \tag{A–44}$$

   where B_{wb} represents water formation volume factor at the bubble point of hydrocarbons of interest. Water compressibility, c_w, is calculated using Equation A–45. Pressure, p, represents pressure of interest at which B_w is desired.

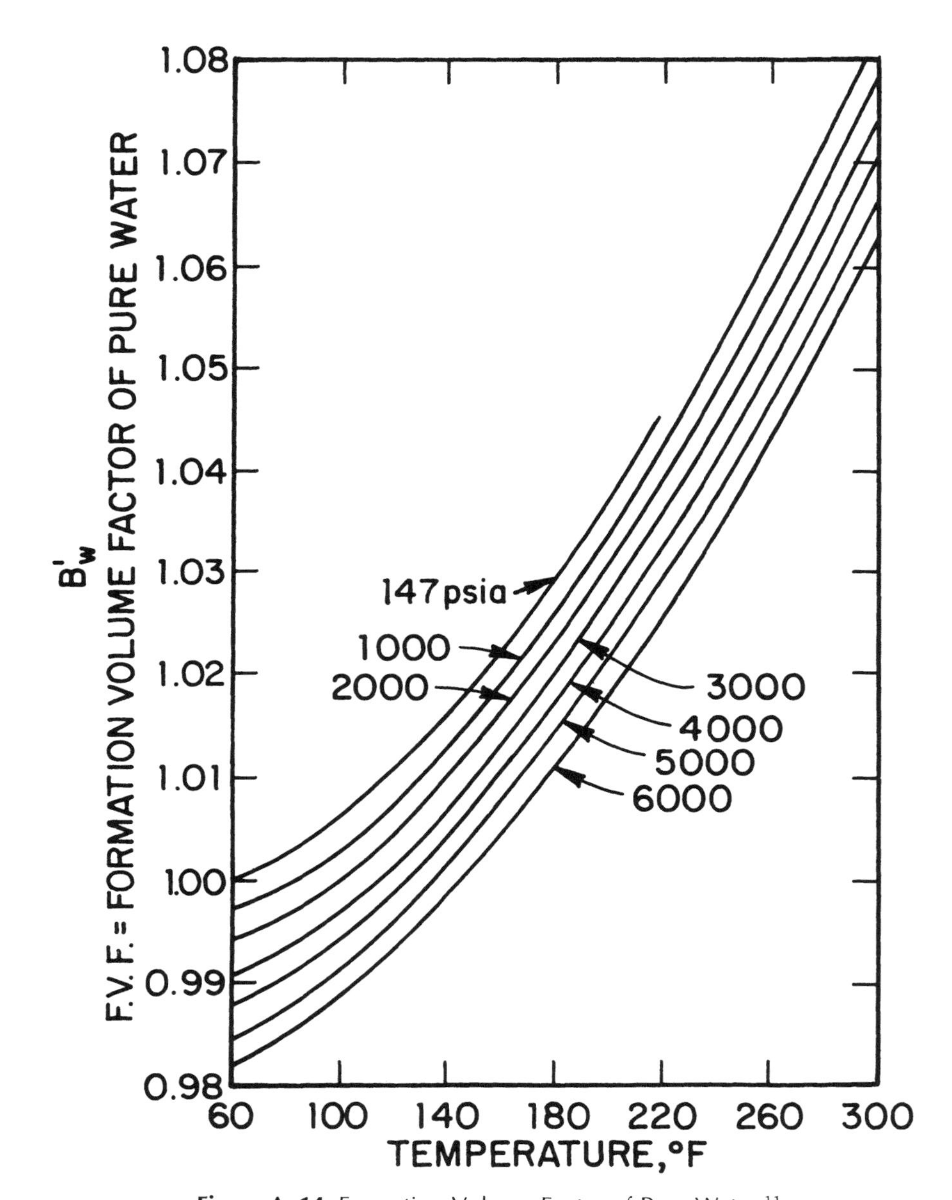

Figure A–14 Formation Volume Factor of Pure Water.[11]

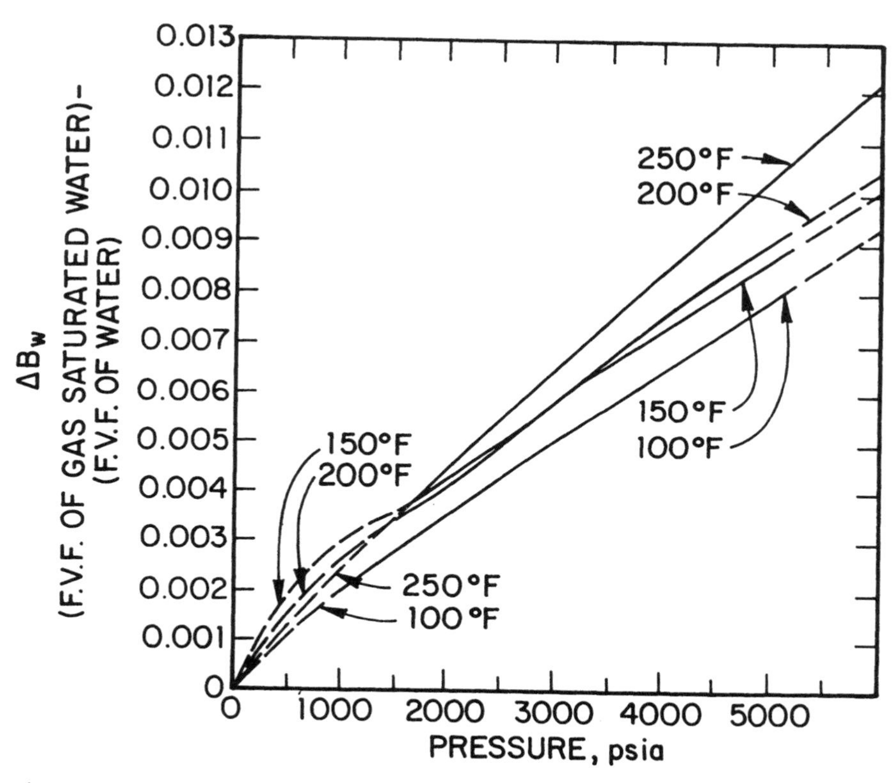

Figure A–15 Difference ΔB_w between B_w Values of Gas-Saturated Pure Water and of Pure Water.[11]

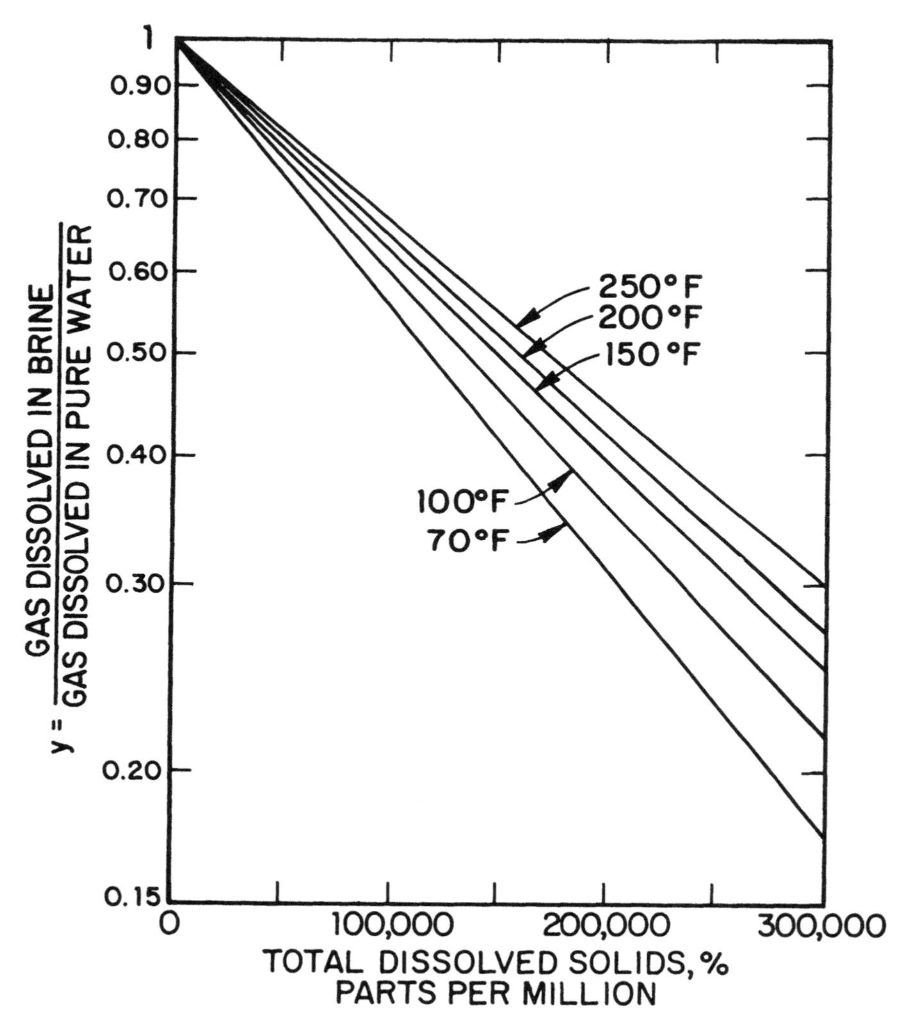

Figure A–16 Effect of Salinity on the Amount of Gas in Solution when Fully Saturated with Gas.[12]

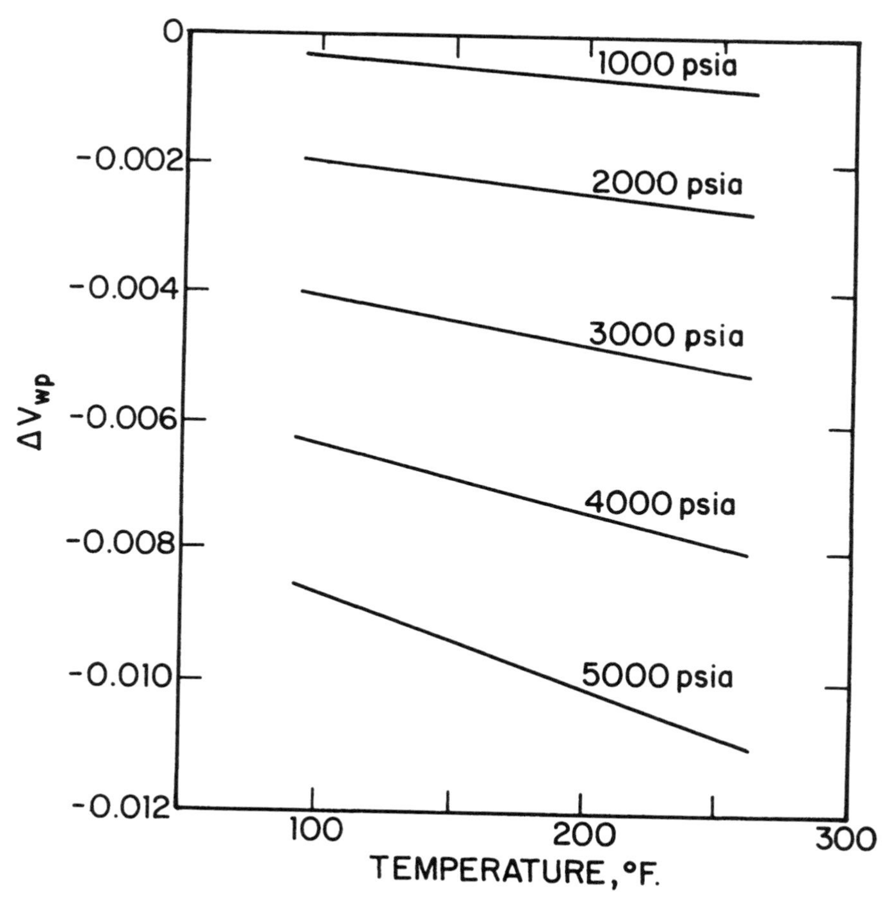

Figure A–17 ΔV_{wp} as a Function of Temperature and Pressure.[13]

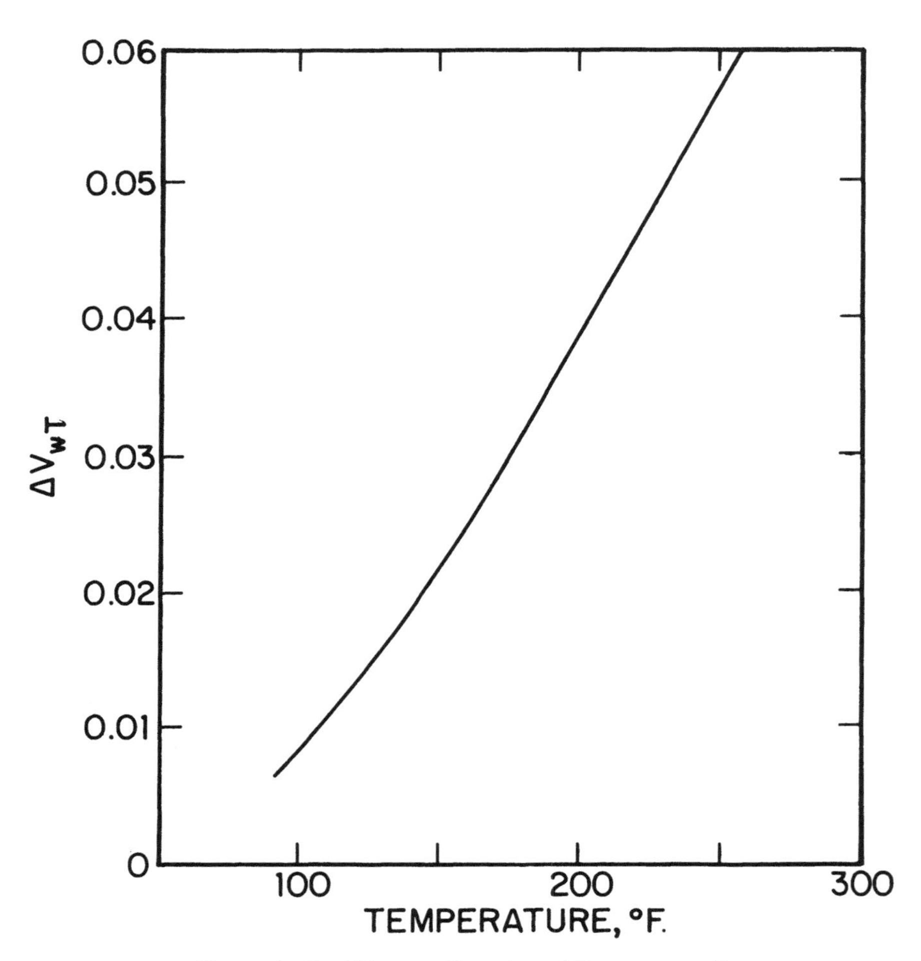

Figure A–18 ΔV_{wT} as a Function of Temperature.[13]

14. WATER COMPRESSIBILITY

Symbol c_w

Units 1/psi

CORRELATION[14]

$$c_w = c_{wp} \times c_{rs} \qquad \text{(A–45)}$$

c_{wp} = c_w for pure water, 1/psi (see Fig. A–19)

c_{rs} = Correction factor for dissolved gas, dimensionless (see Fig. A–20)

Influence of dissolved solids on water compressibility is usually very small and can be neglected for the engineering calculations.

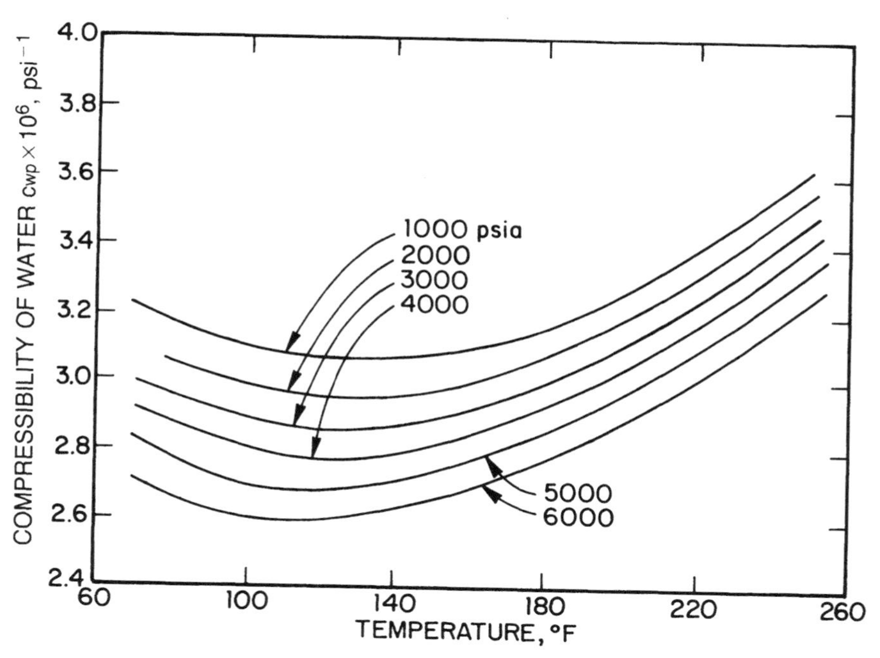

Figure A–19 Compressibility of Pure Water, c_{wp}.[14]

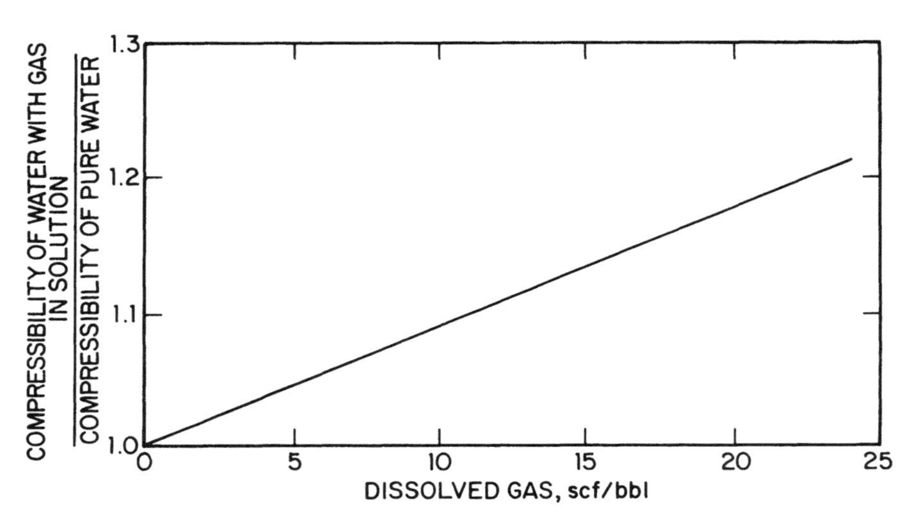

Figure A–20 Compressibility of Pure Water, including Effects of Gas in Solution.[14]

15. WATER DENSITY

Symbol: ρ_w

Units lbm/ft^3

CORRELATION

Water density calculations are similar to those for oil, except that the effects of dissolved gases on water density are small. However, the effect of dissolved solids on water density can be significant. The brine density at standard conditions of 14.7 psia and 60°F, $\rho_{w,sc}$, as a function of total dissolved solids is depicted in Figure A–19.

$$\rho_w = \frac{\rho_{w,sc}}{B_w} \tag{A–46}$$

B_w can be calculated either from Equation A–42, A–43 or A–44.

16. WATER VISCOSITY

Symbol μ_w

Units poise, centipoise, Pascal-sec, lbm/(ft-sec)

In general, water viscosity increases with increased pressure, and also with increasing dissolved solids, and decreases significantly with gas in solution.

CORRELATION

Effect of temperature on water viscosity can be approximated by

$$\ln \mu'_w = 1.003 - 1.479 \times 10^{-2} T + 1.982 \times 10^{-5} T^2 \qquad \text{(A–47)}$$

T is °F and μ'_w is in *cp*.

$$\mu_w = \mu'_w \times c_{ds} \qquad \text{(A–48)}$$

c_{ds} = correction for dissolved solids, dimensionless (see Fig. A–22).

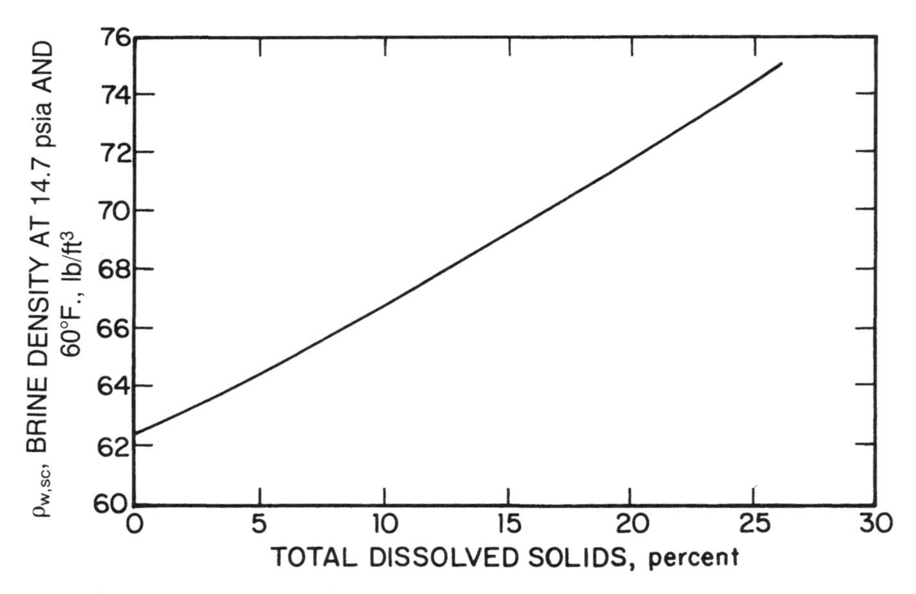

Figure A–21 Density of Brine as a Function of Total Dissolved Solids.[13]

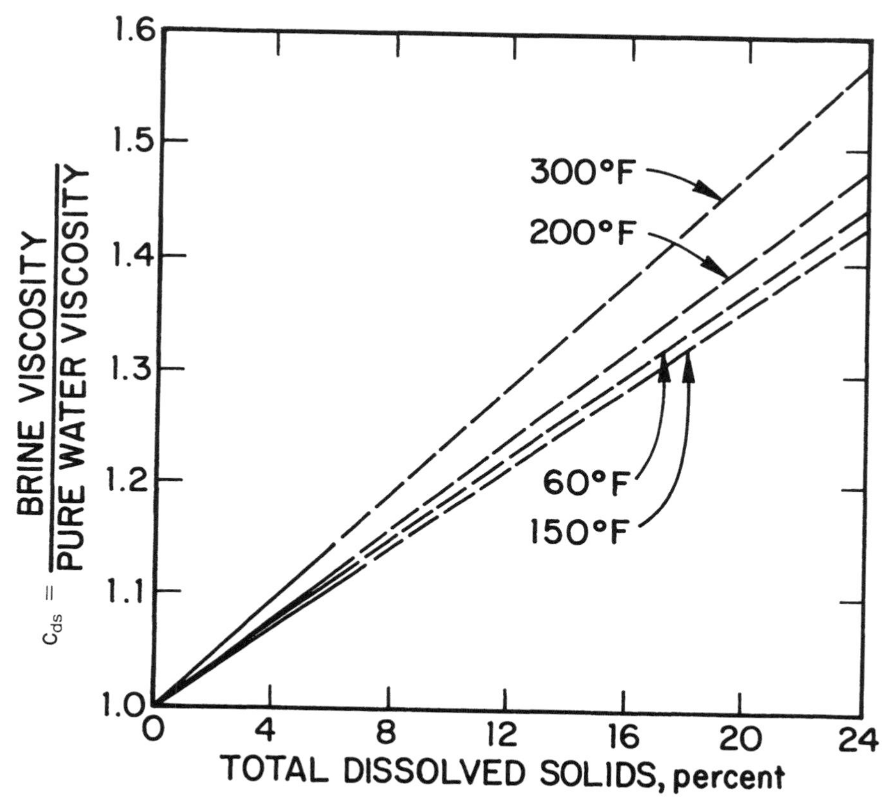

Figure A–22 Variation in Ratio of Brine Viscosity to Pure Water Viscosity over Salinity.[15]

18. SOLUBILITY IN WATER

Symbol R_{sw}

Units SCF/STBW
Solubility of hydrocarbon gases decreases as gas molecular weight increases.

CORRELATION

$$R_{sw} = R'_{sw} \times (y) \qquad \text{(A–49)}$$

R'_{sw} = gas solubility in pure water, SCF/STBW. (see Fig. A–23)
y = correction for water salinity, dimensionless (see Fig. A–16)

EXAMPLE A–10

A well produces water with 15% dissolved solids (150,000 ppm). Determine the following water properties at the bubble-point pressure, 2700 psia, and 250°F. Calculate the following properties.

1. Gas in solution in water at p_b, R_{sw}.
2. Water formation volume factor, B_w.
3. Water Compressibility, c_w.
4. Water Density, ρ_w.
5. Water Viscosity, μ_w.

Solution

1. **Gas in Solution in Water, R_{sw}**
   This is obtained using Equation A–49.

$$R_{sw} = R'_{sw} \times y \qquad \text{(A–49)}$$

From Figure A–23, $R'_{sw} = 17$
From Figure A–16, $y = 0.55$

$$R_{sw} = 17 \times 0.55 = 9.35 \text{ SCF/STBW}$$

2. **Water Formation Volume Factor, B_w**
   B_w can be obtained by two different methods.

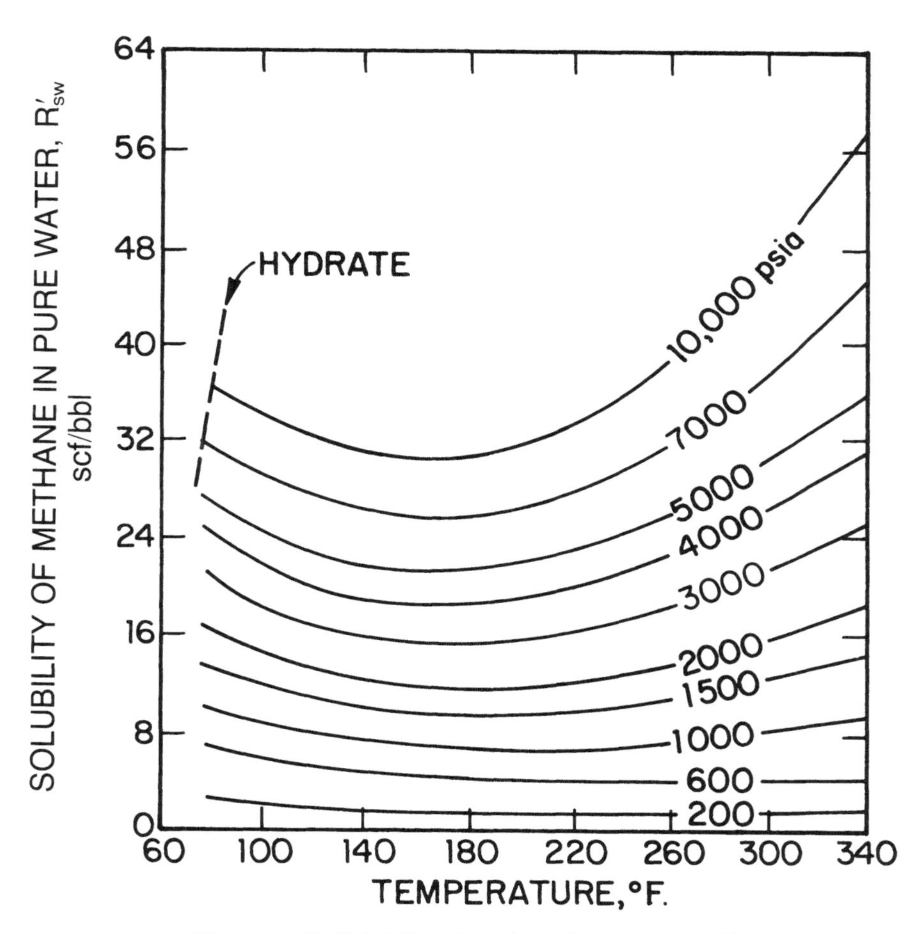

Figure A–23 Solubility of Methane in Pure Water.[16]

Method 1

$$B_w = B'_w + y\,\Delta B_w \qquad \text{(A–42)}$$

Figure A–14 shows $B'_w = 1.050$

Figure A–16 shows $y = 0.55$

Figure A–15 shows $\Delta B_w = 0.006$

$$B_w = 1.050 + 0.55 \times 0.006$$
$$= 1.053 \text{ RB/STBW}$$

Method 2

$$B_w = (1 + \Delta V_{wp})(1 + \Delta V_{wT}) \qquad \text{(A–43)}$$

Figure A–17 shows $\Delta V_{wp} = -0.0045$

Figure A–18 shows $\Delta V_{wT} = 0.057$

$$B_w = (1 - 0.0045)(1 + 0.057)$$
$$= 1.052 \text{ RB/STBW}$$

3. **Water Compressibility, c_w**
Water compressibility is calculated using Equation A–45.

$$c_w = c_{wp} \times c_{rs} \quad \text{(A–45)}$$

Figure A–19 shows $c_{wp} = 3.47 \times 10^{-6}$ 1/psi

$$c_{rs} = 1.08 \quad \text{(Fig. A–20)}$$
$$c_w = 3.47 \times 10^{-6} \times 1.08$$
$$= 3.75 \times 10^{-6} \text{ psi}^{-1}$$

4. **Water Density, ρ_w**
Water density is calculated using Equation A–46.

$$\rho_w = \frac{\rho_{w,sc}}{B_w} \quad \text{(A–46)}$$

Figure A–21 can be used to calculate $\rho_{w,sc}$

$$\rho_{w,sc} = 69.3 \text{ lbm/ft}^3$$
$$\rho_w = \frac{69.3}{1.053} = 65.8 \text{ lbm/ft}^3$$

5. **Water Viscosity, μ_w**
Water viscosity can be calculated using Equation A–48.

$$\mu_w = \mu'_w \times c_{ds}$$
$$\ln \mu'_w = 1.003 - 0.01479\,T + 1.982 \times 10^{-5}\,T^2$$
$$\ln \mu'_w = 1.003 - 0.01479 \times 250 + 1.982 \times 10^{-5} \times 250^2$$
$$= -1.454$$
$$\mu'_w = \exp(-1.454) = 0.234 \text{ cp}$$
$$c_{ds} = 1.33 \text{ (From Fig. A–22)}$$
$$\mu_w = 0.234 \times 1.33 = 0.311 \text{ cp}$$

REFERENCES

1. Chew, J. and Connally, C. A. Jr.: "A Viscosity Correlation for Gas-Saturated Crude Oils," *Trans.* AIME (1959), 23.
2. Beal, C.: "The Viscosity of Air, Water, Natural Gas, Crude Oil and Its Associated Gases at Oil Field Temperatures and Pressures," *Trans.* AIME (1946), 94.

3. Beggs, H. D. and Robinson, J. R.: "Estimating the Viscosity of Oil Systems," *J. Pet. Tech.* (September 1975), 1140–41.
4. Vasquez, M. and Beggs, H. D.: "Correlations for Fluid Physical Property Prediction," *J. Pet. Tech.* (June 1980), 968–70.
5. Lasater, J. A.: "Bubble Point Pressure Correlation," *Trans.* AIME (1958), 379.
6. Standing, M. B. and Katz, D. L.: "Density of Natural Gases," *Trans.* AIME (1942), 140.
7. Brown, G. G. et al: *Natural Gasoline and the Volatile Hydrocarbons.* Natural Gas Association of America, Tulsa, Oklahoma (1948).
8. Trube, A. S.: "Compressibility of Undersaturated Hydrocarbon Reservoir Fluids," *Trans.* AIME (1957), 341.
9. Lee, A. L., et al: "The Viscosity of Natural Gases," *Trans.* AIME (1966), 997.
10. Carr, N. L., et al: "Viscosity of Hydrocarbon Gases under Pressure," *Trans.* AIME (1954), 264.
11. Keenan, J. H. and Keyes, F. G.: *Thermodynamic Properties of Steam,* John Wily & Sons, Inc., New York (1936).
12. Eichelberger, W. C.: *Ind. and Engr. Chem.,* Vol. 47 (1955), 2223.
13. McCain, W. D. Jr.: *The Properties of Petroleum Fluids,* Petroleum Publishing Co., Tulsa, Oklahoma (1990).
14. Dodson, C. R. and Standing, M. B.: "Pressure-Volume-Temperature and Solubility Relations for Natural Gas Water Mixtures," *Drill. and Prod. Prac.,* API (1944), 1973.
15. Frick, T. C.: *Petroleum Production Handbook,* Millet the Printer, Inc., Dallas, Texas, Vol. II (1962).
16. Culbertson, O. L. and McKetta, J. J.: "Solubility of Methane in Water at Pressures to 10,000 Psia," *Trans.* AIME (1951), 223.

APPENDIX B

Gas Compressibility Factor

The ideal-gas law states that the behavior of a gas can be expressed as

$$pV = nRT \tag{B–1}$$

where

p = Pressure, psia
V = Volume, ft^3
n = Number of moles
R = The gas constant, ft^3-psi/(°R-lb-moles)
T = Temperature, Rankine (°R)

As shown below, the ideal-gas law, Equation B–1, can be modified to account for deviation of a real gas from ideal-gas behavior.

$$pV = nzRT \tag{B–2}$$

where z is compressibility factor. The Theorem of Corresponding States demonstrates that real gas mixtures will have the same z- factor for the same values of pseudo-reduced pressure p_{pr} and pseudo-reduced temperature T_{pr} which are defined as

$$p_{pr} = p/p_{pc} \quad \text{(B–3)}$$

$$T_{pr} = T/T_{pc} \quad \text{(B–4)}$$

where p_{pc} and T_{pc} are the pseudo-critical pressure and temperature for hydrocarbon gases, respectively. Table B–1 lists p_{pc} and T_{pc} as a function of gas gravity γ_g for values of γ_g ranging from 0.55 to 1.14.[1] Figure A–8 is Standing and Katz chart for determining z-factors for natural gases as a function of pseudo-reduced pressure p_{pr} and temperature T_{pr}.[2]

Since not all gases produced from a well are pure hydrocarbon gases, deviation from the z-factors given in Figure A–8 occurs if the produced gas contains any impurities. Carbon dioxide, CO_2; Nitrogen, N_2; and Hydrogen sulfide, H_2S, are the most common nonhydrocarbon impurities that are present in a hydrocarbon gas stream. The presence of these impurities cause z-factors to deviate from those given in Figure A–8. Table B–2 lists the effect of the three impurities, CO_2, N_2, and H_2S on the pseudo-critical pressure, p_{pc}, and temperature T_{pc}, based upon percentage volume of the impurities in the gas stream.[1] Once the corrected p_{pc} and T_{pc} values due to the impurities from Table B–2 are obtained, one can calculate the new values of p_{pr} and T_{pr} from Equations B–3 and B–4, respectively, and then obtain the corrected z-factors for new values of p_{pr} and T_{pr} from Figure A–8.

If P-V-T data are not available, then one can use one of the many published correlations in the literature for calculating z-factors. Table B–3 lists thirteen computational methods for calculating z-factor. Table B–3 also lists the deviation of these methods from the Standing and Katz chart in average absolute error percent.[3]

TABLE B–1 PSEUDOCRITICAL PROPERTIES OF HYDROCARBON GASES, p_{pc} AND T_{pc}*

γ_g	p_{pc}	T_{pc}	γ_g	p_{pc}	T_{pc}
0.55	673	336	0.85	664	441
0.56	673	341	0.86	664	444
0.57	672	346	0.87	663	448
0.58	672	350	0.88	663	451
0.59	672	354	0.89	662	454
0.60	671	358	0.90	662	457
0.61	671	362	0.91	662	461
0.62	671	365	0.92	662	464
0.63	670	368	0.93	661	467
0.64	670	372	0.94	661	471
0.65	670	375	0.95	660	474
0.66	670	378	0.96	660	477
0.67	669	382	0.97	659	481
0.68	669	385	0.98	659	484
0.69	669	388	0.99	659	487
0.70	668	392	1.00	658	491
0.71	668	395	1.01	658	494
0.72	668	398	1.02	657	497
0.73	668	401	1.03	656	500
0.74	667	405	1.04	656	504
0.75	667	408	1.05	655	507
0.76	667	411	1.06	655	510
0.77	666	415	1.07	654	514
0.78	666	418	1.08	654	517
0.79	666	421	1.09	653	520
0.80	665	424	1.10	652	524
0.81	665	428	1.11	652	527
0.82	665	431	1.12	651	530
0.83	665	434	1.13	651	534
0.84	664	438	1.14	650	537

Do not interpolate; values are inclusive to the next higher value.

* Used by permission of the Pacific Energy Association (formerly the California Natural Gasoline Association), taken from Bulletin No. TS–461, 1947. Data are for the average of the natural gas mixtures occuring in the western states—primarily California.

TABLE B–2 CORRECTION FOR PSEUDOCRITICAL PROPERTIES OF HYDROCARBON GASES FOR CARBON DIOXIDE, NITROGEN, AND HYDROGEN SULFIDE* (ADD IF POSITIVE OR SUBTRACT IF NEGATIVE)

Volume Percent of CO_2 or N_2 in Gas	Carbon Dioxide, CO_2		Nitrogen, N_2		Hydrogen sulfide, H_2S	
	p_{pc}	T_{pc}	p_{pc}	T_{pc}	p_{pc}	T_{pc}
1	+ 4	− 1	− 1	− 3	+ 2	−2
2	+ 8	− 3	− 3	− 6	+ 4	−2
3	+ 12	− 5	− 5	− 9	+ 7	−3
4	+ 17	− 7	− 7	−11	+10	−3
5	+ 21	− 9	− 9	−14	+15	−4
6	+ 25	−11	−11	−17		
7	+ 30	−12	−12	−20		
8	+ 34	−14	−14	−22		
9	+ 39	−16	−16	−25		
10	+ 44	−17	−17	−28		
11	+ 48	−19	−19	−30		
12	+ 53	−21	−21	−33		
13	+ 57	−22	−22	−36		
14	+ 61	−24	−24	−39		
15	+ 66	−26	−26	−42		
16	+ 70	−27	−27	−44		
17	+ 74	−29	−29	−47		
18	+ 79	−31	−31	−50		
19	+ 83	−32	−32	−52		
20	+ 87	−34	−34	−55		
21	+ 92	−36	−36	−58		
22	+ 96	−37	−37	−60		
23	+100	−39	−39	−63		
24	+104	−41	−41	−66		
25	+109	−42	−42	−68		
26	+113	−44	−44	−71		
27	+117	−46	−46	−74		
28	+122	−47	−47	−77		
29	+126	−49	−49	−79		
30	+130	−51	−51	−82		
31	+134	−52	−52	−85		
32	+139	−54	−54	−87		
33	+143	−56	−56	−90		
34	+147	−57	−57	−93		
35	+152	−59	−59	−95		
36	+156	−61	−61	−98		

* Values for carbon dioxide and nitrogen are based on data from the California Natural Gasoline Association Bulletin No. TS–461. Used by permission of the Interstate Oil Compact Commission. Values for hydrogen sulfide are supplied by the author.

TABLE B–3 COMPARISON OF METHODS FOR COMPUTER CALCULATION OF z-FACTORS

1. Gray-Sims[4]
2. Sarem[5]
3. Leung[6]
4. Papay[7]
5. Hankinson, et al.[8]
6. Carlile-Gillett[9]
7. Hall-Yarborough[10]
8. Brill[1]
9. Dranchuk, et al.[12]
10. Dranchuk-A. Kassem[13]
11. Gopal[14]
12. Burnett[15]
13. Papp[16]

	1	2	3	4	5	6
Average error, %	0.145	−0.043	0.638	−4.889	2.261	−0.052
Average absolute error, %	0.190	0.939	2.115	7.969	2.799	0.208

Isotherm Average Absolute Error in z-Factor, %
(Deviation from Standing and Katz)

T_r	1	2	3	4	5	6
1.2	0.070	2.160	7.689	19.490	11.387	0.334
1.3	0.081	1.370	4.179	12.541	6.176	0.214
1.4	0.075	0.991	2.558	9.518	3.497	0.179
1.5	0.048	0.787	1.704	8.564	2.193	0.115
1.6	0.063	0.742	1.183	8.023	1.469	0.259
1.7	0.437	0.618	0.855	7.067	1.093	0.216
1.8	0.435	0.561	0.479	5.702	0.802	0.182
2.0	0.354	0.434	0.472	3.407	0.553	0.233
2.4	0.324	0.654	0.532	1.209	0.299	0.191
3.0	0.016	1.071	1.500	4.171	0.524	0.153

Isobar Average Absolute Error in z-Factor, %
(Deviation from Standing and Katz)

p_r	1	2	3	4	5	6
0.2	0.055	1.147	1.689	0.235	0.181	0.129
0.5	0.050	0.529	1.185	0.389	0.425	0.118
1.0	0.047	1.176	0.825	0.562	0.782	0.065
1.5	0.015	0.880	0.577	1.372	0.987	0.138
2.0	0.031	0.957	1.608	2.688	0.564	0.160
2.5	0.000	1.567	2.078	3.059	1.199	0.149
3.0	0.046	1.206	1.502	2.132	1.836	0.200
3.5	0.064	0.738	0.880	2.021	2.364	0.161
4.0	0.014	0.696	1.078	3.018	2.903	0.134

TABLE B–3 Cont.

4.5	0.049	0.622	1.091	3.876	1.179	0.115
5.0	0.039	0.861	1.046	4.501	3.598	0.127
5.5	0.255	1.054	0.867	4.776	2.078	0.098
6.0	0.495	1.117	0.638	4.908	2.429	0.110
6.5	0.678	0.934	0.330	4.895	2.876	0.087
7.0	0.854	0.586	0.413	5.185	3.346	0.251
8.0	0.700	0.217	1.506	6.968	4.443	0.659
10.0	0.035	0.756	2.275	20.450	6.395	0.708
15.0	0.000	1.859	18.480	72.409	10.800	0.330

	7	8	9	10	11	12	13
Average error, %	−0.158	−3.423	−0.017	−0.002	0.105	−3.882	0.120
Average absolute error, %	0.512	3.966	0.361	0.304	1.338	4.601	0.539

Isotherm Average Absolute Error in z-Factor, %
(Deviation from Standing and Katz)

T_r							
1.2	0.552	1.627	0.417	0.335	1.177	20.639	1.068
1.3	0.620	1.042	0.402	0.417	1.251	2.251	0.547
1.4	0.425	0.372	0.361	0.308	0.745	2.070	0.790
1.5	0.602	0.857	0.169	0.143	1.008	1.441	0.877
1.6	0.424	0.724	0.313	0.244	1.161	1.379	0.451
1.7	0.341	0.425	0.424	0.302	1.497	1.204	0.350
1.8	0.381	0.447	0.293	0.254	1.588	1.215	0.325
2.0	0.489	1.909	0.281	0.183	2.210	2.156	0.310
2.4	0.682	7.149	0.284	0.221	0.461	4.920	0.287
3.0	0.607	25.104	0.665	0.633	2.279	8.740	0.382

Isobar Average Absolute Error in z-Factor, %
(Deviation from Standing and Katz)

p_r							
0.2	1.301	0.292	0.088	0.079	0.121	0.627	0.230
0.5	1.553	0.591	0.151	0.165	0.190	1.488	0.375
1.0	0.567	1.175	0.299	0.265	0.514	2.842	0.268
1.5	0.422	2.014	0.424	0.323	0.775	3.844	0.387

TABLE B–3 Cont.

2.0	0.583	3.136	0.483	0.369	1.238	3.669	0.891
2.5	0.577	3.635	0.289	0.234	0.730	4.303	0.918
3.0	0.575	3.445	0.337	0.317	1.656	6.680	0.703
3.5	0.447	3.344	0.514	0.424	0.714	8.359	0.658
4.0	0.437	3.547	0.701	0.574	0.899	9.603	0.650
4.5	0.331	3.757	0.665	0.528	0.725		0.532
5.0	0.325	3.983	0.569	0.477	0.684		0.465
5.5	0.310	4.249	0.434	0.372	3.590		0.498
6.0	0.503	4.359	0.242	0.247	3.008		0.579
6.5	0.304	4.495	0.125	0.182	2.522		0.613
7.0	0.242	4.636	0.155	0.146	2.098		0.631
8.0	0.258	4.998	0.202	0.185	1.602		0.616
10.0	0.109	6.100	0.247	0.159	1.301		0.319
15.0	0.381	13.622	0.575	0.424	1.710		0.367

REFERENCES

1. Smith, R. V.: *Practical Natural Gas Engineering,* PennWell Publishing, Tulsa, Oklahoma, 1983.
2. Katz, D. L., Cornell, D., Kobayashi, R., Poettmann, F. H., Vary, J. A., Elenbaas, J. R., Weinaug, C. F.: *Handbook of Natural Gas Engineering,* McGraw-Hill Book Company, Inc., N.Y., 1959.
3. Takacs, G.: "Comparing Methods for Calculating z-Factor," *Oil & Gas Journal,* pp. 43, May 15, 1989.
4. Gray, E. H. and Sims, H. L.: "Z-Factor Determination in a Digital Computer," *Oil & Gas Journal,* pp. 80–81, July 20, 1959.
5. Sarem, A. M.: "Z-Factor Equation Developed for Use in Digital Computers," *Oil & Gas Journal,* p. 118, Sept. 18, 1961.
6. Dranchuk, P. M. and Quon, D.: "A General Solution of the Equations Describing Steady State Turbulent Compressible Flow in Circular Conduits," *Journal of Canadian Petroleum Technology,* pp. 60–65, Summer 1965.
7. Papay, J.: "A Termelestechnologai Parameterek Valtozasa a Gaztelepek Muvelse Soran," *OGIL Musz. Tud. Kozl.,* pp. 267–273, Budapest, 1968.
8. Hankinson, R. W., Thomas, L. K., and Phillips, K. A.: "Predict Natural Gas Properties," *Hydrocarbon Processing,* pp. 106–108, April 1969.
9. Carlile, R. E. and Gillett, B. E.: "Digital Solutions of an Integral," *Oil & Gas Journal,* pp. 68–72, July 19, 1971.
10. Yarborough, L. and Hall, K. R.: "How to Solve Equation of State for z-Factors," *Oil & Gas Journal,* pp. 86–88, Feb. 18, 1974.
11. "Two-phase Flow in Pipes," Intercomp Course, The Hague, 1974.
12. Dranchuk, P. M., Purvis, R. A., and Robinson, D. B.: "Computer Calculations of Natural Gas Compressibility Factors Using the Standing and Katz Correlation," Inst. of Petroleum Technical Series, No. IP 74–008, 1974.

13. Dranchuk, P. M. and Abou-Kassem, J. H.: "Calculations of z-Factors for Natural Gases using Equations of State," *Journal of Canadian Petroleum Technology*, pp. 34–36, July–September 1975.
14. Gopal, V. N.: "Gas z-Factor Equations Developed for Computer," *Oil & Gas Journal*, Aug. 8, 1977.
15. Burnett, R. R.: "Calculator Gives Compressibility Factors," *Oil & Gas Journal*, pp. 70–74, June 11, 1979.
16. Papp, I.: "Uj Modszer Foldgazok Elteresi Tenvezojenek Szamitasara," *Koolaj es Foldgaz*, pp. 345–347, November 1979.

APPENDIX C

Conversion Factors

TABLE C–1 TEMPERATURE CONVERSIONS

	1.	2.	3.	4.
	°F	°C	°R	K
1. degree Fahrenheit (°F)	1.000	(°F-32)/1.8	°F + 459.67	(°F + 459.67)/1.8
2. degree Celsius (°C)	1.8(°C) + 32	1.000	1.8(°C) + 491.67	°C + 273.15
3. degree Rankine (°R)	°R − 459.67	(°R − 491.67)/1.8	1.000	°R/1.8
4. kelvin (K)	1.8(K) − 459.67	K − 273.15	1.8 (K)	1.000

TABLE C–2 LENGTH CONVERSIONS MULTIPLICATION FACTOR

TO CONVERT FROM / TO	1. inches (in)	2. feet (ft)	3. yards (yd)	4. miles (mi)	5. centimeter (cm)	6. meter (m)	7. kilometer (km)
1. inches (in)	1.0	$8.333\,333 \times 10^{-2}$	$2.777\,778 \times 10^{-2}$	$1.578\,282 \times 10^{-5}$	2.540	2.540×10^{-2}	2.540×10^{-5}
2. feet (ft)	12.0	1.0	$3.333\,333 \times 10^{-1}$	$1.893\,939 \times 10^{-4}$	30.480	$3.048\,0 \times 10^{-1}$	$3.048\,0 \times 10^{-4}$
3. yards (yd)	36.0	3.0	1.0	$5.681\,818 \times 10^{-4}$	91.44	$9.144\,0 \times 10^{-1}$	9.144×10^{-4}
4. miles (mi)	$6.336\,003 \times 10^{4}$	5.280×10^{3}	1.760×10^{3}	1.0	$1.609\,344 \times 10^{5}$	$1.609\,344 \times 10^{3}$	1.609 344
5. centimeter (cm)	$3.937\,008 \times 10^{-1}$	$3.280\,84 \times 10^{-2}$	$1.093\,613 \times 10^{-2}$	$6.213\,712 \times 10^{-6}$	1.0	1.000×10^{-2}	1.000×10^{-5}
6. meter (m)	39.370 08	3.280 84	1.093 613	$6.213\,712 \times 10^{-4}$	100.0	1.0	1.000×10^{-3}
7. kilometer (km)	$39.370\,08 \times 10^{3}$	$3.280\,84 \times 10^{3}$	$1.093\,613 \times 10^{3}$	$6.213\,712 \times 10^{-1}$	1.000×10^{5}	1.000×10^{3}	1.0

TABLE C–3 AREA CONVERSIONS MULTIPLICATION FACTOR

TO CONVERT FROM / TO	1. in^2	2. ft^2	3. mi^2	4. cm^2	5. m^2	6. km^2
1. square inches (in^2)	1.0	6.944 444 $\times 10^{-3}$	2.490 977 $\times 10^{-10}$	6.451 600	6.451 600 $\times 10^{-4}$	6.451 600 $\times 10^{-10}$
2. square feet (ft^2)	1.44 $\times 10^{2}$	1.0	3.587 007 $\times 10^{-8}$	9.290 304 $\times 10^{2}$	9.290 304 $\times 10^{-2}$	9.290 304 $\times 10^{-8}$
3. square miles (mi^2)	4.014 489 $\times 10^{9}$	2.787 84 $\times 10^{7}$	1.0	2.589 988 $\times 10^{10}$	2.589 988 $\times 10^{6}$	2.589 988
4. square centimeters (cm^2)	1.550 $\times 10^{-1}$	1.076 391 $\times 10^{-3}$	3.861 022 $\times 10^{-11}$	1.0	1.000 0 $\times 10^{-4}$	1.000 000 $\times 10^{-10}$
5. square meters (m^2)	1.550 $\times 10^{3}$	10.763 91	3.861 022 $\times 10^{-7}$	1.000 0 $\times 10^{4}$	1.0	1.000 000 $\times 10^{-6}$
6. square kilometers (km^2)	1.550 $\times 10^{9}$	1.076 391 $\times 10^{7}$	3.861 022 $\times 10^{-1}$	1.000 0 $\times 10^{10}$	1.000 0 $\times 10^{6}$	1.0

- 1 Acre = 43,560 ft^2, 1 dary = 1000 md
- 1 md = $9.86 \times 10^{-16}\ m^2 = 9.86 \times 10^{-12}\ cm^2 = 1.127 \times 10^{-3}\ \frac{(B/D)\ cp}{ft^2(psi/ft)}$

TABLE C–4 VOLUME CONVERSIONS MULTIPLICATION FACTOR

TO CONVERT FROM / TO	1. in^3	2. ft^3	3. cm^3	4. m^3
1. cubic inches (in^3)	1.0	$5.787\ 035 \times 10^{-4}$	$1.638\ 706 \times 10$	$1.638\ 706 \times 10^{-5}$
2. cubic feet (ft^3)	1.728×10^3	1.0	$2.831\ 685 \times 10^4$	$2.831\ 685 \times 10^{-2}$
3. cubic centimeters (cm^3)	$6.102\ 376 \times 10^{-2}$	$3.531\ 466 \times 10^{-5}$	1.0	$1.000\ 000 \times 10^{-6}$
4. cubic meters (m^3)	$6.102\ 376 \times 10^4$	$3.531\ 466 \times 10$	1.000×10^6	1.0

TABLE C–5 LIQUID VOLUME CONVERSIONS MULTIPLICATION FACTOR

TO CONVERT TO / FROM	1. gal	2. U.K. gal	3. bbl (oil)	4. ft^3	5. L	6. m^3
1. gallons (U.S.) (gal)	1.0	$8.326\,739 \times 10^{-1}$	$2.380\,952 \times 10^{-2}$	$1.336\,805 \times 10^{-1}$	3.785 412	$3.785\,412 \times 10^{-3}$
2. Imperial gallons (U.K. gal)	1.200 95	1.0	$2.859\,406 \times 10^{-2}$	$1.605\,437 \times 10^{-1}$	4.546 092	$4.546\,092 \times 10^{-3}$
3. barrels (oil, 42 gal) (bbl)	42.0	$3.497\,230 \times 10$	1.0	5.614 583	$1.589\,873 \times 10^{2}$	$1.589\,873 \times 10^{-1}$
4. cubic feet (ft^3)	7.480 52	6.228 833	$1.781\,076 \times 10^{-1}$	1.0	$2.831\,685 \times 10$	$2.831\,685 \times 10^{-2}$
5. liters (L)	$2.641\,720 \times 10^{-1}$	$2.199\,692 \times 10^{-1}$	$6.289\,810 \times 10^{-3}$	$3.531\,466 \times 10^{-2}$	1.0	$1.000\,000 \times 10^{-3}$
6. cubic meters (m^3)	$2.641\,720 \times 10^{2}$	$2.199\,692 \times 10^{2}$	6.289 810	$3.531\,466 \times 10$	1.000×10^{3}	1.0

TABLE C–6 DENSITY CONVERSIONS MULTIPLICATION FACTOR

TO CONVERT FROM / TO	1. gm/cm^3	2. lb/ft^3	3. kg/L	4. kg/m^3
1. $grams/cm^3$	1.0	$6.242\,797 \times 10$	1.0	$1.000\,000 \times 10^3$
2. lb/ft^3	$1.601\,846 \times 10^{-2}$	1.0	$1.601\,846 \times 10^{-2}$	$1.601\,846 \times 10$
3. kg/L	1.0	$6.242\,797 \times 10$	1.0	$1.000\,000 \times 10^3$
4. kg/m^3	$1.000\,000 \times 10^{-3}$	$6.242\,797 \times 10^{-2}$	1.0×10^{-3}	1.0

TABLE C–7 PRESSURE CONVERSIONS MULTIPLICATION FACTOR

TO CONVERT FROM / TO	1. atm	2. bar	3. lbf/in^2	4. kgf/cm^2	5. in Hg (32°F)	6. mmHg (32°F)	7. ftH_2O (39.2°F)	8. Pa
1. atmosphere (standard) (atm)	1.0	1.013 250	$1.469\,6 \times 10$	1.033 228	$2.992\,133 \times 10$	7.600×10^2	$3.389\,952 \times 10$	$1.013\,250 \times 10^5$
2. bar	$9.869\,233 \times 10^{-1}$	1.0	$1.450\,377 \times 10$	1.019 716	$2.953\,006 \times 10$	$7.500\,638 \times 10^2$	$3.345\,623 \times 10$	$1.000\,000 \times 10^5$
3. lbf/in^2 (psi)	$6.804\,573 \times 10^{-2}$	$6.894\,757 \times 10^{-2}$	1.0	$7.030\,695 \times 10^{-2}$	2.036 026	$5.171\,507 \times 10$	2.306 73	$6.894\,757 \times 10^3$
4. kgf/cm^2	$9.678\,411 \times 10^{-1}$	$9.806\,65 \times 10^{-1}$	$1.422\,334 \times 10$	1.0	$2.895\,909 \times 10$	$7.355\,613 \times 10^2$	$3.280\,935 \times 10$	$9.806\,650 \times 10^4$
5. inHg (32°F)	$3.342\,097 \times 10^{-2}$	$3.386\,38 \times 10^{-2}$	$4.911\,529 \times 10^{-1}$	$3.453\,147 \times 10^{-2}$	1.0	2.54×10^2	1.132 955	$3.386\,38 \times 10^3$
6. mmHg (32°F) (torr)	$1.315\,789 \times 10^{-3}$	$1.333\,22 \times 10^{-3}$	$1.933\,672 \times 10^{-2}$	$1.359\,506 \times 10^{-3}$	$3.937\,01 \times 10^{-2}$	1.0	$4.460\,451 \times 10^{-2}$	$1.333\,22 \times 10^2$
7. ftH_2O (39.2°F)	$2.949\,894 \times 10^{-2}$	$2.988\,98 \times 10^{-2}$	$4.335\,149 \times 10^{-1}$	$3.047\,912 \times 10^{-2}$	$8.826\,475 \times 10^{-1}$	$2.241\,926 \times 10$	1.0	$2.988\,98 \times 10^3$
8. pascal (Pa)	$9.869\,233 \times 10^{-6}$	1.000×10^{-5}	$1.450\,377 \times 10^{-4}$	$1.019\,716 \times 10^{-5}$	$2.953\,006 \times 10^{-4}$	$7.500\,638 \times 10^{-3}$	$3.345\,623 \times 10^{-4}$	1.0

TABLE C–8 RATE OF FLOW CONVERSIONS MULTIPLICATION FACTOR

TO CONVERT FROM / TO	1. ft^3/min	2. U.S. gal/day	3. bbl/day	4. m^3/s
1. ft^3/min	1.0	1.077 195 $\times 10^{4}$	2.564 749 $\times 10^{2}$	4.719 474 $\times 10^{-4}$
2. gal/day	9.283 374 $\times 10^{-5}$	1.0	2.380 952 $\times 10^{-2}$	4.381 264 $\times 10^{-8}$
3. bbl/day	3.899 017 $\times 10^{-3}$	42.0	1.0	1.840 131 $\times 10^{-6}$
4. m^3/s	2.118 88 $\times 10^{3}$	2.282 477 $\times 10^{7}$	5.434 396 $\times 10^{5}$	1.0

TABLE C–9 MASS CONVERSIONS MULTIPLICATION FACTOR

TO CONVERT FROM / TO	1. ounces (avoir)(oz)	2. pounds (avoir)(lb)	3. metric ton (t)(tonne)	4. kilograms (kg)
1. ounces (oz)(avoir)	1.0	6.25 $\times 10^{-2}$	2.834 952 $\times 10^{-5}$	2.834 952 $\times 10^{-2}$
2. pounds (lb)(avoir)	16.0	1.0	4.535 924 $\times 10^{-4}$	4.535 924 $\times 10^{-1}$
3. metric ton (t)(tonne)	3.527 397 $\times 10^{4}$	2.204 622 $\times 10^{3}$	1.0	1.000 000 $\times 10^{3}$
4. kilograms (kg)	3.527 397 $\times 10$	2.204 622	1.000 000 $\times 10^{-3}$	1.0

TABLE C–10 VELOCITY CONVERSIONS MULTIPLICATION FACTOR

TO CONVERT TO / FROM	1. ft/s	2. ft/min	3. ft/h	4. m/s
1. ft/s	1.0	60.0	3600.0	3.048 000 $\times 10^{-1}$
2. ft/min	1.666 666 $\times 10^{-2}$	1.0	60.0	5.080 000 $\times 10^{-3}$
3. ft/h	2.777 777 $\times 10^{-4}$	1.666 666 $\times 10^{-2}$	1.0	8.466 667 $\times 10^{-5}$
4. m/s	3.280 84	1.968 504 $\times 10^{2}$	1.181 102 $\times 10^{4}$	1.0

TABLE C–11 VISCOSITY (ABSOLUTE) CONVERSIONS MULTIPLICATION FACTOR

TO CONVERT TO / FROM	1. cp	2. poise g/(cm − s)	3. lb/(ft − s)	4. pascal second (Pa − s)
1. centipoise (cp)	1.0	1.00 $\times 10^{-2}$	6.719 689 $\times 10^{-4}$	1.000 000 $\times 10^{-3}$
2. poise g/(cm − s)	100.0	1.0	6.719 689 $\times 10^{-2}$	1.000 000 $\times 10^{-1}$
3. lb/(ft − s)	1.488 164 $\times 10^{3}$	1.488 164 $\times 10$	1.0	1.488 164
4. pascal second (Pa − s)	1.000 $\times 10^{3}$	10.0	6.719 689 $\times 10^{-1}$	1.0

TABLE C–12 KINEMATIC VISCOSITY: ABSOLUTE VISCOSITY IN MASS UNITS DIVIDED BY MASS DENSITY CONVERSIONS MULTIPLICATION FACTOR

TO CONVERT FROM \ TO	1. cS	2. S	3. ft^2/s	4. m^2/s
1. centistokes (cS)	1.0	1.000×10^{-2}	$1.076\ 391 \times 10^{-5}$	$1.000\ 000 \times 10^{-6}$
2. stokes (cm^2/s)(S)	1.000×10^{2}	1.0	$1.076\ 391 \times 10^{-3}$	$1.000\ 000 \times 10^{-4}$
3. ft^2/s	$9.290\ 304 \times 10^{4}$	$9.290\ 304 \times 10^{2}$	1.0	$9.290\ 304 \times 10^{-2}$
4. square meter per second (m^2/s)	1.000×10^{6}	1.000×10^{4}	$1.076\ 391 \times 10$	1.0

TABLE C–13 LAND MEASUREMENT CONVERSIONS MULTIPLICATION FACTOR*

TO CONVERT FROM \ TO	1. ft^2	2. mi^2	3. acre	4. m^2
1. square foot (ft^2)	1.0	$3.587\ 007 \times 10^{-8}$	$2.295\ 675 \times 10^{-5}$	$9.290\ 304 \times 10^{-2}$
2. square mile (mi^2)	$2.787\ 840 \times 10^{7}$	1.0	$6.399\ 974 \times 10^{2}$	$2.589\ 988 \times 10^{6}$
3. acre (U.S. survey)	$4.356\ 018 \times 10^{4}$	$1.562\ 506 \times 10^{-3}$	1.0	$4.046\ 873 \times 10^{3}$
4. square meter (m^2)	$1.076\ 391 \times 10$	$3.861\ 022 \times 10^{-7}$	$2.471\ 044 \times 10^{-4}$	1.0

* 1 league = 3 miles
1 square mile = 640 acres
1 township = 36 square miles
1 section = 1 square mile
1 rod = 16.5 feet

APPENDIX D

The Calculation of Pseudo-Skin Factors

In many oil and gas wells, the observed flow rate is different than that calculated theoretically. As noted in Chapter 2, the concept of skin factor was developed to account for the deviation from the theoretical rate. For example, for an oil well located centrally in the drainage plane, during pseudo-steady state flow, the oil flow rate can be calculated as

$$q = 0.007078kh\,(\bar{p} - p_{wf})/[\ln(r_e/r_w) - 3/4 + s_t] \qquad \text{(D-1)}$$

All parameters in Equation D–1 are in conventional U.S. oil field units and s_t is total skin factor, which includes the effects of partial penetration, perforation density, well stimulation, mechanical skin damage due to drilling and completion, etc. A positive value of s_t would result in a reduction of flow rate while a negative value of s_t would result in a flow enhancement. Damaged wells have positive skin factors and stimulated wells have negative skin factors.

The mechanical skin factor s_m represents well damage caused by drilling and completion fluids. The majority of the drilled wells when completed show mechanical skin damage, and hence, the well is normally acidized before it is put on production. The mechanical damage, denoted as a positive skin factor, would cause a loss in well productivity if it is not removed. In addition to mechanical skin damage, many other parameters may cause either loss or gain in well productivity. These parameters include (1) wells completed in part of the pay zone, i.e., partially penetrating wells; (2) near-wellbore turbulence; (3) perforation density; (4) slant wells, etc. The change in well productivity due to these parameters is described by assigning an equivalent skin factor called *pseudo-skin factor*.

A skin factor of a well is estimated either by drill stem test (DST), or by drawdown or buildup tests. The skin factor calculated from well test analysis is usually a linear combination of a mechanical skin factor and various pseudo-skin factors, for example, for a partially penetrating well,

$$s_t = \frac{s_m}{b'} + s_p \qquad \text{(D–2)}$$

where s_t is the total skin factor (which would be obtained from well test analysis), s_m represents mechanical skin factor, s_p represents the pseudo-skin factor caused by partial penetration and b' is penetration ratio which is defined in Equation D–5. Similarly, for a gas well producing at high flow rate, the total skin factor is given by

$$s_t = s_m + Dq_g \qquad \text{(D–3)}$$

where D is turbulence coefficient and is defined in Equation D–24. It is clear from Equations D–2 and D–3 that the mechanical skin factor caused by damage or stimulation can be obtained if and only if relevant pseudo-skin factors are known. There are a few correlations which can be used to calculate various pseudo-skin factors. These correlations are included in this Appendix.

1. PSEUDO-SKIN FACTOR DUE TO PARTIAL PENETRATION

As noted earlier, many oil and gas wells are completed in a part of the pay zone. This is normally referred to as a partially penetrating well. Some of the correlations that are available to calculate skin factors due to a partial penetration are listed below.

A. BRONS AND MARTING METHOD[1]

In oil and gas wells, the deviation from radial flow due to restricted fluid entry causes an additional pressure drop. Brons and Marting[1] first expressed this effect of partial penetration as a pseudo-skin factor and suggested that it can be correlated by the following equation

$$s_p = \left(\frac{1}{b'} - 1\right) [\ln (h_D) - G(b')] \tag{D–4}$$

where

$$b' = h_p/h, \text{ penetration ratio} \tag{D–5}$$

$$h_D = \frac{h}{r_w}\sqrt{\frac{k_h}{k_v}}, \text{ dimensionless pay zone thickness} \tag{D–6}$$

h_p = perforated interval, ft

h = total pay zone thickness, ft

k_h = horizontal permeability, md

k_v = vertical permeability, md

$G(b')$ is a function of penetration ratio b'. An approximate correlation for $G(b')$ is

$$G(b) = 2.948 - 7.363b' + 11.45(b')^2 - 4.675(b')^3 \tag{D–7}$$

Brons and Marting[1] presented plots for pseudo-skin factor due to partial penetration, s_p, as a function of penetration ratio b' for several h_D values (see Fig. D–1). These plots (Fig. D–1) are for three types of well completion configurations, which are shown in Figure D–2. For each configuration, the dimensionless pay zone thickness, h_D, is defined differently. The definitions of h_D for three well completion configurations are summarized below.

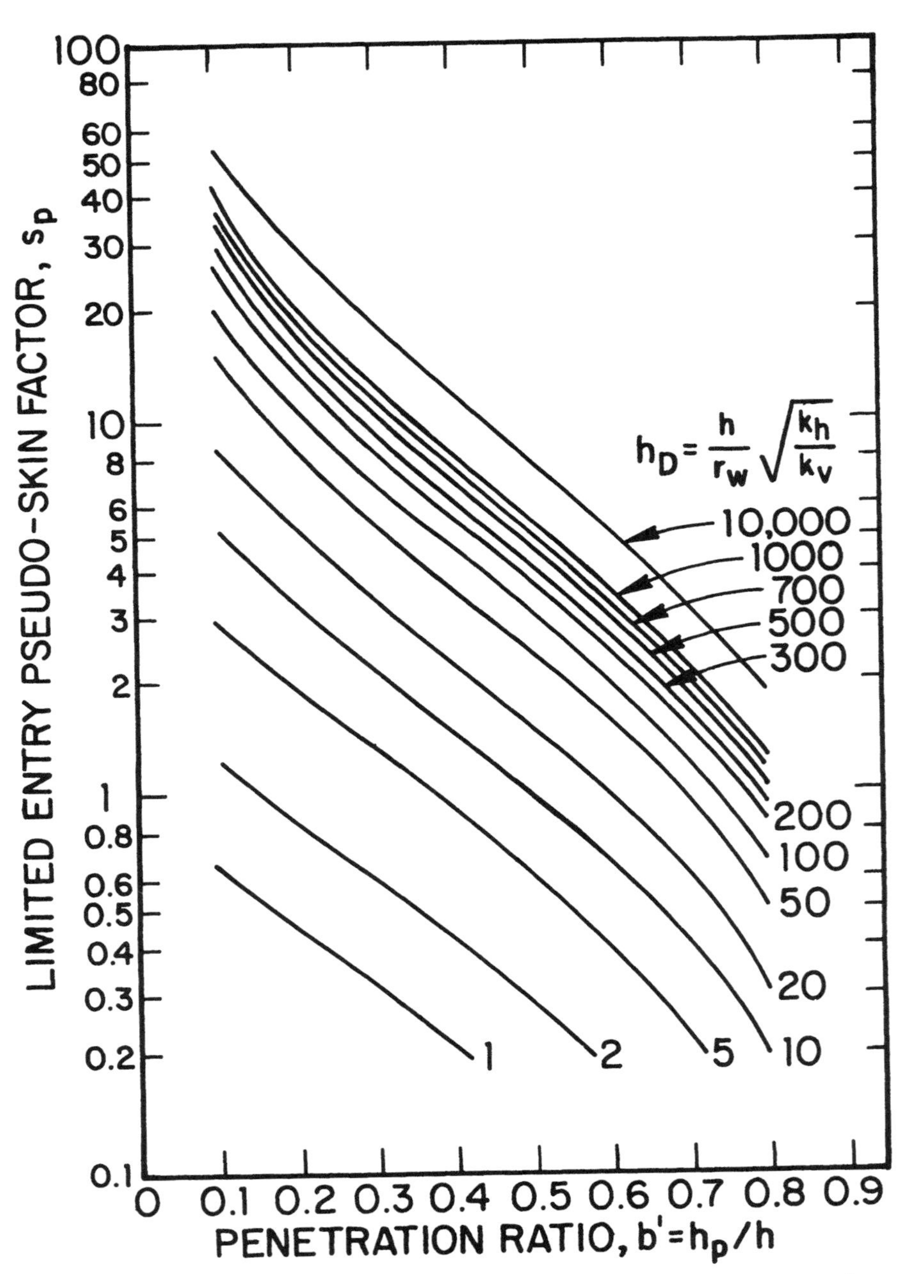

Figure D–1 A Correlation of Pseudo-Skin Factor Due to Partial Penetration[1]

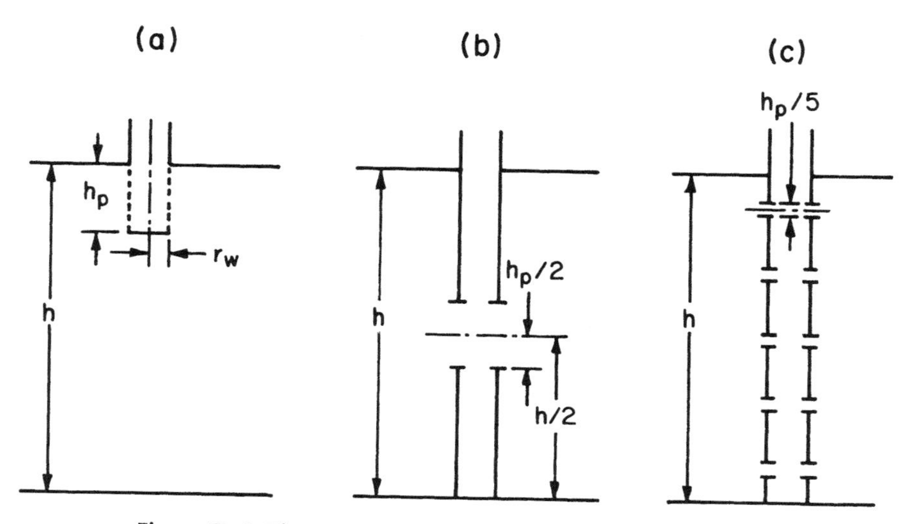

Figure D–2 Three Types of Well Completion Configurations[1]

WELL CONFIGURATION	*DEFINITION OF h_D*
Well producing from the top (or bottom) of the formation, Fig. D–2(a)	$h_D = \frac{h}{r_w}\sqrt{\frac{k_h}{k_v}}$
Well only producing from the central section, Fig. D–2(b)	$h_D = \frac{h}{2r_w}\sqrt{\frac{k_h}{k_v}}$
Well with N intervals open to production, Fig. D–2(c)	$h_D = \frac{h}{2Nr_w}\sqrt{\frac{k_h}{k_v}}$

B. ODEH CORRELATION[2]

Odeh[2] presented an equation for calculating pseudo-skin factor for an arbitrarily location of perforated interval h_p,

$$s_p = 1.35\left(\frac{1}{b'} - 1\right)^{0.825}\Big[\ln(r_w h_D + 7) - 1.95 - \ln(r_{wc})\{0.49 + 0.1\ln(r_w h_D)\}\Big] \qquad \text{(D–8)}$$

where

$$r_{wc} = r_w \exp\left[0.2126\left(2.753 + \frac{z_m}{h}\right)\right] \tag{D-9}$$

$$\text{for } 0 < \frac{z_m}{h} < 0.5,$$

and

$$z_m = h_1 + (h_p/2) \tag{D-10}$$

h_1 = distance from the pay zone top to the top of the open interval, ft

and b' and h_D are defined in Equations D–5 and D–6, respectively. If $h_1 = 0$, i.e., if well perforates at the top of formation, then use

$$r_{wc} = r_w \tag{D-11}$$

If $z_m/h > 0.5$, replace z_m/h in Equation D–9 by $1 - (z_m/h)$.

C. YEH AND REYNOLDS CORRELATION[3]

Yeh and Reynolds[3] found that a pseudo-skin factor caused by partial well completion depends upon (1) the dimensionless wellbore length h_{wD}, (2) the penetration ratio b', and (3) the location of the perforated interval. Their correlation for calculating pseudo-skin factor is

$$s_p = \left(\frac{1 - b'}{b'}\right) \ln (h_{wD}) \tag{D-12}$$

where

$$h_{wD} = \frac{C'b'(1 - b')\,h_D}{\exp(C_1)} \tag{D-13}$$

$$C_1 = 0.481 + 1.01(b') - 0.838(b')^2 \tag{D-14}$$

h_D denotes dimensionless reservoir thickness as defined in Equation D–6. The parameter C' accounts for the location of the open interval and is obtained from Figure D–3, where $\Delta z_D = \min\,[h_1/h,\ h_2/h]$. The definitions of h_1, h_2 and h are shown in Figure D–4. If the open interval is at the top or bottom of the reservoir, then $C' = 2$. This correlation can also be used for multilayered reservoirs.

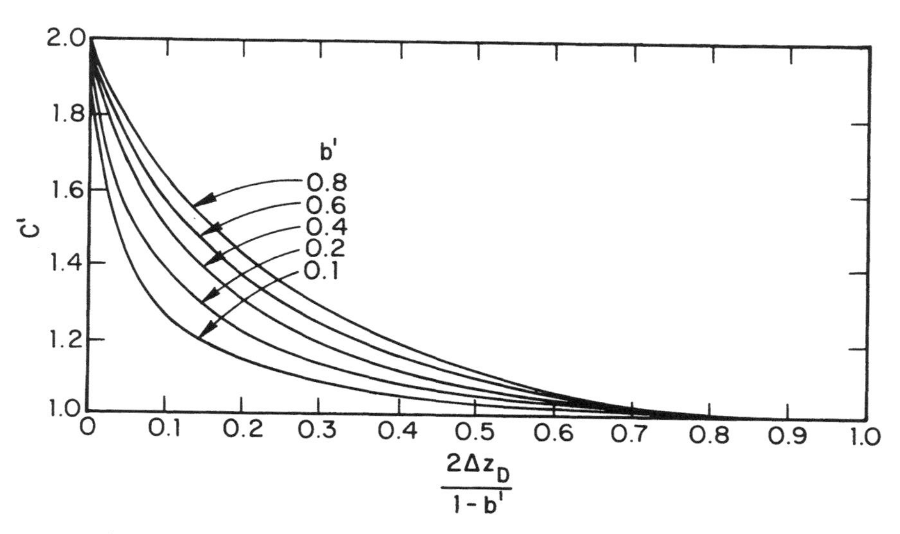

Figure D–3 A Graphical Correlation of Parameter C′ to be used in Equation D–13[3]

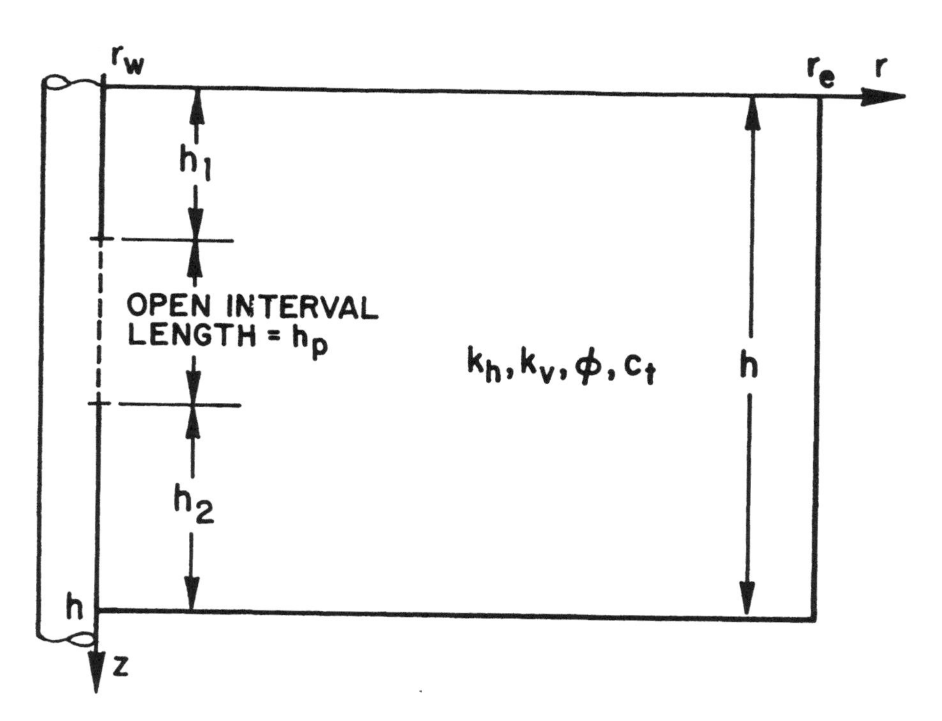

Figure D–4 A Schematic View of a Restricted-Entry Well.

D. PAPATZACOS' CORRELATION[4]

Papatzacos[4] showed that for a single-layer, infinite reservoir, the pseudo-skin factor for partial penetration can be estimated as

$$s_p = \left(\frac{1}{b'} - 1\right) \ln\left(\frac{\pi h_D}{2}\right) + \frac{1}{b'} \ln\left[\frac{b'}{2 + b'}\left(\frac{A-1}{B-1}\right)^{1/2}\right] \quad \text{(D–15)}$$

where

$$A = \frac{h}{h_1 + 0.25\, h_p} \quad \text{(D–16)}$$

and

$$B = \frac{h}{h_1 + 0.75\, h_p} \quad \text{(D–17)}$$

where h_1 is the distance from the top of the reservoir to the top of the open interval (see Fig. D–4).

An important feature of partial penetration is that pseudo-skin factor s_p is always greater than zero, i.e., restricted entry always reduces the well productivity. Initially, at early times, a partially penetrated well behaves as if it is producing from a formation of thickness h_p instead of h. After some time, a transition to flow from the entire formation is observed, after which the steady-state pseudo-skin factor s_p is established. This transient effect of restricted fluid entry may be important for low-permeability reservoirs.

It is important to note that if a partial penetration skin factor is estimated from a well test analysis, then one can estimate effective vertical permeability, k_v of the reservoir using either of the Equations D–4, D–8, D–12 and D–15.

2. PSEUDO-SKIN FACTOR DUE TO PERFORATIONS

Most reservoir engineering equations assume an open-hole well. To ensure mechanical integrity, the majority of the wells are cemented and perforated. Perforations, depending upon their shot density, offer flow restrictions to the wellbore, resulting in a reduced production rate. This loss of productivity due to perforations can also be expressed as a pseudo-skin factor. The pseudo-skin factor due to perforation s_{pf} primarily depends upon perforation geometry and perforation quality. The parameters that determine s_{pf} are: (1) penetration depth, (2) perforation diameter, (3) shot density, and (4) phasing.

In general, a long perforation depth, a large perforation diameter, and a high shot density enhances well performance. Phasing is the angular pattern of shots around the wellbore. Strip shooting (0° phasing), where all shots are along the same side of the wellbore, gives the lowest flow capacity as compared to other phasing angles.

The pseudo-skin factor due to perforation s_{pf} can be calculated from Figures D–5 and D–6, which depict pseudo-skin factors as a function of shot density, phasing, and perforation depth beyond casing.[5,6] These two figures are specifically for ½-in. perforations through cemented casing in 9½-in. drilled wellbore diameter. The depth of penetration is measured from sandface. For a wellbore with a different diameter d, the apparent penetration depth L_{pa} should be used, where L_{pa} is given by

$$L_{pa} = L_p \left(\frac{9.5}{d}\right). \tag{D–18}$$

The effect of perforation diameter (other than ½ in.) on pseudo-skin factor is negligible.

One needs to estimate perforation depth, before using Figures D–5 and D–6 to estimate pseudo-skin factor due to perforation. Standard Berea depths of perforations are usually provided by the perforating companies, which are based on experimental work following procedures and specifications given in the booklet, API RP43. The actual perforation depths may vary from the standard depths due to variations in rock properties, fluid density and casing properties. Thompson[7] presented a correlation for estimating the correction factor which can be applied to standard Berea penetration to estimate penetration L_p in a given formation,

$$L_p = [L_{pB} \times 10^m] - (d_s + d_c) \tag{D–19}$$

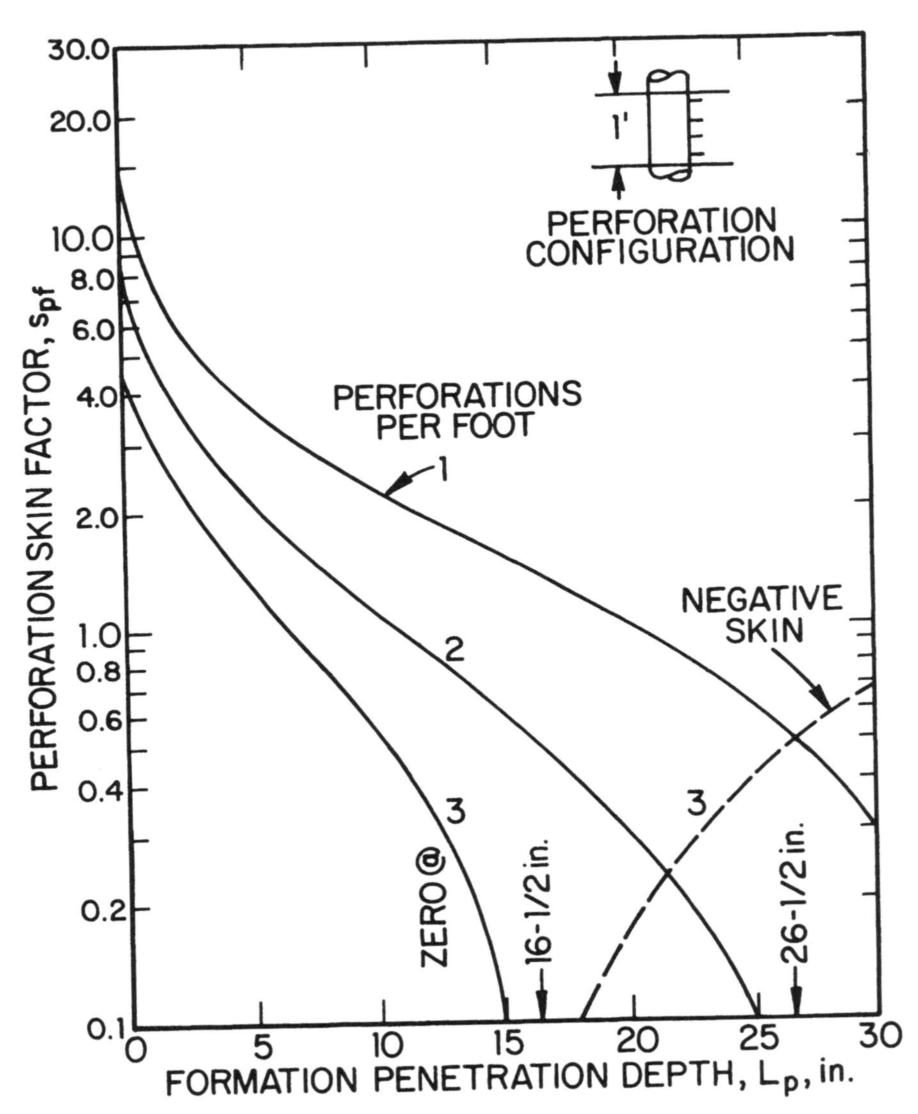

Figure D–5 Perforation Pseudo-Skin Factor, 0° Phasing[5,6]

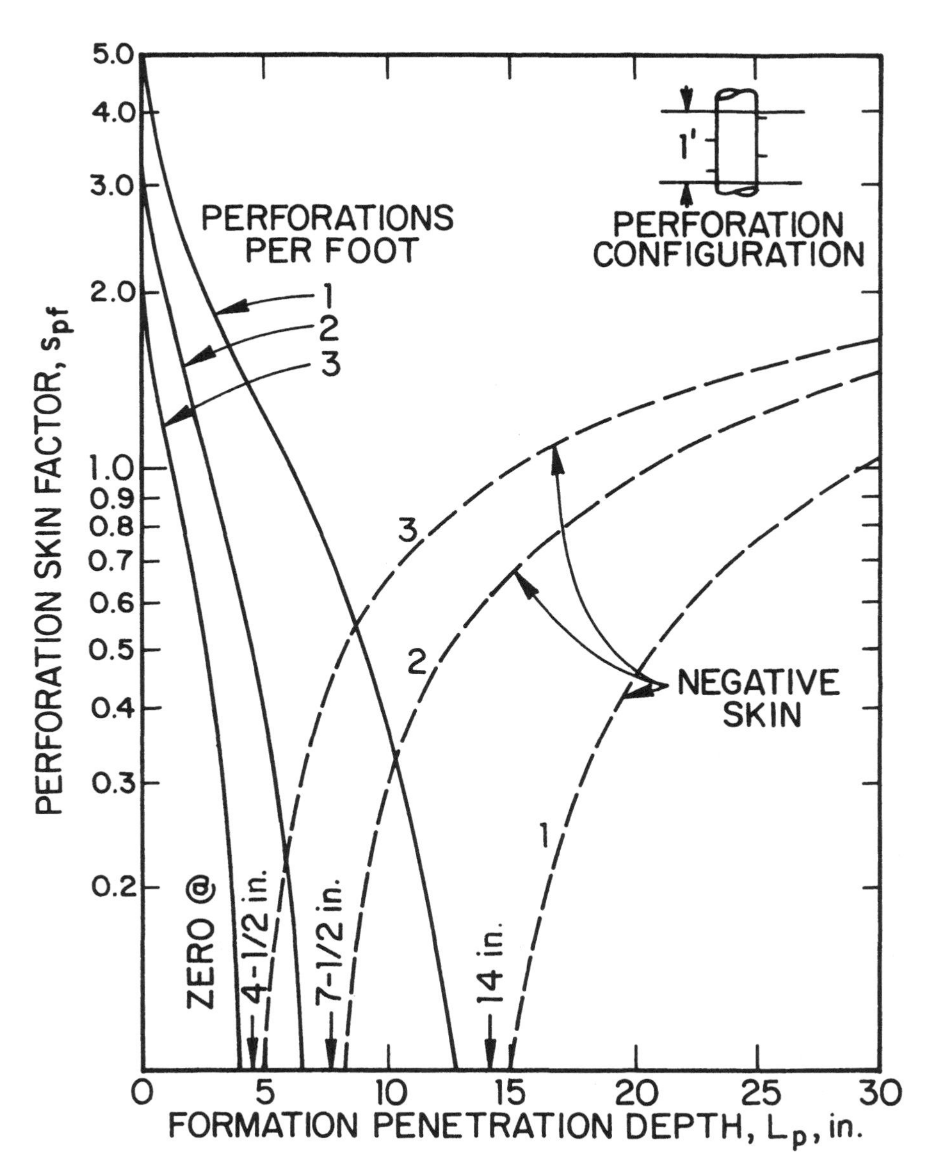

Figure D–6 Perforation Pseudo-Skin Factor, 180° Phasing[5,6]

where

$m = 3.7 \times 10^{-5} (C_B - C)$
L_{pB} = standard Berea depth of penetration, in
C_B = Berea compressive strength ($\approx$ 6500) psia
C = rock compressive strength, psia
d_s = cement sheath thickness, in (usually 0.75 in)
d_c = casing thickness, in (usually 0.375 in)

Thompson's correlation, Equation D–19, applies only to jet guns with shaped charges. According to Saucier and Lands,[8] formation stress is a more suitable parameter to correct API standard penetration depth. Reference 8 includes the following equation for correcting perforation depth.

$$L_p = 0.7\,[L_{pB} - (d_s + d_c)]\,M_p \tag{D–20}$$

where M_p is penetration depth correction factor. Figure D–7 presents penetration depth correction factor, M_p, for three types of formations: Austin limestone, Wasson dolomite and Berea sandstone.

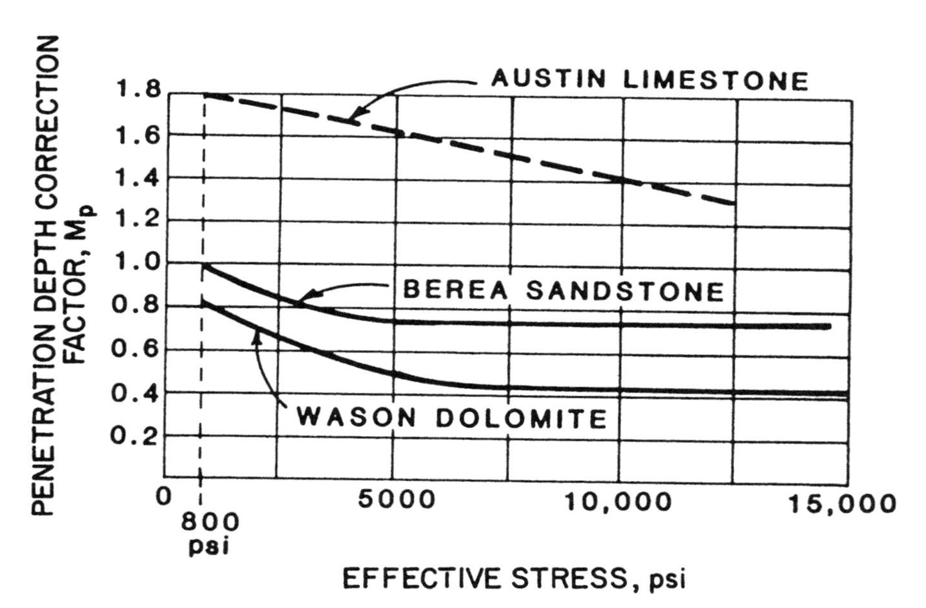

Figure D–7 Penetration Depth Correction Factor, M_p[8]

A more detailed discussion on estimation of perforation depth, L_p, is provided in References 9 and 10. Table D–1 presents the penetration depths for different perforation diameters.[10]

TABLE D–1 PERFORATING GUN DATA[10]

Gun Size, in.	Tubing, Casing, in.	Perforation Diameter, in., avg.	Penetration,* in. avg.	longest
Retrievable through tubing				
1⅜	4½ csg	0.21	3.03	3.30
1 9/16	5½ csg	0.24	4.7	5.48
1 11/16	4½ to 5½ csg	0.24	4.8	5.50
2	4½ to 5½ csg	0.32	6.5	8.15
2⅛	2⅞ tbg to 4½ csg	0.33	7.2	8.15
2⅝	4½ csg	0.36	10.36	10.36
Expendable through tubing				
1⅛	4½ csg	0.19	3.15	3.15
1¼	2⅜ tbg	0.30	3.91	3.91
1⅜		0.30	5.1	5.35
1 11/16	2⅞ tbg to 5½ csg	0.34	6	8.19
2 1/16	5½ to 7 csg	0.42	8.2	8.6
2⅛	2⅞ tbg to 5½ csg	0.39	7.7	8.6
Retrievable casing guns				
2¾	4½ csg	0.38	10.55	10.5
2⅞	4½ csg	0.37	10.63	10.6
3⅛	4½ csg	0.42	8.6	11.1
3⅜	4½ csg	0.36	9.1	10.8
3⅝	4½ & 5½ csg	0.39	8.9	12.8
4	5½ to 9⅝ csg	0.51	10.6	13.5
5	6¾ to 9⅝ csg	0.73	12.33	13.6

* Penetration length measured from casing ID.

3. SKIN FACTOR DUE TO REDUCED CRUSHED-ZONE PERMEABILITY

The effect of a crushed zone can be expressed as a mechanical skin factor. McLeod[11] modelled the perforation surrounding a crushed zone as a "horizontal microwell" with formation damage around it. McLeod's equation[11] for steady-state skin factor due to reduced crushed-zone permeability is

$$s_c = \left(\frac{k}{k_{dp}} - \frac{k}{k_d}\right) \frac{12h_p}{NL_p} \ln\left(\frac{r_{dp}}{r_p}\right) \tag{D–21}$$

where

k = formation permeability, md
k_{dp} = crushed-zone permeability, md
k_d = damage-zone permeability near the wellbore, md
r_{dp} = crushed-zone radius, in.
r_p = perforation radius, in.
L_p = depth of penetration, in.
h_p = perforated interval, ft
N = total number of perforations

Figure D–8 shows the geometry of perforation with crushed zone. The permeability of crushed zone can be approximated as[10]

$k_{dp} = 0.1k$ if perforated overbalanced and
$k_{dp} = 0.4k$ if perforated underbalanced.

The crushed zone radius, r_{dp}, can be estimated as[10]

$$r_{dp} = r_p + 0.5 \tag{D–22}$$

Some engineers believe that Equation D–21 is difficult to use with certainty and thus it is primarily useful for sensitivity analysis.

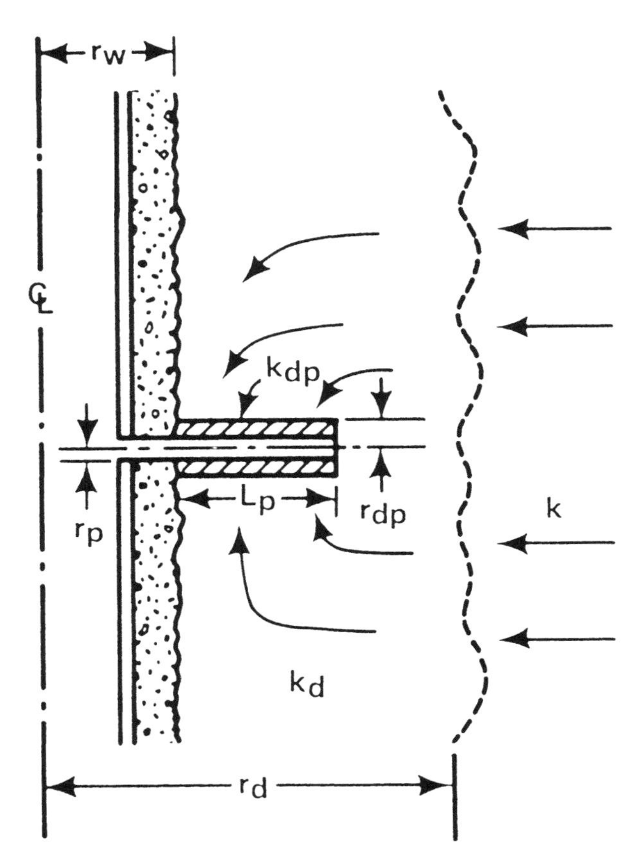

Figure D–8 Geometry of Perforation With Crushed Zone[10]

4. RATE-DEPENDENT SKIN FACTOR DUE TO NEAR-WELLBORE TURBULENCE

Darcy's law of flow through porous media is for laminar flow regime. At high flow velocities, i.e., in a turbulent region, Darcy's law is no longer valid. (This deviation from Darcy's flow is normally more pronounced in gas wells than in oil wells in high-permeability reservoirs.) The deviation from Darcy's law is expressed as a skin factor s_{tu} also called rate-dependent skin factor which is defined as

$$s_{tu} = Dq_g \tag{D–23}$$

where q_g is gas flow rate in Mscfd and D is the turbulence coefficient (1/Mscfd) and is given by[10,12,13]

$$D = \frac{2.226 \times 10^{-15} k_a \beta' \gamma_g h}{\mu_{pwf} r_w h_p^2} \tag{D–24}$$

where

$$\beta' = \frac{2.73 \times 10^{10}}{k_a^{1.1045}} \tag{D–25}$$

or

$$\beta' = \frac{2.33 \times 10^{10}}{K_a^{1.201}} \tag{D–26}$$

β' = high velocity flow coefficient, 1/ft
k_a = effective gas permeability near the wellbore, md
μ_{pwf} = gas viscosity at wellbore conditions, cp
γ_g = specific gas gravity (air = 1)
r_w = wellbore radius, ft
h_p = thickness of perforated interval, ft.

Note that in Equation D–24 μ_{pwf} is a function of pressure, thus the turbulence coefficient D is pressure-dependent. Equations D–25 and D–26 for turbulent coefficient β' are from References 9 and 1, respectively. A somewhat different result will be obtained, depending upon definition of the β' used. References 12 and 13 include detailed discussions on the effect of turbulent flow in porous media.

Chapter 3 includes pseudo-skin factors due to horizontal or slant wells. Additionally, Chapter 5 includes pseudo-skin factors for fractured vertical wells.

REFERENCES

1. Brons, F. and Marting, V. E.: "The Effect of Restricted Fluid Entry on Well Productivity," *Journal of Petroleum Technology*, pp. 172–174, February 1961.
2. Odeh, A. S.: "An Equation for Calculating Skin Factor Due to Restricted-Entry," *Journal of Petroleum Technology*, pp. 964–965, June 1980.
3. Yeh, N. S. and Reynolds, A. C.: "Computation of the Pseudo-Skin Caused by a Restricted-Entry Well Completed in a Multilayer Reservoir," *SPE Formation Evaluation*, pp. 253–263, June 1989.
4. Papatzacos, P.: "Approximate Partial-Penetration Pseudo-Skin for Infinite-Conductivity Wells," *SPE Reservoir Engineering*, pp. 227–234, May 1988, *Trans.*, AIME, vol. 283.
5. Harris, M. N.: "The Effect of Perforating on Well Productivity," *Trans.*, AIME, vol. 237, pp. 518, 1966.
6. Standing M. B.: *Reservoirs,* Course Manual in Reservoir Engineering Continental Shelf Development, Petroleum Industry Courses, NTH Trondheim, Norway, 1980.
7. Thompson, G. D.: "Effects of Formation Compressive Strength on Perforator Performance," *Drilling and Production Practice*, API, 1962.
8. Saucier, R. J. and Lands, J. F.: "A Laboratory Study of Perforations in Stressed Formation Rocks," *Journal of Petroleum Technology*, pp. 1347–1353, September 1978.
9. Golan, M. and Whitson, C. H.: *Well Performance,* International Human Resources Corporation, Boston, Massachusetts, 1986.
10. Brown, K. E.: *The Technology of Artificial Lift Methods*, vol. 4, Tulsa, Oklahoma, PennWell Publishing Co., 1984.
11. McLeod, Jr., H. O.: "The Effect of Perforating Conditions on Well Performance," *Journal of Petroleum Technology*, pp. 31–39, January 1983.
12. Wattenbarger, R. A. and Ramey, Jr., H. J.: "Gas Well Testing with Turbulence, Damage and Wellbore Storage," *Journal of Petroleum Technology*, pp. 877–887, August 1968.
13. Ding, W.: *Gas Well Test Analysis*, MS Thesis, University of Tulsa, Tulsa, Oklahoma, 1986.

APPENDIX E

Recovery Factors

TABLE E–1 PRIMARY RECOVERY IN STOCK TANK BARRELS PER ACRE-FOOT*
(Per Percent Porosity for Depletion Type Reservoirs)†

Oil Solution GOR m^3/m^3 (ft^3/bbl)	Oil Gravity API	Sand or Sandstone			Limestone, Dolomite or Chert		
		Maximum	Average	Minimum	Maximum	Average	Minimum
10.7	15	7.22	4.87	1.44	17.87	2.56	.36
(60)	30	11.95	8.52	4.88	20.87	6.29	1.85
	50	19.20	13.89	9.46	24.78	11.84	5.07
35.6	15	6.97	4.62	1.75	16.33	2.65	.51
(200)	30	11.57	7.90	4.38	19.05	5.75	1.52
	50	19.42	13.73	9.15	23.44	11.40	4.36
106.9	15	7.56	4.76	2.50	12.69	3.29	.90
(600)	30	10.48	6.52	3.61	14.64	4.70	(1.24)
	50	15.05	9.74	5.85	17.30	7.25	(2.06)
178.1	15	—	—	—	—	—	—
(1,000)	30	12.34	7.61	4.52	13.26	5.38	(1.63)
	50	11.96	7.15	4.10	12.79	4.83	(1.24)
356.2	15	—	—	—	—	—	—
(2,000)	30	—	—	—	—	—	—
	50	10.58	6.45	4.04	9.64	(4.26)	(1.47)

* 1 bbl = 0.159 m^3 and 1 Acre-foot = 1233.53 m^3

† Arps, J. J. and Roberts, T. G.: Petroleum Trans of AIME, vol. 204, pp. 120–127, 1955.

TABLE E–2 PRIMARY RECOVERY FACTORS†

Production Mechanism	Lithology	State	Average Primary Recovery Factor, % At Average Value of OOIP
Solution Gas Drive	Sandstones	California	22
		Louisiana	27
		Oklahoma	19
		Texas 7C, 8, 10*	15
		Texas 1–7B, 9*	31
		West Virginia	21
		Wyoming	25
Solution Gas Drive	Carbonates	All	18
Natural Water Drive	Sandstones	California	36
		Louisiana	60
		Texas	54
		Wyoming	36
Natural Water Drive	Carbonates	All	44

* Texas is subdivided into Districts by the Texas Railroad Commission.

† Statistical Analysis of Crude Oil Recovery and Recovery Efficiency, API Bulletin D–14, Second Edition, Dallas, Texas, April 30, 1984.

TABLE E–3 SECONDARY RECOVERY FACTORS†

Secondary Recovery Method	Lithology	State	Primary Plus Secondary Recovery Factor At Average OOIP, %	Ratio of Secondary to Primary Recovery Factor At Average OOIP, %
Pattern Waterflood	Sandstone	California	35	0.33
		Louisiana	51	0.40
		Oklahoma	28	0.62
		Texas	38	0.50
		Wyoming	45	0.89
Pattern Waterflood	Carbonates	Texas	32	1.05
Edge Water Injection	Sandstone	Louisiana	55	0.33
		Texas	56	0.64
Gas Cap Injection	Sandstone	California	44	0.48
		Texas	43	0.23

† Statistical Analysis of Crude Oil Recovery and Recovery Efficiency, API Bulletin D–14, Second Edition, Dallas, Texas, April 30, 1984.

APPENDIX F

Glossary

BUILDUP TEST: This is a test where the well is shut-in and the pressure is allowed to build up. Drawdown and *Buildup tests* are conducted to evaluate reservoir and well properties.

CRITICAL RATE, q_c: *Critical rate* is the maximum rate at which oil is produced without production of water or gas (whenever top gas or bottom water zones are present, along with an oil zone).

DECLINE CURVE: *Decline curve* is a plot of production history against time. There are three types of *decline curves:* Linear, Log-Log, and Semi-Log. Extrapolation of these curves is used to calculate reserves. In general, the reserves calculated by using a semi-log plot are different than those calculated using a log-log plot.

DIMENSIONLESS PRESSURE, p_D: This is used to calculate the change of pressure over time. Normally, the *dimensionless pressure* is based upon initial reservoir pressure minus the present well flowing pressure. Therefore, for a fixed flow rate, as time progresses, the well flowing pressure decreases and *dimensionless pressure* increases.

DIMENSIONLESS TIME, t_D: Dimensionless time is used to describe the variation in different mathematical solutions over time. There are two types of dimensionless time. One is based on either the wellbore radius for a vertical well, the fracture half-length for a fractured vertical well, or half the well length for a horizontal well. The other type of dimensionless time is based upon the drainage area A, and this dimensionless time is the same regardless of the well type. This dimensionless time, based on the drainage area, is normally used to calculate the beginning of pseudo-steady state.

DIMENSIONLESS WELL HALF-LENGTH, L_D: This defines the length of a horizontal well in dimensionless terms. It is defined as a length of a well divided by two times the reservoir height and multiplied by the square root of vertical to horizontal permeability ($L_D = (L/2h)\sqrt{k_v/k_h}$).

DRAINAGE AREA, A: Drainage Area is the area drained by the given well.

DRAINHOLE: Drainholes are normally drilled through an existing vertical well, using either a short or ultrashort radius drilling technique. The length of the drainhole is limited to a few hundred feet.

MULTIPLE DRAINHOLES: In some drilling techniques, such as ultrashort radius, it is possible to drill more than one drainhole at a given elevation. These are called multiple drainholes.

DRAWDOWN TEST: This is a well test where a well is produced at a constant rate. During this test the bottom hole producing pressure decreases over time. Similar to a build-up test, data from the drawdown can be used to calculate reservoir and well properties.

EFFECTIVE WELLBORE RADIUS, r'_w: The effective wellbore radius of a horizontal well or a damaged or a stimulated vertical well is a wellbore radius of an equivalent unstimulated vertical well. This vertical well with a large wellbore will produce at the same rate as that of a horizontal well. The concept of an effective wellbore radius is used to convert production performance of a damaged or a stimulated well or horizontal well to that of an equivalent vertical well.

FLOW REGIMES: A producing well goes through different flow regimes and they are described below:

1. INFINITE ACTING: This state occurs as soon as a well is put on production. During this state, the well has seen no physical or artificial boundaries. Therefore, it is called infinite acting.

2. TRANSITION STATE: The transition state is a state where the well has seen one physical boundary but has not seen the other boundaries. For vertical wells, in the horizontal plane, there are two boundaries: one in the x direction and one in the y direction. However, a horizontal well has three boundaries: a boundary in the x direction in the horizontal plane, a boundary in the y direction in the horizontal plane, and a third boundary in the vertical plane of the reservoir.

3. PSEUDO-STEADY STATE, *pss*: Pseudo-steady state is defined as a state when the pressure front from the well has reached all of its drainage boundaries, and as more and more fluid is withdrawn from the reservoir, average reservoir pressure as well as the pressure at the reservoir boundary decreases over time.
4. TIME TO REACH PSEUDO-STEADY STATE, t_{pss}: Time to reach pseudo-steady state is the time it takes for the reservoir to go through the infinite and transition state. Currently, there is no common industry-accepted definition to calculate the beginning of pseudo-steady state. However, when data are not available, the beginning of pseudo-steady state can be assumed to be $t_{DA} = 0.1$, where t_{DA} is dimensionless time based upon the drainage area. It is important to note that the time to reach the pseudo-steady state is independent of the well configuration. It is fixed for a given drainage volume or drainage area.
5. STEADY STATE: Steady state is a state where one can maintain a constant pressure at the wellbore and also at the boundary of the reservoir. This pressure drop is independent of time. In reality, this boundary condition is rarely achieved. However, steady state is mathematically simple, and therefore, it helps to derive negative skin factors and productivity indices for a given well configuration.

FRACTURE: There are two types of fractures. One is artificially created, while the other is a natural fracture.

DIMENSIONLESS FRACTURE CONDUCTIVITY, F_{CD}: This is defined as a fracture permeability multiplied by the fracture width divided by the product of formation permeability and the fracture half-length. From a mathematical standpoint, when F_{CD} value is greater than 100, the performance of the finite-conductivity fracture is identical to that of an infinite-conductivity fracture. However, from a practical standpoint, all fracture jobs with dimensionless fracture conductivity F_{CD} above 25 would give results very similar to that of an infinite-conductivity fracture. It is also important to note that as the length of the fracture increases, the dimensionless-fracture conductivity decreases. This indicates that for a given formation and proppants, beyond a certain length, increasing fracture length does not give additional productivity.

FINITE-CONDUCTIVITY FRACTURE: Although in all fracture jobs infinite-fracture conductivity is desired, actual fracture jobs such as a water frac, acid frac, and propped fractures, have finite-conductivity. Different types of proppants give different fracture conductivities. For example, bauxite gives the highest fracture flow capacity or conductivity, while Jordan sand gives a little lower fracture flow capacity.

FRACTURE FLOW CAPACITY: Fracture flow capacity is defined as $k_f b_f$, where k_f is fracture permeability and b_f is fracture width. Units of fracture flow capacity are md-ft or md-m. Most service companies (which do fracture jobs), normally have a booklet which will give information on the fracture flow capacities of their various fracture treatments.

Generally, fracture flow capacities of various fractured sands are on the order of thousands of darcy-feet. However, in a reservoir, under high pressure, this flow capacity may decrease considerably. Moreover, some fractures lose their flow capacity as time progresses. Refer to service company booklets for further information.

FULLY PENETRATING FRACTURE: *Fully penetrating fracture* is a fracture that intersects the entire reservoir height.

INFINITE-CONDUCTIVITY FRACTURE: In this fracture, pressure drop within the fracture is zero. This is a mathematically ideal condition, which is desired in most fracture jobs. In low-permeability reservoirs, if a fracture has a significantly higher permeability than the formation, then for all practical purposes the fracture behaves as though it has an infinite-conductivity. For an infinite-conductivity fracture.

$$\text{Effective wellbore radius} = \text{Total fracture length}/4$$

UNIFORM-FLUX FRACTURE: In a uniform-flux fracture, a constant amount of fluid is produced per unit length of the fracture. To achieve this condition, from a mathematical standpoint, pressure drop should exist within the fracture, with minimum pressure at the fracture center and maximum pressure at the fracture tips. This also provides the required pressure drop for flow from fracture tips to the fracture center. Normally, the vertical well intersects the center of the fracture. For a uniform flux-fracture:

$$\text{Effective wellbore radius} = \text{Total fracture length}/(2e)$$

GAS LOCKING AND WATER LOCKING: These situations can occur in a horizontal well due to its well shape. The well itself has a minimum pressure drop along its length. Hence, it is very difficult to remove gas and water locked in "bends" of a horizontal well.

HORIZONTAL WELL: A horizontal well is normally a new well 1000 to 4000 feet long and drilled from the surface. A horizontal well can also be defined as a well parallel to the reservoir bedding plane.

MOBILITY RATIO, *M*: Mobility ratio indicates displacement efficiency in an EOR process. It is the ratio of viscosities and permeabilities of two fluids, one fluid being displaced (such as oil) and the other fluid being injected (such as water). A mobility ratio less than one indicates a favorable mobility ratio; i.e., the displacement is very efficient. A mobility ratio greater than one indicates a less efficient displacement.

MULTIPLE WELLS THROUGH A SINGLE VERTICAL WELL: In some places, several horizontal drainholes at different elevations have been drilled through one single vertical well.

PERMEABILITY, k: Permeability, normally expressed in millidarcies, expresses the formation's flow capacity.

HORIZONTAL PERMEABILITY, k_h: Horizontal permeability is normally measured in the horizontal plane of the reservoir.

EFFECTIVE HORIZONTAL PERMEABILITY, $\sqrt{k_x k_y}$: In the horizontal plane, if permeability in the x direction is not the same as in the y direction, then effective horizontal permeability is $\sqrt{k_x k_y}$. In naturally fractured formations and in certain geological environments, high permeability is observed in one direction as compared to the other.

EFFECTIVE RESERVOIR PERMEABILITY, $\sqrt{k_v k_h}$: If the permeabilities in the horizontal direction and the vertical direction are different, effective reservoir permeability is normally approximated as $\sqrt{k_v k_h}$.

PERMEABILITY RATIO, k_v/k_h: This is the ratio of vertical to horizontal permeability.

VERTICAL PERMEABILITY, k_v: Vertical permeability is measured in the vertical plane of the reservoir.

PRODUCTIVITY INDEX, J: Productivity Index J is defined as flow rate per unit pressure drop. The productivity index is often used in many calculations. Units of productivity index are bbl/day-psi or m^3/day-kPa.

SHAPE FACTOR, C_A: Shape factors are used only during the pseudo-steady state. That is, when the well has seen all of its boundaries and the reservoir pressure has started decreasing as more and more fluid is withdrawn from the reservoir. The shape factors include the influence of well location on well productivity in a given drainage area. A well drilled at a different location within the drainage area would have different productivities, and shape factors account for this effect. Similarly, for a horizontal well in a given drainage area, for different well lengths, and well locations, different shape factors are obtained.

SHAPE-RELATED SKIN FACTOR, s_{CA}: This is an equivalent skin factor for a shape factor. For convenient mathematical use, shape factors are converted into skin factors to calculate well productivity. The shape-related skin factors are generally positive. The influence of the shape skin factor is to reduce the well productivity, therefore it is desirable that the well should have a minimum shape-related positive skin factor, preferably zero. In a limiting case, the horizontal well shape factors are the same as those for infinite-conductivity fully penetrating fractures.

SKIN FACTOR, s: Skin factor is a concept which is introduced to account for the excess pressure drop that occurs around the wellbore. This pressure drop can be caused by several reasons. For example, a partially penetrating well or well damaged due to drilling mud invasion would give an additional skin pressure drop. Normally, a newly drilled well would have a skin factor on the order of $+1$ to $+10$.

NEGATIVE SKIN FACTOR: The negative skin factor term is used to indicate stimulated vertical wells. In a well that is acidized or stimulated, the permeability in the near-wellbore region is higher than that in the rest of the reservoir. This results in a lower pressure drop in the near-wellbore region than those observed with unstimulated wells. To account for this, negative skin factors are used. Horizontal wells can also be represented as highly stimulated vertical wells or wells with high negative skin factors.

SLANT WELLS: Slant well is an inclined well. This well has an inclination angle α with the vertical axis. This slant well is also referred to as a deviated well with a deviation angle of α.

WATER BREAKTHROUGH TIME, t_{BT}: The well is put on production at a certain rate, and the time it takes for the water or gas to break through by coning is called breakthrough time. Before the breakthrough time, only single-phase flow occurs, such as only the flow of oil or gas. After the breakthrough time, more than one fluid will be produced from the wellbore.

WATER AND GAS CONING: Water and gas coning is the situation where undesirable fluids (water or gas) are produced along with the oil. For example, in vertical wells, to avoid water coning, the well is generally perforated only through the top portion of the reservoir. This helps to minimize water coning. This is also called partial penetration and is represented as a positive skin factor.

WELLBORE STORAGE: Wellbore storage indicates a time during the well test when the test is influenced by the volume of the wellbore as well as the compressibility of the wellbore fluid. In a horizontal well where long wellbores are involved, the wellbore storage period could be a significant part of the well test. Therefore, whenever possible, it is recommended to conduct a well test on a horizontal well by using a downhole shut-in device.

WELL ECCENTRICITY, δ: Well eccentricity for a horizontal well is defined as the elevation difference between the horizontal well center and the center of the vertical plane of the reservoir.

WELL LENGTH, L: Well length is the total productive length of a horizontal well.

WELL SPACING: Well spacing is the distance between the wells. Normally it is referred to in terms of acreage such as 40-acre spacing or 80-acre spacing, etc.

INDEX

T

EQUATION INDEX

CHAPTER 2: RESERVOIR ENGINEERING CONCEPTS

CHAPTER 3: STEADY STATE SOLUTIONS

CHAPTER 4: INFLUENCE OF WELL ECCENTRICITY

CHAPTER 5: COMPARISON OF HORIZONTAL AND FRACTURED VERTICAL WELLS

CHAPTER 6: TRANSIENT WELL TESTING SOLUTIONS

CHAPTER 7: PSEUDO-STEADY STATE SOLUTIONS

CHAPTER 8: WATER AND GAS CONING IN HORIZONTAL AND VERTICAL WELLS

CHAPTER 9: HORIZONTAL WELLS IN GAS RESERVOIR

CHAPTER 10: PRESSURE DROP THROUGH A HORIZONTAL WELL